ELECTRON TRANSPORT IN QUANTUM DOTS

ELECTRON TRANSPORT IN QUANTUM DOTS

edited by

Jonathan P. Bird
Associate Professor, Department of Electrical Engineering
Arizona State University
Tempe, Arizona

&

Visiting Professor, Center for
Frontier Elecronics and Photonics
Chiba University
Chiba, Japan

KLUWER ACADEMIC PUBLISHERS
Boston / Dordrecht / London

Distributors for North, Central and South America:
Kluwer Academic Publishers
101 Philip Drive
Assinippi Park
Norwell, Massachusetts 02061 USA
Telephone (781) 871-6600
Fax (781) 871-9045
E-Mail: kluwer@wkap.com

Distributors for all other countries:
Kluwer Academic Publishers Group
Post Office Box 322
3300 AH Dordrecht, THE NETHERLANDS
Telephone 31 786 576 000
Fax 31 786 576 254
E-mail: services@wkap.nl

Electronic Services <http://www.wkap.nl>

Library of Congress Cataloging-in-Publication Data

Electron Transport in Quantum Dots
Edited by Jonathan P. Bird
Includes bibliographical references and index.
ISBN 1-4020-7459 X

Printed on acid-free paper.

Printed in the United States of America.

Image: Scanning-electron micrograph of a split-gate quantum dot. The white spacer bar at the bottom of the image denotes a distance of 1 micron. Image provided courtesy of: Dr. Y. Takagaki, Paul-Drüde Institute, Germany.

To Hinako and Jack
For giving so much, while often getting so little back

Contents

List of Contributors

R. Akis
Department of Electrical Engineering and Center for Solid-State Electronics Research,
Arizona State University, Tempe, AZ 85287–6206, USA
Fax: 480-965-8058
Email: Richard.Akis@asu.edu

Bruce Alphenaar
Department of Electrical and Computer Engineering,
409 Lutz Hall,
University of Louisville, Louisville, KY 40292, USA
Tel: 502-852-1554
Fax: 502-852-1577
Email: brucea@louisville.edu

Robert H. Blick
Department of Electrical and Computer Engineering,
University of Wisconsin-Madison,
1415 Engineering Drive,
Madison, WI 53706, USA
Tel: 608-262-1952
Email: blick@engr.wisc.edu

A. M. Chang*
Physics Department, Purdue University, 525 Northwestern Ave.,
West Lafayette, IN 47907-2036, USA
Tel: 765-494-3012
Fax: 765-494-0706
Email: yingshe@physics.purdue.edu

Susumu Komiyama
Department of Basic Science,
University of Tokyo,
Komaba 3-8-1, Meguro-ku, Tokyo 153-8902, Japan
Email: csusumu@ASone.c.u-tokyo.ac.jp

Craig S. Lent
Department of Electrical Engineering,
University of Notre Dame,
Notre Dame, IN 46556, USA
Email: lent@callisto.ee.nd.edu

Heiner Linke
Department of Physics,
University of Oregon,
Eugene, OR 97403-1274, USA
Tel: 541 346-4583
Fax: 541 346-5861
Email: linke@duoregon.edu

Andy Sachrajda
Bldg. M-23A, Rm. 156, Institute for Microstructural Sciences,
National Research Council of Canada, 1191 Montreal Rd.,
Ottawa, Ontario K1A 0R6, Canada
Tel: 613-993-9773
Fax: 613-952-8701
Email: andy.sachrajda@nrc.ca

S. Tarucha
Department of Physics, University of Tokyo, 7-3-1 Hongo,

*Effective August 1, 2003, new address will be Department of Physics, Duke University, Durham, NC 27708, USA. Email: yingshe@phy.duke.edu

Bunkyo-ku, Tokyo 113-0033,
Japan
Email:
tarucha@phys.s.u-tokyo.ac.jp

Richard P. Taylor
Department of Physics, University of Oregon, 1371 E. 13th Ave.
Eugene, OR 97403-1274, USA
Tel: 541 346-4741
Fax: 541 346-3422
Email: rpt@darkwing.uoregon.edu

Dieter Weiss
Department of Physics, Universität Regensburg,
D-93040 Regensburg, Germany
Tel: +49-941-943-3197
Fax: +49-941-943-3196
Email: dieter.weiss@physik.uni-regensburg.de

Preface

When I was contacted by Kluwer Academic Publishers in the Fall of 2001, inviting me to edit a volume of papers on the issue of electron transport in quantum dots, I was excited by what I saw as an ideal opportunity to provide an overview of a field of research that has made significant contributions in recent years, both to our understanding of fundamental physics, and to the development of novel nanoelectronic technologies. The need for such a volume seemed to be made more pressing by the fact that few comprehensive reviews of this topic have appeared in the literature, in spite of the vast activity in this area over the course of the last decade or so. With this motivation, I set out to try to compile a volume that would fairly reflect the wide range of opinions that has emerged in the study of electron transport in quantum dots. Indeed, there has been no effort on my part to ensure any consistency between the different chapters, since I would prefer that this volume instead serve as a useful forum for the debate of critical issues in this still developing field. In this matter, I have been assisted greatly by the excellent series of articles provided by the different authors, who are widely recognized as some of the leaders in this vital area of research.

The articles collected in this volume cover a wide range of range of aspects of quantum-dot transport, but may broadly be classified into three distinct groups. Chapters 1–4 deal with transport in weakly-coupled quantum dots, in which current flow arises predominantly from tunneling, and in which many-body phenomena dominate the resulting behavior. Chapters 5–8, on the other hand, focus on transport in open dot and antidot systems, in which the Coulomb blockade is suppressed by charge fluctuations, and in which interest lies in the nature of the connection between the quantum-mechanical and semiclassical descriptions of electron transport. Finally, Chapters 9–11 explore some of the emerging applications of quantum dots, as single-photon detectors, in quantum cellular automata, and in carbon-nanotube spintronics.

In preparing a volume such as this, the efforts of many are required to bring the project to completion, and it gives me the greatest of pleasure to acknowledge the many contributions that have made this work possible. While I hesitate to place these in any order of importance, I would, first and foremost, like to express my sincere gratitude to the various authors who contributed the different chapters to this book. I would also like to thank Danette Knopp of Kluwer Academic, who had the confidence in me to complete this project. While Danette has now moved on to the next stage of her career, I hope that she will be proud of the final product of our joint work. More recently, I have worked with Greg Franklin at Kluwer Academic, who has provided a seamless transition of power since the departure of Danette. I would also like to acknowledge the support from my sponsors, who provided me with the freedom to work on this project, in particular Larry Cooper at the Office of Naval Research, and Jerry Smith at the Department of Energy. Special appreciation is also reserved for Richard Akis and David Ferry, of Arizona State University, who have been my partners in my own intellectual journey into the world of quantum dots. I feel lucky to count you both as friends, and look forward to further explorations together in the future. Finally, the last stages of this work were completed during my tenure as a visiting professor in the Center for Frontier Electronics and Photonics at Chiba University. I would especially like to acknowledge the generous support during this time from my friend and colleague, Prof. Yuichi Ochiai, and from Mrs. Yuko Komai of the Venture Business Laboratory at Chiba University.

Chapter 1

Interactions, Spins and the Kondo Effect in Quantum-Dot Systems

S. Tarucha[1,2,3], K. Ono[1], T. Fujisawa[3], W. G. van der Wiel[1,4,5], and L. P. Kouwenhoven[4]

[1] *Department of Physics, University of Tokyo, Hongo, Bunkyo-ku, Tokyo 113-0033, Japan*
[2] *ERATO Mesoscopic Correlation Project, Japan Science and Technology Corporation, Atsugi-shi, Kanagawa 243-0198, Japan*
[3] *NTT Basic Research Laboratories, Nippon Telegraph and Telephone Corporation, Atsugi-shi, Kanagawa 243-0198, Japan*
[4] *Department of Applied Physics, Delft University of Technology, PO Box 5046, 2600 GA Delft, The Netherlands*
[5] *PREST, Japan Science and Technology Corporation, Hongo, Bunkyo-ku, Tokyo 113-0033, Japan*

1. Introduction

Semiconductor quantum dots are often referred to as artificial atoms because the electronic energy spectrum is determined by the effects of quantum mechanical confinement and interactions [1]. Zero-dimensional (0D) energy levels are well defined in a sufficiently small dot that the size of the confining potential is comparable to the Fermi wavelength of electrons. On the other hand, the interaction effect is characterized by a Coulombic energy (single electron charging energy) cost for trapping an additional electron in a dot. The charging energy is simply given by e^2/C (C, total capacitance) when the dot is large enough for containing many electrons. This gives rise to so-called "Coulomb blockade" and is explained using an orthodox theory [2]. However, when the dot is small and contains just a few electrons, the interaction effect as well as the quantum mechanical effect depends strongly on the electronic configuration, and cannot be characterized using a capacitance model [2,3]. Adding an electron to such a small quantum dot costs a certain energy and simultaneously changes the electronic configuration to minimize the total energy, i.e. sum of the quantum mechanical energy and interaction energy. The cost for the quantum mechanical energy is associated with the orbital energy. Each orbital state is spin degenerate, so antiparallel spin filling of the same orbital state is generally favored. However, this is not the case when we consider the cost of interaction energy. Exchange energy is gained when electrons are added with parallel spins as compared to antiparallel spins. Therefore, if we assume a two-electron state formed by the filling of degenerate orbital states, the ground state is usually a spin triplet state. In addition, direct Coulomb energy is larger for two electrons in the same orbital state than those in two different orbital states. This gives rise to certain preference for the filling of degenerate orbital states.

In semiconductor quantum dots there are various parameters to influence the electronic configuration, such as the number of electrons, N, magnetic field, B, and shape and strength of the confining potential [2,3]. The orbital energy is significantly modified using a magnetic field at just a few T, because for the typical size of the confining potential the electron wavelength is not so large as compared to the magnetic confinement. This enables the manipulation of orbital degeneracies in quantum dots in the relatively low magnetic field range [2,3]. In the high magnetic field range the interaction effect becomes more important than the effect of quantum mechanical confinement, because all of the electrons are confined to the lowest Landau level. This gives rise to various transitions in the electronic configuration to a fully spin-polarized state, then to more strongly correlated states such as magic number states and Wigner molecule states [4], though these strongly correlated states have not yet been demonstrated.

Shape of the confining potential is also an important parameter for the electronic configuration. A "good" quantum number, which gives rise to systematic sets of degeneracies of orbital states, is present for a quantum dot confined by a potential with a high degree of symmetry. In a quantum dot formed in a circular mesa of a 2D electron gas (2DEG), atom-like electronic properties such as shell-filling and obeyance of Hund's rule are observed [5]. These properties arise from the filling of systematic sets of degenerate states confined by a 2D harmonic potential. Breaking the rotational symmetry by deforming the confining potential lifts the degeneracy present in a circular quantum dot, and destroys the distinctive shell structure and modifies the way of spin filling [4,6]. In a strongly elliptic quantum dot a novel spin configuration "spin density wave" is predicted to appear [6].

In specific semiconductor quantum dots those parameters that can influence the electronic configuration can be adjusted systematically. This allows us to investigate various kinds of interaction effects in a quantum dot in a controlled manner. In this chapter we describe such controllable interaction effects in quantum dot systems. We mainly discuss the spin-related properties observed in a 2D harmonic quantum dot. The atom-like electronic properties are revealed by measurements of electron transport through the dot, and are first reviewed in section 2. The in-plane 2D harmonic confinement potential leads to maximal 0D level degeneracy of single-particle states in the dot. Consecutive filling of each set of degenerate 0D states is directly responsible for the shell-filling. On the other hand, Hund's first rule is associated with the interaction effects for the filling of electrons amongst nearly degenerate states. Parallel-spin filling is then favored amongst half-filled degenerate states in each shell. The key ingredient for manipulating the way of spin filling is the tuning of orbital degeneracies. The spin filling occurs to form the ground state with a high-spin value. A spin triplet state is the case made out of two electrons: two parallel spin electrons in two different but nearly degenerate orbital states. The excited state is a spin singlet state having the same orbital configuration but with antiparallel spins. We show in section 3 that the two-electron spin state is manipulated between a spin singlet and a spin triplet by magnetically adjusting the orbital degeneracies. Contributions from the direct Coulomb and exchange effects on the spin state are discussed in detail.

In sections 4–6 we describe the spin effects obtained by controlling over a spin singlet and triplet states. The first is the effect of Pauli exclusion principle observed in a coupled two quantum dot system with each dot holding a single 0D state. The Pauli exclusion prevents two electrons of parallel spin from occupying a single spatial orbital. Then single electron transport through the system is only allowed for two singlet states: the first has an antiparallel spin electron in each dot and the second does two antiparallel spin electrons in one of the two dots. The transport is blocked once a spin triplet state having a parallel spin electron in each dot is excited. In section 4 we show that this blockade (by Pauli exclusion) occurs under the condition of nonlinear transport.

For study on the spin effects in quantum dots in terms of electron transport, we generally assume that spin is a stable quantum number in the quantum dot at least on the time scale of electron transport. For quantum dots made in a nonmagnetic semiconductor there are usually no strong sources to relax the spin state. Energy relaxation due to phonons is very strong for spinless electrons in any semiconductor materials, and the phonon emission time is on the order of nsec or shorter. In contrast if the relaxation accompanies a spin flip, the relaxation time should be much longer than this time scale. Many of previous experiments on the relaxation time are carried out using an optical pump and probe technique. This technique is not so suitable for electronic systems in quantum dots because contributions from holes cannot be neglected. We have developed an electrical pump and probe technique to measure the relaxation time from an excited state to the ground state in a quantum dot with and without a spin flip [7]. The spin-flip relaxation is the case from a spin triplet state (excited state) to a spin singlet state (ground state). We show in section 6 that the measured relaxation time is several orders of magnitude longer than that without a spin-flip.

The last topic on the spin effect described in section 7, is the Kondo effect [8]. The Kondo effect is a typical many-body spin effect, and observed in the 1930s for metals containing magnetic impurities [9]. The resistance increases as the temperature is lowered below a characteristic temperature, because higher-order scattering of electrons at Fermi level is promoted by the antiferromagnetic exchange coupling to the local magnetic moment at the magnetic impurity site. The same effect, but between an unpaired electron spin in a quantum dot and the Fermi level in the contact leads, has recently been discovered for a semiconductor quantum dot [10]. The Kondo effect in quantum dots shows quite a few interesting aspects that have never been studied for metals. Here we describe the observation of complete transparency of the dot by the $S = \frac{1}{2}$ Kondo effect ("unitary limit of the conductance") and the Kondo effect enhanced by the degeneracy of spin singlet ($S = 0$)-triplet ($S = 1$) states.

2. Atom-Like Properties of Electrons Confined in a Quantum Dot

2.1. Two-Dimensional Harmonic Quantum Dot

Clean 2D harmonic quantum dots have only recently been fabricated in a semiconductor heterostructure [5,11]. This subsection presents the electronic properties of such 2D harmonic quantum dots containing just a few electrons, which resemble these of real atoms.

Electrons bound to a nuclear potential experience sufficiently strong quantum mechanical confinement and Coulomb interactions that they are well arranged in ordered states. This leads to the well-known ordering of atoms in the periodic table. The ionization energy has large maxima for atomic numbers 2, 10, 18, ... when shells are completely filled with electrons. In addition, for the filling of electrons in similar orbitals, parallel spins are favored until the set of orbitals is half filled as expected from Hund's first rule. This also gives rise to secondary maxima in the ionization energy [12]. A good analogue to the 3D shells in atoms can be realized in quantum dots (artificial atoms) with the shape of a circular disk. The disk-shaped quantum dots we can fabricate, are formed in a laterally gated micron-sized double barrier structure, and contain a tunable number of electrons N starting from $N = 0$.

For the simplest explanation of the atom-like shell structure we ignore, for the moment, the interaction effects and consider the eigen states, i.e. single-particle states confined in a 2D harmonic quantum dot. The eigen-energy for a 2D harmonic oscillator at $B = 0$ becomes $E_{n,\ell} = (2n+|\ell|+1)\hbar\omega_0$. Here $n(= 0, 1, 2, \ldots)$ is the radial quantum number, and $\ell(= 0, \pm 1, \pm 2, \ldots)$ is the angular momentum quantum number of the oscillator and ω_0 is the oscillator frequency. Each state $E_{n,\ell}$ is two-fold spin-degenerate. The eigen-states are systematically degenerate from the lowest level (Fig. 1.1): so including spin degeneracy, the first, second, and third "shells" respectively are two-fold degenerate with $E_{0,0}$ (1s orbital), four-fold degenerate with $E_{0,1} = E_{0,-1}$ (2p orbital), six-fold degenerate with $E_{0,2} = E_{0,-2}$ (3d orbital) $= E_{1,0}$ (3s orbital). Note that the degeneracy with respect to the angular quantum number of $\pm|\ell|$ arises from the rotational symmetry in the lateral confinement, while the additional degeneracy like for the 3s and 3d orbital states is associated with the harmonicity of the confining potential. The square of the eigen-wavefunction is plotted in Fig. 1.1 for these 1s, 2p, 3s and 3d orbital states. The number of nodes of the wavefunction along the radial direction is given by the radial quantum number n. If the angular momentum is non-zero, then an additional node appears at the center. The larger ℓ the more spreads

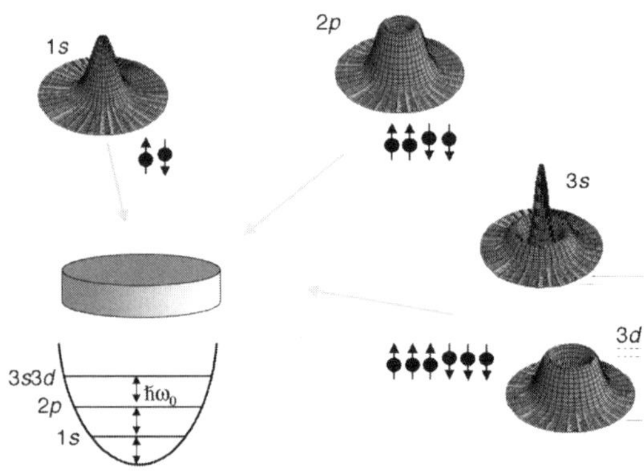

Fig. 1.1. Eigen states ($E_{n,\ell}$), i.e. single-particle states confined in a two-dimensional harmonic quantum dot, and the spatial form of the square of the eigen-wavefunction: 1s ($E_{0,0}$), 2p ($E_{0,\pm 1}$), 3s ($E_{1,0}$) and 3d ($E_{0,\pm 2}$) orbital states.

out the wavefunction. These plots are helpful for understanding the relative strength of Coulomb interactions. For example, an electron in a large angular momentum state suffers from weaker Coulomb interaction with the other electrons.

For non-interacting electrons these $E_{n,\ell}$ states are consecutively filled from the lowest, and complete filling of each shell is attained for special electron numbers of $N = 2, 6, 12, 20, \ldots$ These are the "magic numbers" that characterize the shell structure, resembling that in real atoms.

The eigen-states for a 2D harmonic quantum dot in the presence of magnetic field (B field) perpendicular to the 2D plane of the quantum dot are the Fock–Darwin (FD) states [3,13,14], and expressed by:

$$E_{n,\ell} = -(\ell/2)\hbar\omega_c + (n + 1/2 + |\ell|/2)\hbar(4\omega_0^2 + \omega_c^2)^{1/2} \tag{1}$$

where $\hbar\omega_c = eB/m^*$ is the cyclotron energy (for GaAs $\hbar\omega_c = 1.76$ meV at $B = 1$ T).

Figure 1.2(a) shows $E_{n,\ell}$ versus B calculated for $\hbar\omega_0 = 2$ meV. The Zeeman spin-splitting is neglected since it is usually much smaller than the orbital energy shift with B field. The orbital degeneracy at $B = 0$ T is lifted on initially increasing B, reflecting the first term of Eq. (1): a state having a positive (negative) angular momentum shifts down (up) and the shift is larger for the larger value of $|\ell|$. As B is increased further, new crossings or new degeneracies can occur. The last crossing occurs along the bold line in Fig. 1.2(a). This last crossing is denoted as a filling factor of $\nu = 2$, in analogy to the quantum Hall effect in a large 2DEG. In contrast to the bulk 2D case, degeneracy in the Landau level is lifted by the confinement, so that beyond the bold line the down-going

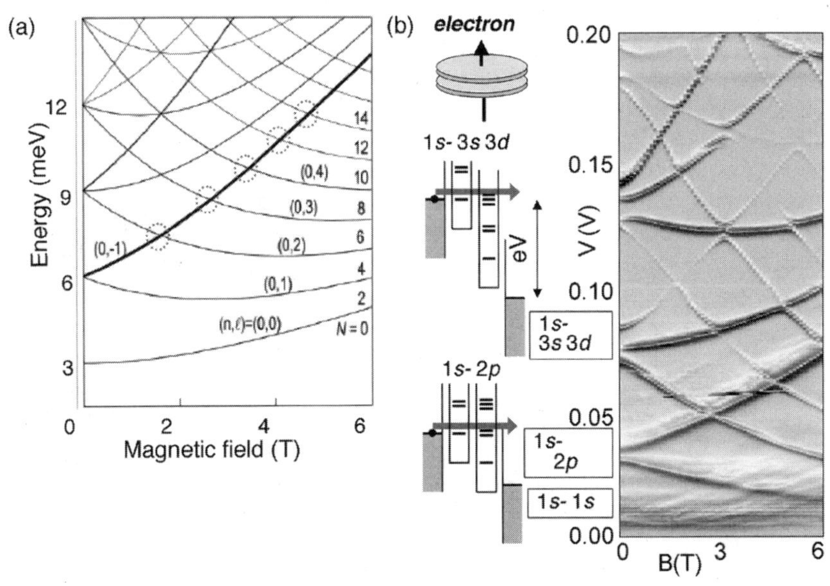

Fig. 1.2. (a) Fock–Darwin states calculated for $\hbar\omega_0 = 2$ meV. Dotted circles indicate the last crossing of the Fock–Darwin states. (b) Single-particle tunneling spectrum measured using a double dot (dot 1 and 2) device (see Section 2.3). Single-particle state resonance between the 1s state in dot 1 and various orbital states in dot 2 are attained by adjusting source-drain voltage.

FD-states merge to form the lowest Landau level. The sequence of magic numbers in the lowest Landau level is simply 2, 4, 6, 8, . . .

2.2. Quantum Dot Devices and Experimental Setup

The device configuration is schematically shown in Fig. 1.3. The scanning electron microscope image shows the actual shape of the device mesa. A single quantum dot is located inside each sub-micron cylindrical mesa (geometrical diameter $= 0.4$ to $0.6\,\mu$m) made from an undoped double-barrier structure (DBS), which consists of an undoped 12 nm-$In_{0.05}Ga_{0.95}As$ well and two undoped $Al_{0.22}Ga_{0.78}As$ barriers of thickness 9.0 and 7.5 nm. Above and below the DBS there is an n-doped GaAs contact (source and drain), and the circular dot is located between the two heterostructure barriers. The third electrode is a Schottky gate, which is wrapped around the mesa. The dot is strongly confined in the vertical direction by the heterostructures, while it is softly confined in the lateral direction by the Schottky gate induced depletion region. This depletion region is approximated by a harmonic potential, and the characteristic energy $\hbar\omega_0$ is evaluated from measurements of the magnetic field dependence of the Coulomb oscillations [3]. Upon application of a negative gate voltage the dot is squeezed inwards, and at the same time the number of electrons, N, is decreased. For the $0.5\,\mu$m-diameter mesa, the dot is pinched-off or $N = 0$ for $V_g < -1.6$ V. Note that the strength of confinement decreases as N is increased, due to the effect of Coulomb screening. We evaluate an electron density of $1.7 \times 10^{15}\,\text{m}^{-2}$ from measurement of Shubnikov–de Haas oscillations for a large area device. From this value we estimate that about 10 electrons are in the dot when the effective diameter is 100 nm.

The current, I, flowing vertically through the dot is measured as a function of gate voltage, V_g, in response to a dc voltage, V, applied between the contacts. For a

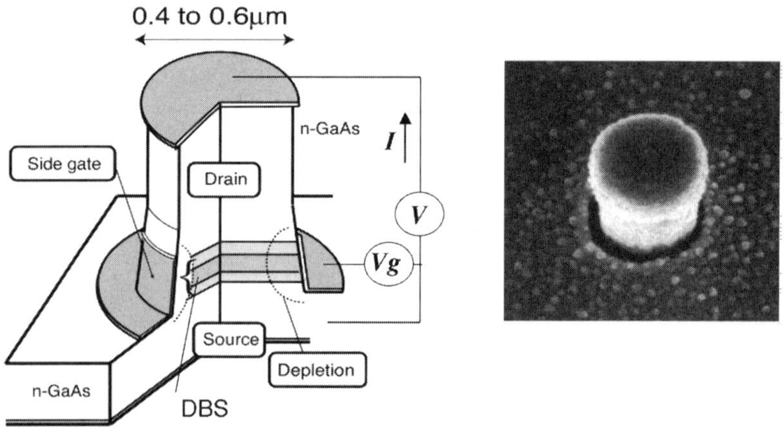

Fig. 1.3. Schematic of a vertical device containing a quantum dot. A two-dimensional harmonic quantum dot is located between the heterostructure barriers. The scanning electron microscope (SEM) image shows the overview of the device in a submicron-sized circular mesa.

sufficiently small V to be in the linear-response regime, the current shows a series of current peaks as a function of gate voltage (i.e. Coulomb oscillations), corresponding to the one-by-one change in the number of electrons in the dot (see section 2.4).

2.3. *Single-Particle Tunneling Spectroscopy*

Here we study resonant tunneling through the single-particle states in the present quantum dot to demonstrate that the lateral confinement has actually the form of a 2D harmonic oscillator. Tunneling through a quantum dot is usually influenced by the Coulomb interaction between electrons in the dot, since electrons are dynamically accumulated in the dot. Single-particle tunneling can be observed when the escape of an electron from the dot is strongly promoted, i.e. when the tunneling rate is much higher for the escape barrier than for the incident barrier. This is for example the case when a large bias voltage of V is applied to adjust the energy alignment between the Fermi level in the source contact and the states confined in a dot. Then the escape tunnel barrier becomes significantly lower than the incident tunnel barrier. The quantum numbers for each single-particle state can be assigned from the B field dependence of the bias voltage for resonance [15]. We actually use a weakly coupled double dot device to measure the resonant tunneling from the $1s$ state in the first dot on the source side to the $1s, 2p, 3s, 3d, \ldots.$ states in the second dot on the drain side as a function of bias voltage V (Fig. 1.2(b)). We can then avoid possible problems such as thermal broading of the Fermi level and influence by a 2DEG accumulated on the source contact at the heterostructure barrier. The device is prepared using a triple barrier structure instead of the DBS for the device shown in Fig. 1.3. Note that V_G is common between the two dots. We apply a sufficiently negative gate voltage to pinch-off the double dot system and apply a large voltage V to adjust energy alignment between the $1s$ state ($E_{0,0}$) in the first dot and the various states ($E_{n,\ell}$) in the second dot. This energy alignment is well identified as a narrow current peak in the intensity plot of $\mathrm{d}I/\mathrm{d}V$ versus B and V (Fig. 1.2(b)). Note that about 10% of V drops across the center barrier. A small peak at $V \sim 0.01$ V is due to $1s$-$1s$ resonance. To the high voltage of the lowest single peak at $B = 0$ T, there are nearly degenerate two peaks at $V \sim 0.03$ V ($1s$−$2p$ resonance) and three peaks at $V \sim 0.08$ V ($1s$−$3s3d$ resonance). These degeneracies are lifted as the B field is increased. The evolution of the peaks with B field is quite similar to the FD diagram in Fig. 1.2(b). Exactly speaking, the peak shift represents the B-dependence of $E_{n,\ell} - E_{0,0}$. So the lowest peak shows no clear shift with B field. Most of the higher lying peaks show the first crossing at the same B field of 2 T. This is well reproduced by the FD diagram calculated for $\hbar\omega_0 = 5$ meV. This value is the same as we previously evaluated for a single dot holding one or two electrons [3].

The two(three) peaks associated with the $2p$ states ($3s3d$ states) are somewhat split even at $B = 0$ T. This indicates lifting of degeneracy for the second(third) shell by about 0.2 meV, probably due to small deformation in the lateral confinement under such a largely biased condition.

2.4. *Shell Filling and Hund's First Rule*

In section 2.3 we confirmed that the single-particle states in our quantum dot are well represented by the FD states. Now, we discuss the filling of these single-particle states with interacting electrons. The single-particle level degeneracy is actually lifted due to the Coulomb interactions. However, when the quantum mechanical confinement energy is comparable to or greater than the interaction energy, each set of degenerate states or each shell is still consecutively filled from the lowest, so that we expect the same series of magic numbers of $N = 2, 6, 12, \ldots$ as for the single-particle case. In addition, for the filling of electrons in the same shell, parallel spins are favored in accordance with Hund's first rule. This leads to another series of magic numbers of $N = 4, 9, 16, \ldots$ corresponding to the half filling of the second, third, fourth shells, . . ., respectively.

The interaction effects can be studied from measurements of Coulomb oscillations for the dot (see section 2.2). Figure 1.4(a) shows the current oscillations (Coulomb oscillations) observed for a device with a geometrical diameter of 0.5 μm [5,11]. A small bias of 0.15 mV is set for V, so that only the ground states (GSs) contribute to the current. The values of N can be identified in Coulomb blockade region between the peaks, starting from $N = 0$, because for $V_g < -1.6$ V no further current peaks are observed, i.e. the dot is empty. When N becomes smaller than 20, the oscillation period depends strongly on N. The current peak to the left of a Coulomb blockade region with N trapped electrons thus measures the N-electron GS electro-chemical potential $\mu(N)$. For example, the first, second and third peaks from the left measure the one, two and three electron GS electro-chemical potentials, respectively. The peak spacing labeled by "N" therefore corresponds to the increment of electro-chemical potential, $\Delta(N) = \mu(N+1) - \mu(N)$. In Fig. 1.4(a) the peak spacing is only given in gate voltage, i.e. $\Delta V_g(N)$, and $\Delta(N) = \alpha \Delta V_g(N)$. α is a conversion factor from gate voltage to energy, and can be determined from measurement of the so-called "Coulomb diamonds" [2]. Thus obtained $\Delta(N)$ is plotted as a function of N in Fig. 1.4(b). In correspondence with the spacings between the Coulomb oscillations, $\Delta(N)$ is unusually large for $N = 2$, 6 and 12, and is also relatively large for $N = 4$, 9 and 16 (not shown for $N = 16$). The values of 2, 6 and 12 arise from the complete filling of the first, second and third shells, respectively, while those of 4, 9 and 16 are due, respectively, to the half filling of the second, third and fourth shells with parallel spins in accordance with Hund's first rule. For example, the GS for $N = 4$ is a spin triplet state having two parallel spin electrons in the orbital degenerate $2p$ shell states: $(n, \ell) = (0, 1)$ and $(0, -1)$. The first excited state (ES) is a spin singlet having

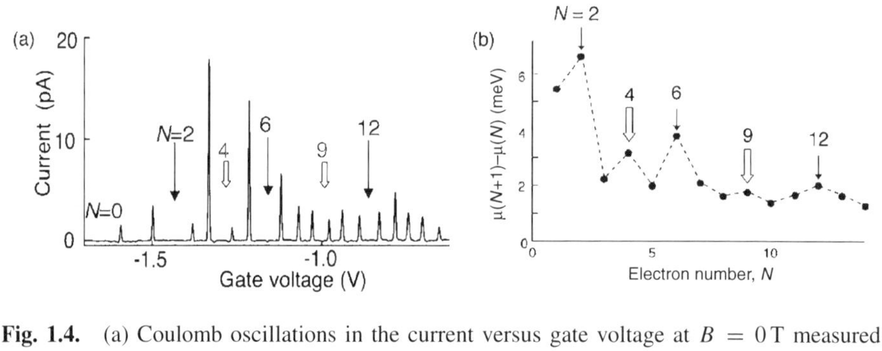

Fig. 1.4. (a) Coulomb oscillations in the current versus gate voltage at $B = 0$ T measured at 50 mK. The excitation voltage $V = 0.15$ meV. (b) Change of electro-chemical potential, $\mu(N+1) - \mu(N)$, as a function of electron number, N.

two antiparallel spin electrons both in one of the $2p$ shell states: $(n, \ell) = (0, 1)$. The excitation energy is 0.7 meV, which is equivalent to the exchange energy between two $2p$ electrons.

Figure 1.5 shows the B field dependence of the position of the current peaks measured for a device prepared in the same way. It is clear to see that the current peaks generally shift in pairs with B. This pairing is due to the consecutive filling of spin degenerate states. So from the shift of the paired peaks on increasing B, we assign quantum numbers to the respective pairs. For example, the lowest, second lowest, and third lowest pairs in the vicinity of $B = 0$ T correspond to the filling of electrons in the FD states $(n, \ell) = (0, 0)$, $(0, 1)$ and $(0, -1)$ with antiparallel spins, respectively. Furthermore, the wiggles or anticrossings between pairs of peaks correspond to the crossings of the FD states. However, from close inspection of the peak pairing, we find that the pairing is modified at each anticrossing too in line with Hund's first rule. This is discussed in detail in section 3. The last wiggle of each pair of peaks appears along the dashed line, which corresponds to the bold line ($\nu = 2$) in Fig. 1.2(a). We note that the $\nu = 2$ line is more vertical in the experiment than in the calculation. This indicates that the lateral confinement becomes weaker as N increases because of the effect of self-screening and screening of the gate metal and leads [3].

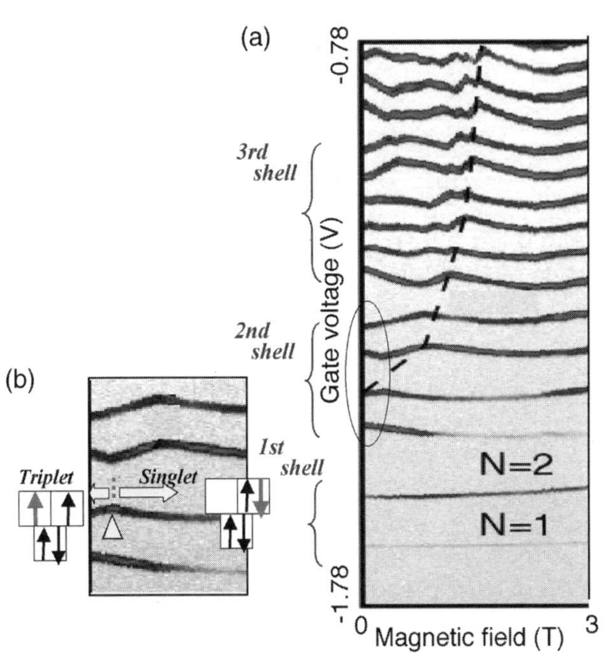

Fig. 1.5. Intensity plot of the current in the plane of magnetic field and gate voltage. The device used for the experiment is not the same but prepared in the same way as that showing the Coulomb oscillations in Fig. 1.4(a). (a) Dark lines indicate the current peak (Coulomb peak) positions. The dashed line corresponds to the dotted line in Fig. 1.2(a). (b) Magnified plot of current peaks corresponding to the filling of the second shell (at $B = 0$ T). For the half filling of the second shell, i.e. $N = 4$, a spin triplet (singlet) state is the ground state on the left (right) of the triangle point. The boxes indicate the electronic configuration for the $N = 4$ ground state.

2.5. Exchange Effect for Hund's First and Second Rules at $B = 0$ T

When we inspect closely the evolution of the first two current peaks on initially increasing B field for the filling of the second, third and fourth sells, respectively in Fig. 1.5, we find for all of the shells that the first (second) peak shifts down (up). Comparing these data to the FD-spectrum, we assign the states corresponding to the first and second peaks such that: $E_{0,-1}$ and $E_{0,1}$ for the $N=3$ and 4 peaks, $E_{0,-2}$ and $E_{0,2}$ for the $N=7$ and 8 peaks, and $E_{0,-3}$ and $E_{0,3}$ for the $N=13$ and 14 peaks [3,16]. These assignments are further confirmed from measurements of the excitation spectrum, which allows us to detect the GS and ESs simultaneously [3,17]. At $B = 0$ T the first two electrons for the filling of each shell occupy the states $E_{0,\ell}$ with opposite signs of $\ell(=\pm 1, 2$ and 3, respectively). This is because the GS is a spin triplet state favored by Hund's rule as compared to a spin singlet state, which is the ES. The triplet state has the same orbital symmetry implying that it is energetically lower than the singlet state, i.e. ES only by the exchange (EX) energy. The singlet is favored as the B field is increased, and finally becomes the GS.

We derive the EX energy for $N = 8$ as illustrated in the inset to Fig. 1.6 from the intersection with the $B = 0$ T axis of the thin line (spin singlet ES) and the measured data (spin triplet GS). Note that the triplet to singlet transition occurs for the $N=8$ GS at the upward kink in the inset. Thus obtained energy is simply associated with the

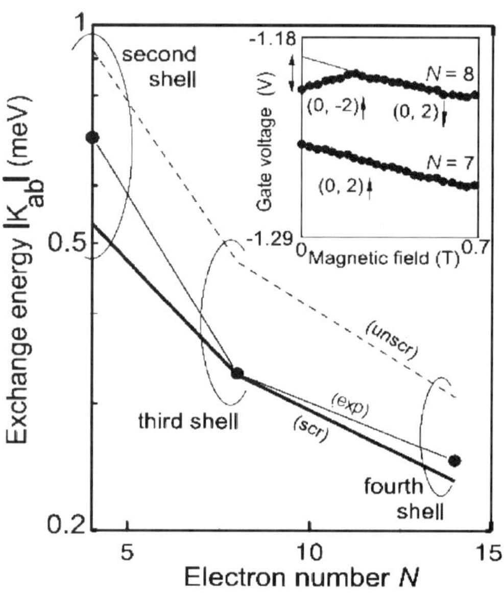

Fig. 1.6. Exchange energy $|K_{ab}|$, on a log scale, associated with spin-triplets formed when each new shell is filled by just two electrons at $B = 0$ T. The inset shows an expansion for the filling of the first two electrons into the third shell (i.e. $N = 7$ and 8). The vertical double-arrow represents $|K_{ab}|$ in units of gate voltage which is then converted to energy. The calculated curves in the main figure are for the unscreened (dashed) and screened (solid) cases (see Section 3).

states (0, 2) and (0, −2). Likewise we derive the EX energy associated with the states (0, 1) and (0, −1) for $N = 4$ and that with (0, 3) and (0, −3) for $N = 14$, and plot the values in Fig. 1.6. This figure shows that the EX energy quickly becomes smaller for higher lying shells, reflecting the more spreading out of the wavefunction for the larger angular momentum state (see Fig. 1.1).

From the electron distributions of the FD states we calculate the EX energy for two electrons occupying two degenerate states $(0, \pm|\ell|)$. We take $\hbar\omega_0 = 3\ (2)$ meV to reproduce the data for $N = 4$ (8 and 14). The dashed curve in Fig. 1.6 is the calculation when we neglect screening of the interactions within the dot by electrons in the leads and in the gate. In this case the Coulomb potential falls off as $1/r$, where r is the distance between the electrons. For the solid curves we have approximated the screening effects by replacing the Coulomb potential by $\exp\{-r/d\}/r$. We have taken $d = 10$ nm which is roughly the thickness of the tunnel barriers. The screened case provides the best quantitative agreement for our realistic choices of the confining energy and the screening distance.

Note that the total angular momentum, L, is $L = 0$, when the third and fourth shell is just filled by two electrons. This means that the present quantum dot does not show Hund's second rule, which favors to maximize L for the shell filling in real atoms. This is because of the deviation of the confining potential from a quadratic form, which is demonstrated by a 3D self-consistent computer simulation [18]. The single-particle degeneracy is then lifted for the shells higher than the second order such that states having the smaller angular momentum of $|\ell|$ are more lifted with those having the same orbital symmetry $(\pm|\ell|)$ unchanged. The energy for the lifting is of the same order as variations in $\Delta(N)$ (see Fig. 1.4(b)) of the third and higher shells, so it can significantly influence the spectrum of $\Delta(N)$.

3. Tunable Spin States with Magnetic Field

3.1. Model: Orbital Degeneracy and Two-Electron States

The GS for two electrons in the dot is either a spin single state or a spin triplet state. If there is only a single orbital state, the GS is always a spin singlet state having two antiparallel spin electrons in the same orbital state. However, if there are more than one orbital states, the GS can be a spin triplet state having two parallel spin electrons in two different orbital states, depending on the degeneracy of states. Hund's first rule for the second shell at $B = 0$ T described in section 2.5 is typically the case. Here we consider the two-electron states in the same way but in the presence of a B field. As typically shown in the FD-spectrum, there are usually a number of degeneracies or crossing of single-particle states as a function of B field (dotted circles in Fig. 1.2(a)). So here we use a simple model to generalize the condition for building up a spin triplet state formed by the first two electrons for the filling of degenerate two FD states.

For the moment we only consider the EX energy to favor a spin triplet state. If the two orbital states are energetically separated by the amount $\Delta\varepsilon$ more than the EX energy $K(<0)$, i.e. $\Delta\varepsilon > |K|$ a spin singlet state having two antiparallel spin electrons in the lower-lying orbital state is always the GS. On the other hand, if the

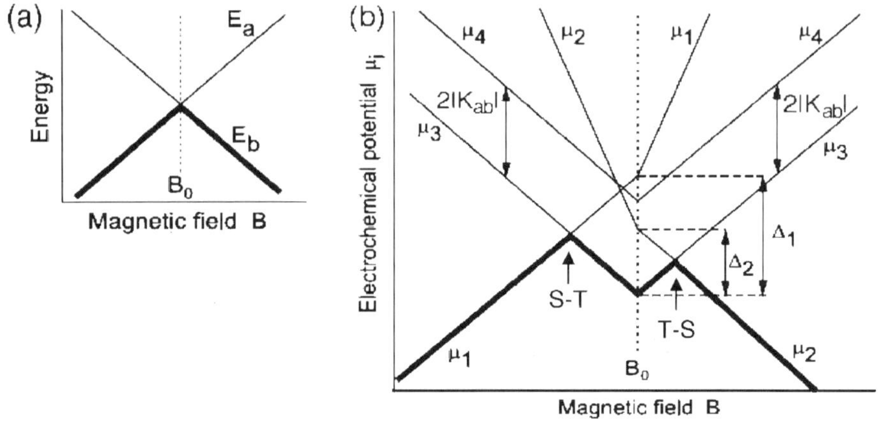

Fig. 1.7. (a) Schematic diagram of two single-particle states with energies E_a and E_b crossing each other at a magnetic field $B = B_0$. (b) Electrochemical potential, $\mu_i(2) = U_i(2) - U(1)$, for two interacting electrons. The thick line represents the ground state whereas the thin lines show the excited states.

two orbital states are sufficiently close to each other to satisfy that $\Delta\varepsilon < |K|$, the GS is a spin triplet state having two parallel spin electrons with one in each orbital state. This can be modeled for two, spin-degenerate single-particle states with energies, E_a and E_b, crossing each other at $B = B_0$ (Fig. 1.7(a)). The GS energy, $U(1)$, for one electron occupying these states, equals E_a for $B < B_0$ and E_b for $B > B_0$ (thick line in Fig. 1.7(a)). For two electrons we can distinguish four possible configurations with either total spin $S = 0$ (spin-singlet) or $S = 1$ (spin-triplet). (We neglect the Zeeman energy difference between $S_z = -1, 0$ and 1.) The corresponding energies, $U_i(2, S)$ for $i = 1$ to 4, are given by:

$$U_1(2, 0) = 2E_\mathrm{a} + C_\mathrm{aa} \tag{2a}$$

$$U_2(2, 0) = 2E_\mathrm{b} + C_\mathrm{bb} \tag{2b}$$

$$U_3(2, 1) = E_\mathrm{a} + E_\mathrm{b} + C_\mathrm{ab} - |K_\mathrm{ab}| \tag{2c}$$

$$U_4(2, 0) = E_\mathrm{a} + E_\mathrm{b} + C_\mathrm{ab} + |K_\mathrm{ab}| \tag{2d}$$

Here, $C_{ij}(i, j = \mathrm{a, b})$ is the direct Coulumb (DC) energy between two electrons occupying states with energies E_i and E_j, and K_ab is the EX energy ($K_\mathrm{ab} < 0$) between two electrons occupying E_a and E_b with parallel spins. The electrochemical potential is defined for a two electron GS as $\mu(2) = U(2, S) - U(1)$. For each $U_i(2, S)$ we obtain potentials: $\mu_i(2) = U_i(2, S) - E_\mathrm{a}$ for $B < B_0$ and $U_i(2) - E_\mathrm{b}$ for $B > B_0$ (see Fig. 1.7(b)). The GS has $S = 0$ away from B_0. Near B_0, the lowest energy is $\mu_3(2)$, so $S = 1$. The downward cusp in the thick line identifies this spin-triplet region. The transition in the GS from $S = 0$ to 1 and $S = 1$ to 0, respectively, occurs when $\mu_1 = \mu_3$ for $B < B_0$ (labeled "S-T") and when $\mu_2 = \mu_1$ for $B > B_0$ (labeled "T-S"). We define two energies, Δ_1 and Δ_2, to characterize the size of the downward cusp in the GS at

$B = B_0$:

$$\Delta_1 = \mu_1 - \mu_3 = C_{aa} - C_{ab} + |K_{ab}| \tag{3a}$$

$$\Delta_2 = \mu_2 - \mu_3 = C_{bb} - C_{ab} + |K_{ab}| \tag{3b}$$

$$\Delta_1 - \Delta_2 = C_{aa} - C_{bb} \tag{3c}$$

If the states with E_a and E_b have the same type of orbitals, EX interactions only contribute to the downward cusp since $C_{aa} = C_{bb} = C_{ab}$. However, FD states at the crossings at finite B fields usually have different types of orbitals, and thus $C_{aa} \neq C_{bb} \neq C_{ab}$, so both of EX and DC interactions can significantly contribute to the downward cusp.

3.2. Observation of Spin Singlet–Triplet Transitions

In Fig. 1.5 we see the wiggles of paired peaks corresponding to the crossings of FD states in the low B field range. To examine in detail the experimental data in the vicinity of such a crossing point we show in Fig. 1.8(a) a magnified plot for the current peaks between $N = 7$ and 16 for $B = 0$ to 3 T. We see clearly the paired peaks indicating the antiparallel spin filling of a single orbital state by two electrons. Modifications to

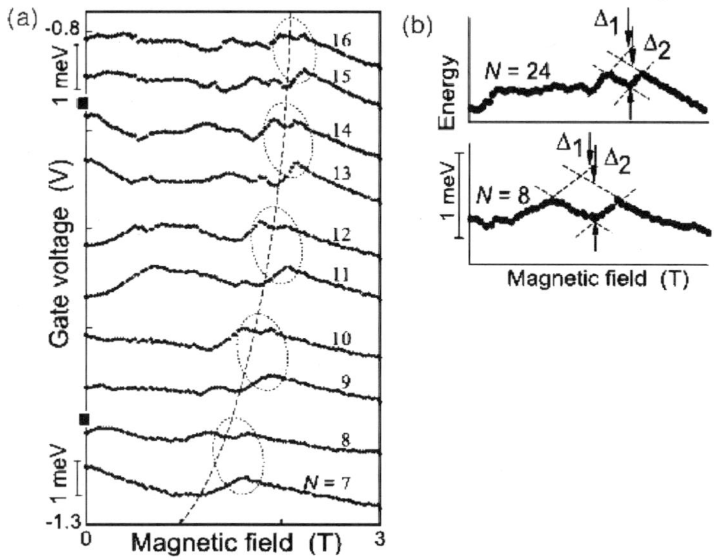

Fig. 1.8. (a) Magnified plot of the magnetic field dependencies of current peak positions from $N = 7$ to 16 shown in Fig. 1.5. The bars along the gate voltage axis show 1 meV energy scales calibrated at -1.26 and -0.85 V. The dotted curve indicates the last crossing between single-particle states. Dashed ovals group pairs of ground states for odd and even electron numbers. Spin transitions in the ground states are indicated by ■ at $B = 0$ T and occur inside the ovals for $B \neq 0$ T. (b) Magnified plots of the $N = 8$ and 24 current peaks versus magnetic field. The dashed lines illustrate how the interaction-energy parameters, Δ_1 and Δ_2 are determined (explained in the text).

the pairing are observed for the peaks labeled by ■ at 0 T, and in each of the dotted ovals connecting pairs of peaks at non-zero field. These are all signatures of Hund's first rule; i.e. spin-polarized filling. We show expansions of the evolution of the $N = 8$ and $N = 24$ peaks in Fig. 1.8(b) to demonstrate the resemblance between the data and the model of Fig. 1.7(b). The downward cusps are clearly seen. The dashed lines form a parallelogram, from which we obtain parameters Δ_1 and Δ_2 in Eqs 3(a)–(c).

3.3. *Derivation of Direct Coulomb and Exchange Energy*

To compare the two-electron model with larger electron numbers, we assume that other states are far away in energy so that they can be neglected. Then, the downward cusps in Fig. 1.7(b) should occur for all higher *even*-electron numbers, whereas they should be absent for all higher *odd*-electron numbers. This is clearly observed from the ovals in Fig. 1.8(a). For instance, the B field dependence of the 9th peak compares well to the thick line in Fig. 1.7(a), i.e. a transition between two single-particle states occupied by one electron. The B field dependence of the 10th peak compares well to the thick line in Fig. 1.7(b), i.e. an extra downward cusp where the GS has two electrons with parallel spins ($S = 1$), occupying two different orbital states which are nearly degenerate. Other pairs of even and odd numbered peaks show the same behavior. This justifies our assumption that the other electron states can be neglected and that we can simplify the many-electron system to a system of just one or two electrons.

As illustrated in Fig. 1.8(b), we can derive the experimental values for Δ_1 and Δ_2 for different N. These values are plotted as solid symbols in Fig. 1.9. We find that Δ_1 is larger than Δ_2 for all electron numbers, implying that $C_{\mathrm{aa}} > C_{\mathrm{bb}}$. As N increases from 6 to 12, Δ_1 first increases and then slowly decreases, whilst Δ_2 slightly decreases. For comparison we show the exchange energy for $N = 8$ in the same figure.

The above discussion has been kept general for crossings between any type of single-particle states. We calculate DC and EX energies using FD states. As can be seen from Fig. 1.2(a) the last crossing is always a crossing between just two FD states. The up-going state is always $(n, \ell) = (0, -1)$, whereas the down-going state, $(0, \ell > 1)$, has an increasing angular momentum for states with higher energy. The relation with Fig. 1.7(a) is: $E_{\mathrm{a}} = E_{0,-1}$ and $E_{\mathrm{b}} = E_{0,\ell>1}$.

We take $\hbar\omega_0 = 2$ meV for the calculation to reproduce the data for $N > 8$ and obtain Δ_1 and Δ_2 with Eqs 3(a)–(c). For comparison we show the calculated screened and unscreened values by the dashed and solid curves, respectively. Figure 1.9 shows that screening considerably reduces Δ_1 to values much closer to the experimental values. For all N, we find that $\Delta_1 > \Delta_2$ and thus $C_{\mathrm{aa}} > C_{\mathrm{bb}}$. Since the average radius of the wavefunctions increases with angular momentum, two electrons are closer together when they both occupy $(0, -1)$ compared to when they both occupy $(0, \ell = N/2 - 1)$ for even-$N > 4$ (or $\ell > 1$), so the DC interaction is stronger in the former than the latter. This explains our observation that $C_{\mathrm{aa}} > C_{\mathrm{bb}}$ for all N. The overlap between *different* wavefunctions, $(0, -1)$ and $(0, \ell = N/2 - 1)$, decreases for even-$N > 4$ (or $\ell > 1$). This results in a decrease in both C_{ab} and $|K_{\mathrm{ab}}|$ with N. Note that $|K_{\mathrm{ab}}|$ decreases more rapidly with N than Δ_1 and Δ_2. This means that the downward cusp observed in Fig. 1.8 arises mainly from the direct Coulomb effect for $N > 8$.

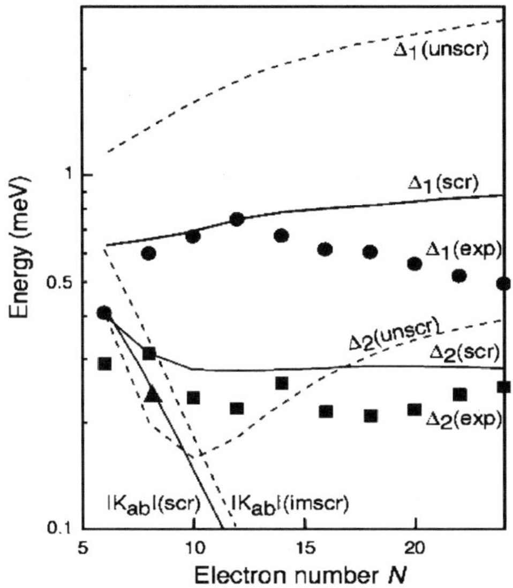

Fig. 1.9. Experimental values for the energy parameters $\Delta_1(\bullet)$ and $\Delta_2(\blacksquare)$, on a log scale, versus electron number derived from data as shown in Fig. 1.8(a). The exchange energy $|K_{ab}|(\blacktriangle)$ is also shown for $N = 8$ (Fig. 1.6). The dashed and solid curves are calculated from the FD-wavefunctions at 2 T, for an unscreened (unscr) and screened (scr) Coulomb interaction.

Our general model used for the calculations in Figs 1.6 and 1.9 provides a clear identification of effects due to EX and DC interactions. We note that there are more advanced calculations supporting our analyses [19].

4. Pauli Blockade in Single Electron Tunneling

4.1. Many-Body Effect and Pauli Exclusion

Coulomb blockade discussed to date solely reflects the charge discreteness. Regulated by the Coulomb blockade, the number of electrons in the dot is varied one-by-one (single electron tunneling), and this means that the N electron GS can only be constructed by adding (subtracting) one electron to (from) the $N-1(N+1)$ electron GS. However, in some cases the GS cannot be constructed in this way because of the many-body effect as well as the Pauli exclusion. In the presence of many-body interactions, particularly internal ferromagnetic or antiferromagnetic spin couplings, the total spin for the N electron GS, $S(N)$, can be different from that for the $N-1$ electron GS, $S(N-1)$ by more than $\frac{1}{2}$: $|\Delta S(N)| = |S(N) - S(N-1)| > \frac{1}{2}$. Then the transition between the N and $N-1$ GS cannot be undertaken by an event of single electron tunneling, because the single event can only change the total spin by $\pm\frac{1}{2}$. This leads to blockade of single electron tunneling, which is called "spin blockade" [20,21]. This blockade is usually not so strong as Coulomb blockade, because it can be lifted by spin relaxation in the quantum dot. Recently a couple of experiments on the spin blockade have been reported [21,22]. In those experiments, however, it is difficult to assign the spin effect because the many-body state is not well defined.

On the other hand, Pauli exclusion leads to a simple spin effect to block the single electron tunneling because it arises from the Fermionic nature of electrons. Pauli exclusion states that a single quantum state cannot be occupied by more than one electron. In other words two parallel spin electrons cannot occupy the same orbital state. This can lead to suppression of single electron tunneling, for example, in such a system that electrons in the source contact are fully spin polarized, because parallel spin electrons cannot be consecutively filled into the same orbital state in the quantum dot even when $|\Delta S| = \frac{1}{2}$. Here the many-body effect in the quantum dot is not important at all. The 2DEG in the contact lead can be fully spin-polarized in the presence of a sufficiently high B field for reaching the $\nu = 1$ quantum Hall region. However, in such a strong B field the states in the quantum dot can also be confined to the spin-polarized lowest Landau state, and then accommodate only spin-polarized electrons with no suppression of single electron tunneling. This makes difficult or complicated to facilitate the experiment. Instead, we have recently used a weakly coupled double quantum dot to demonstrate the Pauli effect [23]. In the following subsections we show that in transport through a series of 0D sites, with symmetry under inversion appropriately broken, the Pauli effect, in combination with the Coulomb blockade, can be used to block current altogether in one direction while permitting it to flow in the opposite direction.

Besides the Pauli effect described above, suppression of single electron tunneling arising from the formation of $\nu = 2$ edge channels in the contact leads has been observed [24]. The dot-lead tunneling probability is very small for down spin electrons in the inner edge channel than up spin electrons in the outer edge channel. Then single electron tunneling for adding a down spin electron to the dot is significantly reduced.

4.2. *Weakly Coupled Double Dot*

Suppose we have two sites, site 1 and site 2, weakly coupled together, and one electron is permanently localized on site 2 (see Fig. 1.10). Now consider transport of a second electron through the system between two contact leads. The number of electrons on site 1 (2), N_1 (N_2), varies between 0 and 1 (1 and 2). An electron can only be transported through the system when the transition between the two-electron states $(N_1, N_2) = (0, 2)$ and $(1, 1)$ is allowed. This condition is met when the necessary energy cost to add one more electron to the system is compensated for by the action of a nearby plunger gate voltage, or the voltage between the leads, which is a familiar single-electron tunneling phenomenon. Crucially, spin effects also markedly influence the electron transport. Because the tunnel coupling between the two sites is sufficiently weak, the $(N_1, N_2) = (1, 1)$ spin-singlet and spin-triplet states are practically degenerate. Additionally, for $(N_1, N_2) = (0, 2)$, only a spin singlet is permitted because of Pauli exclusion. Therefore, electron transport is only allowed for a channel made from the $(1, 1)$ and $(0, 2)$ singlet states. This always holds true for reverse bias when the chemical potential of the left lead, μ_ℓ, nearest site 1, is lower than that of the right lead, μ_r, nearest site 2, because only an antiparallel spin electron can be injected onto site 2 from the right lead (Fig. 1.10). On the other hand, for forward bias, $\mu_\ell > \mu_r$, either the $(1, 1)$

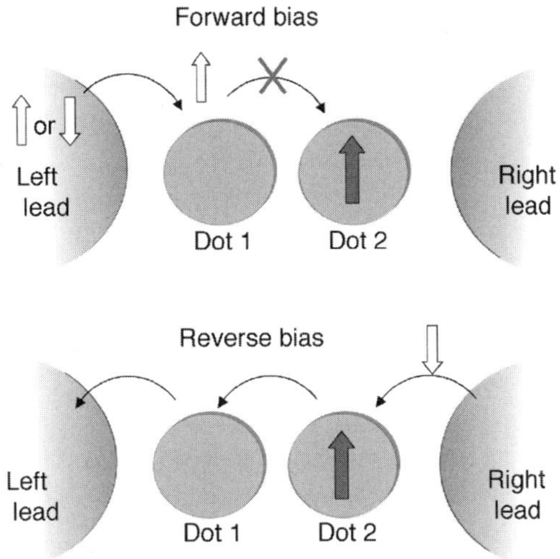

Fig. 1.10. Model for current rectification by Pauli effect in a series of two quantum dots, dot 1 and 2. Each quantum dot holds just one spin degenerate orbital state, and an up spin electron is permanently trapped in dot 2. Single electron tunneling occurs via the two-electron states in the double dot: from the left (right) to the right (left) lead for forward (reverse) bias.

singlet or triplet can be populated with more or less the same probability by injection of an electron on to site 1 from the left lead. If the (1, 1) singlet is populated, this subsequently generates a single electron tunneling current flowing through the singlet state. Once the triplet is populated, however, subsequent electron transfer from site 1 to 2 is blocked by Pauli exclusion (Pauli blockade). Note that an electron arriving on site 1 cannot go back to the left lead because of the fast relaxation of the hole state left behind in the lead. When the forward bias is sufficiently small compared to the coupling energy between the left lead and site 1, or to the effective electron temperature, the electron on site 1 of the (1, 1) triplet can go back to the left lead, in which case current suppression does not occur. This is the case only for linear transport near zero bias where we can actually expect to observe a single electron tunneling current. Thus the (1, 1) triplet will sooner or later be occupied on a time scale sufficiently longer than the electron tunneling time between the leads, and this should lead to clear current suppression, for example, in dc measurement.

The device used for the experiment is made by vertically coupling two circular quantum dots, which is similar to those used for the experiment of single-particle tunnelling spectroscopy (see section 2.3). The specific device we discuss is a 0.6 μm cylindrical mesa and has an AlGaAs (8 nm)/InGaAs (12 nm)/AlGaAs (6 nm)/InGaAs (12 nm)/AlGaAs (8 nm) triple barrier structure (see Fig. 1.11(a)). The tunnel coupling energy between two dots is estimated to be 0.3 meV [25]. This is much smaller than the 2D harmonic potential energy ($\hbar\omega_0 \sim 4$ meV), and the charging energy for each dot ($U \sim 4$ meV). Again note that in this device configuration, the total number of electrons in the whole double dot system, $N (= N_1 + N_2)$, can be varied one-by-one as a function of gate voltage, V_G, starting from $N = 0$ [25,26]. Here, we label the two dots dot 1 and

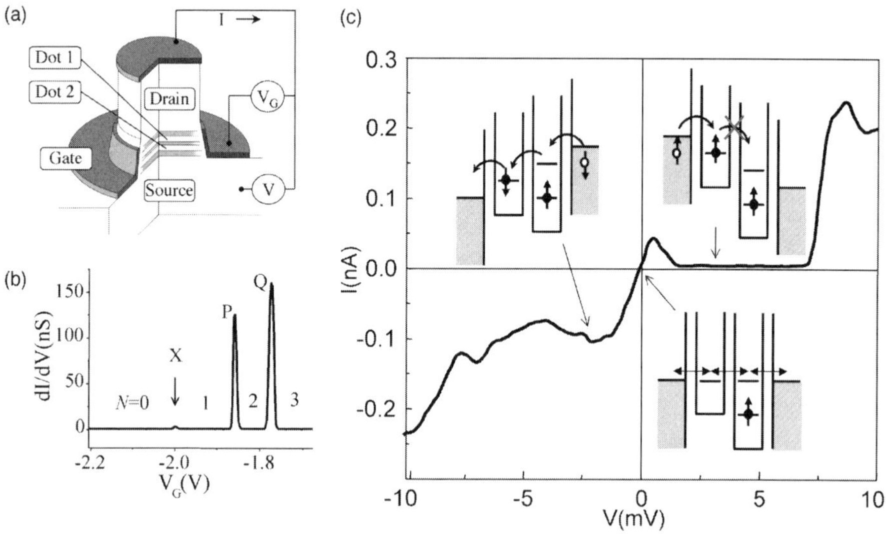

Fig. 1.11. Blockade of single electron tunneling current by Pauli exclusion in a weakly coupled double dot system: I versus V curve measured for V_G fixed at the second Coulomb peak ($V_G = -1.854$ V, upper left inset).

dot 2, and they correspond to the site 1 and site 2 in Fig. 1.10, respectively. Similarly $N_1(N_2)$ is the number of electrons in dot 1(dot 2). The transmission coefficients for all the tunnel barriers (dot–electrode and dot–dot) are sufficiently weak that electron transport can be discussed just in terms of sequential tunneling between the source and drain.

The general situation of Fig. 1.10 can be reproduced in our device (see Fig. 1.11(c), potential diagrams), if there is an appropriate potential offset, 2δ, between the two quantum dots at $V = 0$ V [27]. Then, just one electron is trapped in the $1s$ orbital state of dot 2, and the two-electron GS is either $(N_1, N_2) = (1, 1)$ or $(0, 2)$. For $(1, 1)$, the singlet and triplet states are nearly degenerate because the coupling between the two dots is very weak. For $(0, 2)$, only the singlet GS is initially relevant. A $(0, 2)$ triplet ES can be formed by putting two parallel spin electrons in the $1s$ and $2p$ states in dot 2. However, its energy is normally much higher than that of the $(1, 1)$ and $(0, 2)$ GS, so it does not influence the transport in our discussion for the moment. When viewing Fig. 1.10, we can see that electrons can be transported by the $(0, 2)$ and $(1, 1)$ singlet states for reverse bias, whereas for forward bias, the $(1, 1)$ triplet can be populated, leading to the blockade of electron transport. We now define the electro-chemical potential of the (N_1, N_2) GS to be $\mu(N_1, N_2)$, and the chemical potential of the source (drain) contact to be μ_r (μ_ℓ). Pauli blockade in Fig. 1.10 can appear when electrons are injected from the drain to the $(1, 1)$ triplet in the nonlinear transport.

We assume that spin is conserved in the electron tunneling throughout the double dot, and also that there are no spin flips on a time scale sufficiently longer than the electron tunneling time. This condition actually holds for our experiment, which we will explain later.

We measure the dc current, I, flowing vertically through the two dots as a function of V and V_G to study the linear and nonlinear electron transport. V_G and V, respectively, are used to change the electrostatic potential of the two dots together, and the potential offset between the two dots. By adjusting these voltages and using a double dot sample with an appropriate potential offset 2δ, we can realize the situation in Fig. 1.10. All of the data is taken at an effective electron temperature of about 0.2 K, and about 10% of the applied voltage V is dropped between the two dots. Note that the potential offset 2δ ranging from ~0.5 to a few meV, has been observed in all of our double dot samples. This is probably due to small decrease in the effective electron density in the cylindrical mesa just above the two dots relative to that below [27]. We select a double dot sample having an appropriate value of 2δ such that the condition $\mu(1, 1) \cong \mu(0, 2)$ is satisfied at $V = 0$ V.

Figure 1.11(b) shows the first three Coulomb peaks in the dI/dV versus V_G measured for $V \sim 0$ V. The first peak (X) is very small but definitely present at $V_G \sim -2$ V, and this indicates the transport through the double-dot system for N fluctuating between 0 and 1. The second peak (P) and the third peak (Q) are much larger. This implies that tunneling is elastic between the source and drain leads for $N = 1 \Leftrightarrow 2$ and $2 \Leftrightarrow 3$. However, because of the potential offset 2δ, this is not the case for $N = 0 \Leftrightarrow 1$. Electrons must tunnel through the lowest state in dot 2, but since there is no available state for tunneling in dot 1, second-order tunneling through dot 1 is necessary [28]. We then study nonlinear transport through the two-electron states by measuring the $I - V$ for V_G fixed at the second Coulomb peak (X) (Fig. 1.11(c), main curve). The data reveals a large, nearly constant dI/dV at $V \sim 0$ V for both bias polarities, and this is due to elastic tunneling in the linear response region. Tunneling here occurs via the (1, 1) and (0, 2) singlet states, which are aligned at $V \sim 0$ V when we take into account the effect of the potential offset. Transport is linear for $V \sim 0$ V, and is nonlinear for $|V| \geq 1$ mV (see potential diagrams in Fig. 1.11(c)). The linear transport conductance (dI/dV) is pretty large, reflecting elastic tunneling though the double dot system. For the nonlinear transport, the current is clearly suppressed in forward bias because of Pauli exclusion, whereas a large current flows in reverse bias because of inelastic tunneling via the singlet states. The Pauli blockade is lifted for $V > 7$ mV when an electron can be ejected from the lowest state of dot 2 to the source. The Pauli blockade is further confirmed from measurements of nonlinear transport for various V_G values [23].

From the level of the current ($I = 1 \sim 2$ pA) in the Pauli blockade region, we estimate the lifetime of the spin triplet to be $e/I \sim 100$ ns. This is markedly longer than the time taken for just single-electron inelastic tunneling between the contact leads ($e/I \sim 1.6$ ns) obtained from the current level for $V < -1$ mV. From Fig. 1.11 we also find that the magnitude of the current in the Pauli blockade region is more or less the same as that observed for the second order tunneling through the lowest state in dot 2. This suggests that the current in the Pauli blockade region is also caused by second order tunneling processes [28], and that the intrinsic spin-flip relaxation time is longer than 100 ns. An even longer relaxation time for spin has recently been reported for a single quantum dot, which is described in section 5. As well as the second order tunneling processes, a small current can also be induced by spin-flips inside the dot. This, for example, can arise from spin-orbit scattering, however, its contribution is not yet well understood. Note that the current level in the Pauli blockade region is much higher than that in the region of Coulomb blockade (<100 fA). This clearly demonstrates that unknown current leakage is negligible in our system.

4.3. Lifting of Pauli Blockade

To explore further the spin blockade region in Fig. 1.11(c), we study the influence of a magnetic field perpendicular to the 2D plane of the dot. In our model, Pauli blockade in forward bias can be lifted if a $(N_1, N_2) = (0, 2)$ triplet is available, and it can be made lower in energy than the (1, 1) triplet (Fig. 1.12(a)). Then, a spin-triplet channel is opened and current can flow. The (0, 2) triplet at low B field is the ES with two parallel spin electrons in dot 2, one in the $1s$ state and the other in the $2p$ state. At $B = 0\,\mathrm{T}$, the $2p$ single-particle state has a twofold orbital degeneracy: $2p_+$ for $\ell = 1$ and $2p_-$ for $\ell = -1$. When the B field is initially increased, the $2p_+$ state approaches the $1s$ state (see Fig. 1.2(a)). In addition, the Coulomb energy of the two-electron states is increased due to magnetic confinement of the wave functions, but it occurs at a slower rate for the (0, 2) triplet than for both of the (1, 1) triplet and (0, 2) singlet, because the $2p$ state is more spatially extended than the $1s$ state (Fig. 1.1). Therefore, the (0, 2) triplet becomes energetically favored relative to the other two-electron states. To track the B field-induced change in energy of the (0, 2) triplet relative to the (1, 1) triplet, we investigate the related resonance, which gives rise to a current peak in forward bias. Figure 1.12(b) shows the B field-induced downshift of the $1s$–$2p_+$ resonance, which is identified by the current peak labeled ▼ in I–V that cuts across the Pauli blockade region. As the B field is increased toward 5 T, the $1s$–$2p_+$ current peak approaches the $V = 0\,\mathrm{V}$ axis as it progressively enters the Pauli blockade region. Pauli blockade is then lifted, and we can see enhanced current on and to the right of the current peak.

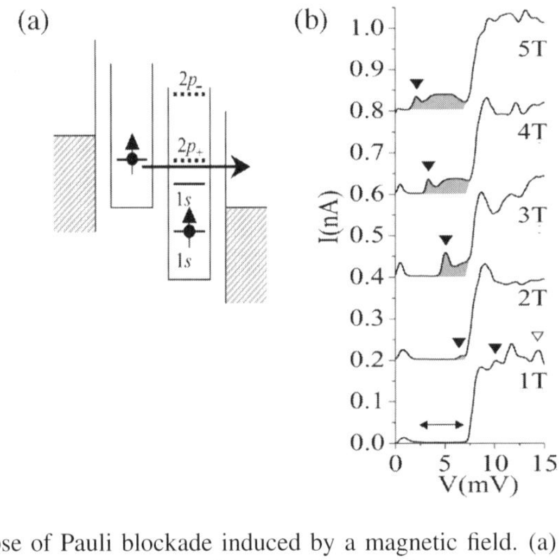

Fig. 1.12. Collapse of Pauli blockade induced by a magnetic field. (a) Schematic potential diagram showing how Pauli blockade is lifted by the B field induced lowering of the $2p$ state in dot 2. (b) Shift of the $1s$ (dot1)–$2p_+$ (dot 2) resonance peak (▼) in the I–V cutting across the Pauli blockade region with increasing B field from 1 to 5 T. Each I–V curve is vertically offset by 0.2 nA. For the $1s - 2p_-$ resonance, a peak (▽) is only indicated for 1 T.

We note that for plots like Fig. 1.12(b), but different V_g, the $1s$–$2p_+$ resonance peak actually enters the $N = 2$ Coulomb blockade region and immediately disappears. This also demonstrates that the spin blockade observed in Figure 1.11(c) is spin-related rather than charge-related.

5. Energy Relaxation with and without Spin-Flip

5.1. Allowed and Forbidden Transitions in Quantum Dots

Strength of transition from one state to another strongly depends on their quantum states, known as selection rules for radiative transitions in atoms [29]. Spectroscopic studies of allowed transitions in hydrogen and helium atoms provided crucial evidence for the Bohr's model of an atom. "Forbidden" transitions, which are usually allowed by higher-order processes or other mechanisms, indicate how well the quantum numbers describe the system. Here we discuss the strength of the transitions and their selection rules in $N = 1$ and 2 electron quantum dots. Understanding the transitions in quantum dots is an important issue for dynamics of quantum states. We focus on spontaneous transitions from a higher state to a lower state (energy relaxation), which measure how strongly the quantum states are coupled to their environments. Some coupling mechanisms for spin and orbital degrees of freedom in quantum dots are discussed.

5.2. Electrical Pump and Probe Measurement

In the Coulomb blockade regime, N is fixed at an integer. When a gate voltage is immediately changed from one Coulumb blockade region at $N = N_0(N_0 + 1)$ to another at $N = (N_0 + 1)\ (N_0)$, one electron is injected (extracted) into (from) the quantum dot. These injection/extraction can be done with a time accuracy approximately given by the inverse of their tunneling rates, Γ (Γ^{-1} can be varied in a wide range from 1 ps to 1 μs for typical quantum dots). Moreover, an electron can be (sometimes selectively) injected into, or extracted from, GS or ESs by adjusting external voltages. Therefore, the external voltages can be designed to measure the energy relaxation time from an ES to the GS (electrical pump and probe measurement). We repeatedly apply voltage pulses to push the system out of equilibrium, and measure the time-integrated non-equilibrium transient current in the following two ways.

A very simple scheme is to apply a rectangular-shaped voltage-pulse to a gate electrode, as schematically shown in Fig. 1.13(a) [7]. During the steady state before the pulse, both the ES and the GS for a particular number of electrons are located above the Fermi energy of the contact leads μ_L and μ_R for a sufficiently long period to ensure that these states are completely empty (see Fig. 1.13(b)). Then the positive pulse lowers the potential of the dot so that only the ES is located in the transport window (see Fig. 1.13(c)). Transport through the ES continues until the GS becomes occupied. First, an electron tunnels into either the ES or the GS with probabilities, $\Gamma_{L,e}$ and $\Gamma_{L,g}$, respectively, but only one electron can enter the dot. If an electron is injected into the ES, it can relax to the GS, or tunnel to the right lead to give a net current. However, the transport is blocked once the GS is occupied. By choosing strongly asymmetric barriers, $\Gamma_L \gg \Gamma_R$, an electron can stay in the ES for a long time. Thus, the decay

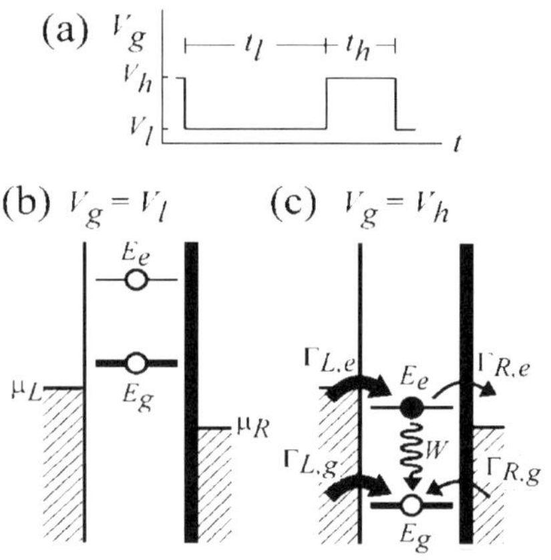

Fig. 1.13. (a) Schematic of pulse waveform used for the relaxation time measurement. (b) and (c) Schematic energy diagrams for low (b) and high (c) pulse situations. The thick and thin vertical lines denote the asymmetric tunneling barriers. Tunneling processes are shown by arrows.

rate, D, of the transient current should reflect the energy relaxation rate, W, from the ES to the GS.

We can relate D and W from the rate equations,

$$\begin{aligned}\frac{d}{dt}\rho_e &= \Gamma_{L,e}(1-\rho_e-\rho_g) - \Gamma_{R,e}\rho_e - W\rho_e \\ \frac{d}{dt}\rho_g &= (\Gamma_{R,g}+\Gamma_{L,g})(1-\rho_e-\rho_g) + W\rho_e\end{aligned} \tag{4}$$

by taking into account all the tunneling processes shown schematically in Fig. 1.13(c). Here ρ_e and ρ_g are the average electron numbers in the ES and the GS, respectively. $\rho_e = \rho_g = 0$ at the beginning of the pulse $t = 0$, and $0 \leq \rho_e + \rho_g \leq 1$ is satisfied due to Coulomb blockade. We obtain a form $\rho_e(t) = A(1-e^{-Ft})e^{-Dt}$, assuming $\Gamma_L \gg \Gamma_R$ and $W < \Gamma_{L,e} + \Gamma_{L,g}$ (valid for the visible current). $A \simeq \Gamma_{L,e}/(\Gamma_{L,e} + \Gamma_{L,g})$ is the injection efficiency into the ES. The filling rate, $F \simeq \Gamma_{L,g} + \Gamma_{L,e}$, is so large that $e^{-Ft} \sim 0$ in our time domain. The decay rate,

$$D = W + \Gamma_{R,e}(1-A) \tag{5}$$

contains information about W. The average number of tunneling electrons, $\langle n_e \rangle$, can be written as

$$\langle n_e \rangle \simeq \int \Gamma_{R,e}\rho_e dt = A\Gamma_{R,e}(1-e^{-Dt})/D \tag{6}$$

and can be compared with the experiments. In this scheme, small number, which can be much smaller than 1, of electrons per pulse contribute to the current. By repeating the pulse by a frequency, f_{rep}, a reasonable current, $\langle n_e \rangle e f_{rep}$, is obtained. Therefore, the decay rate D of the transient current gives the relaxation rate W, provided that

$$\Gamma_{R,e} \ll W < \Gamma_{L,e} + \Gamma_{L,g} \tag{7}$$

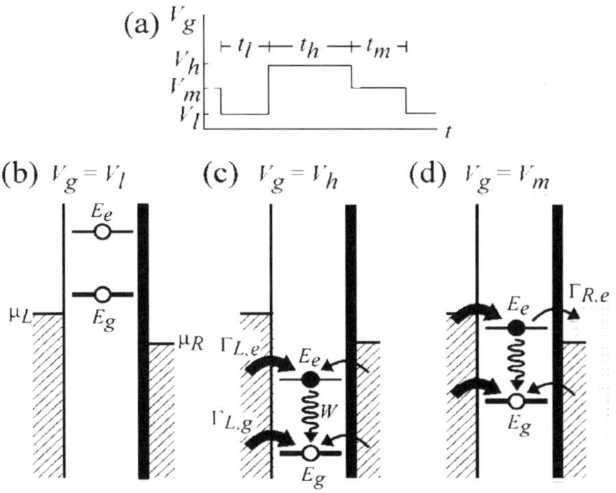

Fig. 1.14. (a) Schematic of double-step pulse waveform to measure extremely long relaxation times. (b)–(d) Schematic energy diagram showing low (b), high (c) and intermediate (d) pulse situations.

This condition is satisfied when the saturation value of $\langle n_e \rangle$ is much smaller than unity and $\langle n_e \rangle$ can be well approximated by a single exponential curve.

More improved method for a longer relaxation time, i.e. $W < \Gamma_{R,e}$ is to apply a double-step voltage-pulse, in which V_g is changed between three voltages, V_ℓ, V_h and V_m as shown in Fig. 1.14(a) [30]. First, when $V_g = V_\ell$ (Fig. 1.14(b)), an empty quantum dot is prepared during a sufficiently long period. When V_g is suddenly increased to V_h (Fig. 1.14(c)) so that both ES and GS are located below the Fermi energies, an electron enters to either ES or GS. Once an electron enter either states, neither it can leave the dot, nor another electron can enter the dot. If it is in the ES, the ES may relax to the GS during this time. Therefore, the average electron numbers in the ES after this period, t_h, is approximately given by $\rho_{e,h} = A \exp(-Wt_h)$ for $t_h \gg \Gamma_L^{-1}$, where $A \simeq \Gamma_{L,e}/(\Gamma_{L,e}+\Gamma_{L,g})$. When V_g is changed to V_m (Fig. 1.14(d)), an electron in the ES can tunnel out to the drain, if the ES is not yet relaxed to the GS. This tunneling process can be described by Eq. (4) with initial conditions, $\rho_e = \rho_{e,h}$ and $\rho_g = 1 - \rho_{e,h}$. With a sufficiently long duration, $t_m \gg \Gamma_{R,e}^{-1}$, for complete readout, the average number of tunneling electrons, $\langle n_e \rangle$, can be written as

$$\langle n_e \rangle = A' \exp(-Wt_h), \tag{8}$$

where $A' = \Gamma_{R,e}/\Gamma_{R,g}$. We repeatedly apply the double-step pulse (effectively -10^7 times) to obtain a reliable current I_p, and evaluate the average number of tunneling electrons, $+n_{e'} = I_p(t_\ell + t_h + t_m)/e$. Provided that $W < \Gamma_R \ll \Gamma_L$, the relaxation rate W can be measured with this method.

These two techniques were used to measure the relaxation time in quantum dots. Practically, high-speed rectangular voltage-pulse (or double-step pulse) is introduced to a gate electrode through a low-loss coaxial cable. Due to a parasitic capacitance of the electrode, the rise time of the pulse at the sample was $\tau_{rise} = 0.7 - 2$ ns, which is monitored by the reflected signal (time-domain reflectometry). The rise time of the applied pulse must be comparable to or shorter than Γ_R^{-1}, otherwise electron cannot be injected into the ES. The rectangular waveform may contain a small ringing structure and fluctuations (less than 10% of the pulse height in the experiments), which slightly broaden the current peak and make energy selectivity worse.

5.3. Momentum Relaxation in $N = 1$ Quantum Dots

We apply electrical pump and probe measurements to quantum dots defined in sections 2 and 3. The most simple relaxation process is the transition from the $2p$ state (1st ES) to the $1s$ state (GS) in $N = 1$ quantum dot, i.e. artificial hydrogen atom. This transition changes the electron's orbital momentum but preserves the spin, as schematically shown in Fig. 1.15(a). The $2p$ state in a 2D harmonic quantum dot has two-fold orbital degeneracy at $B = 0$ T, which might complicates the analysis. Instead an elliptic quantum dot is used to lift the orbital degeneracy of $2p$ state. The quantum dot used for the experiment is elliptic in the lateral confining potential, and we approximate the lateral confinement by two orthogonal harmonic potentials in the x and y plane (lateral directions) [31]. The characteristic confinement energies of the quantum dot studied here are $\hbar\omega_x \sim 2.5$ meV and $\hbar\omega_y \sim 5.5$ meV. This non-circularity does not affect much our discussion or the underlying physics. The Zeeman splitting of the electron spin is small ($E_Z \sim 0.1$ meV at $B = 5$ T), and is unresolved in our experiments.

Single rectangular pulse technique was used to measure the momentum relaxation time in one-electron system. Figure 1.15(b) shows typical transient current signals, which are converted into $\langle n_e \rangle$, obtained at various pulse length, t_h. This behavior can be fitted well with the exponential function of Eq. (6) (solid lines in Fig. 1.15(b)), and the small saturation value of $\langle n_t \rangle$, ~ 0.02 indicates very efficient relaxation. This ensures the condition (Eq. (6)) for the relaxation time measurement. The relaxation time estimated from the rise time of $\langle n_t \rangle$ is $\tau_{1s-2p} \sim 10$ ns at $B = 1$ T and 3 ns at

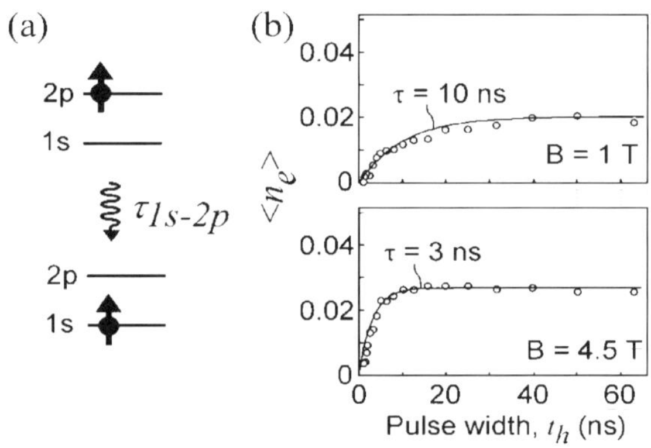

Fig. 1.15. (a) The electron configurations of $N = 1$ quantum dot before and after relaxation. (b) The average number of tunneling electrons per pulse, n_t, measured at $B = 1$ T and 4.5 T. The relaxation time is obtained from the exponential curve (solid line) fitted to the data.

$B = 4.5\,\mathrm{T}$ for the case in Fig. 1.15(b). The magnetic field dependence of is τ_{1s-2p} is summarized in Fig. 1.16(c).

For a QD system whose energy spacing is meV region, the electron–phonon interaction, rather than photon, is significant, and the energy spacing coincides with the acoustic phonon energy in the linear dispersion regime [32]. Only the spontaneous emission can contribute to the energy relaxation process at low temperature (thermal energy $\sim 10\,\mu\mathrm{eV} \ll$ the energy spacing between the $1s$ and $2p$ states, $\varepsilon_{1s-2p} \sim 2\,\mathrm{meV}$). Because of the discrete energy of the states, the relaxation involves an emission of a phonon of energy equal to ε_{1s-2p}. The corresponding phonon wavelength is given by $\lambda_{1s-2p} = hv_{phonon}/\varepsilon_{1s-2p}$, where $v_{\mathrm{phonon}} = 5100\,\mathrm{m/s}$ is the phonon velocity in GaAs. Figure 1.16(a) and (b) show how ε_{1s-2p} and λ_{1s-2p} change in the magnetic field along the z direction.

Theoretically, the strength of the electron–phonon interaction is expected to be suppressed for phonon wavelengths smaller than the characteristic size of the quantum dot. This suppression has been discussed in the context of inefficient optical transitions (phonon bottleneck effect) [33]. However most optical characteristics do not agree with this theory, probably because of interactions with the optical phonon or the Auger mechanism [34]. Our experiments on a simple one-electron quantum dot reveal intrinsic phonon emission characteristics for moderate acoustic phonon energies. The λ_{1s-2p} dependence of the phonon emission rate can be seen in the B field dependence. In our experiment, λ_{1s-2p} increases with B field as shown in Fig. 1.16(b), while the characteristic lateral size of the quantum dot in the x/y direction, ℓ_x and ℓ_y, decrease with increasing B field. The characteristic size is given by $\ell_{x/y} = \sqrt{\hbar/m^*}(\omega_{x/y}^2 + \omega_c^2/4)^{-1/4}$. Therefore, with decreasing B field, λ_{1s-2p} becomes shorter than the size of the quantum dot, and so phonon emission is suppressed, i.e. τ_{1s-2p} increases (see Fig. 1.16(c)). This qualitatively indicate the suppression of phonon emission and the signature of the phonon bottleneck effect.

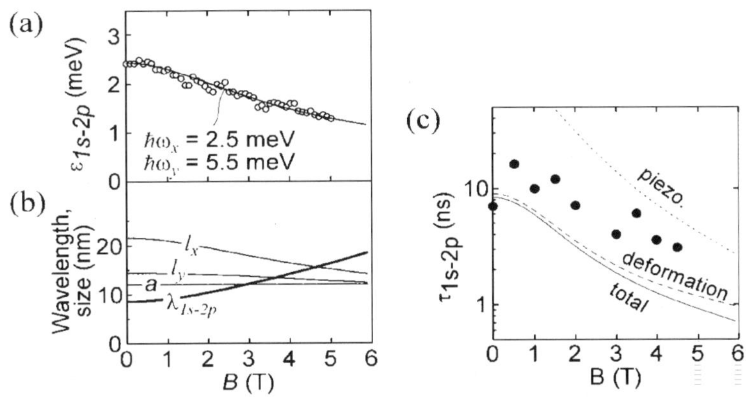

Fig. 1.16. Magnetic field dependence of (a) the energy spacing between the $2p$ excited state and the $1s$ ground state, $\varepsilon_{1\mathrm{s}-2\mathrm{p}}$, (b) the longitudinal acoustic phonon wavelength, λ, and characteristic sizes of the quantum dot (a, λ_x and λ_y), and (c) the energy relaxation time, $\tau_{1\mathrm{s}-2\mathrm{p}}$. The solid line in (c) is calculated for spontaneous emission of an acoustic phonon.

In order to be more quantitative, we calculate the phonon emission rate from Fermi's golden rule including both deformation and piezoelectric types of coupling with standard GaAs material parameters [35,36]. The deformation coupling gives a higher (lower) rate than piezoelectric coupling for an energy higher (lower) than $\sim 0.6\,\text{meV}$ for GaAs. For simplicity, the calculation is done for a circular dot, whose effective confinement energy is $\hbar\omega_{\text{eff}} = \hbar\sqrt{\omega_x\omega_y(1+\omega_c^2/(\omega_x+\omega_y)^2)}$. This assumption is reasonable because Coulomb interactions in an elliptic quantum dot just scale with $\hbar\omega_{\text{eff}}$ [31]. As shown by the solid line in Fig. 1.16(c), we find good agreement with the data. The difference by about a factor of 2 or 3 might come from the assumptions about the confinement potential and uncertainty in the material parameters. Thus, the fast energy relaxation in the $N = 1$ quantum dot can clearly be understood by spontaneous emission of an acoustic phonon. It should be noted that the electron–phonon interaction is maximized when half of the phonon wavelength is comparable to the size of the quantum dot. Therefore, the relaxation time may become longer if quantum dots are properly designed.

5.4. *Spin Relaxation in $N = 2$ Quantum Dot*

In contrast, the relaxation time is remarkably different if it involves a spin-flip. We discuss a spin-flip energy relaxation that appears in two-electron quantum dot, i.e. artificial helium atom. At low magnetic fields, the many-body GS is a spin-singlet (labeled S) having two antiparallel spin electrons in the $1s$ state, while the first ES is a spin-triplet (labeled T) having two parallel-spin electrons, one each occupying the $1s$ and $2p$ orbitals [5,17]. Figure 1.17(a) schematically shows the electronic configurations. Another spin singlet state with two antiparallel spin electrons occupying the $1s$ and $2p$ orbitals, has a higher energy, so we neglect it. Because of Coulomb interactions, the energy spacing between the two states, $\varepsilon_{\text{S-T}}$ ($\sim$0.6 meV at $B = 0\,\text{T}$), is smaller than ε_{1s-2p}. Energy relaxation from the first ES (T) to the GS (S) not only involves the same change in orbital momentum as that in the $N = 1$ quantum dot, but also requires a spin-flip because of Pauli exclusion (see section 4). A simple

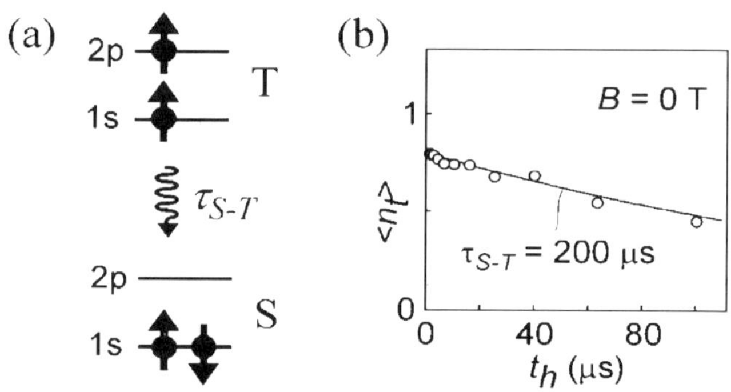

Fig. 1.17. (a) The electronic configurations of $N = 2$ quantum dot before and after relaxation. (b) The average number of tunneling electrons per pulse, n_e measured at 0 T. The relaxation time, $\tau_{1s2p} = 200\,\mu\text{s}$, is obtained from the exponential decay (solid line).

phonon–emission transition from the spin-triplet to the spin singlet is forbidden by spin conservation.

We now discuss to what degree this transition is "forbidden". Since very long relaxation time is expected, a double step pulse technique was used to investigate the relaxation time. Figure 1.17(b) shows a typical measurement of $\langle n_t \rangle$ at 0 T, indicating a relaxation time of $\tau_{S-T} \sim 200\,\mu s$. This relaxation time is 4 to 5 orders of magnitude longer than that observed in the $N = 1$ quantum dot. We find no clear B field dependence of τ_{S-T} (always longer than $100\,\mu s$), at least for the energy spacings between 0.6 meV at $B = 0$ T and 0.24 meV at $B = 2$ T.

First, we study thermal effects, which always enhance the inelastic processes. Shown in Fig. 1.18(a) is a typical temperature dependence of τ_{S-T} measured at $B = 0$ T. No clear change is observed up to 0.5 K, while τ_{S-T} decreases above 0.5 K. This behavior can be understood by the thermal excitation from the triplet state to the contact leads. Then another electron, which has an opposite spin to the left electron, tunnels into the singlet state. This sequential tunneling process exchange spin with leads. Considering the Fermi–Dirac distribution, the thermal-activated scattering time, τ_{th}, is given by

$$\tau_{th}^{-1} \sim \Gamma_{tot,e}/(\exp(\Delta_1/k_B T) + 1) \qquad (9)$$

where $\Gamma_{tot,e} = \Gamma_{L,e} + \Gamma_{R,e}$ is the total tunneling rate for the excited state, and Δ_1 is the activation energy from the triplet state to the lead. This yields $\tau_{th} = 100\,\mu s$ at 1 K, and explains well the change at $T > 0.5$ K. However, τ_{S-T} at low temperature is dominated by other mechanisms.

τ_{S-T} strongly depends on the high gate-pulse voltage V_h, during which relaxation takes place, as shown in Fig. 1.18(b). Although V_h is always in the $N = 2$ Coulomb blockade region, τ_{S-T} strongly depends on V_h. This V_h dependence implies an influence of the source and drain leads. Even though the Coulomb blockade is robust in the suppression of transport, higher-order tunneling processes can contribute to the relaxation and change τ_{S-T}. An electron in the dot can be replaced with an electron of opposite spin from the leads. This results in energy loss in the quantum dot, whereas the lead gains the same energy. This inelastic cotunneling rate, τ_{CO}^{-1}, is estimated by

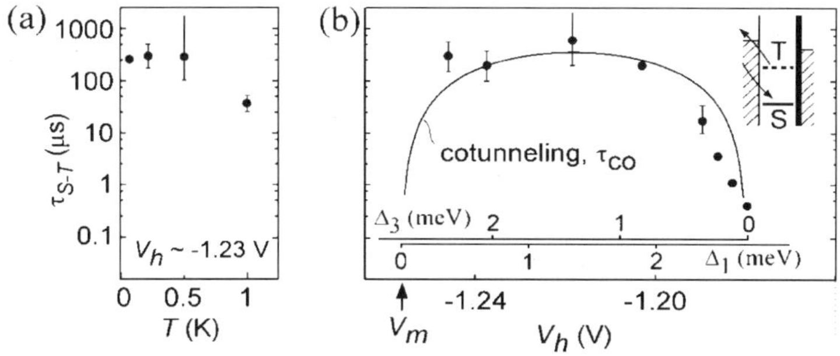

Fig. 1.18. (a) Temperature, dependence of the relaxation time τ_{ST} at $B = 0$ T. (b) The gate voltage, V_h, dependence of τ_{ST}. V_h is also converted into Δ_1 and Δ_3 energy scales. The solid line is calculated for cotunneling processes. The inset shows a schematic of the inelastic cotunneling processes.

considering second-order tunneling processes [28,37]. For the relaxation mechanisms considered here, the $N = 2$ quantum dot can relax virtually through $N = 1$ or $N = 3$ intermediate states. Note that this process does not cause a net current even at a finite voltage of $e|V_{sd}| < \varepsilon_{S-T}$ [38]. Assuming $V = 0$ V and zero temperature for simplicity, we obtain

$$\tau_{CO}^{-1} = \varepsilon_{S-T}(\hbar\Gamma_L + \hbar\Gamma_R)^2(\Delta_1^{-1} + \Delta_3^{-1})^2/h. \tag{10}$$

Here, Δ_1 and Δ_3, respectively are the energies required to excite the initial $N = 2$ triplet state to the $N = 1$ and 3 intermediate states. We can extract Δ_1 and Δ_3 from V_h, and the values are shown in Fig. 1.18(b). The solid line shows τ_{CO}, the relaxation time due to cotunneling, calculated with experimentally deduced parameters ($\varepsilon_{S-T} = 0.6$ meV, and $(\Gamma_s + \Gamma_d)^{-1} = 7$ ns). Clearly the observed relaxation time can be well understood by inelastic cotunneling.

It should be noted that the cotunneling process can be easily weakened by using thicker tunneling barriers. Then we expect longer relaxation time, which would be dominated by intrinsic processes, like spin-orbit interactions. Spin-orbit interaction, which mixes the spin and orbital degrees of freedom, is known to be a dominant spin-flip mechanism in 2DEG in semiconductors. Causes of spin-orbit interaction include the lack of crystal inversion symmetry, the electric field of the confinement potential, and impurities and interfaces of the structures. Spin-orbit interactions are predicted to be a dominant spin-flip mechanism in GaAs quantum dots [39], although this is still an extremely small effect. According to the theoretical calculation based on the bulk GaAs band structure, i.e. the lack of crystal inversion symmetry, the spin-flip energy relaxation time for the quantum dot we studied is expected to be $\tau_{SO} = 600\,\mu$s. Although our observation of $\tau_{S-T} \sim 200\,\mu$s is dominated by the cotunneling process, it might be close to the upper limit of the spin-flip relaxation time.

For simplicity, we consider the spin-orbit interaction energy, Δ_{SO} only for coupling between the $1s$ and $2p$ orbitals, but including all effects which mix spin and orbital degrees of freedom. Simple perturbation theory [40] predicts that the relaxation time from the triplet to the singlet is given by

$$\tau_{S-T,SO} \sim (\varepsilon_{S-T}/\Delta_{SO})^2 \tau_{phonon}(\varepsilon_{S-T}) \tag{11}$$

This first-order perturbation is obtained for the two triplet states with non-zero z-components of spin, by approximating the two-electron wavefunction with the Slater determinant of the one-electron $1s$ and $2p$ wavefunctions. Here, $\tau_{phonon}^{-1}(\varepsilon_{S-T})$ is the phonon emission rate at the phonon energy, ε_{S-T}, and we know that τ_{phonon} is well accounted for by the electron-phonon interaction. Therefore, we can deduce an upper bound of $\Delta_{SO} < 4\,\mu$ eV from our observations ($\tau_{S-T} > 200\,\mu$s). This value is close to the spin-orbit induced spin splitting energy ($\sim$2.5 μeV) observed in a GaAs 2DEG system [41]. Note that spin-orbit interactions are significantly enhanced in nanoparticles, e.g. copper, probably because of impurities or interfaces [40], but quantum dot is largely free of these undesirable factors.

Our experiments indicate that the spin degree of freedom in quantum dots is well isolated from the orbital degree of freedom. This is particularly attractive for applications to spin memories and spin quantum bits (qubits) [42,43]. For a simple scheme involving just a single-electron spin in a magnetic field, the spin–orbit interactions can degrade the energy relaxation time (T_1) of a spin qubit. We estimate the dominant contribution, $T_{1,\mathrm{so}}$, using a perturbative approach,

$$T_{1,so} \sim (\varepsilon_{1s-2p}/\Delta_{SO})^2 \tau_{phonon}(\varepsilon_Z) \tag{12}$$

Since $\Delta_{\mathrm{SO}} < 4\,\mu\mathrm{eV}$, this yields $T_{1,\mathrm{so}} > 1\,\mathrm{ms}$ for a Zeeman splitting $\varepsilon_{\mathrm{Z}} = 0.1\,\mathrm{meV}$ and $\varepsilon_{\mathrm{1s-2p}} \sim 1.2\,\mathrm{meV}$ at $B = 5\,\mathrm{T}$($T_{1,\mathrm{so}} > 100\,\mu\,\mathrm{s}$ at $B = 9\,\mathrm{T}$). This $T_{1,\mathrm{so}}$ is thus comparable to that obtained by electron-spin-resonance for donor states in GaAs [44], and is much longer than the time required for typical one- and two-qubit operations [45]. Note that small spin-orbit interactions are also desirable with respect to the dephasing time (T_2) of a spin qubit [46]. Our results therefore encourage further research in the use of the spin degree of freedom in quantum dots.

5.5. *Comparison with Real Atoms*

Our observations for $N = 1$ and 2 quantum dots can be compared with real atoms [29]. The transition from the $2p$ state to the $1s$ state in atomic hydrogen is allowed by photon emission (the Lyman α transition line with a lifetime of 1.6 ns), while that in the $N = 1$ quantum dot or artificial hydrogen is allowed by *phonon* emission. The transition lifetime is accidentally more or less the same, although the mechanism and the energy are quite different. The transition from the spin triplet state to the spin singlet state is forbidden by conservation of the total spin for both atomic helium and $N = 2$ quantum dots or artificial helium. The lifetime of the spin-triplet helium atom is actually extremely long (7860 s), because it is also forbidden by the parity of the wavefunction, i.e. transition from the $2s$ orbital to $1s$ orbital is forbidden. Moreover, these "forbidden" transitions can occur by collisions with electrons for the helium atom, and by cotunneling for the $N = 2$ quantum dot. The difference between the allowed and forbidden transitions leads to more than 12 orders of magnitude difference in the relaxation times for real hydrogen and helium atoms. The ratio decreases with increasing atomic number, because spin–orbit interaction increases and the Russel–Saunders approximation, becomes worse. The ratio appeared in potassium and calcium atoms, which are located in the same row as the gallium and arsenide in the periodic table, is $1.5\mathrm{X}10^5$. This ratio is comparable to the ratio in a GaAs quantum dot ($> 3\mathrm{X}10^4$). This crude comparison implies that the spin-orbit interaction in the nanostructure can be almost comparable to the atomic property.

6. The Kondo Effect in Quantum Dots

6.1. *The Kondo Physics: Metal and Quantum Dot System*

Normally, the electrical resistance of a pure metal decreases as its temperature is lowered (solid curve in Fig. 1.19). The electrons can travel through a metallic crystal more easily when the vibrations of the atoms are small. The resistance typically saturates around 10 K, due to static effects in the metal. The value of the low-temperature resistance depends on the number of defects in the material. However, the temperature

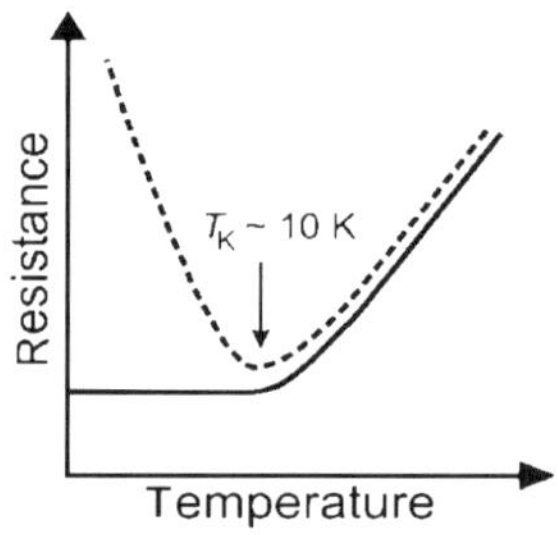

Fig. 1.19. Schematic temperature dependence of the resistance of pure metals (solid curve) and metals with a small concentration of magnetic impurity atoms (dashed curve). The latter curve shows a minimum around the Kondo temperature, T_K.

dependence changes considerably when a small concentration of magnetic atoms, such as cobalt, is added to the metal. Rather than saturating at low temperature, the resistance increases as the temperature is lowered further (dashed curve in Fig. 1.19). Since the early 1930s there have been many observations of this anomalous temperature dependence of the resistance of metals [9]. This behavior, however, was not explained earlier than in 1964, when the Japanese theorist Jun Kondo provided an explanation [7,10]. Since then, the phenomenon has been referred to as the Kondo effect.

The Kondo effect in metals occurs because below a certain temperature, which is called the Kondo temperature, T_K, the mobile electrons in the host metal tend to screen the non-zero total spin of the electrons in the magnetic impurity atom. In the simplest model of a magnetic impurity, introduced by Anderson in 1961 [47], there is only one electron level with energy ε_0 and the impurity spin is $\frac{1}{2}$. The z-component is fixed as either "spin up" or "spin down". However, so-called exchange processes can effectively flip the impurity spin, while simultaneously creating a spin excitation in the Fermi sea. This spin exchange changes the energy spectrum of the system. When many such processes are added coherently, a new state—the Kondo resonance—is generated with the same energy as the Fermi level. Such a resonance is very effective at scattering electrons with energies close to the Fermi level. Since the same electrons are responsible for the low-temperature conductivity of a metal, the strong scattering contributes greatly to the resistance. As many electrons need to be involved, the Kondo effect is a many-body phenomenon. The whole system—i.e. the magnetic impurity atom plus its surrounding electrons—forms a spin singlet. The energy scale for this singlet state is the Kondo temperature.

A quantum dot connected to source and drain leads, can nicely mimic the above situation of a localized spin impurity in a Fermi sea. Therefore, the Kondo effect was expected to occur in quantum dot systems as well [48]. However, there is an important difference between metal and quantum dot systems. In a metal, electrons are described by plane wave functions. Scattering from impurities mixes electron waves with different momenta. This momentum transfer increases the resistance. In a quantum dot, on the contrary, all the electrons have to travel through the device, as there is no electrical path around it. In this case, the Kondo resonance makes it easier for states belonging to the two opposite leads to mix. This mixing process increases the conductance, that is, the Kondo effect produces the *opposite* behavior in a quantum dot to that of

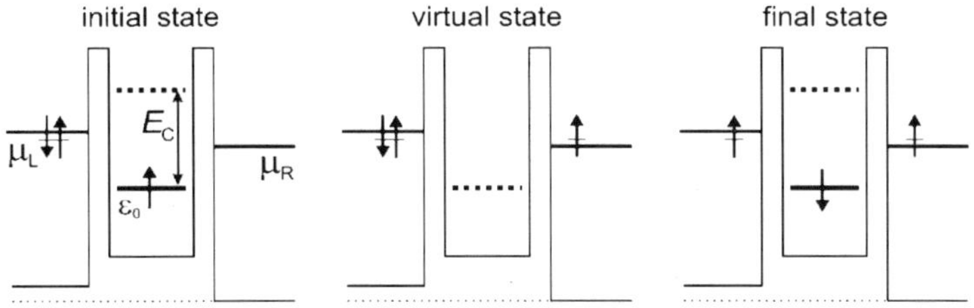

Fig. 1.20. The Anderson model of a magnetic impurity applied to a single level quantum dot connected to source and drain leads. The level has an energy $-|\varepsilon_0|(\mu_L - \mu_R \approx 0)$.

a bulk metal. Figure 1.20 shows the Anderson model of a magnetic impurity, applied to a single level quantum dot connected to source and drain leads. The level has an energy ε_0 below the Fermi energy of the leads and is initially occupied by one spin-up electron (see leftmost diagram of Fig. 1.20). Adding another electron is prohibited by the charging energy, E_C, while it would at least cost $|\varepsilon_0|$ to remove the electron. By virtue of quantum uncertainty, the spin-up electron can momentarily tunnel out of the dot, leaving the dot-lead system in a classically forbidden virtual state (middle diagram of Fig. 1.20). If the spin-up electron is replaced by a spin-down electron from the leads, as in the rightmost diagram of Fig. 1.20, the dot spin has effectively been flipped. Many such events combine to produce the Kondo effect in quantum dots, which leads to the formation of an extra resonance at the Fermi level of the leads.

The above spin-flip processes, and hence the Kondo effect, are favored by a strong coupling between the dot and the leads. The Kondo temperature is related to the parameters of the Anderson model by [49]

$$T_K = \frac{\sqrt{\Gamma E_C}}{2} e^{\pi \varepsilon_0(\varepsilon_0 + E_C)/\Gamma E_C} \tag{13}$$

where Γ is the width of the dot level, which is broadened by electrons tunneling to and from it. So the parameters that characterize the single-level dot system—E_C, ε_0 and Γ—can be replaced by a single energy scale, T_K. The advantage of quantum dots in studying the Kondo effect is that the parameters that determine the Kondo temperature, can be easily changed by adjusting the voltages on the gates.

The main characteristics of electron transport through a quantum dot with spin-degenerate levels, coupled strongly to the leads, are schematically depicted in Fig. 1.21. For an odd number of electrons on the dot, the total spin of the dot, S, is necessarily non-zero and in the simplest case $S = \frac{1}{2}$, like in Fig. 1.20. However, for an even electron number on the dot—again in the simplest scenario—the upper level is occupied by one spin-up electron and one spin-down electron. A dot with an even number of electrons therefore has $S = 0$ and the Kondo effect is not expected to occur. This "even–odd-asymmetry" results in the temperature dependence of the linear conductance, G, as shown in Fig. 1.21(a). In the "odd" or "Kondo" Coulomb valleys the conductance increases as the temperature is lowered, due to the Kondo effect. In the "even" valleys, on the contrary, the conductance decreases, due to a decrease of thermally excited transport through the dot. The temperature dependence of the conductance in the

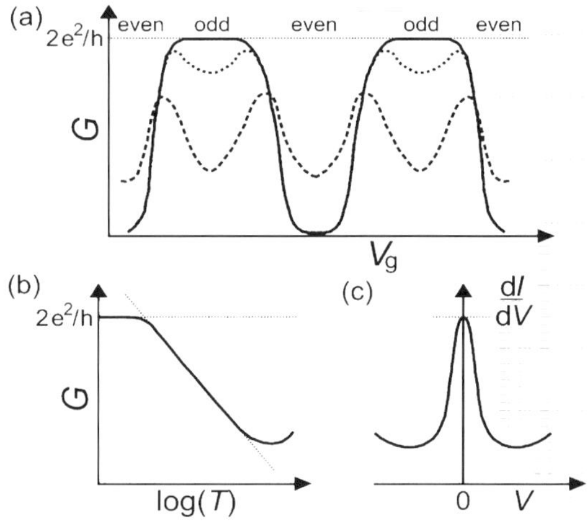

Fig. 1.21. Schematic representation of the main characteristics of electron transport through a quantum dot strongly coupled to the leads, showing the Kondo effect. (a) Linear conductance, G, versus gate voltage, V_g. The solid curve is for $T \ll T_K$, the dotted curve for $T \approx T_K$ and the dashed curve for $T > T_K$. The Kondo effect only occurs for odd electron number, resulting in the odd-even asymmetry between the different Coulomb valleys. (b) In the Kondo valleys the conductance increases logarithmically with lowering temperature, saturating at $2e^2/h$. (c) The Kondo resonance leads to a zero-bias resonance in the differential conductance, dI/dV, versus bias voltage, V.

Kondo valleys is shown in Fig. 1.21(b). The conductance increases logarithmically with decreasing temperature [48], and saturates at a value $2e^2/h$ at the lowest temperatures [50]. Although the dot has two tunnel barriers and the charging energy tends to block electrons from tunneling into or out of it, the Kondo effect makes that electrons are transmitted perfectly through the dot. This complete transparency of the dot has led to the name "unitary limit" of conductance. The first experimental observation of this limit in quantum dots is reported in section 6.2. The Kondo resonance at the Fermi level of the leads manifests itself as a zero-bias resonance in the differential conductance, dI/dV, versus V, as shown in Fig. 1.21(c). The full width at half maximum of this resonance gives an estimate for the Kondo temperature.

In section 6.2 we show that a quantum dot having an odd number of electrons well coupled to the leads and a spin $S = \frac{1}{2}$, can be described in terms of the Anderson impurity model. However, in section 6.3 it is found that quantum dots can also push research into the Kondo effect in new directions, where artificial structures can be exploited in regimes that are inaccessible with magnetic impurities in metals.

6.2. *The Kondo Effect in the Unitary Limit*

The first observations of the Kondo effect in quantum dots were reported in 1998 [51]. The results of these experiments followed reasonably well the expectations from the Anderson impurity model [47]. However, in none of these studies the theoretically predicted unitary limit of conductance at $2e^2/h$ was found. Here we discuss an experiment in which for the first time this unitary limit is observed [52]. A strong Kondo effect is observed in a semiconductor quantum dot when a small magnetic field is applied and the Coulomb blockade for electron tunneling is overcome completely by the Kondo effect.

The device of [52] was fabricated from an AlGaAs/GaAs heterostructure grown by molecular beam epitaxy on semi-insulating GaAs. A modulation-doped AlGaAs/GaAs heterojuction was used to create a 2DEG with a density of $3X10^{15}$ m^{-2}, 100 nm below the surface. The quantum dot was formed inside a $1D$ wire defined in the 2DEG by dry etching (top-left inset to Fig. 1.22(a)). Quantum confinement was established by means of two transverse gates. We indicate by $V_{\rm gl}$ and $V_{\rm gr}$ the corresponding voltages. All measurements were performed in a dilution refrigerator with a base temperature $T = 15$ mK, using a standard lock-in technique with an ac voltage excitation between source and drain contacts of 3 μV.

Figure 1.22 contains some essential results from [52]. In Fig. 1.22(a) the conductance, G, is plotted as a function of gate voltage $V_{\rm gl}$ for different temperatures. All traces were taken in a magnetic field $B = 0.4$ T perpendicular to the 2DEG. At base temperature, conductance in the valleys around $V_{\rm gl} = -413$ mV and -372 mV reaches the value of $2e^2/h$. In fact, the valleys tend to disappear. When the temperature is increased, two separate Coulomb peaks develop with growing peak spacing. The adjacent Coulomb valleys show an opposite T-dependence. This even–odd asymmetry suggests a pairwise filling of the dot levels, i.e. an unpaired spin in a valley with

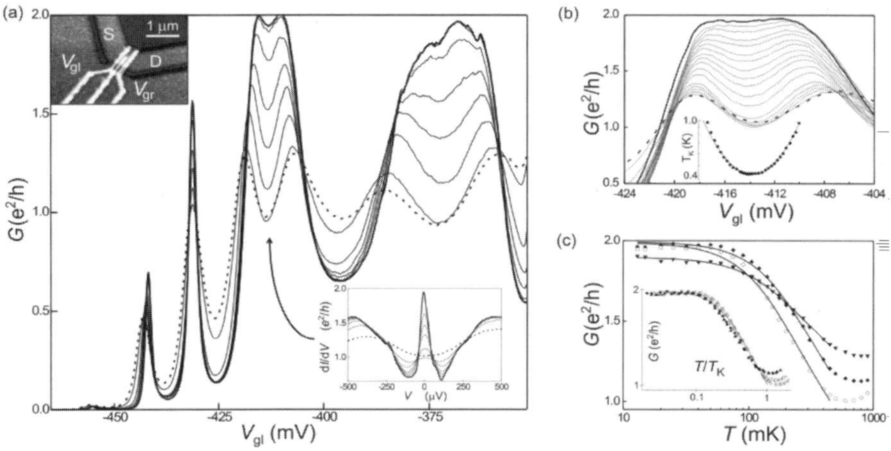

Fig. 1.22. (a) Coulomb oscillations in G versus $V_{\rm gl}$ at $B = 0.4$ T for different temperatures, $T = 15$ mK (thick solid trace) up to 800 mK (thick dashed trace). $V_{\rm gr}$ is fixed at -448 mV. Top-left inset: atomic force microscope image of the device. Bottom-right inset: differential conductance, $\mathrm{d}I/\mathrm{d}V$, versus dc bias voltage, V, for $T = 15$ mK (thick solid trace) up to 900 mK (thick dashed trace), at $V_{\rm gl} = -413$ mV and $B = 0.4$ T. (b) $G(V_{\rm gl})$ around the Kondo plateau in the case of optimized symmetric tunnel barriers. T ranges between 15 mK (thick solid trace) up to 800 mK (thick dashed trace). Inset: Kondo temperature, $T_{\rm K}$, at the Kondo plateau as obtained from many fits as in (c); the solid line is a parabolic fit of $\log(T_{\rm K})$ versus $V_{\rm gl}$. (c) $G(T)$ at fixed gate voltage for $V_{\rm gl} = -411$ (solid diamonds), -414 (open circles) and -418 (solid triangles) mV. Inset: G versus normalized temperature, $T/T_{\rm K}$, for six different gate voltages.

N = odd, where we observe the Kondo effect, and a spin singlet for N = even. In the right inset to Fig. 1.22(a) we show the differential conductance, dI/dV, versus source-drain bias, V, for different T in the middle of the Kondo plateau at $V_{gl} = -413\,\text{mV}$. The pronounced peak around $V = 0\,\text{V}$ reflects the Kondo resonance at the Fermi energy. We note that since the Zeeman splitting for $B = 0.4\,\text{T}$ is several times smaller than $k_B T_K$, the Kondo resonance is not split by the magnetic field. The peak height (i.e. the linear conductance) has a logarithmic T-dependence with a saturation at $2e^2/h$ for low T denoting the achievement of the unitary limit. The unitary limit implies that the transmission probability through the quantum dot is equal to one. Although U is an order of magnitude larger than $k_B T_K$, the Kondo effect completely determines electron tunneling at low energies (i.e. $T \ll T_K$ and $eV \ll k_B T_K$).

These measurements were taken after optimizing the two barrier gate voltages, V_{gl} and V_{gr}, in order to obtain nearly equal tunnel barriers. However, sweeping V_{gl}, as in Fig. 1.22(a), changes the left barrier much more effectively than the right one and hence the barriers cannot be symmetric over the whole V_{gl}-range. For a quantitative comparison to theory, we chose V_{gr} such that, upon sweeping V_{gl}, we could obtain a flat plateau close to $2e^2/h$ (Fig. 1.22(b)). The two discernable Coulomb oscillations at the highest temperatures have completely merged together at low T. In Fig. 1.22(c) we fit G versus T for different gate voltages to the empirical function

$$G(T) = G_0 \left(\frac{T_K'^2}{T^2 + T_K'^2} \right)^s \tag{14}$$

This function is an analytical approximation to numerical renormalisation-group (NRG) results for the Anderson impurity model [53]. $T_K' = T_K/\sqrt{2^{1/s} - 1}$ and s is a fit parameter that should be close to 0.2 for a spin-$\frac{1}{2}$ impurity. Indeed we find $s = 0.29$ and good agreement between experimental data and theoretical curves. The inset to Fig. 1.22(b) shows T_K versus V_{gl} as obtained from many fits as in Fig. 1.22(c). Log(T_K) follows a quadratic dependence on V_{gl} (i.e. ε_0) with a minimum in the middle of the conductance plateau. From the parabolic fit we estimate $U = 0.5\,\text{meV}$ and $\Gamma = 0.23\,\text{meV}$ (see [52] for more details). All over the plateau, G is a universal function of the normalized temperature, T/T_K, regardless of the other energy scales, U, ε_0 and Γ see inset to Fig. 1.22(c).

6.3. *Singlet–Triplet Kondo Effect*

In this section we focus on a more exotic Kondo phenomenon that cannot be explained within the spin-$\frac{1}{2}$ Anderson model. A novel Kondo effect is observed in a few-electron vertical quantum dot (described in sections 2–4) [54]. The effect occurs for an even number of electrons at a spin-singlet/spin-triplet (ST) transition [16,55], which is tuned by a magnetic field.

Not only for a quantum dot with an odd number of electrons it is possible to have a finite net electron spin. It is also possible that a quantum dot with N = even has a total spin $S = 1$. This occurs for example when the last two electron spins are parallel. If we ignore the remaining $N - 2$ electrons, this corresponds to a triplet state. Parallel spin filling is a consequence of Hund's first rule (sections 2 and 3). A ST transition can be induced by adjusting a magnetic field as described in section 3.

The spin of the triplet state can also be screened by cotunneling events. This process is illustrated in Fig. 1.23. In contrast to single-particle states that are considered in the spin-$\frac{1}{2}$ Kondo problem, the spin triplet consists of three degenerate two-particle states. Cotunneling exchanges only one of the two electrons with an electron from the leads. The total spin of the many-body Kondo state depends on how many modes in the leads couple effectively to the dot [56]. If there is only one mode, the screening is not complete and the whole system does not reach a singlet state. In this case the Kondo effect is called "underscreened". Calculations show that also for $S = 1$ a narrow Kondo resonance arises at the Fermi energy, however, the corresponding T_K is typically lower than in the case of $S = \frac{1}{2}$ [57].

In the experiment described in this section, we actually have the situation of a quantum dot with even N, where the spin singlet and spin triplet states for the last two electrons ($N = 5, 6$) are degenerate. The singlet state is represented by $|S, S_z\rangle = |0, 0\rangle$, where S_z is the z-component of the total spin on the dot. The triplet consists of three degenerate two-electron states (the Zeeman is negligible at the magnetic field

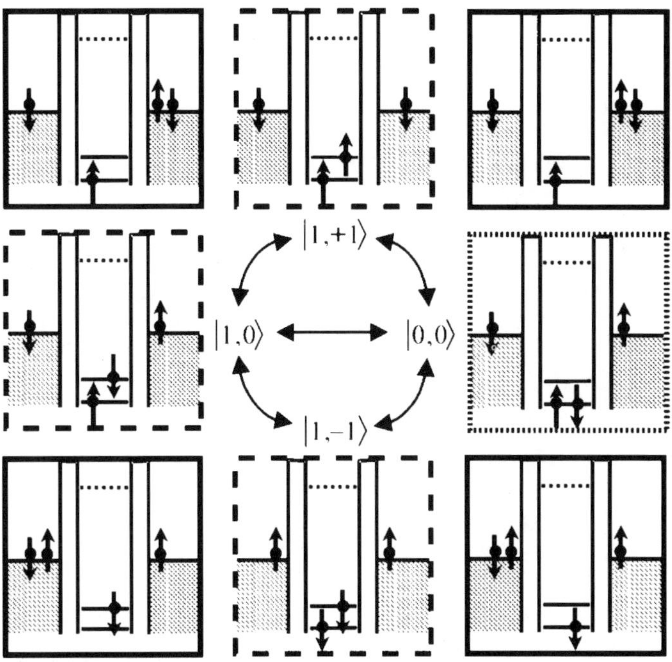

Fig. 1.23. Cotunneling in an integer-spin quantum dot for N = even at a singlet–triplet degeneracy. Two electrons can share the same orbital with opposite spins (singlet state in the dotted line box) or occupy two distinct orbitals in one of the three spin triplet configurations (dashed line boxes). The different spin states are coupled by virtual states (solid line boxes). Similar to the spin-$\frac{1}{2}$ case, spin-flip events can screen the local magnetic moment. Note that an $S = 1$ Kondo effect only involves $|1, +1\rangle$, $|1, 0\rangle$ and $|1, -1\rangle$.

considered here): $|S, S_z\rangle = |1, -1\rangle, |1, 0\rangle$ and $|1, 1\rangle$. Figure 1.23 illustrates the different cotunneling processes relevant at the ST degeneracy. Starting from $|1, 1\rangle$ cotunneling via a virtual state $|\frac{1}{2}, \frac{1}{2}\rangle$, can lead either to the triplet state $|1, 0\rangle$, or to the singlet state $|0, 0\rangle$. Via a second cotunneling event the state $|1, -1\rangle$ can be reached. As for the $S = 1$ case, the local spin can fluctuate by cotunneling events. By coupling to all triplet states, the singlet state enhances the spin exchange interaction between the dot and the leads, resulting in a higher rate for spin fluctuations. This particular situation yields a strong Kondo effect, which is characterised by an enhanced T_K. Recent scaling calculations indeed indicate a strong enhancement of T_K at the singlet–triplet degeneracy [58].

The quantum dot device is shown in the inset to Fig. 1.24(a) and has the shape of a rectangular pillar. The internal confinement potential is close to a 2D ellipse [6]. The tunnel barriers between the quantum dot and the source and drain leads are thinner than in devices studied before [6,16] such that cotunneling processes are enhanced. Figure 1.24(a) shows the linear response conductance (dc bias voltage $V = 0\,\mathrm{V}$) versus V_g and B. Regions shown black have low conductance and correspond to the regimes of Coulomb blockade for $N = 3$–10. In contrast to the experiment described in the previous section, performed on a lateral dot with unknown electron number, here the number of confined electrons is precisely known.

Gray stripes represent Coulomb peaks as high as $\sim e^2/h$. The pairing of the first two lower peaks reflects the GS evolution for $N = 3$ and 4, indicating the anti-parallel filling of the same orbital for the third and fourth electron. However, a modification to

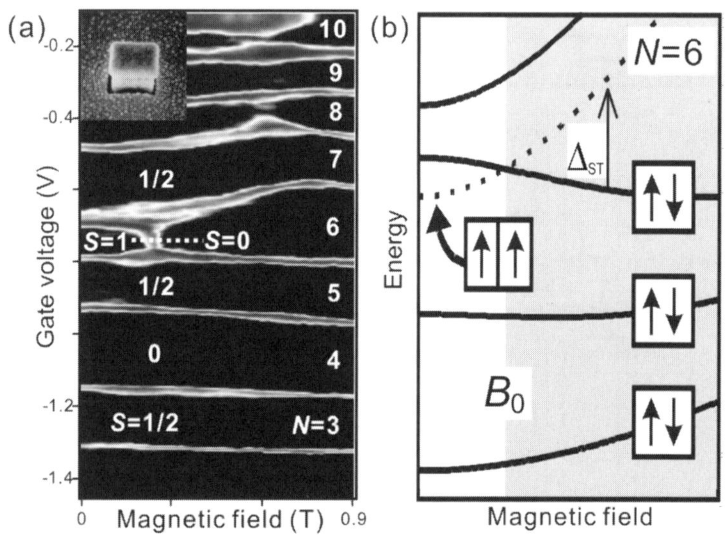

Fig. 1.24. (a) Gray-scale representation of the linear conductance versus V_g and B. Gray stripes bounded by white denote conductance peaks of height $\sim e^2/h$. Black regions of low conductance indicate Coulomb blockade. Inset: Scanning electron micrograph of a quantum dot with dimensions $0.45 \times 0.6\,\mu\mathrm{m}^2$ and height of $\sim 0.5\,\mu\mathrm{m}$. (b) Schematic energy spectrum evolving with magnetic field. Solid and dashed lines are explained in the text. Δ_{ST} represents the energy difference between the triplet and the singlet states.

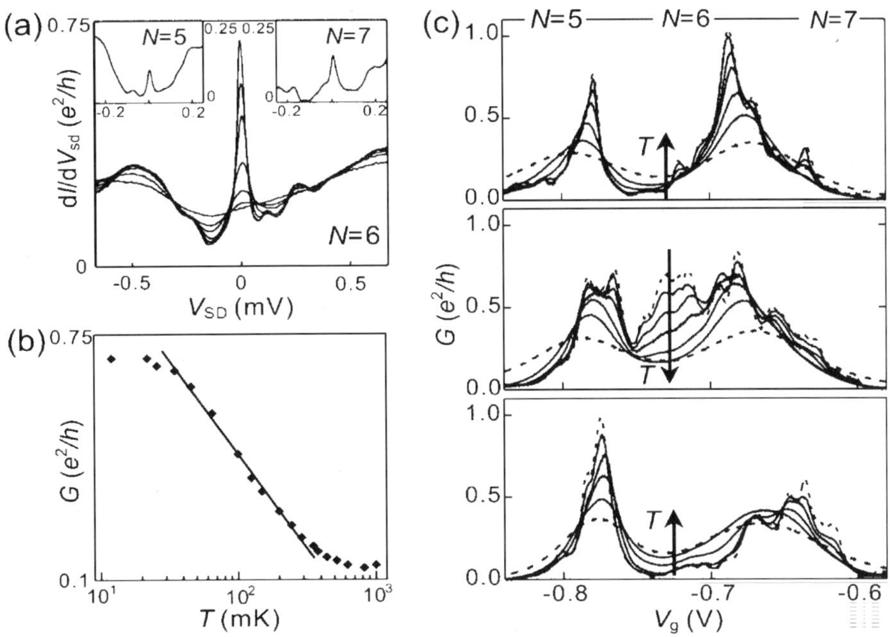

Fig. 1.25. (a) Temperature dependence of the differential conductance, $\mathrm{d}I/\mathrm{d}V$, versus source drain voltage, V, at the singlet–triplet degeneracy ($V_g = -0.72\,\mathrm{V}$, $B = 0.21\,\mathrm{T}$) and for $T = 14$, 65, 100, 200, 350, 520, and 810 mK. The left (right) inset shows the Kondo resonance for $N = 5$ ($N = 7$), measured at $V_g = -0.835\,\mathrm{V}$ ($V_g = -0.625\,\mathrm{V}$), and for $B = 0.11\,\mathrm{T}$, $T = 14\,\mathrm{mK}$. (b) Peak height of zero-bias Kondo resonance versus T as obtained from (a) (solid diamonds). The solid line demonstrates a logarithmic Temperature dependence. (c) Temperature dependence of the linear conductance versus V_g for $B = 0.12\,\mathrm{T}$ (spin-triplet ground state), $B = 0.22\,\mathrm{T}$ (singlet–triplet degeneracy), and $B = 0.32\,\mathrm{T}$ (spin-singlet ground state). Each panel shows 7 traces at $T = 20$ (dot-dashed line), 35, 70, 120, 260, 490 (solid lines), 1050 (dashed line) mK.

the pairing is found for the $N = 5$ and 6 GS lines. The $N = 5$ state has $S = \frac{1}{2}$, and the corresponding Coulomb peak shows a smooth B evolution. Instead, the stripe for $N = 6$ has a small upward cusp at $B \approx 0.22\,\mathrm{T}$ (which is referred to as B_0). From earlier analyses [6] and from measurements of the excitation spectrum [17,54], this cusp is identified as a transition in the GS from a spin triplet to a spin singlet. At the transition a strong enhancement of the conductance is observed. In fact, over a narrow range around 0.22 T, the Coulomb gap for $N = 6$ completely disappears. Figure 1.24(b) shows the ST transition in a schematic energy spectrum. Solid lines represent the B-evolution of the first four orbital levels in a single-particle model. The dashed line is obtained by subtracting the two-electron exchange coupling from the fourth level. At the crossing between this dashed line and the third orbital level at $B = B_0$ the ground state for $N = 6$ undergoes a triplet-to-singlet transition. In Fig. 1.25(a) differential conductance measurements, $\mathrm{d}I/\mathrm{d}V$ versus V, are shown taken at $B = B_0$ and V_g, corresponding to the dotted line in Fig. 1.24(a). At $T = 14\,\mathrm{mK}$ the narrow resonance around $V = 0$ has a full-width-at-half-maximum, FWHM $\approx 30\,\mu\mathrm{V}$. The height of the zero-bias resonance decreases logarithmically with T (see Fig. 1.25(b)). These are typical "fingerprints" of the Kondo effect. The saturation at low T is ascribed to electronic noise. From the FWHM ($\approx k_B T_K$), $T_K \approx 350$ mK is estimated. As stated before, the Zeeman spin splitting is neglected, since $g\mu_B B_0 \approx 5\,\mu\text{V} \ll k_B T_K$, implying that the spin triplet is in fact three-fold degenerate at $B = B_0$. This condition is essential to this particular Kondo effect. Alternative schemes have been proposed for a Kondo effect where the degeneracy of the triplet state is lifted by a large magnetic field [59]. Recently, also this kind of ST Kondo effect has been observed [60].

For $N = 6$ we only find an anomalous T-dependence at the ST degeneracy. Figure 1.25(c) shows G versus V_g for different T. The two Coulomb peaks correspond to the transition from $N = 5$ to 6 and from $N = 6$ to 7. The small, short-period modulations superimposed on the Coulomb peaks are due to a weak charging effect in the upper part of GaAs pillar above the dot [51]. We will however focus on the general T dependence. The upper panel is at $B = 0.12$ T. Upon increasing T (arrow), the valley conductance for $N = 6$ goes up due to thermally activated transport. A similar behavior is observed in the lower graph for $B = 0.32$ T. In contrast, at the ST transition for $B = 0.22$ T we find an opposite T-dependence, again indicating the formation of a Kondo resonance. At the lowest T, the valley conductance is as high as $0.7\,e^2/h$, which is close to the height of the Coulomb peaks.

The T dependence for $N = 5$ and 7 is clearly different from that in the non-Kondo valley for $N = 6$ (lower panel of Fig. 1.25(c)). Such a difference is a manifestation of the ordinary spin-$\frac{1}{2}$ Kondo effect expected for $N =$ odd. Indeed the corresponding zero-bias resonances are clearly observed (see insets to Fig. 1.25(a)). Their height, however, is much smaller than for the singlet–triplet Kondo effect. There is also some indication for a triplet Kondo effect in the T-dependence for $N = 6$ at $B = 0.12$ T, although the associated zero-bias anomaly is not as clear. We note that in a lateral quantum dot a similar ST degeneracy can lead to a two-stage Kondo effect, which is characterized by two Kondo temperatures [61,62]. For such an effect to occur, single-channel leads are prerequisite, which is usually not the case for a vertical dot device.

7. Summary

A semiconductor quantum dot is a small solid state device equipped with various functions to modify the effects of quantum mechanical confinement and interactions in quantum dots. These functions can be used to probe the physics of 0D electronic systems that is not well established in the case of many-body interactions, strong correlations, coherent or incoherent processes, etc, and also to exemplify rules and principles in quantum mechanics that are fundamental, but difficult to test in conventional material systems. Topics reviewed in this chapter are all related to these features of quantum dot devices with a special emphasis on the spin effects observed in the electron transport.

The key ingredient here is the manipulation of the spin configuration: between a spin singlet state ($S = 0$) and a spin triplet state ($S = 1$) and between an unpaired spin state ($S = \frac{1}{2}$) and a paired spin state ($S = 0$). The singlet–triplet configuration in a single quantum dot is associated with the filling of degenerate (or non-degenerate) orbital states. Parallel spin filling is usually favored for degenerate orbitals in line with

Hund's first rule. We use a 2D harmonic quantum dot containing just a few electrons to demonstrate the manipulation of the spin configuration, because in this type of quantum dots the orbital degeneracies are magnetically adjusted in the diagram of Fock–Darwin states and also because the number of electrons N in the dot is varied one-by-one starting from $N = 0$.

We apply techniques to control over the spin configuration in quantum dots to study various spin effects on the electronic properties: Pauli exclusion on spin transport, energy relaxation with and without a spin-flip, and the Kondo effect. The two-electron spin state is very simple, but can significantly influence the electronic properties of quantum dot systems. Pauli exclusion principle is a typical example. Because of this principle, electron transport is only allowed for spin singlet states in a series of quantum dots. This is directly demonstrated by the observation of Pauli blockade in the single electron tunneling through a weakly coupled double dot system when two same-spin electrons occupy the lowest energy state in each dot, i.e. a spin triplet state. To study the dynamics of spin relaxation in quantum dots we develop an electrical pump and probe technique to measure the energy relaxation time from a higher state to a lower state. The measured relaxation time from a spin triplet state to a spin singlet state is very long, indicating that the spin degree of freedom in quantum dots is well isolated. The Kondo effect has long been observed for metal including magnetic impurities and understood in the framework of the Anderson impurity model. The observation of the Kondo effect in quantum dots, however, invokes reviving interests in the Kondo physics, because in quantum dots various parameters can be manipulated to modify the exchange coupling. The unitary limit of conductance due to the $S = \frac{1}{2}$ Kondo effect and the enhancement of the Kondo effect due to the singlet–triplet degeneracy described here are such examples.

Electron spin is a well defined quantum number in solid state material, and when implemented in quantum dot devices, it can be manipulated to modify the electronic properties, as illustrated in this chapter. These features will make quantum dots attractive not only for study of fundamental physics but also for applications to novel spin devices like spin memories, filters and quantum bits for quantum computation as well.

Acknowledgments

The authors thank Y. Tokura, D. G. Austing, S. Sasaki, S. De Franceschi, and L. I. Glazman for valuable discussions. Part of the authors acknowledge financial support from the Specially Promoted Research, Grant-in-Aid for Scientific Research, from Ministry of Education, Science, and Culture in Japan and from CREST-JST PREST-JST, and the DARPA grant number DAAD19-01-1-0659 of the QuIST program.

References

1. M. Reed, *Sci. Am.* **268**, 118 (1993); M.A. Kastner, *Phys. Today* **46**, 24 (1993); R.C. Ashoori, *Nature* **379**, 413 (1996).
2. For reviews, *Mesoscopic Phenomena in Solids* (eds. B.L. Altshuler, P.A. Lee, and R.A. Webb) Elsevier (1991); U. Meirav and E.B. Foxman, *Semicond. Sci. Technol.* **11**,

255 (1996); *Proceedings of the NATO Advance Study Institute on Mesoscopic Electron Transport* (eds. L.L. Sohn, L.P. Kouwenhoven, and G. Schoen) Kluwer Series E345 (1997).

3. S. Tarucha, D.G. Austing, S. Sasaki, Y. Tokura, W. van der Wiel, and L.P. Kouwenhoven, *Appl. Phys.* A **71**, 367 (2000).
4. For reviews, S. Tarucha, D.G. Austing, T. Fujisawa, and L.P. Kouwenhoven, in *Optical and Electronic Process of Nano-Matters* (ed. M. Ohtsu), Kluwer Academic Publishers, 57–93 (2001); See P.A. Maksym and T. Chakraborty, *Phys. Rev. Lett.* **65**, 108 (1990) for magic number states, and A.A. Koulakov and B.I. Shklovskii, *Phil. Mag.* B **77**, 1235 (1998) for Wigner molecules.
5. S. Tarucha, D.G. Austing, T. Honda, R.J. van der Hage, and L.P. Kouwenhoven, *Phys. Rev. Lett.* **77**, 3613 (1996).
6. D.G. Austing, S. Sasaki, S. Tarucha, S.M. Reimann, M. Koskinen, and M. Manninen, *Phys. Rev.* B **60**, 11514 (1999); S.M. Reimann, M. Koskinen, J. Kolehmainen, M. Manninen, D.G. Austing, and S. Tarucha, *Eur. J. Phys.* D **9**, 105 (1999).
7. T. Fujisawa, Y. Tokura, and Y. Hirayama, *Phys. Rev.* B **63**, 081304 (2001); T. Fujisawa, D.G. Austing, Y. Tokura, Y. Hirayama, and S. Tarucha, *Phys. Rev. Lett.* **88**, 236802 (2002).
8. J. Kondo, *Prog. Theor. Phys.* **32**, 37 (1964).
9. W.J. de Haas, J.H. de Boer, and G.J. van den Berg, *Physica* **1**, 1115 (1934).
10. See for reviews, L.P. Kouwenhoven and L.I. Glazman, *Phys. World* **14**, 33 (2001).
11. D.G. Austing, T. Honda, and S. Tarucha, *Semicond. Sci. Technol.* **11**, 388 (1996).
12. M. Alonso and E.J. Finn, *Quantum and Statistical Physics*. Addison-Wesley (1968).
13. V. Fock, *Z. Phys.* **47**, 446 (1928); C.G. Darwin, *Proc. Camb. Phil. Soc.* **27**, 86 (1930).
14. M. Ciorga, A. Wensauer, M. Pioro-Ladriere, M. Korkusinski, J. Kyriakidis, A.S. Sachrajda, and P. Hawrylak, *Phys. Rev. Lett.* **88**, 256804 (2002).
15. K. Ono, D.G. Austing, Y. Tokura, and S. Tarucha, *Physica* B, **314** 450 (2002).
16. S. Tarucha, D.G. Austing, Y. Tokura, W.G. van der Wiel, and L.P. Kowenhoven, *Phys. Rev. Lett.* **84**, 2485 (2000).
17. L.P. Kouwenhoven, T.H. Oosterkamp, M.W.S. Danoesastro, M. Eto, D.G. Austing, T. Honda, and S. Tarucha, *Science* **278**, 1788 (1997).
18. P. Matagne, J.P. Leburton, D.G. Austing, and S. Tarucha, *Phys. Rev.* B **65**, 085325 (2002).
19. There are various calculations consistent with our experiment; exact calculation for $N < 7$ by A. Wojs and P. Hawrylak, *Phys. Rev.* B **53**, 10841 (1996) and also by M. Eto, *Jpn. J. Appl. Phys.* **36**, 3924 (1997), Hartree–Fock calculations for $N > 8$ by A. Natori et al., *Jpn. J. Appl. Phys.* **36**, 3960 (1997), H. Tamura,*Physica* B **249–251**, 210 (1998) and M. Rontani et al., *Phys. Rev.* B **59**, 10165 (1999), and spin density-functional theory at $B = 0$ T by In-Ho Lee et al., *Phys. Rev.* B **57**, 9035 (1998), and M. Koskinen, M. Manninen, and S.M. Reimann, *Phys. Rev. Lett.* **79**, 1817 (1997), and also at non-zero B by O. Steffen, U. Rossler, and M. Suhrke, *Europhys. Lett.* **42**, 529 (1998).
20. For single dot, D. Weimann, H. Hausler, and B. Kramer, *Phys. Rev. Lett.* **74**, 984 (1995); Y. Tanaka and H. Akera, *Phys. Rev.* B **53**, 3091 (1996).
21. For double dot, H. Imamura, H. Aoki, and P.A. Maksym, *Phys. Rev.* B **57**, R4259 (1998); Y. Tokura, D.G. Austing, and S. Tarucha, *J. Phys.: Condens. Matter* **11**, 6023 (1999).
22. L.P. Rokhinson, L.J. Guo, S.Y. Chou, and D.C. Tsui, *Phys. Rev.* B **63**, 035321 (2001); A.H. Huettel, H. Qin, A.W. Holleithner, R.H. Blick, K. Neumaier, D. Weimann, K. Eberl, and J.P. Kotthaus, cond-mat/0109104.
23. K. Ono, D.G. Austing, Y. Tokura, and S. Tarucha, *Science* **297**, 1313 (2002).
24. M. Ciorga, A.S. Sachrajda, P. Hawrylak, C. Gould, P. Zawadzki, S. Jullian, Y. Feng, and Z. Wasilewski, *Phys. Rev.* B **61**, R16315 (2000).

25. D.G. Austing, T. Honda, K. Muraki, Y. Tokura, and S. Tarucha, *Physica* B **249–251**, 206 (1998).
26. S. Amaha, D.G. Austing, Y. Tokura, K. Muraki, K. Ono, and S. Tarucha, *Solid State Commun.* **119**, 183 (2001).
27. M. Pi, A. Emperador, M. Barranco, F. Garcias, K. Muraki, S. Tarucha, and D.G. Austing, *Phys. Rev. Lett.* **87**, 066801 (2001).
28. D.V. Averin and Y.V. Nazarov, Single Charge Tunneling, in *Coulomb Blockade Phenomena in Nanostructures* (eds. by H. Grabert and M.H. Devoret), Plenum Press and NATO Scientific Affairs Division, 217–247 (1992).
29. H.A. Bethe and E.E. Salpeter, *Quantum Mechanics of One- and Two-Electron Atoms*. Springer, Berlin 1957.
30. T. Fujisawa, D.G. Austing, Y. Tokura, Y. Hirayama, and S. Tarucha, *Nature* **419**, 278 (2002).
31. Y. Tokura, S. Sasaki, D.G. Austing, and S. Tarucha, *Physica* B **298**, 260 (2001).
32. T. Fujisawa, T.H. Oosterkamp, W.G. van der Wiel, B.W. Broer, R. Aguado, S.Tarucha, and L.P. Kouwenhoven, *Science* **282**, 932 (1998).
33. H. Benisty, C.M. Sotomayer-Torres, and C. Weisbuch, *Phys. Rev.* B **44**, 10945 (1991).
34. U. Bockelmann, P. Roussignol, A. Filoramo, W. Heller, G. Abstreiter, K. Brunner, G. Böhm, and G. Weimann, *Phys. Rev. Lett.* **76**, 3622 (1996).
35. K. Seeger, *Semiconductor Physics: An Introduction*. Springer-Verlag, Berlin, 153–213 (1985).
36. U. Bockelmann, *Phys. Rev.* B **50**, 17271 (1994).
37. M. Eto, *Jpn. J. Appl. Phys.* part 1 **40**, 1929 (2001); E.V. Sukhorukov, G. Burkard, and D. Loss, *Phys. Rev.* B **63**, 125315 (2001).
38. S. De Franceschi, S. Sasaki, J.M. Elzerman, W.G. van der Wiel, S. Tarucha, and L.P. Kouwenhoven, *Phys. Rev. Lett.* **86**, 878 (2001).
39. A.V. Khaetskii and Y.V. Nazarov, *Phys. Rev.* B **61**, 12639 (2000).
40. W.P. Halperin, *Rev. Mod. Phys.***58**, 533 (1986).
41. Y.A. Bychkov and E.I. Rashba, *JETP Lett.* **39**, 78 (1984).
42. D. Loss and D.P. DiVincenzo, *Phys. Rev.* A **57**, 120 (1998).
43. P. Recher, E.V. Sukhorukov, and D. Loss, *Phys. Rev. Lett.* **85**, 1962 (2000).
44. M. Seck, M. Potemski, and P. Wyder, *Phys. Rev.* B **56**, 7422 (1997).
45. J.A. Gupta, R. Knobel, N. Samarth, and D.D. Awschalom, *Science* **292**, 2458 (2001).
46. K.V. Kavokin, *Phys. Rev.* B **64**, 075305 (2001).
47. P.W. Anderson, *Phys. Rev.* **124**, 41 (1961).
48. L.I. Glazman and M.E. Raikh, *JETP Lett.* **47**, 452 (1988); T.K. Ng and P.A. Lee, *Phys. Rev. Lett.* **61**, 1768 (1988).
49. F.D.M. Haldane, *Phys. Rev. Lett.* **40**, 416 (1978).
50. A. Kawabata, *J. Phys. Soc. Jpn.* **60**, 3222 (1991).
51. D. Goldhaber-Gordon, H. Shtrikman, D. Mahalu, D. Abusch-Magder, U. Meirav, M.A. Kastner, *Nature* **391**, 156 (1998); S.M. Cronenwett, T.H. Oosterkamp, and L.P. Kouwenhoven, *Science* **281**, 540 (1998); J. Schmid, J. Weis, K. Eberl, and K. von Klitzing, *Physica* B bf256–258, 182 (1998).
52. W.G. van der Wiel, S. De Franceschi, J.M. Elzerman, T. Fujisawa, S. Tarucha, and L.P. Kouwenhoven, *Science* **289**, 2105 (2000).
53. D. Goldhaber-Gordon, J. Göres, M.A. Kastner, H. Shtrikman, D. Mahalu, and U. Meirav, *Phys. Rev. Lett.* **81**, 5225 (1998); T.A. Costi, A.C. Hewson, and V. Zlatic, *J. Phys. Condens. Matter* **6**, 2519 (1994).
54. S. Sasaki, S. De Franceschi, J.M. Elzerman, W.G. van der Wiel, M. Eto, S. Tarucha, and L.P. Kouwenhoven, *Nature* **405**, 764 (2000).

55. W.G. van der Wiel, T.H. Oosterkamp, J.W. Janssen, L.P. Kouwenhoven, D.G. Austing, T. Honda, and S. Tarucha, *Physica* B **256–258**, 173 (1998).
56. D.C. Mattis, *Phys. Rev. Lett.* **19**, 1478 (1967); P. Nozières and A. Blandin, *J. Physique* **41**, 193 (1980).
57. Y. Wan, P. Phillips, and Q. Li, *Phys. Rev.* B **51**, 14782 (1995); W. Izumida, O. Sakai, and Y. Shimizu, *J. Phys. Soc. Jpn.* **67**, 2444 (1998).
58. M. Eto and Y.V. Nazarov, *Phys. Rev. Lett.* **85**, 1306 (2000); M. Eto and Y.V. Nazarov, *Phys. Rev.* B **64**, 085322 (2001); M. Eto and Y.V. Nazarov, *Phys. Rev.* B **66**, 153319 (2002).
59. M. Pustilnik, Y. Avishai, and K. Kikoin, *Phys. Rev. Lett.* **84**, 1756 (2000); D. Giuliano and A. Tagliacozzo, *Phys. Rev. Lett.* **84**, 4677 (2000).
60. J. Nygård, D.H. Cobden, and P.E. Lindelof, *Nature* **408**, 342 (2000).
61. W. Hofstetter and H. Schoeller, *Phys. Rev. Lett.* **88**, 016803 (2002).
62. W.G. van der Wiel, S. De Franceschi, J.M. Elzerman, S. Tarucha, L.P. Kouwenhoven, J. Motohisa, F. Nakajima, and T. Fukui, *Phys. Rev. Lett.* **88**, 126803 (2002).

Chapter 2

Microwave Spectroscopy on Single and Coupled Quantum Dots

R. H. Blick*, A. W. Holleitner, and H. Qin

Center for NanoScience and Sektion Physik, Ludwig-Maximilians-Universität, Geschwister-Scholl-Platz 1, 80539 München, Germany

1. Introduction

The overwhelming strength of quantum dot physics is found in the experimenter's ability to devise and test a broad variety of artificial molecular structures. This has been widely used to study the electron–electron interaction in terms of Coulomb blockade of single electron transport [1]. Moreover, the atomic and molecluar characteristics of quantum dots [2] is most prominently expressed in transport spectroscopy on excited N-electron states and the corresponding magnetic field dispersion. It is this similarity to real atoms and molecules which led to considering quantum dots for quantum information processing [3, 4]. A number of proposals have been introduced how to integrate these artificial molecules (as shown in Fig. 2.1) for building such quantum computing devices [5, 6]. The straightforward realization of a quantum bit (qubit) is given, e.g. by coupling two of these quantum dots and hence forming the most simple form of an artificial hydrogen molecule. Fabricated with great accuracy in a whole variety of circuits the molecular binding mechanisms can be probed [7–9]. Apart from these computational aspects quantum dots allow probing the foundations of quantum mechanics, since by now not only single charges can be manipulated but also the spin degree of freedom.

The question arising in this context is how information in quantum dots stored as a phase coherent electronic state is interacting with its environment. Here we will mainly consider photonic fields interacting with the discrete electronic states in the quantum dots. Single photons allow to manipulate single electrons tunneling on and off the quantum dots, which is termed photon assisted tunneling (PAT). This mechanism enables switching information stored in a dot in the range of nsec down to some

*Current address: University of Wisconsin-Madison, Department of Electrical and Computer Engineering, Madison, WI 53705, USA. Email: blick@engr.wisc.edu

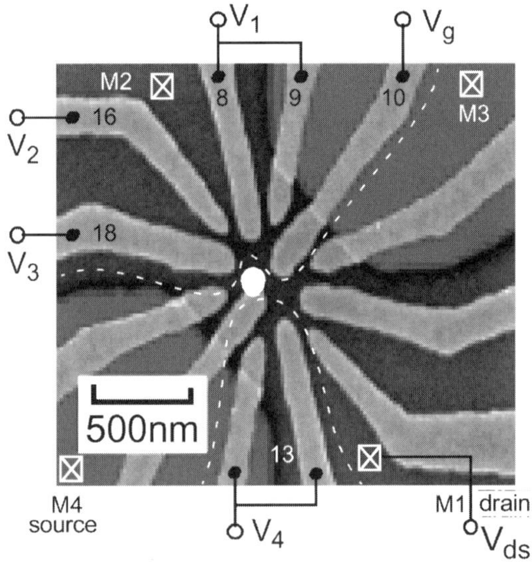

Fig. 2.1. Top view taken with a scanning electron beam microscope of a typical coupled quantum dot: a variety of gating electrodes confines two dots. In this gate geometry source and drain contacts are located beneath a displacer layer, allowing usage of additional electrodes [8].

psec [10]. As for real molecules microwave spectroscopy is a helpful tool in studying the electronic degrees of freedom. The main difference lies in the fact that in the following we are dealing with a single "solid state molecule" and simultaneously apply transport and microwave spectroscopy. This molecule is defined by coupling two quantum dots, leading to the formation of coherent electronic states in this molecule [11–21]. In the following we want to first give an outline of the fabricational procedure, then a variety of methods of photon generation are introduced. This is extended by a summary of the microwave experiments on coupled dots which is the best example for probing PAT as well as coherent modes. In addition data on single quantum dots are presented discussing dynamic conductance.

2. Aspects of Fabrication

The material system of choice are AlGaAs/GaAs-heterostructures containing a high-mobility two-dimensional electron gas with a phase coherence length of typically 10–30 μm at low temperatures. Definition of quantum dots is usually achieved by electron beam lithography and adjacent deposition of Schottky field-effect gates. Quantum dots contain roughly 1–100 electrons and are attached to quasi metallic leads by tunneling barriers. The main advantage of heterostructures is the high degree of perfection with which its electronic and photonic properties can be tailored. These materials already enabled ground breaking work which demonstrated that quantum dots in the few electron limit show not only charge quantization, but reveal a discrete energy spectrum similar to real atoms [1]. While it might not be possible with laterally integrated quantum dots to build a full scale quantum computer, these dot circuits are the ideal model systems for studying the principles of operation, which later on might be implemented in molecular devices.

The devices we use for microwave spectroscopy are realized in a two-dimensional electron gas (2DEG) being 90 nm below the surface of an AlGaAs/GaAs heterostructure. At a bath temperature of some 10 mK the electron mobility and the density are usually found to be $\mu = (8\text{–}12) \times 10^5\ \mathrm{m^2/Vs}$ and $n_s = (1\text{–}3) \times 10^{15}\ \mathrm{m^{-2}}$, respectively. By pinching off the tunneling barriers of one of the dots, we first characterize each dot individually. From transport spectroscopy we find charging energies for the dots to be of the order of $E_C = e^2/2C_\Sigma = 1\text{–}2\ \mathrm{meV}$. Taking into account the electron density the number of electrons in the dots is estimated to be 10–50 depending on the dot radii. Minimum feature size we have achieved are of the order of 7 nm [22].

3. Measurement Techniques

In order to study the dynamics of the few-electron systems properly new experimental measurement techniques have to be realized. In this work several approaches to probing the dynamics in the GHz-range were taken, which will be outlined in the following [23]: The "classical" means of spectroscopy of quantum dots are transport and capacitance spectroscopy. This allows to obtain data on the system under non-equilibrium conditions, however, direct detection of coherent processes in the time or frequency domain requires more intricate techniques. The experimental tools reported here give the techniques at hand to investigate such questions.

Starting point for the measurements on quantum dots presented in this work is transport spectroscopy: The high impedance quantum dots ($R_{\mathrm{dc}} \cong 1\text{–}100\ \mathrm{M\Omega}$) are mounted on a chip carrier and placed in a dilution refrigerator operating at 20 mK. At temperatures of $\sim$1 K Coulomb blockade kicks in and single electron tunneling can be monitored in a two-port conductance measurement. It is important to reduce all possible sources of noise: The most common are ground loops, 50/60 Hz interference, coupling of radio frequencies, and microphonic noise. The current noise has to be suppressed to some 10 fA in order to reduce the effective electron temperature in the quantum dots below 100 mK. On the hand by using ultra-small quantum dots fabricated in silicon the onset of Coulomb blockade is found even above 100 K.

3.1. Classical Microwave Spectroscopy

Microwave spectroscopy on single and coupled quantum dots is understood as measuring the induced photo current. The microwave sources easily available are broad band tunable synthesizers, operating in the range of several MHz up to 100 GHz. Microwave radiation is coupled via impedance matched coaxial lines to the electrodes of the nanostructure. According to conventional wisdom coupling is modelled capacitively by assuming the high frequency current will modulate the potentials sufficiently. The microwave signal $f = 10\ \mathrm{MHz}\text{–}50\ \mathrm{GHz}$ is coupled into the cryostat using coaxial lines or via a circular waveguide and irradiated onto the sample using an antenna formed out of a conducting loop. Above 50 GHz wave guides are utilized to coax the radiation to the sample—for our setup the lower cut-off is given by the wave guide diameter, while the upper one is determined by the absorption of the stain-less steel wave guide walls.

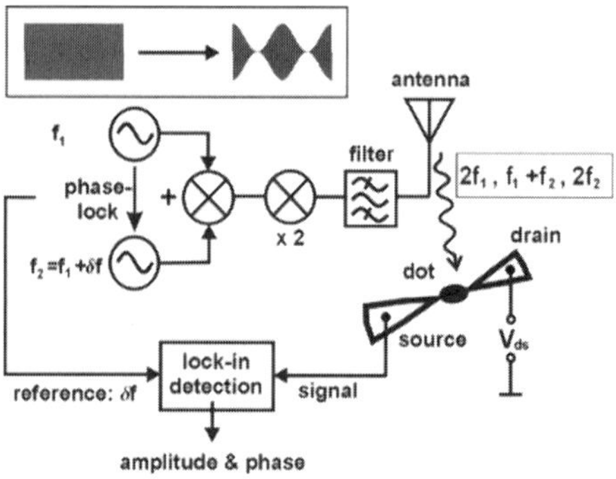

Fig. 2.2. Standard setup for transport and microwave spectroscopy: Either a single source is used or two sources for amplitude and phase sensitive measurements can be combined. For the two source setup two millimeter waves with a slight frequency-offset generated by two phase-locked microwave synthesizers are added, doubled and filtered. The resulting modulated signal is irradiated on the quantum dot by means of an antenna. With a highly sensitive preamplifier and a lock-in amplifier both amplitude and phase of the photo-conductance are detected at the modulation frequency δf. In the inset the microwave signal before and after modulation is schematically depicted.

In addition to applying only a single microwave source or an on-chip source, the combination of two phase-locked synthesizers enables measurements similar to pump-probe experiments known from optics (see Fig. 2.2) [24]: Two microwave synthesizers are phase-locked and tuned to slightly different frequencies, e.g. $f_1 = 18.08\,\text{GHz}$ and $f_2 = 18.08\,\text{GHz} + \delta f$ with $\delta f = 2.1\,\text{kHz}$. The two signals are added, frequency-doubled and filtered with a band pass. Due to the band pass only microwaves with frequencies $2f_1$, $f_1 + f_2$ and $2f_2$ are irradiated upon the quantum dot. As these frequency components have a rigid phase relation, their superposition leads to a modulated microwave signal with modulation frequency δf. We have thus produced a flux of photons with energy $2hf_1 = 0.15\,\text{meV}$ whose intensity varies periodically in time with frequency 2.1 kHz. Electronic transport induced by these photons can be detected with a lock-in amplifier at the frequency of the microwave beat. Thus, the detected signal is solely due to the irradiation and contains no dc contribution. It is therefore possible to observe PAT even in the non-linear regime, where the broadening of the ordinary conductance resonances normally masks the photon-induced features. Another advantage of this technique is the possibility of heterodyne detection which allows for determination of both amplitude and relative phase of the signal (see below). This is not possible using a single microwave source and a simple modulation technique with a modulator-diode. Hence our method enables the determination of phase relaxation rates. In both cases—single and double source—best results were achieved by attaching a Helmholtz antenna terminating the coax line, radiating onto the dot structures.

3.2. *Non-Linear Transmission Lines*

An even more advanced setup is the ultra wide band spectrometer shown in Fig. 2.3 consisting of two nonlinear transmission lines (NLTLs), generating trains of short

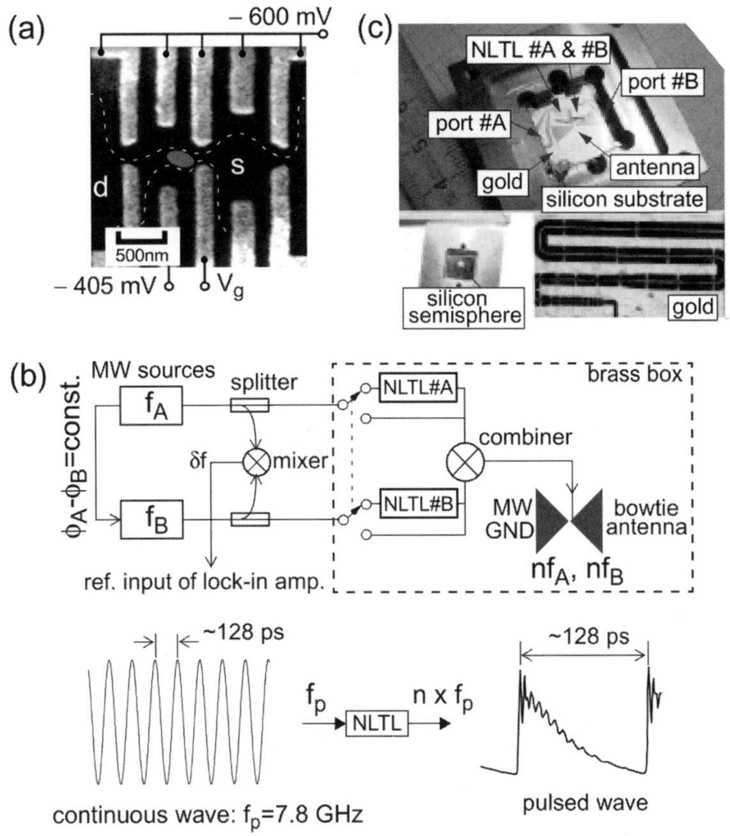

Fig. 2.3. Ultra wide band millimeter wave spectrometer set-up by combining two non-linear transmission-lines (NLTLs) [16]: (a) Top view of the heterodyne spectrometer circuit: the circuit is in the center of a brass box with two SMA ports feeding the two NLTLs. The NLTLs radiate into a power combiner, which is terminated in the apex of the bow-tie antenna. (b) View of the silicon hemisphere on the back side of the box, which emits the radiation. (c) Magnification of one of the meander-shaped transmission lines—the intersections are the positions of the Schottky diodes. (d) Extended setup of Fig. 2.2 with the NLTLs included.

pulses with harmonic frequency contents up to 400 GHz [25, 26]. In contrast to an earlier version of the spectrometer [23, 27] now both NLTLs are integrated in a single brass box, which is mounted directly on top of a cylindrical wave guide (lower cut-off at 80 GHz) of the sample holder in a dilution refrigerator. Infrared radiation is blocked by a black polyethylene window. This allows us to probe complex photoconductance in the 80–400 GHz range, corresponding to energies of 320 μeV–1.6 meV. The typical charging energies of small dots are of the order of 500 μeV–1.5 meV, while the energies of the excited states in these dots are around 100 μeV–0.5 meV.

A top view is shown in Fig. 2.3(a): The NLTLs are fed with two synthesizers (HP 83711A, $f = 7.8$ GHz) connected on the left and right ports. The output of the two sources is then added by a power combiner and sent to a bow-tie antenna. The combiner and antenna circuit are defined by evaporating gold (Au) onto a silicon (Si) substrate. The impedance of the whole circuit is matched to the 50 Ω antenna. The superimposed radiation of the NLTLs is finally radiated into the cylindrical wave guide through a silicon hemisphere attached to the antenna on the back side of the circuit (see Fig. 2.3(b)). A close-up of one of the NLTLs is given in Fig. 2.3(c). The meander shaped coplanar wave guide is intersected by voltage tunable Schottky diodes, leading to the pulse-forming response of the transmission line. Although the output power of the sources is low to minimize heating, we observe a well-pronounced photoconductance signal, as will be discussed in the following.

3.3. On-Chip Microwave Sources

As seen the classical methods of millimeter wave spectroscopy on dots allow to induce single electron tunneling by the absorption of single photons. Nevertheless, this can viewed as a fairly classical picture, since the wave length of 1–3 mm considerably exceeds the dots diameter of $d \sim 300$–600 nm. In an attempt to verify whether the inhomogeneity of the near-field radiation on the excitations in quantum dots, the radiation source is brought closer to the quantum dots. This is achieved by fabricating a weak link as a Josephson oscillator with a modified atomic force microscope tip directly into the Al-contacts forming the Schottky gates [28]. This is of great advantage for probing the microwave response of quantum dots in the absolute near-field limit, since the photon source is located only 100 nm apart from the tunnel barriers.

An example of such a gate structure is depicted in Fig. 2.4: Two different polarizations of how the radiation emitted from the junction is impinging on the quantum dots can be chosen by placing the junction accordingly. The main disadvantage in the

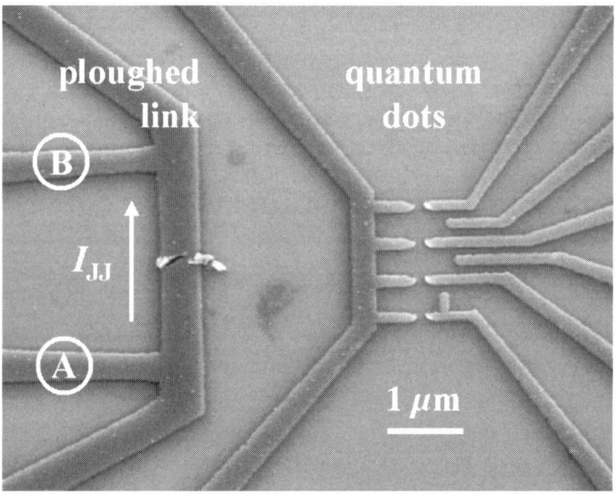

Fig. 2.4. For the ultimate near field: top view of a plowed Josephson junction with the weak link in the center, allowing near field microwave radiation to couple to the quantum dots on the right hand side.

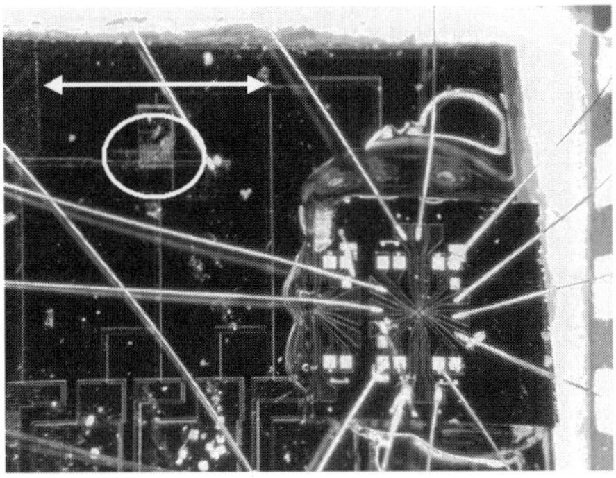

Fig. 2.5. Current state of the art on-chip spectrometer: the quantum dot chip is glued on top of the Josephson junction [21]. The output port of the superconducting microwave source (circle) is fed into an antenna on whose arms the dot chip is placed.

course of the measurements was not the limited emission power of the nano-plowed Josephson junctions, but the complex emission characteristic of such a radiation source. This is indicated by the dc IV-characteristics and the altered IV-characteristics under microwave absorption.

A more direct way of accessing the near field regime is by using hybrid spectrometers set up by combining an on-chip Nb/Al-AlO_x/Nb Josephson junction with the quantum dot circuit. In this case the spectrometers—built by Alexey Ustinov, Universität Erlangen—possess a well defined emission characteristic and can easily be tuned into the desired frequency range [29]. These junctions radiate commonly in the range from few GHz up to 600 GHz. In Fig. 2.5 a typical setup is shown: the spectrometer is fixed in the chip carrier and dc-contacts are attached; on top the quantum dot chip is glued and the contacts are bonded. Best coupling is achieved when the output of the Josephson junction is emitting via broad band antennas (Vivaldi or V-coupler). Most radiation, however, is transduced through the silicon chip the junction is fabricated on, into the GaAs heterostructure. In the measurements tuning of the frequency was possible by varying the Josephson tunnel junction (JTJ) critical current with the applied magnetic field and selecting the appropriate bias point at a resonant state. When operated in the flux-flow regime, the JTJ radiation frequencies $f = 2\mathrm{eV}/h$ are of the order of 100–500 GHz. We employ a finite magnetic field to operate the junction at a Fiske step of the current/voltage characteristic, a self-resonant state. The fundamental cavity resonance frequency of the junction is $f = \bar{c}/(2\ell)$, where $\bar{c}$ is the Swihart velocity and ℓ is the junction length.

The main advantage of this approach is the well understood emission characteristic and the ability to heavily filter all supply lines. As seen before in conventional millimeter wave experiments the existing coaxial lines or waveguides always carry low frequency components and IR radiation, acting as sources of noise. The on-chip spectrometer prevents these noise contributions and will enable more sensitive transport and millimeter spectroscopy. Moreover, the Josephson junction is perfectly suited to investigate electron dynamics in coupled dots in the time domain. This is achieved by operating the source not in cw-mode, but in the pulse-mode. Choosing excitation

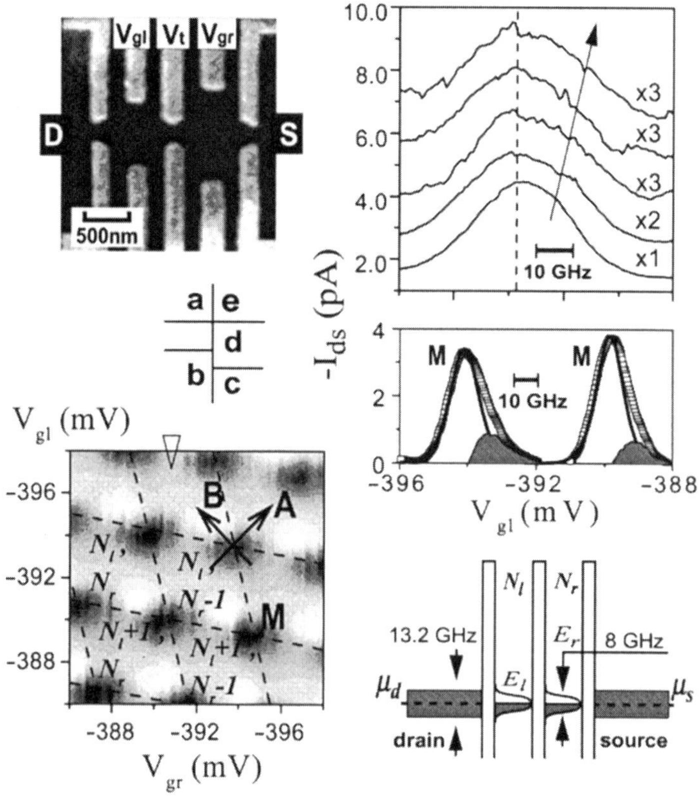

Fig. 2.6. (a) Scanning electron microscope graph of the double dot used in these experiments. (b) Charging diagram of drain-source current with weak tunnel coupling ($G_c \sim 0.08\ e^2/h$) (white: $I_{ds} \sim 0$ pA, black: $I_{ds} \sim -4$ pA). (c) The level diagram illustrates the resonant tunneling and the temperature broadening in the leads and dots. (d) A single trace extracted from (b) at a constant V_{gr} marked by the open triangle. (e) Temperature dependence of the peak asymmetry in the tunneling through the double dot. The bath temperatures are 276, 320, 468, 564, and 634 mK for the curves from bottom to top, respectively.

frequency and delay time appropriately, a synchronized tunneling current in quantum dot systems can be probed.

4. Coherent Modes in Quantum Dots

The basis of the experimental approaches of this work is the understanding of coupled quantum dots. Such a two dot system naturally can be used for the definition of a qubit. The energy scales involved in this setup are the Coulomb interaction E_C of N electrons captured in a quantum dot, the electron spin interaction E_S, excited states ϵ^* of the N-electron system, and the tunnel coupled molecular or coherent state ϵ_t. In the following experiments the foremost goal is to achieve large Coulomb energies

$E_C \sim$ 1–3 meV in order to establish a sufficient signal/noise ratio. Connected to this energy scale are the excited states of the N-electron systems: Empirically it is found that these scale according to $\epsilon^* \sim E_C/5 - E_C/3$. This finally enables observation of tunnel coupled states ϵ_t of the order of 50–200 μeV. The advantage of using discrete states of two quantum dots instead of ground and excited states of the same artificial atom lies in the enhanced flexibility of gating two individual dots. At the same time variation of the coupling strength easily allows for activating or deactivating single qubits. A typical sample and measurements are shown in Fig. 2.6, which will be discussed in detail in the following sections.

4.1. Capacitive Interaction

As for single dots the dominant energy for coupled dots is the charging energy: this can be analyzed by writing the total electrostatic energy of the system, using a capacitance matrix C_{ij}. The diagonal terms C_{AA}, C_{BB} give the single dot's capacitances and $C_{AB} = C_{BA}$ describe the electrostatic interaction [23]. The total energy is given by

$$E = \tfrac{1}{2} Q_i C_{ij}^{-1} Q_j \tag{1}$$

with Q_A, Q_B being the charges $N_A e$ and $N_B e$ of dot A and B ($e = -|e|$). The capacitive coupling of the gate electrodes gives an additional term. As seen partly in Fig. 2.2 two gate voltages are used to span a charging diagram. In the following we use (Θ) and (Π) for the two gates ramped. This gives for the total electrostatic energy

$$E \equiv E(N_A, N_B) = \tfrac{1}{2} Q_i C_{ij}^{-1} Q_j + (V_\Pi \Pi_i + V_\Theta \Theta_i) Q_i \tag{2}$$

with (Π_A, Π_B) and (Θ_A, Θ_B) being the capacitive coupling values of the single dots to the two gates with voltages V_Π und V_Θ applied. This yields for the charges on dots A and B:

$$Q_A = \sum_n C_{An} V_n = C_{AA} V_A + C_{AB} V_B + \sum_{n=\Pi,\Theta} C_{An} V_n, \tag{3}$$

$$Q_B = \sum_n C_{Bn} V_n = C_{BA} V_A + C_{BB} V_B + \sum_{n=\Pi,\Theta} C_{Bn} V_n. \tag{4}$$

From this the voltages V_A and V_B are obtained

$$V_A = \frac{1}{\det C_{ij}} \left(C_{BB} Q_A - C_{AB} Q_B + \sum_{n=\Pi,\Theta} (C_{AB} C_{Bn} - C_{BB} C_{An}) V_n \right), \tag{5}$$

$$V_B = \frac{1}{\det C_{ij}} \left(C_{AA} Q_B - C_{AB} Q_A + \sum_{n=\Pi,\Theta} (C_{AB} C_{An} - C_{AA} C_{Bn}) V_n \right) \tag{6}$$

After integration

$$E(Q_A, Q_B) = \int_0^{Q_A} V_A(Q'_A, Q_B = 0)dQ'_A + \int_0^{Q_B} V_B(Q_A, Q'_B)dQ'_B$$
$$= \frac{1}{\det C_{ij}}\left(\frac{1}{2}C_{BB}Q_A^2 + \sum_{n=\Pi,\Theta}(C_{AB}C_{Bn} - C_{BB}C_{An})V_nQ_A\right.$$
$$\left. + \frac{1}{2}C_{AA}Q_B^2 - C_{AB}Q_AQ_B\right)$$
$$+ \frac{1}{\det C_{ij}}\sum_{n=\Pi,\Theta}(C_{AB}C_{An} - C_{AA}C_{Bn})V_nQ_B \quad (7)$$

For the capacitively coupled gate potentials this results in

$$\Theta_i = \frac{1}{\det C_{ij}}\begin{pmatrix} C_{AB}C_{B\Theta} - C_{BB}C_{A\Theta} \\ C_{AB}C_{A\Theta} - C_{AA}C_{B\Theta} \end{pmatrix} \quad (8)$$

$$\Pi_i = \frac{1}{\det C_{ij}}\begin{pmatrix} C_{AB}C_{B\Pi} - C_{BB}C_{A\Pi} \\ C_{AB}C_{A\Pi} - C_{AA}C_{B\Pi} \end{pmatrix} \quad (9)$$

The conditions for charging an electron onto and off the two dots are for $\mu_S = \mu_D = \mu$:

$$(1) \qquad E(N_A + 1, N_B) - E(N_A, N_B) = \mu \quad (10)$$
$$(2) \qquad E(N_A + 1, N_B + 1) - E(N_A + 1, N_B) = \mu \quad (11)$$
$$(3) \qquad E(N_A, N_B + 1) - E(N_A, N_B) = \mu \quad (12)$$
$$(4) \qquad E(N_A + 1, N_B + 1) - E(N_A, N_B + 1) = \mu \quad (13)$$

The conditions (1)–(4) define the states in the charging diagram from which the phase boundaries can be derived. The only difference of the conditions (1) and (4) is the charging state of the whole system, the same holds for conditions (2) and (3). With the notation

$$\tilde{C}_{ij} = C_{ij}^{-1} \quad (14)$$

condition (1) gives:

$$Q_i\tilde{C}_{ij}\begin{pmatrix} e \\ 0 \end{pmatrix} + \frac{1}{2}\begin{pmatrix} e \\ 0 \end{pmatrix}\tilde{C}_{ij}\begin{pmatrix} e \\ 0 \end{pmatrix} + (V_\Pi\Pi_i + V_\Theta\Theta_i)\begin{pmatrix} e \\ 0 \end{pmatrix} = \mu \quad (15)$$

which leads to

$$V_\Theta^{(1)} = -V_\Pi\frac{\Pi_A}{\Theta_A} + \frac{1}{\Theta_A}\left(\frac{\mu}{e} - \frac{e}{2}\tilde{C}_{AA} - Q_i\tilde{C}_{ij}\begin{pmatrix} 1 \\ 0 \end{pmatrix}\right) \quad (16)$$

and in the same way for condition (4)

$$V_\Theta^{(4)} = -V_\Pi\frac{\Pi_A}{\Theta_A} + \frac{1}{\Theta_A}\left(\frac{\mu}{e} - \frac{e}{2}\tilde{C}_{AA} - Q'_i\tilde{C}_{ij}\begin{pmatrix} 1 \\ 0 \end{pmatrix}\right) \quad (17)$$

with $Q_i' = Q_i + \binom{0}{e}$. The distance of these linear traces is given by

$$\frac{1}{\Theta_A}\binom{0}{e}\tilde{C}_{ij}\binom{1}{0} = \frac{e}{\Theta_A}\tilde{C}_{AB}. \tag{18}$$

From condition (3) it is found:

$$V_\Theta^{(3)} = -V_\Pi \frac{\Pi_B}{\Theta_B} + \frac{1}{\Theta_B}\left(\frac{\mu}{e} - \frac{e}{2}\tilde{C}_{BB} - Q_i\tilde{C}_{ij}\binom{0}{1}\right). \tag{19}$$

and finally for condition (3):

$$V_\Theta^{(2)} = -V_\Pi \frac{\Pi_B}{\Theta_B} + \frac{1}{\Theta_B}\left(\frac{\mu}{e} - \frac{e}{2}\tilde{C}_{BB} - Q_i''\tilde{C}_{ij}\binom{0}{1}\right). \tag{20}$$

with $Q_i'' = Q_i + \binom{e}{0}$ and the offset of the traces

$$\frac{e}{\Theta_B}\tilde{C}_{AB}. \tag{21}$$

These traces define charging diagrams with charges N_A, N_B on the dots. At the crossing points of the traces for (1) and (3), respectively (2) and (4) one finds:

$$E(N_A, N_B + 1) = E(N_A + 1, N_B) \tag{22}$$

and resonant tunneling occurs. The total charge is conserved, but only with the coupled dot system, i.e. the electrons are distributed over the whole system. From these conditions the following equations are obtained:

$$V_\Theta = -V_\Pi \frac{\Pi_A - \Pi_B}{\Theta_A - \Theta_B} - \frac{1}{\Theta_A - \Theta_B}\left(Q_i\tilde{C}_{ij}\binom{-1}{+1} - \frac{e}{2}\tilde{C}_{AA} + \frac{e}{2}\tilde{C}_{BB}\right). \tag{23}$$

These define the charging diagram. Neglecting the interdot capacitance C_{AB} this results in the rectangular upper resonance pattern of Fig. 2.2. With a non-zero coupling C_{AB} conditions (1)–(4) cannot be satisfied at the same time and the degeneracy of the resonances is partially lifted. The offset is determined by $e\tilde{C}_{AB}/\Theta_A$.

4.2. Tunnel Coupling

In addition to the pure capacitive interaction the two discrete resonances in two individual quantum dots can overlap depending on the coupling. This leads to the tunnel splitting. Following Pfannkuche [30] this can be derived from the Hamiltonian of the coupled dot system. In a simplified model the center of mass excitations with the quantum numbers k_A, k_B of N_A, N_B electrons in dot A and B are written

$$[|N_B k_B\rangle|N_A k_A\rangle]_{N_B,N_A,k_B,k_A}. \tag{24}$$

For decoupled dots the Hamiltonian is diagonal

$$H_0^{DQP}|N_B k_B\rangle|N_A k_A\rangle$$
$$= \left[\frac{e^2}{2C_A}N_A^2 + \frac{e^2}{2C_B}N_B^2 + \frac{e^2 C_{AB}}{C^2}N_B N_A + eV_\Pi(\Pi_A N_A + \Pi_B N_B)\right.$$
$$\left. + eV_\Theta(\Theta_A N_A + \Theta_B N_B) + (k_B+1)\Omega_B + (k_A+1)\Omega_A\right]|N_B k_B\rangle|N_A k_A\rangle \tag{25}$$
$$= [E(N_A, N_B) + E^{cm}(k_B, k_A)]|N_B k_B\rangle|N_A k_A\rangle \tag{26}$$

where $\Omega_{B/A}$ characterizes the confinement potential. The coupling of states $N = N_B + N_A$ through the barrier is defined by

$$H_t^{DQP}|N_B k_B\rangle|N_A k_A\rangle = t\sum_{k_1,k_2}\big[|N_B+1, k_B+k_1\rangle|N_A-1, k_A+k_2\rangle$$
$$+ |N_B-1, k_B+k_1\rangle|N_A+1, k_A+k_2\rangle\big]. \tag{27}$$

For simplicity it is assumed that the tunneling matrix element t is constant. If the states $|N_B+1,0\rangle|N_A,0\rangle$ and $|N_B,0\rangle|N_A+1,0\rangle$ are degenerated the two tunnel split states are a linear combination of both with a valence electron. When the electrostatic energy of the two dot system is termed $|N_B,0\rangle|N_A,0\rangle$ with $E(N_A, N_B)$, the energy of the tunnel split state is

$$\epsilon_t = \tfrac{1}{2}(E(N_A, N_B-1) + E(N_A-1, N_B))$$
$$\pm \tfrac{1}{2}\sqrt{(E(N_A, N_B-1) - E(N_A-1, N_B))^2 + 4t^2}. \tag{28}$$

The aim in most experiments on coupled dots is to introduce a finite tunnel splitting for the definition of a quantum bit. The beauty of this approach is the relative high degree of control given at hand. The tunnel splitting can be tuned in by varying the coupling strength of the center gate electrode as will be discussed in the following. In Fig. 2.7(a) measurements on the coupled dot system of Fig. 2.1 are shown. The two dots define a classical charging diagram pattern with clear periodicities being an indication for large quantum dots with similar charging energies. A slight variation of the interdot coupling results in a qualitative change of the diagram as seen in Fig. 2.7(b).

In the regime of strong coupling the two discrete states in the individual dots interact via tunneling $2t$ leading to an altered charging diagram shown in Fig. 2.8(a). The pattern now reveals traces of conductance reflecting the strong wave function overlap between the dots. Following the individual resonance lines a Lorentzian line shape is recovered which is a direct measure for the overlap [14]. The level diagram in Fig. 2.8(b) indicates how the splitting into binding E_B and anti-binding E_A states break up the charging diagram. The splitting is of the order of $\delta E = 2t$ and is found

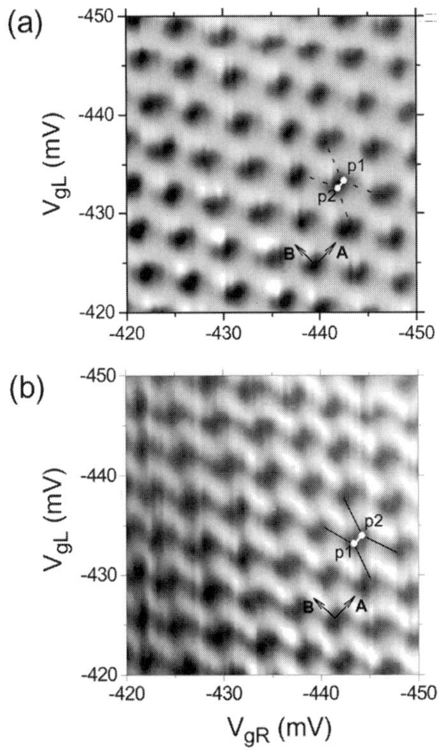

Fig. 2.7. Charging diagram of the coupled dot system from Fig. 2.6 in the regime of weak (a) and strong coupling (b). The dark regions correspond to non-zero conductance, while the white parts indicate Coulomb blockade.

at every crossing point in the diagram. In Fig. 2.8(c) a number of traces are plotted from Fig. 2.8(a): the variation in peak shape is clearly visible. In Fig. 2.8(d) trace 3 is evaluated: the background at s_{2+} and s_{1+} stems from the anti-binding states, while the energy E_{el} is given by the interdot capacitance. It has to be noted that a superposition of Coulomb interaction and direct tunnel coupling is observed in these measurements.

As referred to in Fig. 2.8 the tunnel splitting is even more pronounced when smaller dots are used: in the charging diagram of a coupled dot system shown in Fig. 2.9(a) such a measurement is shown in log-scale representation. In Fig. 2.9(b) the details of the diagram are sketched with the notation given. From the charging diagram single line plots are taken in Fig. 2.10, crossing the charge states $(N_1 - 1, N_2 - 1) \rightarrow (N_1 - 1, N_2)$ and $(N_1, N_2 - 1) \rightarrow (N_1, N_2)$. As seen we find a resonance trace which comprises not only the ground states, but additional peaks corresponding to the anti-bonding states. Evidently, such a coherent mode can be monitored in a coupled dot system. The coupling can be increased by using smaller quantum dots in order to achieve separate peaks for the bonding states. Taking a number of traces from the charging diagram in Fig. 2.9 at the resonance points indicated by A and B, we find a clear tunnel splitting in Fig. 2.10(a). Under variation of a magnetic field applied perpendicular

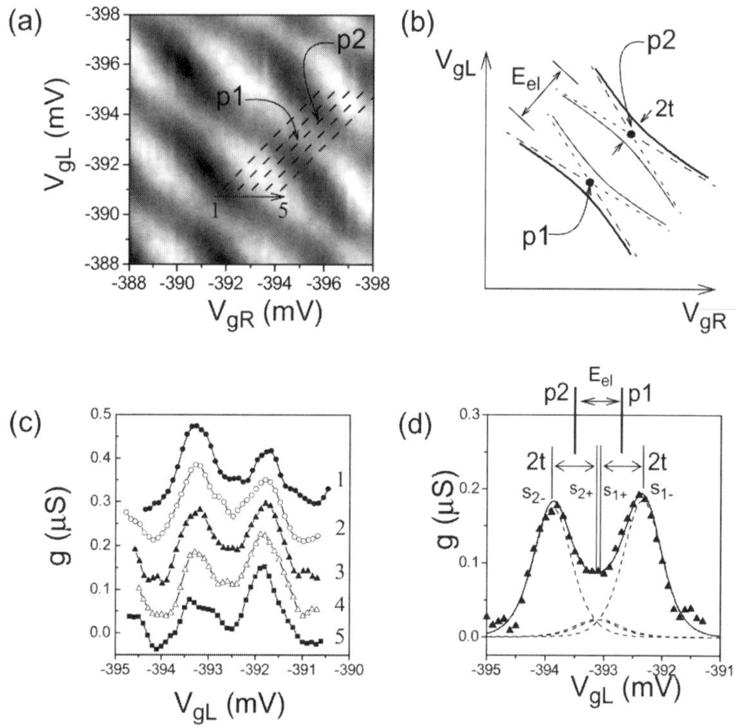

Fig. 2.8. In the regime of strong coupling the two discrete states in the individual dots interact via tunneling $2t$ leading to an altered charging diagram (a). The level diagram (b) indicates how the splitting into binding E_B and anti-binding E_A states break up the charging diagram. The splitting is of the order of $\delta E = 2t$. In (c) a number of traces are plotted from (a): the variation in peak shape is clearly visible. In (d) trace 3 is evaluated: the background at s_{2+} and s_{1+} refer to the anti-binding states, while the energy E_el is given by the interdot capacitance.

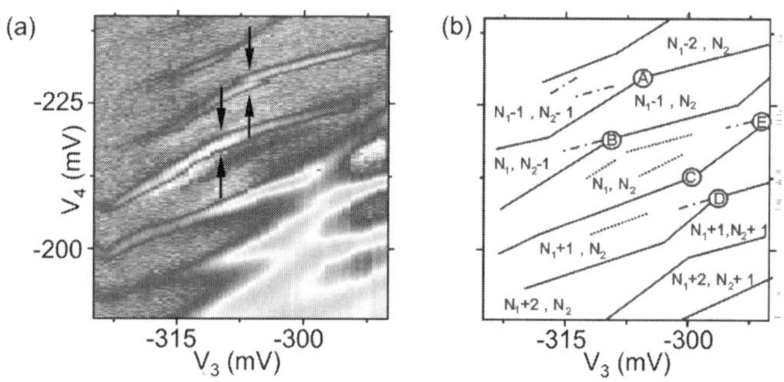

Fig. 2.9. As referred to in Fig. 2.8 the tunnel splitting is even more pronounced when smaller dots are used: in the charging diagram of a coupled dot system shown in (a) such a measurement is shown in log-scale representation at a bath temperature of 50 mK. In (b) the details of the diagram are sketched with the notation.

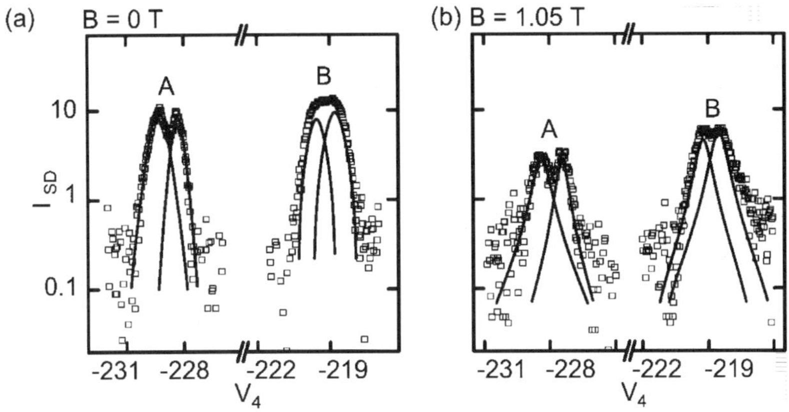

Fig. 2.10. Taking a number of traces from the charging diagram in Fig. 2.9 at the resonance points indicated by A and B, we find a clear tunnel splitting (a). Under variation of a magnetic field applied perpendicular to the plane of the dots these splitted resonances can be adjusted, indicating spin and orbital momenta of the artificial molecule. The tunnel splitting in this case is of the order of ~70 μeV.

to the plane of the dots these splitted resonances can be adjusted, indicating spin and orbital momenta of the artificial molecule. The tunnel splitting in this case is of the order of ~70 μeV, which will be monitored in the following by microwave radiation.

5. Photon Assisted Tunneling in Quantum Dots

5.1. Photon Assisted Tunneling in Single Dots

Placing a quantum dot in a time-dependent electromagnetic field the interaction of single electrons and photons of the electromagnetic field can help to overcome Coulomb blockade. The coupling of radiation to electrons is commonly divided into two regimes: in the adiabatic regime the microwave frequency is below the tunneling rate $f \ll \Gamma$, i.e. the electronic states in the leads and the quantum dots are following the external perturbation adiabatically. In the regime $f \gg \Gamma$ electrons can absorb quanta of the field, which enables tunneling through the dot. The dots considered here reveal charging energies of the order of 0.2–2.0 meV and mean level spacing of 0.05–0.5 meV. The first approach to understanding charge transport under microwave radiation was developed by Tien and Gordon [31] for superconductor junctions. Bruder and Schoeller [18] later on extended this formalism for quantum dots. It is mainly stated that the transition rates of electrons Γ are modified under radiation to $\tilde{\Gamma}(E)$. The dot states are assumed to be discrete while the density of states in the leads is described by Fermi's golden rule, i.e. will be treated as a continuum in the following. The modified tunneling rates are given by

$$\tilde{\Gamma}(E) = \sum_{n=-\infty}^{n=+\infty} J_n^2(x)\Gamma(E + nhf) \tag{29}$$

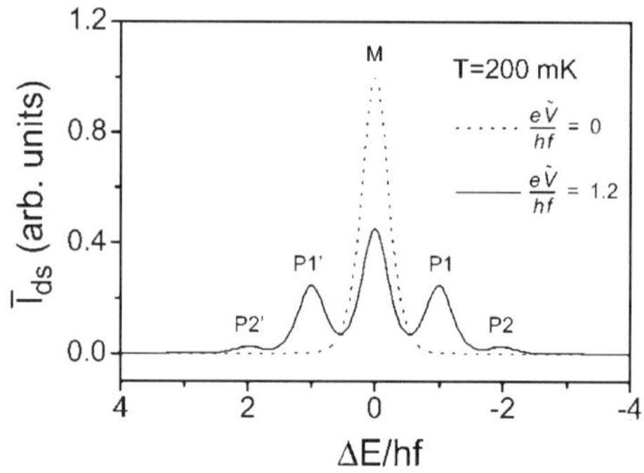

Fig. 2.11. Calculation of photo current $\bar{I}_{\mathrm{ds}}$ versus gate voltage $\mathrm{eV}/hf = \Delta E/hf$ with and without applied microwave radiation. As seen the main resonance is attenuated under radiation while sidebands at multiples of the photon energies $\Delta E/hf = \pm 1, \pm 2, \ldots$ appear.

here $J_n(x)$ is the nth order Bessel function with the argument $x = e\tilde{V}/hf$ where $\tilde{V}$ represents the ac voltage of the radiation. As seen this equation describes absorption $E + nhf$ and emission $E - nhf$ of field quanta. The weight of higher order processes follows the amplitude modulation given by the Bessel functions, as found in the measurements below (see Fig. 2.14). In other words the modification of the transition rates is to be understood as a generation of virtual states. The total photo current is finally given by

$$\bar{I}_{\mathrm{ds}} = \sum_{n=-\infty}^{+\infty} J_n^2\left(\frac{e\tilde{V}}{hf}\right) I_{\mathrm{ds}}\left(V_{\mathrm{g}} + \frac{nhf}{e\alpha}\right) \tag{30}$$

With the static $I_{\mathrm{ds}} - V_{\mathrm{ds}}$-trace without radiation being given by

$$I_{\mathrm{ds}} = g_{\max} V_{\mathrm{ds}} \cosh^{-2}\left(\frac{-e\alpha\left(V_{\mathrm{g}} - V_{\mathrm{g}}^{\Delta}\right)}{2k_{\mathrm{B}}T}\right) \tag{31}$$

the resulting photo current can be easily calculated. Such a calculation is given in Fig. 2.11 where the radiation strength is increased up to the absorption of two photons **P1** and **P2**. The satellite peaks are centered at $\Delta E = V_{\mathrm{g}} - V_{\mathrm{g}}^{\Delta} = nhf$. In Fig. 2.12 the two main absorption processes are sketched: Absorption of a photon in the dot and subsequent tunneling into the lead is analogous to the internal photo effect, while absorption of photons in the leads corresponds to the external photo effect. In order for these transitions to occur a non-zero bias has to be applied, which is usually of the order of some $10\,\mu\mathrm{eV}$. Additionally, transitions through excited states ϵ^* can be induced by the radiation, which are otherwise not accessible due to Coulomb blockade. A photo current through an excited state can be compared to photo ionization of atoms or molecules.

All these features can be recovered in the measurements shown in Fig. 2.13: a single quantum dot is defined and radiation of 35.8 and 40.2 GHz applied. Clearly the main resonance amplitude **M** is lowered and satellites **P1** and **P2** appear. In addition we observe an excited state **PX** resonance with a smaller amplitude. Since the tunneling

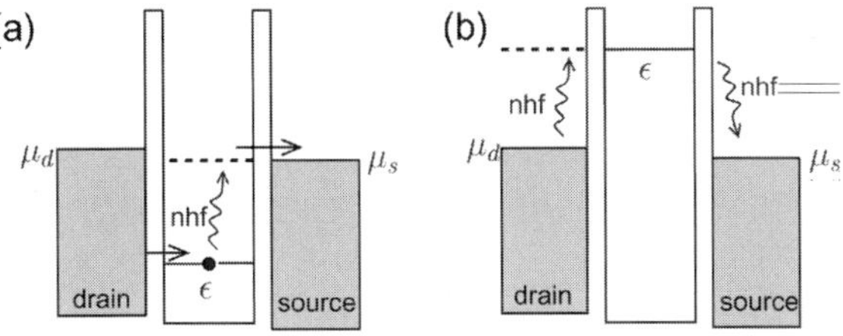

Fig. 2.12. Level diagrams illustrating photon assisted tunneling: (a) a photon is absorbed within the quantum dot and the electron leaves to the source contact. A small bias voltage $V_{\mathrm{ds}} = \mu_{\mathrm{d}} - \mu_{\mathrm{s}}$ is applied. In (b) quanta are absorbed in the drain contact allowing electrons to overcome Coulomb blockade.

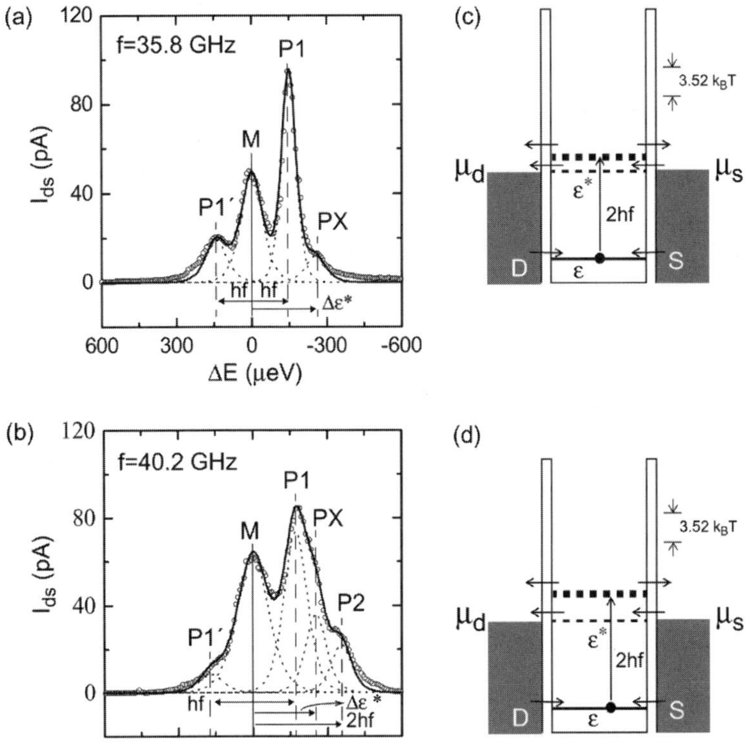

Fig. 2.13. Direct photo current through a single quantum dot under continuous wave radiation at (a) 35.8 GHz and (b) 40.2 GHz. The ground state is noted by **M** while single photon absorption is **P1**, two photon absorption **P2**. Transitions through an excited state ϵ^* are denoted by PX. $T = 270\,\mathrm{mK}$, $B = 1\,\mathrm{T}$, and $V_{\mathrm{ds}} = 30\,\mu\mathrm{eV}$.

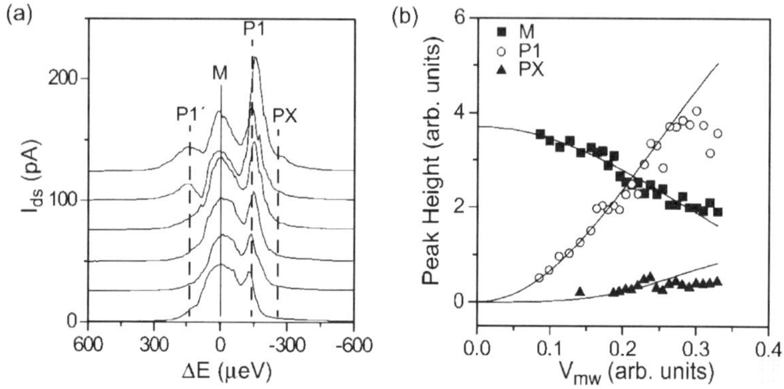

Fig. 2.14. (a) Photo current at 35.8 GHz under increasing radiation power, most prominently seen in the single photon resonance **P1**. In (b) ground, excited and the one-photon resonance are compared in amplitude: as expected the variation follows the Bessel functions of zeroth (**M**) and first order (**P1**), J_0^2 and J_1^2.

barriers for this particular measurement are fairly similar we observe absorption **P1** as well as emission **P1**$'$ of photons. The bias applied in these measurements was of the order of $V_{ds} = +30\,\mu$eV. In varying the microwave power it is possible to gather a more detailed understanding of the different transition states. This is presented in Fig. 2.14: in (a) microwave power at 35.8 GHz is steadily increased. As derived earlier the amplitude of **M** is reduced while the photon resonances rise—of course the energy difference remains constant, e.g. $E_{\mathbf{M}} - E_{\mathbf{P1}} =$ const. In Fig. 2.14(b) the power dependence is plotted in detail, following the the Bessel functions J_0^2 and J_1^2 for resonance **M** and **P1**. The excited state shows a somehow slower rise in amplitude due to the reduced wave function overlap. It has to be noted that the graph is calibrated via the relation $\Delta E = -e\alpha V_g = hf$. The calibration factor α is obtained by tracing the drain/source dependence of the single dot.

In changing the frequency over a broad range from 1 to 50 GHz we obtain the full frequency dependence of the photon sidebands as shown in Fig. 2.15. The best resolution is obtained for the absorption of a single photons. A lower bound is set by the thermal broadening introduced by the radiation, which is estimated to be of the order of 20 GHz considering the full-width-half-maximum is given by $3.52k_BT$. As expected from the relation $E = nhf$ a linear extrapolation can be made yielding the correct result. The same result is found for the two photon absorption resonances. However, the induced excited state transitions appear not to fit into this scheme. Further work is required to clarify the exact nature of these "ionized" states $\Delta\epsilon^*$.

5.2. *Photon Assisted Tunneling in Coupled Dots*

In extending this scheme to coupled quantum dots the same description applies with the essential difference that transitions between the two dots have to be considered, while absorption of electrons in the leads and adjacent tunneling through the dots is reduced (see Fig. 2.16). This has the advantage that thermal broadening is effectively

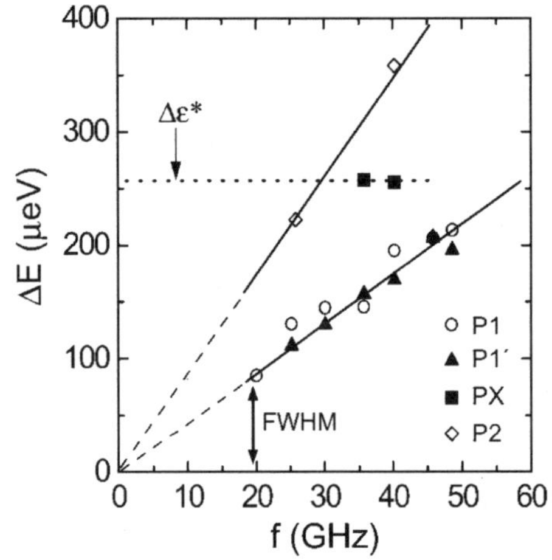

Fig. 2.15. Plotted is the peak distance between **M** and **P1**, **P2**, and **PX**. As expected a linear dependence is found except for the excited state resonance. The thermal broadening given by $\sim 3.52\, k_B T$ as an effective full-width-half-maximum (FWHM) dominates the onset of photon assisted tunneling for single dots.

reduced to below 100 mK corresponding to about 1 GHz. The charging diagram of a weakly coupled dot system is modified under irradiation as exemplified in the scheme shown in Fig. 2.17. Apart from a photo current through the ground state, two processes cause additional resonance lines to appear. Absorption of an electron in the left dot leads to a current through the right dot and vice versa. The transition energy is defined by the energy difference of the two discrete states in the left and right dot ϵ_l and ϵ_r by $\delta E = \epsilon_l - \epsilon_r$. As mentioned this picture holds for weakly coupled dots, while under an increased coupling the transition energy is modified to

$$\Delta E = E_+ - E_- = \sqrt{\delta E^2 + 4t^2} \tag{32}$$

where t gives the tunneling coupling. Accordingly a single electron will slosh back and forth between two dots at the Rabi frequency $\Omega_R = 2t/\hbar$. Hence, a change in coupling will lead to an altered photon assisted response. When we note that

$$nhf = E_+ - E_- = \sqrt{\delta E^2 + 4t^2} \tag{33}$$

we find a direct relation between coupling t and photon energy. Of course in addition to varying the magnitude of the radiation the drain/source voltage can be adjusted to supply the appropriate energy.

A weakly coupled dot system is characterized as follows (see Fig. 2.6): For tuning the dots into the tunnel coupling regime [32–34], the tunneling rates of two outer tunnel barriers are kept fixed around $\Gamma_l^{-1} \sim \Gamma_r^{-1} \sim 2$ nsec and the tunnel conductance G_c of the center barrier is controlled by V_t (see Fig. 2.6(a)). The dots contain around 200 electrons each and show excited states at $\Delta\epsilon^* \cong 120\,\mu$eV above the ground state.

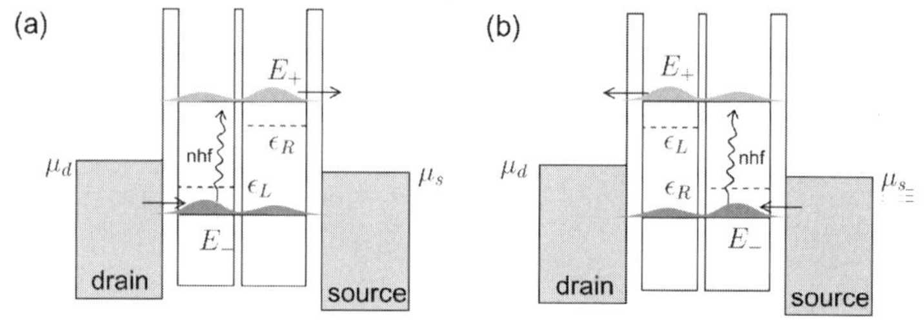

Fig. 2.16. Level diagrams for a coupled quantum dot under irradiation: (a) photon absorption in the left dot, tunneling through the right dot and vice versa (b). In contrast to single dots no thermal broadening is present for coupled dots.

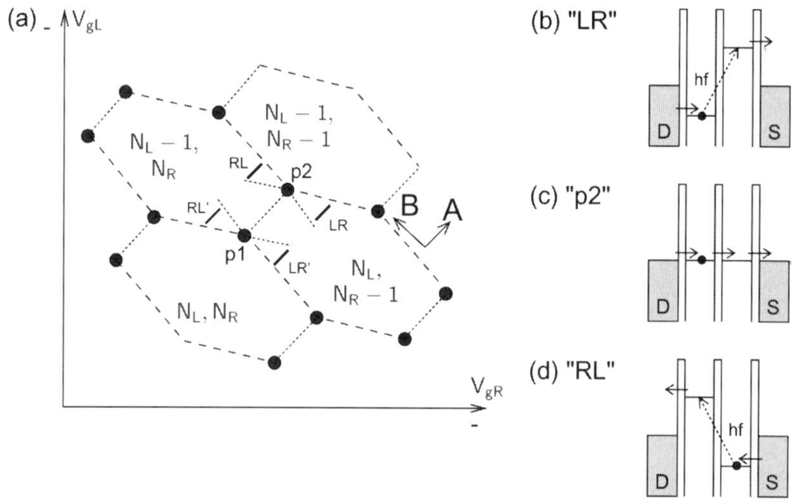

Fig. 2.17. Extending the measurements from single to coupled dots the charging diagram is modified as seen in (a), the coupling is considered to be weak. The different transitions are indicated in (b), (c), and (d): the absorption processes left-to-right **LR** and right-to-left **RL** occur in the Coulomb blockade region at the lines indicated. The ground state states are given by **p1** and **p2**.

The effective calibration factors (α_l and α_r) are 6.8×10^{-2} and 5.2×10^{-2} for the left and right dot, respectively. From the charging diagrams of the drain-source current I_{ds} the charging energies are determined (U_l and U_r) to be 391 μeV and 318 μeV. The drain-source bias V_{ds} is fixed at around $-10\,\mu V \ll \Delta\epsilon^*/e$. In Fig. 2.6(b) the charging diagram of the coupled dot is plotted in a grayscale representation in the weak coupling regime ($G_c \sim 0.08\, e^2/h$). The discrete conductance peaks in Fig. 2.6(b) are main peaks **M** resulting from resonant tunneling through the double dot (see Fig. 2.6(c)). Within the diamond shaped regions enclosed by dashed lines Coulomb blockade prevails and the number of electrons in the dots is well defined.

In the direction marked by arrow **A**, the relative level distance $E_{lr} = -(E_l - E_r)$, is kept constant, while the average $(E_l + E_r)/2$ is increased relative to the chemical potentials (μ_d and μ_s) in the drain and source leads. In direction of **B**, where $(E_l+E_r)/2$ is kept fixed but E_{lr} is varied, the broadening found is dominated by the electron temperature within the dots but not the intrinsic life time of the quantum levels. The temperature broadening in the dots ($h\Gamma$) is estimated to be about $h \times 8$ GHz, which is slightly lower than the electron temperature in the leads. In Fig. 2.6(c), the level diagram takes into account the temperature broadening in the leads and the dots. A typical single trace extracted from the charging diagram is shown in Fig. 2.6(d). On the left side of the peak, where E_l is higher than E_r, it is well approximated by $\cosh^{-2}(\epsilon/2k_B T_e)$ (see the solid line in Fig. 2.6(d)). However, on the right side where ϵ_l is lower than ϵ_r, an additional contribution to the tunneling current exists, which has a maximum at 12 GHz away from the main peak (see dark region in Fig. 2.6(d)). By increasing the bath temperature, the position of this maximum shifts further to the right side, as shown in Fig. 2.6(e). Due to the background radiation and the finite electron temperature electrons absorb acoustic phonons via deformation potential coupling, i.e. at a bath temperature of $T_b = 170$ mK a corresponding frequency of acoustic phonons $f_{ph} \sim 3k_B T_b/h \approx 10.7$ GHz is obtained.

Microwave spectroscopy on the weakly coupled dots ($G_c \sim 0.08e^2/h$) is then studied by taking charging diagrams and applying microwave radiation in the range of 1–20 GHz. The electron temperature is slightly increased to be $T_e \approx 210$ mK in the leads, but the peak position of the sidebands is independent of microwave power applied as noted before for single dots. In Fig. 2.18 a typical charging diagram under irradiation by 20 GHz photons is presented for reverse (a) and forward (b) bias. The satellite peaks for each ground state (**M**) change their position as seen in dependence of bias. We can conclude that the absorption for the present outer tunneling barrier opacity suppress the detection of photon emission signatures. In Fig. 2.19 single lines traces from both diagrams in reverse and forward bias are given. Clearly not only single (**P1**) but double (**P2**) photon absorption is identified.

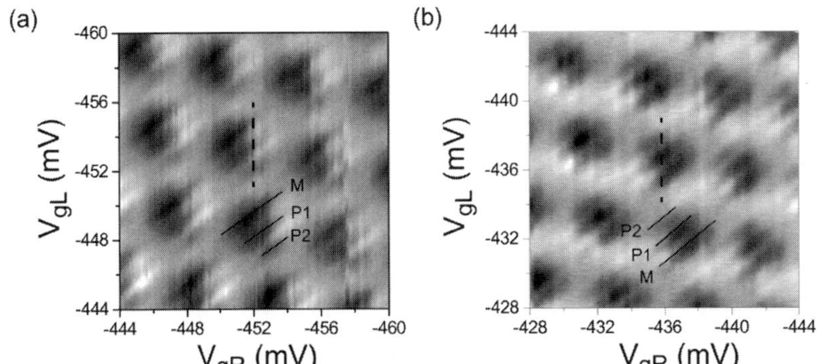

Fig. 2.18. Charging diagram under radiation at 20 GHz: (a) reverse bias of $V_{ds} = -30\,\mu$eV is applied and (b) forward bias $+30\,\mu$eV. The photon induced satellite peaks indicating one (**P1**) and two (**P2**) photon absorption shift accoring to the bias applied. The gray scale is defined as follows: black $I_{ds} < 1.6$ pA and white $I_{ds} > -0.55$ pA. The tunnel barriers are set at $G \sim 0.08\,e^2/h$.

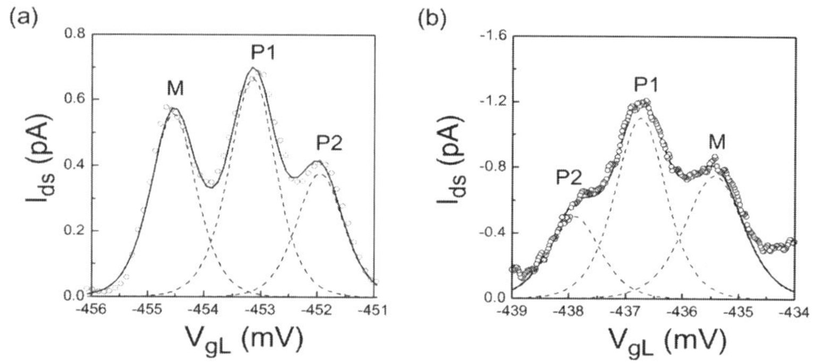

Fig. 2.19. Line traces from the charging diagram under radiation of Fig. 2.18 in forward and reverse current mode.

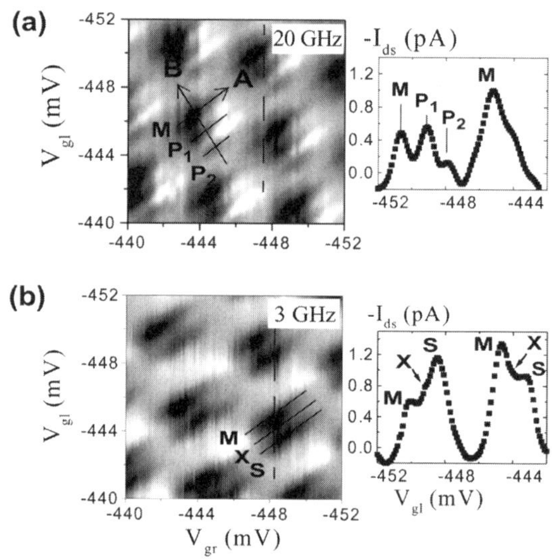

Fig. 2.20. Charging diagrams with microwave radiation at (a) 20 GHz and (b) 3 GHz (white $I_{ds} \sim 0$ pA, black $I_{ds} \sim -4$ pA). The single traces from the grayscale plot show the details. See text for details on symbols **P1**, **P2**, **M**, **S**, and **X**.

A comparison of the charging diagrams at 20 and 3 GHz is shown in Fig. 2.20: in the grayscale plot of Fig. 2.20(a) the double dot shows a regular pattern of resonances. Peaks bound to the main peak **M** are the microwave induced sidebands at 20 GHz. For clarification a single trace taken out of the charging diagram (position indicated by the dashed line) is plotted in the right panel of Fig. 2.20(a). Clearly, we observe one- and two-photon absorption processes marked by **P1** and **P2**. Below 10 GHz only one sideband **S** is clearly observed (see Fig. 2.20(b), only the data for 3 GHz are shown) and most strikingly the distance between **M** and **S** is comparable to that at 20 GHz or even larger. Furthermore, under microwave radiation at 3, 4.5, and 5.9 GHz an additional

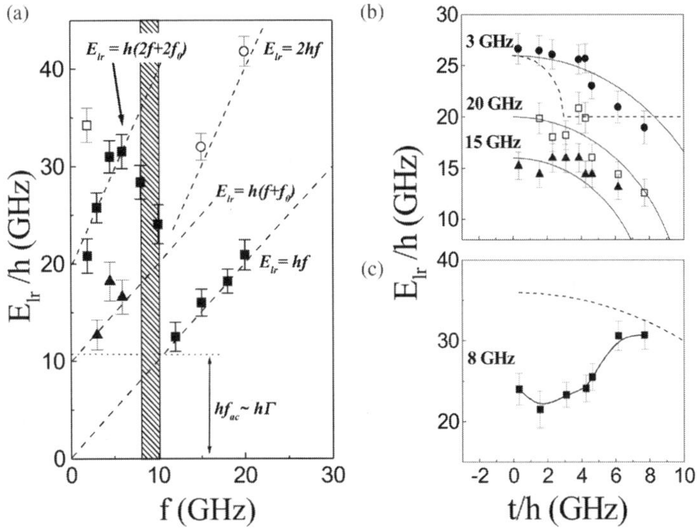

Fig. 2.21. (a) The distance between the side band and the main peak versus the microwave frequency in the weak tunnel coupling regime ($G_c \sim 0.08\ e^2/h$). $f > 10$ GHz: Solid squares and open circles are for sidebands $\mathbf{P}_1$ and $\mathbf{P}_2$, respectively. $f < 8$ GHz: Solid circles and solid triangles stand for sidebands **S** and **X**, respectively. In the hatched region (8–10 GHz), the open diamonds corresponds to sidebands **S**. (b) The coupling dependence of the distance between the side bands and the main peak at 20, 15, and 3. For 20 and 15 GHz, the solid lines corresponds to Eq. (36). For 3 GHz, the dashed line is from Eq. (37), the dash/dotted line is fitted with Eq. (38), and the solid line is based on Eq. (39). (c) At 8 GHz, the dashed curve is calculated from Eq. (39).

tunneling resonance **X** located in between **M** and **S** is observed, as shown in the single traces in Fig. 2.20(b).

In Fig. 2.21(a), the distance between the sidebands (**P1**, **P2**, **X** and **S**) and the main peak (**M**), converted from gate voltages $\Delta V_{\mathrm{gl,gr}}$ to energies E_{lr} using the factors $\alpha_{\mathrm{l,r}}$, are plotted over the microwave frequency applied. Above 10 GHz the distance is equal to the corresponding microwave photon energies hf and $2hf$ as described by the Tien-Gordon theory [31]. Below 8 GHz, however, the distance for sidebands **X** and **S** follow

$$E_{\mathrm{lr}}^{\mathbf{X}} = h(f + f_0) \tag{34}$$

$$E_{\mathrm{lr}}^{\mathbf{S}} = 2h(f + f_0) \tag{35}$$

respectively, with an offset $f_0 \approx 10$ GHz. Apart from the offset, the slopes of the traces correspond to the absorption of one and two photons. Hence, an additional absorption mechanism leading to the offset hf_0 and $2hf_0$ has to be considered. Moreover, it has to be noted that the hatched area in Fig. 2.20(a) marks a transition region between 8 GHz and 10 GHz. As schematically shown in Fig. 2.22, the two states involved in transport are, e.g. the N_lth ground state in the left dot and the N_rth ground state in the right dot.

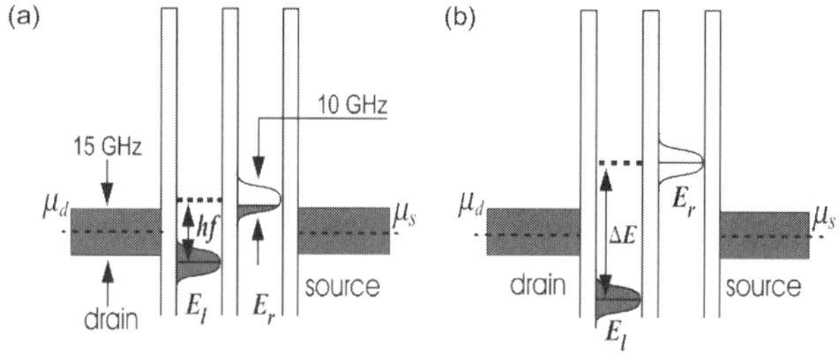

Fig. 2.22. Level diagrams for (a) one-step PAT with microwave frequencies above 10 GHz, and (b) transport with microwave frequencies below 8 GHz.

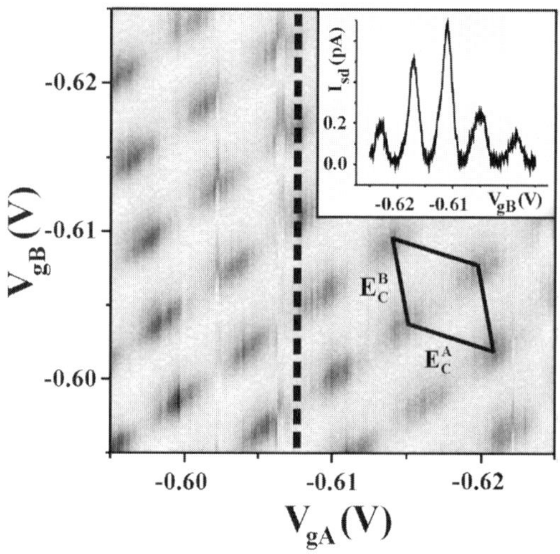

Fig. 2.23. Charging diagram of the double quantum dot as a grayscale plot without microwave radiation (white: $I < 0$ pA, black: $I > 0.5$ pA). A small forward bias of $V_{ds} = 19\,\mu$V is applied to monitor the current. The two dots are weakly coupled by the tunnel barrier and produce a periodic lattice with E_C^A and E_C^B denoting the charging energies, as marked by the diamond. The two gate voltages V_{gA} and V_{gB} span the charging diagram. Inset gives a line plot of the direct-current through the double dot. The range of voltage V_{gB} in this graph coincides with the broken line in the gray-scale plot.

In order to distinguish the different nature of these sidebands and the offset hf_0 and $2hf_0$, the coupling strength between the dots is varied through adjusting the center gate voltage and measuring the energetic distance of sidebands to the main peak. The data is shown in Fig. 2.21(b, c): The variation of the tunnel coupling enables to trace the transition of the artificial molecule from "ionic" to "covalent" binding by determining the peak splitting E_{lr}. At 20 and 15 GHz the peak distance from **P1** to **M** can be roughly fitted by

$$E_{lr}^{\mathbf{P1}} = \sqrt{(hf)^2 - 4t^2} \tag{36}$$

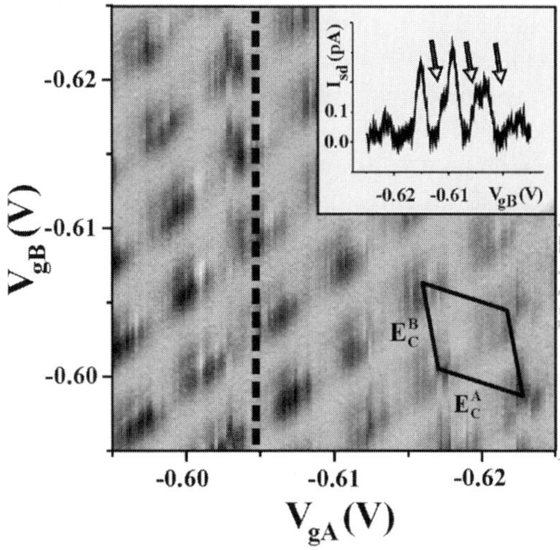

Fig. 2.24. Charging diagram with an identical scan range as in the previous figure, but under microwave radiation by a far-field source at 10 GHz. The sidebands are found on only one side, depending on the tuning of the tunnel barriers. Inset: Photo current with the sidebands induced by the frequency dependent absorption. Pumping of the electrons leads to a reduction of the absolute current value. The gray lines denote the range of voltage V_{gB} which correlates to the frequency of 10 GHz applied.

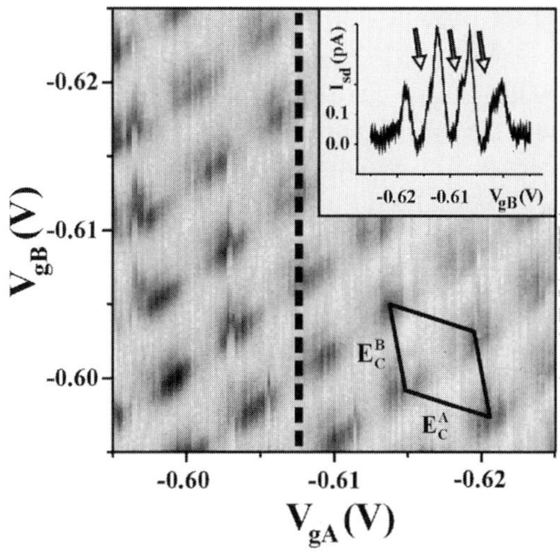

Fig. 2.25. Charging diagram under microwave radiation with the on-chip Josephson oscillator as a source (at 10 GHz). Inset gives the photon-induced current through the dot system with the photon assisted tunneling signal. The peak heights and positions are identical to the plots obtained with the far-field source.

where t is simply assumed to be proportional to the coupling conductance: $t \propto G_c$. Equation (36) is a signature of a coherently coupled two-level system. This implies that the two dots are coupled coherently and form an artificial molecule. The resonance of this molecular state is driven by microwave radiation, which matches the resonance frequency. Commonly, one expects a similar photon coupling below 10 GHz as described by Eq. (36). The coupling dependence obtained turns out to be quite different: An increase of the tunnel coupling at $f = 8$ GHz enhances the splitting (see Fig. 2.21(c)). At $f = 3$ GHz, the coupling then again follows similar traces as those at 20 and 15 GHz (see Fig. 2.21(b)). Hence, an additional energy offset $2hf_0$ being involved below 8 GHz has to be assumed.

Since the coupled dots are operated in the regime of linear response at a bias voltage of only $-10\,\mu$V, electronic excitations from the drain/source leads supplying this energy can be excluded. As mentioned before, the single particle excitation energies of the individual dots $\Delta\epsilon^* \cong 120\,\mu$eV ≈ 30 GHz is larger than $2hf_0$. Hence, intrinsic excitations in the dots will not interfere with molecular states and PAT below 8 GHz. The remaining source of this offset energy quantity $2hf_0$ is the crystal lattice in which the quantum dots are embedded.

As discussed earlier by Fujisawa et al. [35] acoustic phonons—neither plasma excitations nor optical phonons—effectively contribute to electron transport through such a quantum dot device at low temperatures. Compared to this work, the double dot molecule is probed under much lower drain-source bias. As a result, no phonon emission but phonon absorption is observed. The GaAs/AlGaAs lattice is cooled by the ^{3}He/^{4}He mixture to a bath temperature of 170 mK. This yields an acoustic phonon energy of $hf_{\text{ph}} \approx h \times 10.7$ GHz. Assuming a sound velocity $v_{3\text{D}}$ of 4800 m/s in the GaAs host crystal, this 3D phonon mode has a wavelength of about 480 nm, which is comparable to the effective dot diameter d and the distance L between the centers of the two dots ($d, L \sim 500$ nm). This phonon energy also matches the level broadening within the dots, which confirms that the double-dot molecule is strongly heated by the phonon bath. Since the two quantum dots form cavities for 10 GHz phonons, the electrons strongly interact with phonons. Therefore, the energy quantum $2hf_0$ most likely consist of two phonons: $2hf_0 \approx 2 \times hf_{ph}$.

Another source of phonons is the gate structure itself: Similar to interdigitated transducers (IDTs), which are commonly used to launch surface acoustic waves [36], the gates not only confine the electrons in the quantum dots but moreover form a piezoelectrical resonator. Due to the limited number of gates in comparison to conventional IDTs this "Schottky-gate" transducer possesses a large bandwith, i.e. exciting this transducer far off resonance still generates phonons.

Assuming photon and phonon excitations, transport through the double dot at 3 GHz could be coherent as well as sequential. The first process to be considered is the following: a single electron is lifted into an intermediate state by absorption of phonons in one dot and then coupled to the other dot via two-photon absorption, i.e.

$$E_{\text{lr}}^{\text{S}} = \sqrt{(2hf)^2 - 4t^2} + 2hf_0 \tag{37}$$

which is depicted by the dashed curve in Fig. 2.21(b), which is obviously not observed. The other sequential process is given by the absorption of photons in one dot and then the phonon absorption between the dots. The corresponding coupling dependence is written as

$$E_{\mathrm{lr}}^{\mathbf{S}} = \sqrt{(2hf_0)^2 - 4t^2} + 2hf \tag{38}$$

depicted by the dashed/dotted trace in Fig. 2.21(b), which is in reasonable agreement with the experimental data. On the other hand, an electron tunneling coherently through the dots can absorb both photons and phonons. The coupling dependence is then given by

$$E_{\mathrm{lr}}^{\mathbf{S}} = \sqrt{(2h(f + f_0))^2 - 4t^2} \tag{39}$$

illustrated by the solid curve in Fig. 2.21(b). Within the error bars also this process can be assumed, although the sequential absorption of phonons and photons (Eq. (38)) obviously fits better. However, in any case it has to be pointed out that the "molecular state" at 3 GHz is not dephased by sequentially or coherently absorbing phonons and photons. The two quantum dots can be coupled coherently by either microwave photons or acoustic phonons. In general, this can be understood as a superposition of photon *and* phonon assisted tunneling. Above 10 GHz conventional PAT prevails, since the absorption of more than two phonons is much less likely than that of photons. In the transition region from pure PAT to P+PAT, between 8 and 10 GHz, transitions induced by photons and phonons are strongly mixed and no simple superposition of phonons and photons is found. The importance of this results lies in the fact that the coherent electronic mode can interact by absorbing photons and phonons without loss of phase coherence.

6. Dynamic Response of Single Quantum Dots

Spectroscopy on quantum dots is commonly performed either by non-linear transport or by microwave measurements. Ordinary linear transport, i.e. under a small forward bias eV_{ds} between source and drain contacts and without microwave irradiation, only involves quantum dot ground states. In the non-linear case, by applying a finite bias across the artificial atom, also excited quantum dot states can participate in transport [15].

Alternatively, in the presence of a microwave field electrons can absorb or emit photons and thus reach excited quantum dot states otherwise not available in linear transport. In a combination of the two methods described, two coherently coupled microwave sources with a slight frequency offset are used for detecting the complex photoconductance signal (microwave induced tunneling current) at the difference frequency [24]. This enables resolution of PAT in the non-linear regime. Moreover, the detected photoconductance contains the in-phase part (conductance) and out-of-phase part (susceptance). The variation of these two different responses indicates the different dynamics of the involved transport processes.

For the observation of PAT through excited states the size of the quantum dot system is crucial: First, the dot has to be small enough to have a mean energy level spacing $\bar{\Delta}$ large compared to the intrinsic or thermal broadening of the conductance resonances, i.e. $\Gamma, k_{\mathrm{B}}T < \bar{\Delta} = 2\hbar^2/m^*r^2$, where Γ denotes the intrinsic level broadening, T the

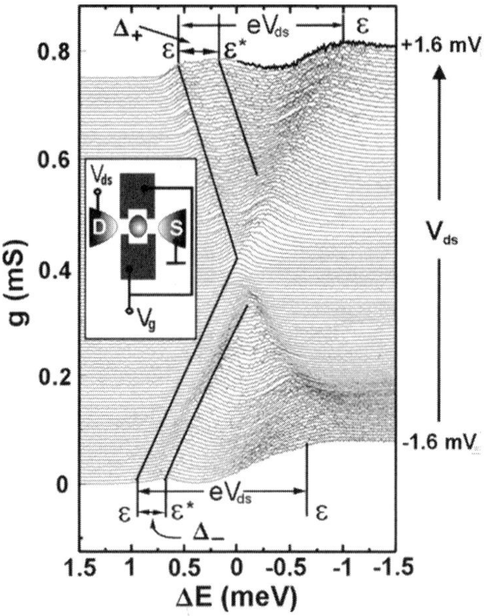

Fig. 2.26. Bias dependence of the quantum dot conductance in the vicinity of a single resonance: The drain-source bias is varied from $V_{ds} = -1.6\,\text{mV}$ (bottom) to $V_{ds} = +1.6$ mV (top) in 120 steps. The horizontal axis is the distance $\Delta E = -e\alpha\Delta V_g$ from the zero-bias ground state resonance (see text). The single resonance at $V_{ds} = 0$ is split into two resonances separated by eV_{ds}. Additional resonances stem from excited states of the quantum dot Δ_+ and Δ_- above the ground state, for positive and negative bias, respectively. The inset shows the gate structure which is used to define the quantum dot.

temperature, r the radius of the dot and $m^* \approx 0.067m_e$ the effective electron mass. Second, the excited state must be attainable via absorption of one or a few photons, i.e. $hf \approx \bar{\Delta}$, where f is the microwave frequency.

The resulting standard direct current (dc) transport is displayed in Fig. 2.26, where the differential conductance dI_{ds}/dV_{ds} in the vicinity of a resonance is shown as a function of forward bias V_{ds} and gate voltage V_g. For convenience, the gate voltage is rescaled to $\Delta E = -e\alpha\Delta V_g$, which is the energetic distance from the ground state resonance at $V_{ds} = 0$ ($\alpha = C_g/C_\Sigma$). The transformation to ΔE allows for a direct extraction of excitation energies from the conductance plot.

From the zero-bias distance between adjacent conductance peaks the total capacitance of the quantum dot is determined to be $C_\Sigma = 85\,\text{aF}$. The quantum dot radius is thus estimated to be $r = 70\,\text{nm}$, i.e. the quantum dot contains only about 20 electrons. As expected, for non-zero bias the ground state resonance (marked ϵ in Fig. 2.26 for comparison with the excited state resonance ϵ^*) splits by eV_{ds}. For $V_{ds} > 0$ an additional conductance resonance due to an excited state at ϵ^* develops, which is $\Delta_+ = (\epsilon^* - \epsilon) = 390\,\mu\text{eV}$ above the ground state. Correspondingly, for $V_{ds} < 0$ a resonance is detected at a distance $\Delta_- = 280\,\mu\text{eV}$ from the ground state. These excitation energies are in good agreement with the mean level spacing $\bar{\Delta} \approx 465\,\mu\text{eV}$

estimated from the dot radius. Hence, two different excited states take part in transport for $V_{ds} < 0$ and $V_{ds} > 0$. Furthermore, as can be seen from Fig. 2.26, the ground state resonance for $\Delta E > 0$ is almost suppressed for $V_{ds} < 0$, whereas the excited state resonance for $\Delta E > 0$ is much stronger. The origin of these "$\Delta E > 0$"-resonances are the alignment of the dot's ground state or the excited state with the chemical potential of the drain reservoir. The strength of these resonances are related to the overlap between the wavefunction of the corresponding quantum dot state and the wavefunctions in the reservoirs. Hence, the variation in conductance indicates that the coupling of the ground state to the drain reservoir is much smaller than that of the excited state. The coupling of the excited state to the reservoirs is about 5.3 times the coupling of the ground state.

Two different techniques are applied to study the transport properties under microwave irradiation. For low forward bias $V_{ds} \approx 0$, the direct current through the dot is measured using a single microwave source. Alternatively, we employ two phase-locked microwave sources which are slightly offset in frequency. This second technique allows to detect photon-induced transport also in the nonlinear regime $|V_{ds}| > 0$. Furthermore, the relative phase of the photon-induced current with respect to the incoming microwave beat can be determined. Results obtained with the first technique are shown in Fig. 2.27, where the current through the quantum dot for small bias ranging from $-19.3\,\mu$V to $+19.3\,\mu$V, is displayed under microwave irradiation at frequency $f = 36.16$ GHz.

For small positive bias the original main peak from the ground state resonance (**M**) as well as a sideband (**G**) in a distance $hf \approx 0.15$ meV are detected. This sideband in the current signal is due to PAT through the ground state. Quite differently, for negative bias additional features in the current signal are induced by the microwaves.

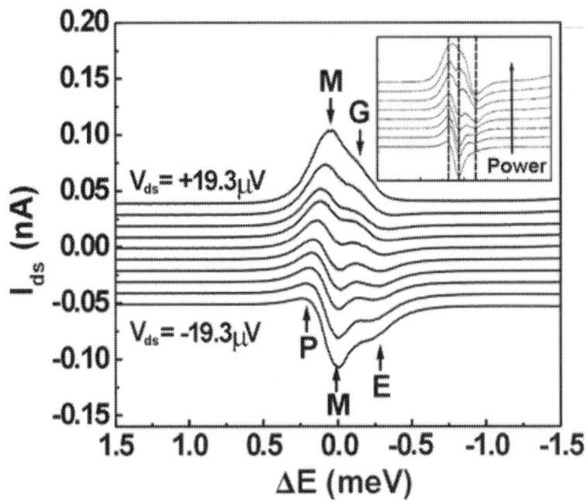

Fig. 2.27. Direct current through the quantum dot for small bias values under irradiation with microwaves of frequency $f = 36.16$ GHz. Next to the main ground state resonance (M) additional features (G, P, E) appear, which can be ascribed to photon-assisted tunneling through the ground state, photon-induced pumping and tunneling through an excited state. In the inset the power dependence of these features is shown for $V_{ds} = -5.1\,\mu$V. The output power of the microwave source is increased from bottom to top in steps of 0.5 dBm.

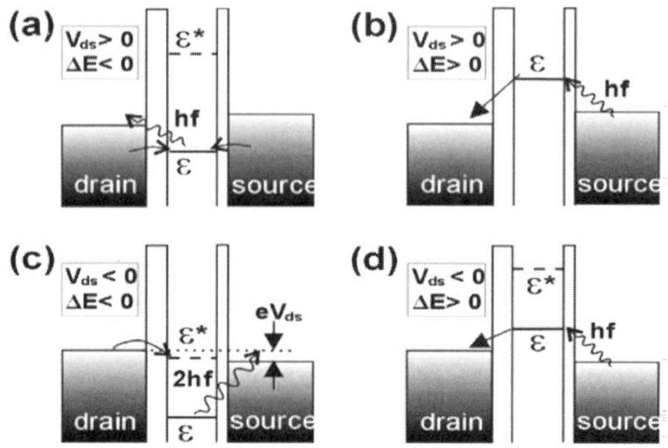

Fig. 2.28. Schematic representation of the photon-induced processes: In the case of positive bias (a, b), only photon-assisted tunneling through the ground state is possible. For $V_{\text{ds}} < 0$, the excited state ϵ^* is approximately two photon energies above the ground state and can be accessed by absorption of two photons (c). Furthermore, due to the asymmetry in microwave absorption across the tunneling barriers, a pumping current can occur (d) for $V_{\text{ds}} < 0$ (see text).

These features can be attributed to photon-induced pumping (**P**) and resonant tunneling through an excited quantum dot state (**E**).

The processes involved are schematically depicted in Fig. 2.28: At low positive bias only the ground state transition Fig. 2.28(a) occurs. As found in the preceding paragraphs, the first excited state for this bias direction is too far above the ground state to be accessible by a one- or two-photon process. The other possible photon-induced ground state transition ($\Delta E > 0$) shown in Fig. 2.28(b) is not detected in the low-bias current signal. However, it is resolved for larger bias values applying the two-source detection scheme (see below). For negative bias $V_{\text{ds}} < 0$ the excited state at $\epsilon^* = \epsilon + \Delta_-$ can participate in transport when the ground state is depopulated by a two-photon absorption process (Fig. 2.28(c)) ($2hf > \Delta_-$), a process analogous to photo ionization.

Normally, this process has a much smaller probability than the one- and two-photon PAT processes. In the present case, however, since the coupling of the excited state to the reservoirs is more than four times stronger than the coupling of the ground state, this process might turn out to be comparable to the pure two-photon PAT in amplitude. Furthermore, a pumping current flows opposite to the bias direction for $\Delta E > 0$, where the ground state is hf above the chemical potentials in the source reservoir (Fig. 2.28(d)). This only happens when the microwave absorption across the right tunnel barrier is larger than that of the left tunneling barrier. In this case, the ground state ϵ is permanently populated with electrons from the source contact which then partly decays into the drain region. From the power dependence (see below), it is confirmed that this pumping current results from PAT.

To ensure that the observed features are not adiabatic effects of the microwave irradiation (e.g. rectification effects) commonly both their frequency and power dependence are determined. The inset of Fig. 2.27 shows the power dependence of the photon-induced features for $V_{ds} = -5.1\,\mu V$. The output power of the microwave synthesizers is changed in steps of 0.5 dBm from trace to trace. Over this wide power range the microwave-induced features do not change in position showing that they are indeed induced by single photons. It is observed that the peak heights increase with microwave power, i.e. a Bessel function behavior: The tunneling current induced by absorbing/emitting n photons is proportional to $J_n^2(x)$, where $x = eV_{ac}/hf$ and V_{ac} is the microwave amplitude across the tunnel barriers. For even higher microwave powers the PAT-like features considerably broaden due to heating effects until they finally vanish.

Studying the power dependence only is not sufficient to reveal the origin of peak (**E**). With the lock-in amplifier the in-phase and out-of-phase photoconductance signals $\gamma_0, \gamma_{\pi/2}$ with respect to the reference are measured. From these the total photoconductance amplitude is obtained: $|A| = \sqrt{\gamma_0^2 + \gamma_{\pi/2}^2}$ and the relative phase Φ which equals $\arctan(\gamma_{\pi/2}/\gamma_0)$ for $\gamma_0 \geq 0$ and $\pi + \arctan(\gamma_{\pi/2}/\gamma_0)$ for $\gamma_0 < 0$, respectively. In Fig. 2.29 the photoconductance amplitude at $f_2 \sim f_1 = 36.16$ GHz and $\delta f = 2.1$ kHz

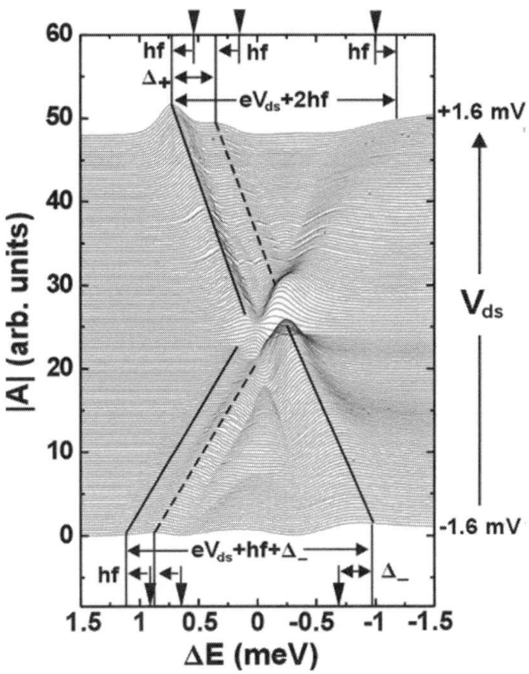

Fig. 2.29. Amplitude $|A|$ of the photoconductance measurement obtained with the two-source setup of Fig. 2.28. The drain–source bias is varied. The position of the resonances found in the dc measurement of Fig. 2.26 are indicated with triangles. With respect to the dc resonances, most photoconductance resonances are shifted by the photon energy hf, as is expected for photon-assisted processes. The resonance for $V_{ds} < 0$ and $\Delta E < 0$, however, is shifted by Δ_-, indicating two-photon PAT through a ground state and the resulting tunneling through an excited state as sketched in Fig. 2.28(c).

is displayed for the same parameter region as the dc measurement shown in Fig. 2.26. With respect to Fig. 2.26, for $V_{ds} > 0$ the conductance window is enlarged by $2hf$. The resonances are each shifted by the photon energy hf which can readily be explained by photon assisted tunneling processes as in Fig. 2.28(a) and (b). This is also the case for the "$\Delta E > 0$"-conductance resonances for negative bias. However, the resonance for $\Delta E < 0$ and small negative bias is clearly shifted by Δ_-, thus enlarging the conductance window to $eV_{ds} + hf + \Delta_-$. The process involved is depicted in Fig. 2.28(c): An electron leaves the quantum dot's ground state for the source reservoir via absorption of two photons. Now, electrons can either refill the ground state or tunnel through the excited state as long as the ground state is depopulated. Transport through the excited state stops when an electron decays to the ground state, or an electron enters the quantum dot's ground state from the leads. With $\Delta E < 0$ and larger negative bias, the photoconductance peak is apparently broadened. The broadening is partly due to other tunneling processes possible at large bias, e.g. one-photon PAT through the ground state. In fact, even at small negative bias there is small tunneling current in between the peak (**M**) and (**E**), which is most probably the one-photon PAT. In the present case, the tunneling process for $\Delta E < 0$ and negative bias is more intricate than the ideal PAT.

In Fig. 2.30, phase traces as well as their respective amplitude signals are displayed for small positive and negative bias ($V_{ds} = +10\,\mu V$ and $V_{ds} = -30\,\mu V$, respectively), corresponding to the central region of Fig. 2.29. For $V_{ds} > 0$ the phase signal remains approximately constant at $\Phi = 0$, which means that the out-of-phase photoconductance $\gamma_{\pi/2}$ is equal to zero. The response of the quantum dot to the microwaves is similar for both of the tunneling processes (**G**). In fact, the two peaks (**G**) in the amplitude signal stem from ground state resonances as depicted in Fig. 2.28(a) and (b). The situation is considerably different for $V_{ds} < 0$ where a strong pumping signal (P) is observed which is caused by a process shown in Fig. 2.28(d). At the position where the photocurrent changes its direction, the amplitude drops to zero and the phase changes trivially by π (this corresponds to crossing zero in the $\gamma_0 - \gamma_{\pi/2}$-plane). The second peak (**E**) stems from the photon-induced tunneling through the excited state as in Fig. 2.28(c). Moving away from this second resonance to more negative ΔE, the phase continuously returns to its original value.

This continuous phase change shows that this transport process results in a finite out-of-phase signal $\gamma_{\pi/2}$. In contrast to the other transport scenarios described above (only the ground state is involved), photon-induced tunneling through the excited state is not a purely conductive transport process but also has capacitive and inductive contributions. This behavior is due to the complicated charging dynamics of the quantum dot for this particular process. The processes involved are PAT events from the ground state to the source reservoir, resonant tunneling through the excited state, recharging of the ground state by the drain reservoir and relaxation from the excited state to the ground state. All these processes have different time constants which additionally depend on the gate voltage (i.e. ΔE). The interplay of these processes results in the observed phase lag. Thus one has a method at hand to determine the admittance of a mesoscopic system [37–39] in the PAT regime which is related to the average relaxation time of the system. In the current setup, for $V_{ds} < 0$ the ground state broadening, due to the coupling to the drain reservoir, is about 400 MHz, while the level broadening from the

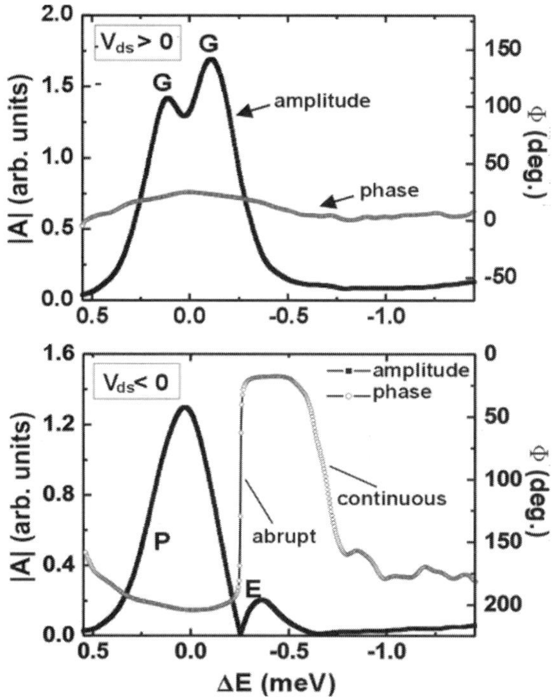

Fig. 2.30. Phase and amplitude of the photoconductance signal for $V_{ds} = +10\,\mu V$ (top) and $V_{ds} = -30\,\mu V$ (bottom). For positive bias, photon-assisted tunneling through the ground state (G) is observed, as schematically depicted in Fig. 2.28(a) and (b). The phase signal remains constant on either side of the resonances. For negative bias, pumping (P) and tunneling through an excited state (E) are induced (cf. Fig. 2.28 (d) and (c)). Vanishing of the amplitude signal between (P) and (E) is accompanied by a trivial jump of π in the phase signal. For more negative values of ΔE the phase signal continuously falls back to its original value.

coupling to the source reservoir is around 2 GHz. The broadening of the excited state coupling to the reservoirs is found to be of the same width of 2 GHz. Hence, the bare tunneling time through the ground state, excluding other time constants, would be less than 2.5 ns. However, the inverse modulation frequency $1/\delta f \approx 500\,\mu s$, which is the time separation between two microwave beat minima, is much larger than the tunneling time. In the few-electron limit, this indicates that it takes the electron a much longer time to relax within the dot than to tunnel through the barriers.

An extension of the measurements to modulation frequencies on the order of 10–100 MHz corresponding to a time scale of 10–100 ns would therefore be desirable. With a shorter microwave beat period it will be possible to probe both the fast tunneling event and the slow relaxation process. It can be concluded that with the frequency f the photon energy hf for the photon-induced process can be adjusted, whereas the modulation frequency δf determines the time scale on which the electronic dynamics of the quantum dot is probed. This phase is related to the susceptance of the quantum dot at very high frequencies. Nontrivial values for this quantity can be attributed to the long charge relaxation times in the quantum dot, this can be exploited for an accurate determination of the relaxation times of excited quantum dot states.

7. The On-Chip Spectrometer

Similar results on coupled dots can be obtained by applying an on-chip microwave source. The source employed is a Josephson oscillator formed by a long Nb/Al–AlO$_x$/Nb junction. In contrast to the excitation spectrum of real atoms or molecules, the spectrum of single or even coupled quantum dots reveals a striking difference in the discrete level structure. For quantum dots it has been shown in a whole variety of experiments that Kohn's theorem [40] prevails. The theorem states that only the center-of-mass (CM) degree of freedom couples to a spatially homogeneous electromagnetic field. In the experiments above only CM excitations were found. Here, the radiation source is placed close to the quantum dots, in an attempt to verify whether the inhomogeneity of the near-field radiation affects the electronic excitations.

The dots used in this experiment are rather classical systems with about ~900 electrons in each dot. In the current measurements the interdot coupling is chosen to be weak ($C_{\mathrm{lr}} = 2\,\mathrm{aF}$), i.e. a hexagonal array of resonance points for the charging diagram is found (see Fig. 2.23), corresponding to the "ionic" coupling limit. All results were obtained in the linear regime (drain/source bias $V_{\mathrm{ds}} = 19\,\mu\mathrm{V}$).

In order to characterize the microwave response of the coupled quantum dots first a conventional microwave source is used for spectroscopy. The charging diagram of the double dot with microwave excitation is shown as a grayscale plot in Fig. 2.24. The sidebands are found on only one side, depending on the tuning of the barrier transmission coefficients. However, since we are interested in the alteration of the microwave coupling by the Josephson tunnel junction (JTJ) we chose deliberately to maintain the tuning with only one sideband clearly appearing in the inset of Fig. 2.24.

The inset shows one of the typical traces from the grayscale plot with the sidebands induced by the frequency dependent absorption. These sidebands are marked by arrows—the peak height modulation is due to the specific trace taken out of the charging diagram (marked by the dashed line). A trace along one of the two periodic resonance lines would yield peaks of similar amplitude. The net pumping of electrons leads to a reduction of the absolute current value down to only some 100 fA. Also the noise floor is slightly enhanced by coupling the radiation.

When the JTJ is operated as a source with a typical emission frequency of $f \cong 10\,\mathrm{GHz}$, we observe charging diagrams as the one shown in Fig. 2.25. The JTJ emission frequency was determined by taking the IV-characteristics. Biasing the JTJ with a current of $I = 1\,\mu\mathrm{A}$ sidebands are recorded which resemble the ones induced by the far-field source. The power emitted is then on the order of $P_{\mathrm{dc}} \sim 20\,\mu\mathrm{V} \times 1\,\mu\mathrm{A} \approx 20\,\mathrm{pW}$, where $20\,\mu\mathrm{V}$ is the voltage drop over the JTJ at 10 GHz. Moreover, the peak broadening is almost identical to the one determined before. As seen, the resonances of the current (marked by lines in the plot) and the induced sidebands (marked by arrows) possess a long term stability. Since the observed resonances for the on-chip source as well as the far-field source are almost identical, it can be concluded that the photon absorption process only depends on the shape of the local electrostatic potential. Varying the frequency of the radiation for such a tunnel coupled dot system results in the well-known linear relation between the position of the sideband relative to the ground state and the frequency for the case of weak coupling.

8. Non-Linear Transmission-Lines for Probing Single Dots

Finally, photoconductance measurements are discussed on a single laterally gated quantum dot in the few electron limit using a wideband, i.e. pulsed, millimeter-wave spectrometer. The underpinning idea in these experiments is to address the temporal evolution of electron wave packets in single and coupled dots. Especially, coupled dot systems should enable the clear detection of Rabi-oscillations [41], as discussed in first measurements earlier [14].

By downconverting a pulsed signal coherently via a beatnote between two microwave sources, the amplitude and phase of the photoconductance can be determined. The spectrometer allows the observation of excited states of the quantum dot and traces these states at non-zero bias. It further enables probing the variation of the quantum dot's capacitance. It is particularly important to note that since the induced photocurrent spectrum is filtered with a lock-in amplifier, broadband spectroscopy on the dot system under a constant thermal load can be performed, in contrast to conventional tuned sources, such as Gunn diodes, whose power variation with frequency complicates broadband measurements.

Although the output power of the spectrometer is low to minimize heating, we observe a well-pronounced photoconductance signal in the dot. Using a digital lock-in amplifier, we are able to probe the amplitude and phase of the photoconductance in a wide-band from 80 to 400 GHz ($10 < n < 50$), corresponding to energies of 0.3–1.6 meV, which is comparable to the charging energy and the mean level spacing of the dot (see below). It is particularly important to note that since we filter the induced photocurrent spectrum with a lock-in amplifier, we can perform broadband spectroscopy on the dot system under a constant thermal load, in contrast to conventional tuned sources, such as Gunn diodes, whose power variation with frequency complicates broadband measurements. Without NLTLs, two phase-locked microwave sources generate a continuous microwave flux, which has been applied to a similar single quantum dot in [24].

First we will focus on the dynamic response by pulsed excitation: We measure both the direct tunnel current (I_{ds}) and the differential conductance ($g = dI_{ds}/dV_{ds}$) of the quantum dot; the dashed lines in Fig. 2.31(a) obtained from the conductance measurement reveal Coulomb blockade (CB) diamonds. Within the diamonds, electron transport is blocked by the charging energy $E_C = e^2/2C_\Sigma$, according to the "orthodox model" of CB. The tunnel barrier connecting the dot with the source contact (thus source-related) is tuned to be less opaque than the drain-barrier, i.e. $\Gamma_d < \Gamma_s$, as schematically shown in the level diagrams in Fig. 2.32. Consequently, the upper-right and the lower-left Coulomb diamond boundaries (along which the dot energy is aligned with the chemical potential in the drain, thus drain-related) are sharper and higher than the upper-left and the lower-right source-related boundaries.

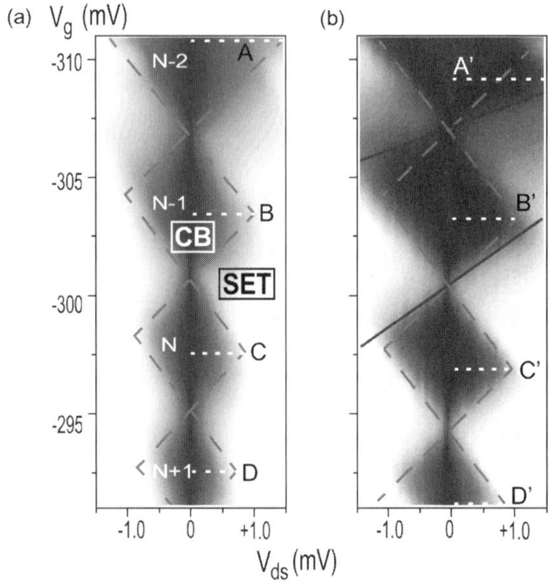

Fig. 2.31. Left: Logarithmic plot of the dc tunnel current (I_{ds}) in the plane of drain-source bias and gate voltage of the quantum dot, bright: $I_{\mathrm{ds}} > 3\,\mathrm{nA}$, dark $< 1\,\mathrm{pA}$. The dashed lines for the enclosed diamonds are obtained from the plot of "dc" conductance, where the diamond boarders are clearer to be determined. Right: Logarithmic plot of the dc tunnel current measured under microwave radiation from the NLTLs (bright $> 0.7\,\mathrm{nA}$, dark $< 0.1\,\mathrm{pA}$). The solid lines indicate the different diamond borders which have smaller slopes. In both plots, the horizontal white dashed lines labeled with A(a), B(b), C(c) and D(d) mark the positions at which $I_{\mathrm{ds}} - V_{\mathrm{ds}}$ curves shown in Fig. 2.31 are extracted.

The varying size of the diamonds indicates that the dot's total capacitance ($C_\Sigma = C_{\mathrm{d}} + C_{\mathrm{s}} + C_{\mathrm{g}}$) changes upon adding or removing a single electron in the few-electron limit. For the two CB diamonds in the center of Fig. 2.31(a), the capacitance of the dot to the drain ($C_{\mathrm{d}} \sim 98\,\mathrm{aF}$) and source ($C_{\mathrm{s}} \sim 67\,\mathrm{aF}$) lead and to the gate ($C_{\mathrm{g}} \sim 25\,\mathrm{aF}$) are denoted respectively. We estimate an average charging energy of $E_{\mathrm{C}} \sim 0.4\,\mathrm{meV}$. The drain-related and source-related slopes of the diamond are $s_1 = \delta V_{\mathrm{g}}/\delta V_{\mathrm{ds}} = 1 + C_{\mathrm{s}}/C_{\mathrm{g}} = 3.66$ and $s_2 = \delta V_{\mathrm{g}}/\delta V_{\mathrm{ds}} = -C_{\mathrm{d}}/C_{\mathrm{g}} = -3.93$, respectively. The slopes yield a scaling factor $\alpha = C_{\mathrm{g}}/C_\Sigma = 0.13$ that relates the gate voltage (V_{g}) to an energy scale ($\Delta E = -e\alpha\Delta V_{\mathrm{g}}$). Although no dc transport through excited states is resolved in the SET regime (Fig. 2.31(a)), we determine the mean level spacing of the dot $\Delta\epsilon^* \sim 200\,\mu\mathrm{eV}$ using transport measurements under pulsed microwave radiation (see below).

We find empirically that millimeter-wave radiation changes the slopes of the diamond boundaries in Fig. 2.31(b) compared to those in Fig. 2.31(a). We observe the drain-related slope s_1 to be decreased to 3.30, and the source-related slope s_2 to about -0.45 (corresponding to the upper-most diamond in Fig. 2.31(b)). Applying the "orthodox model" neglecting the pulsed microwave radiation, the tunnel barrier

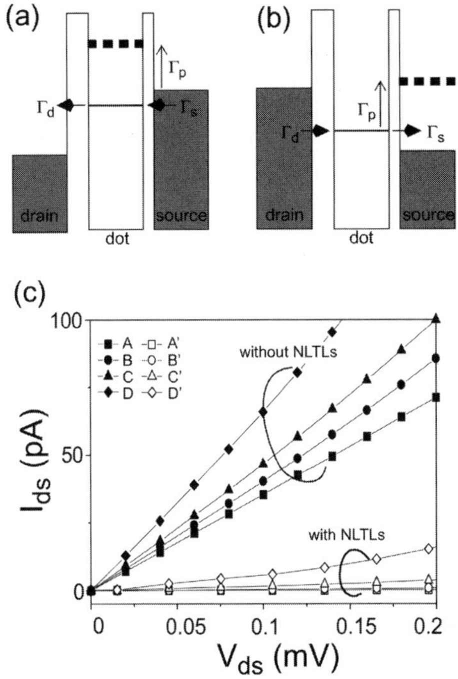

Fig. 2.32. The $I_{ds} - V_{ds}$ curves in the Coulomb blockade regime at positions labeled with A(a), B(b), C(c) and D(d) as shown in Fig. 2.3. The range of $|V_{ds}|$ is from $0\,\mu$V to $200\,\mu$V, in which $e|V_{ds}| < \Delta E^*$, i.e., elastic cotunneling dominates. Beyond this range, more than one quantum dot state is involved and inelastic cotunneling dominates. Inset: The level diagram for $V_{ds} \approx 0\,\mu$V where elastic cotunneling occurs. The electrons in the leads are excited up to virtual states by broadband microwaves. Elastic cotunneling of electrons is strongly suppressed by wideband microwave radiation. (a) At positive drain-source bias, the effective electron number in the dot is increased by pumping electrons into the dot. (b) At negative drain-source bias, the effective electron number in the dot is reduced by pumping electrons out of the dot. (c) The I_{ds}–V_{ds} curves in the Coulomb blockade regime at positions labeled with A(A′), B(B′), C(C′) and D(D′) as shown in Fig. 2.2. The range of $|V_{ds}|$ stretches from 0 to $200\,\mu$V, in which $e|V_{ds}| < \Delta\epsilon^*$, i.e. elastic cotunneling dominates. Beyond this range, more than one quantum dot state is involved and inelastic cotunneling dominates.

capacitances C_d and C_s are reduced by 89% and 18%, respectively; the reduction of C_g is only 3%. Although the slopes under higher gate voltages are closer to the ones in the dc limit, the capacitance of the left contact C_d maintains a clearly different response under pulsed radiation.

We infer from the measurements that this reduction in slopes arises from the effect of fast pumping by pulsed microwave radiation. For the source-related resonance under a positive bias, the level diagram can be illustrated as shown in Fig. 2.32(a). Without microwave radiation, electrons tunnel into the ground state from the source lead at a rate of Γ_s and escape out of the dot at a rate of Γ_d. The total tunneling rate through the dot is found to be $\gamma = \Gamma_d\Gamma_s/(\Gamma_d + \Gamma_s) < 4\,$GHz. When pulsed microwave radiation is present, electrons in the source lead are excited and pumped into excited or virtual states in the dot at a rate of $\Gamma_p = f_p$. Since $\Gamma_d < 8\,\text{GHz} < \Gamma_s$ and then $\Gamma_d < \Gamma_p$, the average electron number $\hat{N}$ in the dot is increased compared to the static value $\hat{N} \sim \Gamma_d\Gamma_s/(\Gamma_d + \Gamma_s)$, i.e. the diamonds' slopes are reduced.

Hence, the slope change observed reflects the dynamic pumping of electrons by pulsed microwave radiation and the related relaxation processes. Furthermore, since the source-related tunnel barrier is more transparent than the drain-related one and the dot can be filled or emptied by electrons from the source, the drain-related resonance experiences a smaller change by pulsed excitation. At higher gate voltage ($V_g > -298\,\text{mV}$ in Fig. 2.31(b)) where both the drain- and source-related barriers are more transparent, the boundary slopes have a smaller reduction under pulsed microwave excitation. By varying the pulse repetition rate f_p and measuring the slope change, it is possible to obtain the tunneling rates through the tunnel barriers. In comparison, earlier work on a similar single quantum dot [24] showed that the slopes of the diamond boundaries are not shifted when the applied microwave radiation is continuous rather than pulsed. Here we emphasize that it is the change of tunneling by pulsed microwave excitation that causes the slope change and hence a phenomenological reduction of the dot capacitance in the frame of the "orthodox model" [42].

The second effect which has to be considered is suppression of cotunneling: In contrast to the "orthodox model" of CB, the finite tunneling current within the CB diamonds (Fig. 2.31(a)) is evidence of cotunneling [43, 44], indicating that the quantum dot is strongly coupled to the leads. Figure 2.31(b) shows the total induced direct tunnel current through the quantum dot under pulsed radiation of all harmonics. Compared to the current without radiation (Fig. 2.31(a)), we see that the finite cotunneling current is strongly suppressed within the CB diamonds. To quantitatively determine the suppression, $I_{ds} - V_{ds}$-curves within the CB diamond are extracted from Fig. 2.31 in Fig. 2.32(c). Every curve is selected at a specific gate voltage (V_g) which is in the center of the CB diamond. In Fig. 2.32(c), the symbols labeled with A (A′), B (B′), C (C′) and D (D′) correspond to the same labels of the dashed horizontal lines in Fig. 2.31. Both with and without radiation the finite tunneling current is proportional to the small bias V_{ds} within the CB regime, indicating elastic cotunneling in which only one quantum dot state participates. Elastic cotunneling dominates when $eV_{ds} < \Delta\epsilon^* \sim 0.2\,\text{meV}$ and $T_e < 2\sqrt{E_C\Delta\epsilon^*}/k_B \sim 4.6\,\text{K}$ [43, 44], which is the case in this experiment.

Without radiation pulses the net elastic cotunneling current I_{ds}^{co} is approximately $I_{ds}^{co} \cong e\Gamma_{co}$ [45]. While elastic cotunneling is enabled by tunneling through virtual states of the dot, it is suppressed under pulsed microwave radiation by a factor of about $I_{ds}^{co}/I_{ds}^{pulsed} \sim 25$, as seen in Fig. 2.32(c). The reason for this lies in the fact that the pulse repetition rate is of the order of $\Gamma_p^{-1} \simeq f_p^{-1} \simeq 128\,\text{psec}$, which is larger than the direct barrier transmission of the dot $\gamma^{-1} = (\Gamma_d\Gamma_s/(\Gamma_d + \Gamma_s))^{-1} > 250\,\text{psec}$. In a simplified picture this can be interpreted in the following way: an electron is trying to pass through the dot at the rate γ, while it is effectively kicked out of the dot at the rate Γ_p. This reduction of dwell time appears as a suppression of cotunneling. Under the condition $\Gamma_p > \gamma$ cotunneling dominates again and the classical Coulomb diamond of Fig. 2.31(a) is restored.

As the third clear result we discuss detection of excited states by the NLTL-spectrometer: Even in the absence of microwave radiation, hardly any fine structure indicating electron transmission through excited states is found in the SET regime (Fig. 2.31(a)). This stems from inelastic tunneling when the dot is strongly coupled to the leads. On the other hand, since the spectrometer emits about 40 harmonics of the fundamental microwave frequency simultaneously, resulting in some thermal broadening and extensive overlap of PAT sidebands, we can not resolve individual sidebands in the measurement shown in Fig. 2.31(b) as is done in single-frequency PAT on quantum dots [13, 16]. We can, however, select or filter the dot's response to a given millimeter-wave harmonic using our phase sensitive detection method, given linear response, which we verify by checking the dot's proportional response to slight changes in spectrometer power [12].

The dual-source spectrometer emits picosecond pulses that have a beat frequency $f_{\mathrm{bn}} = n \times \delta f$ (Fig. 2.3(d)), since it is driven by two phase-locked synthesizers with $f_{\mathrm{A,B}} \approx f_{\mathrm{p}} = 7.8\,\mathrm{GHz}$ offset by $\delta f \equiv 21\,\mathrm{Hz}$. This allows filtering of the corresponding harmonics (f_n) with a digital lock-in amplifier whose reference is δf. Each harmonic introduces a small voltage modulation $\delta V_{\mathrm{ds}}(f_{\mathrm{bn}}) \exp(-i 2\pi f_{\mathrm{bn}} t)$ and $\delta V_{\mathrm{g}}(f_{\mathrm{bn}}) \exp(-i 2\pi f_{\mathrm{bn}} t)$ in the drain-source bias V_{ds} and in the gate voltage V_{g}, respectively. Taking into account all of the harmonics, the corresponding voltage modulation is $\tilde{V}_{\mathrm{ds}} \approx \sum_n \delta V_{\mathrm{ds}}(f_{\mathrm{bn}}) \exp(-i 2\pi f_{\mathrm{bn}} t)$ and $\tilde{V}_{\mathrm{g}} \approx \sum_n \delta V_{\mathrm{g}}(f_{\mathrm{bn}}) \exp(-i 2\pi f_{\mathrm{bn}} t)$. At small V_{ds}, the photoconductance signal yields

$$\tilde{g} = A e^{i\Phi} = X + iY \sim \left(g + V_{\mathrm{ds}} \frac{\partial g}{\partial V_{\mathrm{ds}}} \right) \tilde{V}_{\mathrm{ds}} + V_{\mathrm{ds}} \frac{\partial g}{\partial V_{\mathrm{g}}} \tilde{V}_{\mathrm{g}}. \tag{40}$$

With a two-channel lock-in amplifier we are able to monitor the complex in-phase and out-of-phase (X and Y) components of the photoconductance resulting from the corresponding millimeter-wave harmonics.

In contrast to the total induced current under radiation of Fig. 2.31(b), two photoconductance spectra taken at the 12th harmonic of the beat frequency (corresponding to $f_{12} = 93.6\,\mathrm{GHz}$ (a)) and the 18th harmonic (corresponding to $f_{18} = 140.4\,\mathrm{GHz}$ (b)) are given in Fig. 2.33. The diamond shaped CB regions resemble those observed in Fig. 2.31(b). Furthermore, additional features are observed in the SET regime. We extract single traces from Fig. 2.33(a) and (b) as shown in Fig. 2.33(c) and (d), respectively. The solid lines indicate the ground-state resonances corresponding to the CB diamond borders in Fig. 2.33(a) and (b). The dashed lines reveal fine structures in the SET region. We find that those fine structures observed with positive drain–source bias depend weakly on the harmonic at which the photoconductance is detected. Hence, we interpret the fine structure as transport through excited states but not PAT sidebands (we find $\Delta\epsilon^* \sim 200\,\mu\mathrm{eV}$). Although, PAT sidebands are weakly evident at positive bias, we do observe stronger frequency dependent features, such as two excited-state resonances at the 18th harmonic f_{18} (marked by triangles in Fig. 2.33(b)), while these resonances are not found at the 12th harmonic f_{12} in the SET regime (Fig. 2.33(a)). The ability to resolve excited states under pulsed microwave excitation benefits from the fact that only the response of a single channel is detected by the lock-in amplifier.

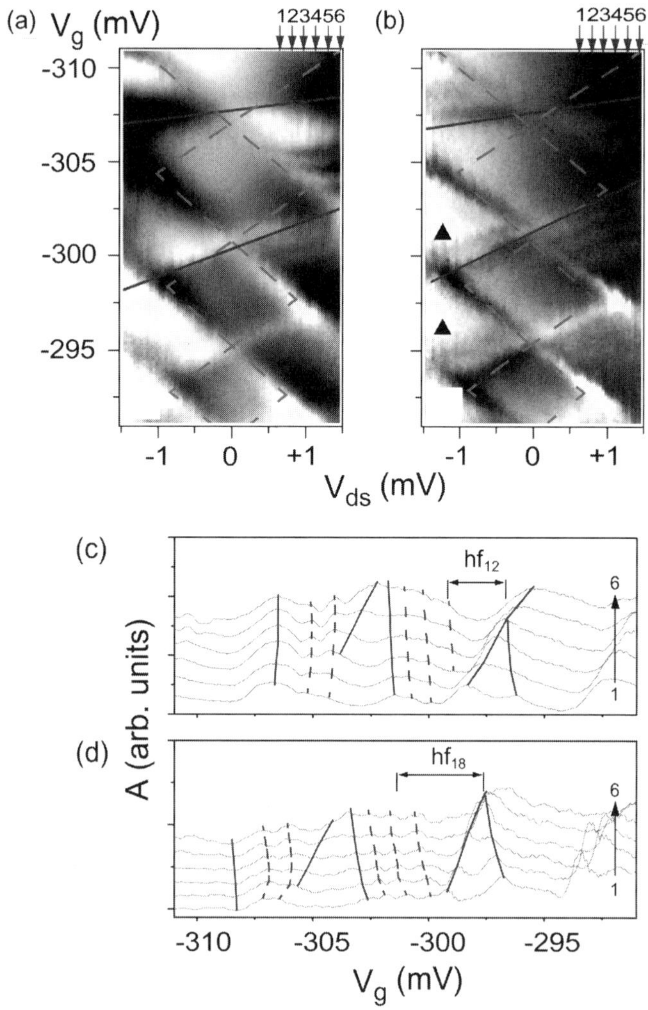

Fig. 2.33. Linear plot of the high frequency conductance at the 12th (a) and 18th (b) harmonic ($f_{12} = 93.6\,\text{GHz}$, $f_{18} = 140.4\,\text{GHz}$). The plotted data is in arbitrary units: bright regions represent negative values while black is positive. In (a) and (b), each dashed diamond boundary is parallel to its counterpart in Fig. 2.2(a). The solid lines show the actual diamond boundaries. Two solid triangles mark resonances from excited states of the dot under negative bias. In (c) and (d), the solid lines trace the ground-state resonances, while the dashed lines represent the excited-state resonances as shown in (a) and (b). The traces in (c) and (d) are extracted from (a) and (b) at six specific gate voltages, respectively.

The phase signal obtained reflects the sign change of the photon-induced tunnel current (otherwise not shown). In general, when the time scale of the slowest tunneling process is comparable to the offset frequency δf, it is possible to deduce this time scale from the phase signal. As studied in [24], when a continuous microwave flux is applied the phase signal reflects the slow charge relaxation processes between the excited and ground states in the dot.

9. Summary

In summary we have shown how to probe quantum dots by microwave radiation with a varitey of different techniques. This comprises measurements on single as well as coupled dots. Latter can be regarded as a single qubit, defined in two dots via tunnel splitting of the discrete states. This tunnel split state was traced in transport measurements and by microwave spectroscopy. Hence, this work describes key aspects of handling quantum dot qubits in solid state systems. A pulsed microwave spectrometer for probing the dynamics of electrons confined in quantum dots was introduced. This setup allowed us to determine the high frequency conductance in a fashion similar to the quasi-static limit. Although, the spin degree of freedom in quantum dots was not discussed here, it is clear that control and manipulation by microwaves will be part of future research. First steps in this direction were taken in early measurements on electron spin resonance [46] and more recently in coupling electron and nuclear spins [47]. Finally, the interaction of electrons with phonons instead of photons will be another important aspect of quantum dot research [48]. This will enable the tailoring of dissipation in these nano-structures.

Acknowledgments

During the last years working on quantum dots I had the pleasure to collaborate closely with Friedrich Simmel, Bernd Irmer, Andreas Hüttel, Karl Eberl, Max Bichler, Dieter Schuh, Alexey Ustinov, Werner Wegscheider, Klaus von Klitzing and Jorg Kotthaus. In particular I want to thank Dan van der Weide without whom realization of this work would not have been possible. I want to acknowledge financial support by the Deutsche Forschungsgemeinschaft through the Schwerpunkt "Quanteninformationsverarbeitung" (Bl/487-2-2), the Sonderforschungsbereich SFB-348, by the Bundesministerium für Forschung und Technologie (BMBF—project 01BM/914 and the German-Israeli-Project DIP), and the Defense Advanced Research Projects Agency (DARPA—contract number F61775-01-WE037).

References

1. L.P. Kouwenhoven, C.M. Marcus, P.L. McEuen, S. Tarucha, R.M. Westervelt, and N.S. Wingreen, in Mesoscopic Electron Transport, Series E, Vol. 345, ed. L.L. Sohn, L.P. Kouwenhoven and G. Schön. 105, Kluwer Dordrecht, Netherlands (1997).
2. R. Ashoori, *Nature* **379**, 413 (1996).
3. R. Landauer, *Phys. Lett.* A **217**, 188 (1996).
4. M.A. Nielsen and I.L. Chuang, *Quantum Computation and Quantum Information.* Cambridge University Press (2000).

5. D. Loss and D.P. DiVincenzo, *Phys. Rev.* A **57**, 120 (1998); P. Recher, E.V. Sukhorukov, and D. Loss, *Phys. Rev. Lett.* **85**, 1962 (2000); Xuedong Hu and S. Das Sarma, cond-mat/0101102 (2001).
6. R.H. Blick and H. Lorenz, *Proc. IEEE Int. Symp. Circ. Sys.* **II-245**, 28 (2000).
7. R.H. Blick, D. Pfannkuche, R.J. Haug, K. von Klitzing, and K. Eberl, *Phys. Rev. Lett.* **80**, 4032 (1998).
8. A.W. Holleitner, H. Qin, C.R. Decker, K. Eberl, and R.H. Blick, *Phys. Rev. Lett.* **87**, 256802 (2001).
9. A.W. Holleitner, R.H. Blick, A.K. Hüttel, K. Eberl, and J.P. Kotthaus, *Science* **297**, 70 (2002).
10. Y. Nakamura, Yu A. Pashkin, and J.S. Tsai, *Nature* **398**, 786 (1999).
11. L.P. Kouwenhoven, S. Jauhar, J. Orenstein, P.L. McEuen, Y. Nagamune, J. Motohisa, and H. Sakaki, *Phys. Rev. Lett.* **73**, 3443 (1994).
12. R.H. Blick, R.J. Haug, D.W. van der Weide, K. von Klitzing, and K. Eberl, *Appl. Phys. Lett.* **67**, 3924 (1995).
13. T.H. Oosterkamp, L.P. Kouwenhoven, A.E.A. Koolen, N.C. van der Vaart, and C.J.P.M. Harmans, *Nature* **78**, 1536 (1997).
14. R.H. Blick, D.W. van der Weide, R.J. Haug, and K. Eberl, *Phys. Rev. Lett.* **81**, 689 (1998).
15. T.H. Oosterkamp, T. Fujisawa, W.G. van der Wiel, K. Ishibashi, R.V. Hijman, S. Tarucha, and L.P. Kouwenhoven, *Nature* **395**, 873 (1998).
16. H. Qin, A.W. Holleitner, K. Eberl, and R.H. Blick, *Phys. Rev.* B **64**, R241302 (2001).
17. A.W. Holleitner, F. Simmel, B. Irmer, R.H. Blick, J.P. Kotthaus, A. Ustinov, and K. Eberl, *N. J. Phys.* **2**, 2 (2000).
18. C. Bruder and H. Schoeller, *Phys. Rev. Lett.* **72**, 1076 (1994).
19. C.A. Stafford and N.S. Wingreen, *Phys. Rev. Lett.* **76**, 1916 (1996).
20. T.H. Stoof and Yu.V. Nazarov, *Phys. Rev.* B **53**, 1050 (1996).
21. A.W. Holleitner, F. Simmel, B. Irmer, R.H. Blick, J.P. Kotthaus, M. Bichler, and W. Wegscheider, Terahertz Spectroscopy and Applications II in Proceedings of SPIE (ed. J.M. Chamberlain) **3828**, 194 (2000).
22. A. Tilke, M. Vogel, F. Simmel, A. Kriele, R.H. Blick, D.A. Wharam, and J.P. Kotthaus, *J. Vac. Sci. Technol.* **B 17**, 1594 (1999).
23. R.H. Blick, PhD Thesis, *Künstliche Atome und Moleküle: Gekoppelte Quantenpunkte und Mikrowellenspektroskopie an Quantenpunkten*, Max-Planck-Institut für Festkörperphysik und Universität Stuttgart (December 1995); *Physics* **57**. Publishing Company Harri-Deutsch, Frankfurt am Main, Germany (1996), ISBN 3-8171-1507-5.
24. H. Qin, F. Simmel, R.H. Blick, J.P. Kotthaus, W. Wegscheider, and M. Bichler, *Phys. Rev.* B **63**, 035320 (2001).
25. U.S. patent 5,748,309, van der Weide et al., May 5 (1998).
26. H. Qin, D.W. van der Weide, J. Truitt, K. Eberl, and R.H. Blick, *Nanotechnology* (2002) 14, 60–64 (2003).
27. R.H. Blick, D.W. van der Weide, R.J. Haug, and K. Eberl, *Phys. Rev. Lett.* **81**, 689 (1998).
28. B. Irmer, F. Simmel, R.H. Blick, H. Lorenz, J.P. Kotthaus, M. Bichler, and W. Wegscheider, Special Issue on Mesoscopic Superconductivity, Superlattices and Microstructures **25**, 785 (1999).
29. A.V. Ustinov, *Physica* D **123**, 315 (1998).
30. D. Pfannkuche, Habilitation Thesis, *Aspects of Coulomb Interaction in Semiconductor Nanostructures*, Max-Planck-Institut für Festkörperforschung and Universität Karlsruhe, (September 1999).
31. P.K. Tien and J.P. Gordon, *Physical Review* **129**, 647 (1963).

32. F.R. Waugh, M.J. Berry, D.J. Mar, R.M. Westervelt, K.L. Campman, and A.C. Gossard, *Phys. Rev. Lett.* **75**, 705 (1995).
33. R.H. Blick, R.J. Haug, J. Weis, D. Pfannkuche, K. von Klitzing, and K. Eberl, *Phys. Rev.* B **53**, 7899 (1996).
34. C. Livermore, C.H. Crouch, R.M. Westervelt, K.L. Campman, and A.C. Gossard, *Science* **274**, 1332 (1996).
35. T. Fujisawa, T.H. Oosterkamp, W.G. van der Wiel, B.W. Broer, R. Aguado, S. Tarucha, and L.P. Kouwenhoven, *Science* **282**, 932 (1998).
36. F.W. Beil, A. Wixforth, and R.H. Blick, *Physica* E **13**, 473 (2002).
37. Y. Fu and S.C. Dudley, *Phys. Rev. Lett.* **70**, 65 (1993).
38. M. Büttiker, A. Prêtre, and H. Thomas, *Phys. Rev. Lett.* **70**, 4114 (1993).
39. M. Büttiker, A. Prêtre, and H. Thomas, *Phys. Rev. Lett.* **71**, 465 (1993).
40. W. Kohn, *Phys. Rev.* **123**, 1242 (1961).
41. T. Brandes, F. Renzoni, and R.H. Blick, *Phys. Rev.* B **64**, R 035319 (2001).
42. H. Grabert and M.H. Devoret (eds.) *Single Charge Tunneling: Coulomb Blockade Phenomena in Nanostructures*, No. 294 in Series B: Physics. NATO Advanced Study Institute, Plenum, New York (1992).
43. D.V. Averin and Y.V. Nazarov, *Phys. Rev. Lett.* **65**, 2446 (1990).
44. A.A. Odintsov, V. Bubanja, and G. Schön, *Phys. Rev.* B **46**, 6875 (1990).
45. S. DeFranceschi, S. Sasaki, J.M. Elzerman, W.G. van der Wiel, S. Tarucha, and L.P. Kouwenhoven, *Phys. Rev. Lett.* **86**, 878 (2001).
46. R.H. Blick, V. Gudmundsson, R.J. Haug, K. von Klitzing, and K. Eberl, *Phys. Rev. Rap. Comm.* **57**, R12685 (1998).
47. A.K. Hüttel, J. Weber, A.W. Holleitner, D. Weinmann, K. Eberl, and R.H. Blick, submitted to *Phys. Rev. Lett.* (2002).
48. E.M. Höhberger, F.W. Beil, R.H. Blick, W. Wegscheider, M. Bichler, and J.P. Kotthaus, *Physica E* **12**, 487 (2002); J. Kirschbaum, E.M. Höhberger, R.H. Blick, W. Wegscheider, and M. Bichler, *Appl. Phys. Lett.* **81**, 280 (2002).

Chapter 3

Nano-Spintronics with Lateral Quantum Dots

A. Sachrajda[1], P. Hawrylak[2], and M. Ciorga[1]

[1] *Quantum Physics Group,* [2] *Quantum Theory Group, Institute for Microstructural Sciences, National Research Council of Canada, Ottawa, Ontario, Canada K1A 0R6*

1. Introduction

Over the last decade there has been a tremendous increase in the research devoted to nanotechnology. This trend is driven both by the potential for new paradigms and technological applications as well as by the fundamental science suggested by new quantum regimes. More recently a similar explosion of effort has commenced in the field of spintronics and, in particular, in semiconductor spintronics [1]. The attraction of exploiting the electrons *other* quantum degree of freedom in semiconductors is clear, making it possible to examine novel device functionalities, achieve combined memory and logic functions and to explore the feasibility of spin-based qubits for quantum information applications. In this chapter we combine these two fields and consider a field we have termed *Nano-spintronics*. Nano-spintronics is related to spintronics in the same way that nano-electronics is to electronics—it is spintronics on the small scale and, ultimately, at the single spin level (single spintronics).

It has already been demonstrated that nano-devices down to the atomic length scale can be fabricated [2]. In addition, there has been a lot of success recently in the controlled self-assembly of these nanostructures [3,4]. The main question is, therefore, not so much the size that devices can be made or their operating temperature (which roughly depends on their size) but other issues, such as operating paradigms, architecture, functionality, and techniques for manipulation. For investigations into functionality and manipulation laterally gated devices [5] provide a useful initial choice. The ability to tune all the relevant parameters controllably through gate action is extremely powerful. Since lateral devices are defined in a two-dimensional electron gas (2DEG), the quantum properties of the 2DEG can also be utilized to create leads with non-trivial properties. In this chapter we summarize some of the theoretical and experimental work which we have performed with our collaborators in this direction. More details and specifics can be obtained in the papers that cover specific areas summarized here.

The chapter is divided as follows. In section 2 we provide a theoretical framework for the results we will present in the following sections. In section 3 we describe the devices used in the experiments as well as basic experimental techniques employed in these studies. We describe how to isolate a single electron by purely electrostatic means and how we can thereby confine a known number of electrons in the quantum dot—a necessary requirement for a successful comparison of the experimental results with theoretical calculations. In section 4 we describe how we can use the properties of the 2DEG leads (electron reservoirs) to achieve spin selectivity for both spin injection and spin detection. A new type of spectroscopy—spin blockade spectroscopy will be described. In section 5 we present results from Coulomb and spin blockade spectroscopy of a quantum dot containing up to 45 electrons. It will be immediately clear that spin and Coulomb blockade spectroscopy provide complementary information. In section 6 we concentrate on the scenario of two electrons. We demonstrate that the read-out of the spin state of a two-electron system (i.e. the conversion of spin information to electrical information), combined with a read-in capability, could provide a non-trivial read-out of a single electron for spin qubit applications. In section 7 we move to a different regime where several more electrons are confined and yet the results are closely related to the one- and two-electron scenarios. This is the regime of filling factor $\nu = 2$ in the dot. We demonstrate spin transitions which had not been predicted, and that originate from the finite range of existence (as a function of electron number) of the $\nu = 2$ spin singlet phase for an even number of electrons. In section 8 we focus on the transition of the $\nu = 2$ phase to the $\nu = 1$ phase via the spin flip regime. In section 9 we demonstrate a novel nano-spintronic functionality—the use of spin to generate negative differential resistance "on demand."

2. Theoretical Framework

In this section we review the theoretical framework required for understanding the electronic properties of gated quantum dots as a function of the number of confined electrons N, with an emphasis on the role of a spin. The starting point is the single particle Hamiltonian of a quantum dot and its energy spectrum. The Hamiltonian describes a quasi-two dimensional electron moving in the (x, y) plane in a perpendicular magnetic field $\boldsymbol{B}$. The lateral confining potential produced by electrostatic gates is a smooth function of position, with a minimum in the center of the dot. Expanding around the minimum and retaining only the lowest terms gives a parabolic confining potential. The Hamiltonian now reads as:

$$H_0 = 1/2m^*[\mathbf{p} + (e/c)A(\mathbf{r})]^2 + 1/2m^*\omega_0^2(x^2 + y^2) \tag{1}$$

where m^* and ω_0 are the effective mass of the electron and level spacing of the confining potential, respectively, and $A(\mathbf{r})$ is the vector potential. The single particle Hamiltonian can be diagonalized using a transformation to a set of harmonic oscillator creation and annihilation operators a and b, respectively, and written as [6]

$$H_0 = \Omega_-(b^+b + 1/2) + \Omega_+(a^+a + 1/2) \tag{2}$$

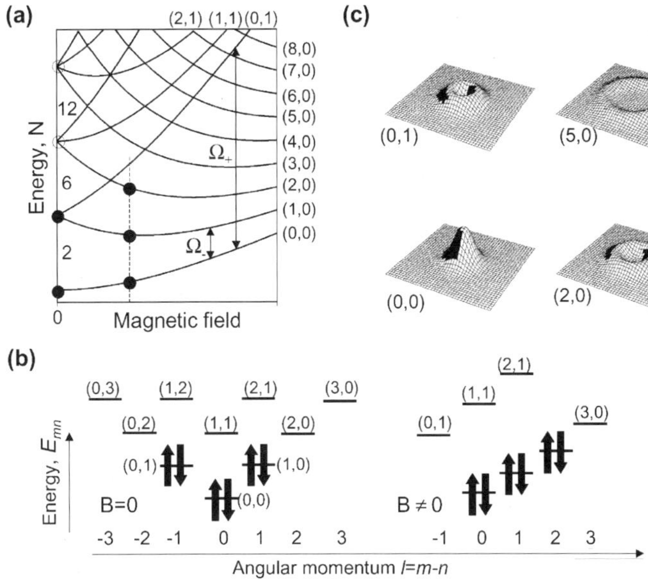

Fig. 3.1. (a) Single particle Fock–Darwin (FD) spectrum of a quantum dot with a parabolic confinement. Numbers in brackets indicate pairs of quantum numbers (m, n). Circles at $B = 0$ indicate shell structure. Full circles indicate levels occupied when the dot contains six electrons. Examples of electronic configurations of ground states for such a dot are shown in (b). Left panel shows $B = 0$ configuration with two shells fully occupied and right panel shows the dot at $\nu = 2$ state, when all electrons reside on orbitals belonging to the spin unresolved lowest Landau level. Examples of wavefunctions of some of those orbitals are shown in (c) together with a wavefunction of the (0, 1) orbital—the lowest orbital of the second Landau level.

The resulting single particle spectrum,

$$E(m, n) = \Omega_{-}(m + 1/2) + \Omega_{+}(n + 1/2) + \frac{1}{2} g\mu B\sigma \tag{3}$$

including the Zeeman energy $E_z = g\mu B\sigma$, is that of two harmonic oscillators with frequencies

$$\Omega_{-/+} = \tfrac{1}{2}\left[(\omega_c^2 + (2\omega_0)^2)^{1/2} \mp \omega_c\right] \tag{4}$$

where ω_c is the cyclotron energy. The single particle spectrum is often called the Fock–Darwin (FD) spectrum. The FD spectrum was first established by Fock and Darwin using a differential equation approach to the solution of the Schrödinger equation. The harmonic oscillator approach [6,7] provides an elegant alternative. The FD spectrum and its magnetic field dependence, with Zeeman energy not included, is shown in Fig. 3.1(a). The FD spectrum at $B = 0$ consists of equally spaced shells with spacing ω_0 and degeneracies monotonically increasing in energy. Increasing the magnetic field decreases the energy spacing between the intra-Landau level m oscillators and increases the energy between inter-Landau level n oscillators. In the high magnetic field limit the low energy states are harmonic oscillator states

$$E(m, 0) = \Omega_-(m + 1/2) + \Omega_+(1/2) \tag{5}$$

which evolve toward the lowest Landau level states with equal spacing, as shown in Fig. 3.1(b) for a six-electron configuration. These specific states can be visualized as rings with an increasing radius, as shown in Fig. 3.1(c). The second lowest states are those evolving toward the second Landau level. As the levels evolve from the shell structure at $B = 0$ to the Landau level structure at high B, many crossings of levels are observed in Fig. 3.1(a). At several special values of B secondary shells appear. The last crossing of states with different angular momenta corresponds to the intersection of the $|m = 0, n = 1\rangle$ state of the second Landau level with states $|m, n = 0\rangle$ of the first Landau level. The first such crossing takes place for the $|m = 0, n = 1\rangle$ (p-shell) and $|m = 2, n = 0\rangle$ (d-shell) orbitals. The FD spectrum to the right of this crossing involves only states of the lowest Landau level. The ladder of $|m, n = 0\rangle$ states can be populated by filling up all states with spin up/down electrons up to the Fermi level as shown in the right panel of Fig. 3.1(b). These electrons form the $\nu = 2$ quantum dot state, that is a spin singlet phase for even number of electrons N. When the level spacing $\Omega\_$ is much smaller then the Zeeman energy E_Z, the electrons with spin down will fill up all levels up to the Fermi level forming the spin-polarized filling factor $\nu = 1$ droplet.

This summarizes the picture of non-interacting electrons in a parabolic quantum dot with a circular symmetry. This symmetry may be perturbed by the asymmetry caused by a gate layout. As will be shown later (section 3), the lateral dot with controlled number of electrons results in a confining potential which consists of a parabolic component

$$V_0 = 1/2m^*\omega^2(x^2 + y^2) \tag{6a}$$

and a perturbation

$$V_1(r) = \frac{1}{2}m\omega^2\gamma^2\left(x - \frac{1}{2D}y^2\right)^2 \tag{6b}$$

This potential is controlled by a strength γ and a length D. The potential affects the shell structure of the quantum dot at $B = 0$ but its effect in a strong magnetic field and at large electron numbers is small [8].

After we have established the single particle spectrum $|m, n, \sigma\rangle = |i\rangle$, we now turn to the discussion of the problem of interacting electrons. The effect of the electron–electron (e–e) interactions is described by the Hamiltonian conveniently written in the language of second quantization as

$$H = \sum_i E_i c_i^+ c_i + \tfrac{1}{2}\sum_{ijkl}\langle ij|V_{ee}|kl\rangle c_i^+ c_j^+ c_k c_l. \tag{7}$$

The operators $c_i^+(c_i)$ create (annihilate) the electron $|i\rangle$ with the single-particle energy E_i, and Coulomb matrix elements $\langle ij|V_{ee}|kl\rangle$ scatter electrons from states $|kl\rangle$ to states $|ij\rangle$. The key difficulty is the size of Hilbert space: for N_{sp} single particle states and N_e electrons there are $\binom{2N_{sp}}{N_e}$ configurations. Even for 10 orbitals and 7(10) electrons this number is 77,520 (184,756) and grows exponentially with the size of the system.

To handle the interacting electron problem it is useful to exploit the symmetries of the problem. In fact, the symmetries, the single particle spectrum, and the form of two-body matrix elements are what differentiates one electronic system from another (the form of the Hamiltonian in second quantization is the same for all systems interacting via two-body interactions). We have already identified angular momentum as a good quantum number for a parabolic dot with a circular symmetry, with the corresponding total angular momentum operator $L = \sum_{nm\sigma}(m-n)c^{+}_{nm\sigma}c_{nm\sigma}$. Many electron states can, therefore, be labeled by their total angular momentum L. The Hamiltonian does not explicitly depend on spin, hence both total spin projection on z-axis

$$S_Z = 1/2\sum_{nm\sigma}\sigma c^{+}_{nm\sigma}c_{nm\sigma} \tag{8a}$$

and total spin

$$S^2 = N/2 + S_Z^2 - \sum_{nmn'm'} c_{n'm'}\uparrow^{+} c_{nm}\downarrow^{+} c_{n'm'}\downarrow c_{nm}\uparrow \tag{8b}$$

are good quantum numbers. These operators and underlying symmetries form a set of "obvious" symmetries. In addition to these obvious, or well-known, symmetries there are "not-so-obvious," or "hidden," symmetries. Consider operator $Q^{+} = N^{-1/2}\sum_{m=0}(m+1)^{1/2}c^{+}_{m+1}c_m$ that operates on the states of the lowest Landau level. It is interesting to note that the commutator of this operator with the Hamiltonian is $[H, Q^{+}] = \Omega_{-}Q^{+}$. This operator (and an analogous operator for inter-Landau level transitions) excite center-of-mass (CM) excitations. These excitations do not depend on mutual interactions among electrons, so the spectrum of the CM operator is just the single particle spectrum. Operator Q^{+} generates exact eigenstates of the electron system. All states can be classified by the number of CM excitations. This is a hidden symmetry of the quantum dot. This symmetry is "hidden," i.e. not "obvious," in the notation of the second quantization but has been demonstrated as a generalized Kohn's theorem in the first quantization [6]. As a result, the CM motion decouples from the relative motion of electrons in parabolic quantum dots and so the *e–e* correlations are independent on the confining potential.

The knowledge of all symmetries implies that the Hilbert space can be divided into blocks corresponding to a total angular momentum L, total spin projection S_z, total spin S, and the number of CM excitations. Not all these symmetries are easily incorporated into a viable computational technique. We employed a number of different approaches in the past: Hartree and Hartree–Fock (HF) calculations [9], HF calculations coupled with multi-pair expansion [10], Local Spin Density Approximation (LSDA) [11], and exact diagonalization (configuration interaction) techniques [12]. These techniques are described in a number of papers [11,13] and the results of these calculations are referred to throughout the text.

Since we focus on spin in this chapter, it is useful to remember how spin affects the electronic properties of quantum dots. This effect is indirect, since the Hamiltonian does not depend on spin, and hence, it is related to the form of the wavefunction. If we write the wavefunction for a given total spin S as $\Psi_{S,S_z}(1, 2, \ldots, N) = \sum_i \Phi_i(1, 2, \ldots, N)\chi_i$, it is a product of spin functions χ, which are the eigenstates of total spin operator, and spatial functions Φ, which are functions of electron coordinates $(1, 2, \ldots N)$. Since the wavefunction has to transform under the permutation group, the permutation properties of spin functions χ determine the form of spatial functions Φ. This form of Φ, in turn, affects charge density and Coulomb interaction among electrons. We can easily see this in a simple example of two electrons on two orbitals a and b. The spin singlet $S = 0$ (triplet $S = 1$) state is anti-symmetric (symmetric) under permutation and requires the orbital wavefunction to be symmetric (anti-symmetric). When the Coulomb energy is evaluated, the symmetric (anti-symmetric) spatial wavefunction leads to Coulomb energy given by $\langle a, b|V|b, a\rangle + (-)\langle b, a|V|a, b\rangle$, i.e. direct and repulsive (attractive) exchange term. Therefore, the total spin, through the symmetry of the wavefunction, changes the interaction energy of two electrons. The total energy difference $J = 2\langle a, b|V|a, b\rangle$ between singlet and triplet will be discussed later in details.

3. Experimental Devices and Techniques

Lateral quantum dots owe their existence to the split gate technique invented in Cambridge in the mid 1980s [5]. The key feature of the method is the use of patterned surface gates to define structures in a two-dimensional electron gas (2DEG), placed typically around 100 nm below the surface. It has been used to study a variety of structures, e.g. quantum dots [14] and antidots [15]. The beneficial property of this technique is the electrostatic control of device parameters and the resultant flexibility. Quantum dots, e.g., can be defined with tunable entrance and exit barrier heights, electron occupation number, shape and size. The split gate technique has led to more discoveries in mesoscopics (e.g. quantum conductance in small constrictions [16]) than any other technique.

The Coulomb blockade (CB) effect in semiconductors was discovered by Scott-Thomas et al. [17] in their measurements on quantum wires. It soon became apparent that this technique could form the basis of an extremely powerful technique for spectroscopic transport measurements—Coulomb blockade spectroscopy. Indeed, it is interesting to note, that while single dot spectroscopy has recently become a very important technique in optical studies [18], single dot transport spectroscopy, using Coulomb blockade techniques, has been utilized for over a decade [19]. Whereas excellent detailed reviews of this technique are available elsewhere [20], here we describe the basic concepts, the knowledge of which is required for understanding problems discussed in this and the following sections.

Consider the *generic* lateral quantum dot illustrated in Fig. 3.2(a). It consists of six gates. When a sufficiently negative gate voltage (greater than the depletion value, typically ~ -0.4 V) is applied to the gates, a small puddle of electrons, a quantum dot, is isolated in the depleted region within the 2DEG. The gates marked "T" are used to define tunneling barriers into and out of the dot. The gates marked "P", so called "plunger" gates, are used to control the number of electrons confined in the dot. When the voltages applied to gates are made more negative, the size of the dot shrinks due to the expansion of the depletion region beyond the lithographic regions of each gate. Figure 3.2(b) shows the cartoon of the potential landscape through a quantum dot. The energy gap between occupied and empty levels in the dot is dominated by a charging

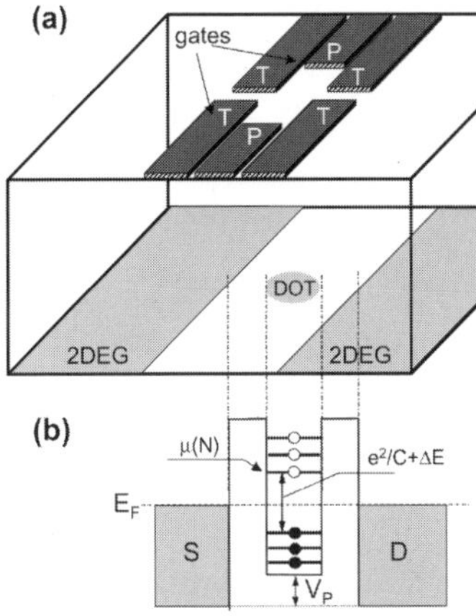

Fig. 3.2. (a) A schematic representation of a typical lateral quantum dot. Shown are six metallic gates deposited above two-dimensional electron gas (2DEG) confined in AlGaAs/GaAs heterostructure. The white area in the plane of 2DEG represents the region depleted of electrons as a result of the negative voltage applied to the gates. A small island of electrons—quantum dot—is formed in the middle of this region. (b) A schematic representation of energy levels in the quantum dot with a charging energy e^2/C dominating the gap between occupied (full circles) and unoccupied (empty circles) levels. For details see text.

energy e^2/C, where C is a total capacitance of the dot. Within the orthodox model of a Coulomb blockade spectroscopy a simple capacitance expresses the interactions of the electrons in the quantum dot with themselves and with the outside world. Current can flow through the dot (one electron at a time) only when the electrochemical potentials of the quantum dot $\mu(N)$, defined as

$$\mu(N) = E(N) - E(N-1) \tag{9}$$

where $E(N)$ is the energy of the ground state of the N-electron dot, becomes aligned with the chemical potential of the source (S) and drain (D) leads, also defined in 2DEG. When this condition is not fulfilled, the current is blockaded. The plunger gate, through its voltage V_P, is used to continuously change the electrochemical potential of the dot, i.e. to change the number of electrons confined in the dot. The peak in the measured current through the dot is observed at V_P values for which above mentioned condition occurs.

A typical single V_P scan in CB spectroscopy, performed at zero magnetic field, is shown in Fig. 3.3(a). Four CB peaks are shown. On a Coulomb blockade peak single electrons are repeatedly added to and then removed from the dot, one electron at a time, resulting in a finite current. Typical magnetic field scan is shown in Fig. 3.3(b). The spacing between the peaks are typically dominated by charging energy e^2/C, together with a component ΔE that is related to an addition spectrum. The latter forms the basis for the quantum dot spectroscopy. Several other interesting features can be seen in the figure. The large step observable in all of the peaks, labeled with a star-symbol, is due to a Landau level depopulation event in the 2DEG. Since one is matching electrochemical

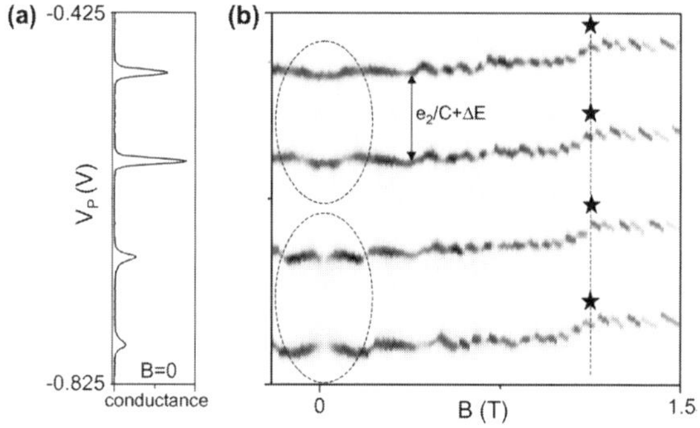

Fig. 3.3. Typical Coulomb blockade (CB) spectrum of a lateral quantum dot showing four CB peaks. Plotted is the conductance of the dot containing around 40 electrons. (a) Single plunger voltage V_P scan at zero magnetic field; (b) Magnetic field dependence of the same peaks as in (a). The spacing between peaks is dominated by a charging energy e^2/C. Dashed ovals indicate peaks paired together near zero magnetic field. Star symbol marks a magnetic field independent step in a peak position, related to Landau level depopulation in 2DEG.

potentials of the dot and leads one is simultaneously performing spectroscopy on the leads. The leads in lateral devices, as mentioned above, are also defined in 2DEG. The quantum dot is then, in a sense, a local probe of the 2DEG edge. The low field "magneto-fingerprint" appears to be similar for pairs of peaks as might be expected if the electrons populate the same state of the dot differing only in their spin state. Such a pairing, however, is rather rarely observed in lateral dots for more than a few peaks at a time.

The data shown in Fig. 3.3 were obtained in a linear regime of CB spectroscopy, i.e. with a very small (typically a few μeV) source-drain bias voltage V_{SD} applied across the sample. In this regime only the ground states of quantum dots are being probed. By applying a larger voltage across the quantum dot (greater than the level spacing, but still less than the charging energy, e^2/C) the electrons can tunnel into the dot through excited states in addition to the ground state. On exiting the dot, the electron may leave the dot in one of the excited states. The transport is still, of course, one electron at a time. Important information about the excited states spectrum of a quantum dot can be obtained through analysis of so called Coulomb diamonds—structures in a (V_P, V_{SD}) plane evolving from single peaks as a result of increased V_{SD}—and current stripes—magnetic field sweeps with a high source-drain voltage applied [21].

The quantum dot depicted in Fig. 3.2(a) cannot easily be used to perform spectroscopy on few electron dots. Trivially, a sufficiently negative voltage applied to the gates will empty the dot of electrons. The tunneling barriers, however, will be totally pinched off at the same time due to the indirect effect of the plunger gates on the tunneling barrier. To eliminate this problem it was necessary to change the layout of the dot to that illustrated in Fig. 3.4(a) [22]. It consists of four gates: a top gate ("T"), plunger gate ("P") and two finger gates ("F"). The most important feature of this layout is the definition of each of the tunneling barriers with two gates—one with a small

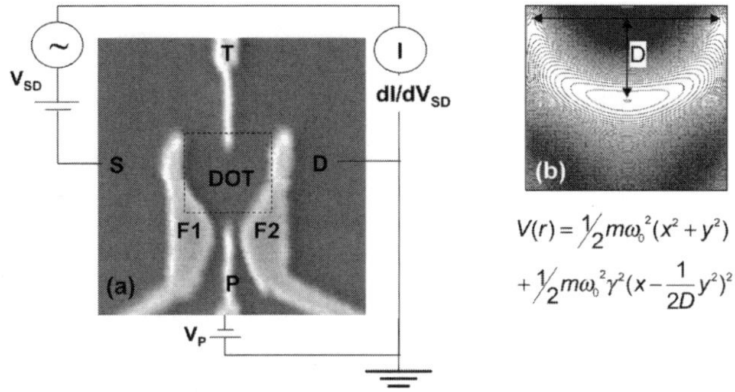

Fig. 3.4. (a) Scanning electron microscope (SEM) picture of a lateral quantum dot device with a small tunable number of electrons. Shown are four metallic gates defining a dot. Shown also is a schematic of the experimental circuit used to perform Coulomb blockade transport experiments. (b) Deformed parabolic potential $V(r)$ created by the gates in the region squared in (a).

perimeter in the quantum dot and one with a large perimeter. The same tunneling barrier height can be achieved for different ratios of gate voltages applied to the T and F gates, but these will have very different effects on the size of the dot. The 2DEG in the wafer used to built devices used in experiments described in this chapter is about 90 nm below the wafer surface, the bulk wafer mobility is 2×10^6 cm^2/Vs while the density is 1.7×10^{11}cm^{-2}. Shown in the figure is also a simple schematic of the circuit used in the experiments. Both dc and ac measurements could be performed, enabling us to measure either the total current I flowing through the device or a differential conductance dI/dV_{SD}.

The effect of the gates on the potential at the GaAs/GaAlAs heterojunction can be evaluated [8]. Figure 3.4(b) shows the potential profile experienced by electrons. We see that the potential inside the dot resembles a slightly deformed parabolic potential of the form discussed in section 2. It consists of a parabolic minimum described by Eq. (6a) and a weak wire potential described by Eq. (6b), which connects the two quantum point contacts.

Figure 3.5 illustrates the procedure for emptying a quantum dot. Initially a sufficient voltage is applied to the F gates to deplete the 2DEG underneath them. A sufficiently negative voltage is then applied to the T gate to create operational tunneling barriers. Under these conditions around 50 electrons are isolated. The voltage applied on the F gates is then made more negative, which reduces the size of the dot but which also has the tendency to pinch off the tunneling barriers. It is therefore necessary to compensate this effect by gradually reducing the negative voltage applied to the T gate. The procedure is then repeated. The plunger gate is simultaneously used to sweep through the different occupation numbers N. Voltage on the T gate is also swept, in a direction opposite to V_P, however, to compensate for the effect of pinching off the barriers by the increased negative voltage on P gate. The range of that sweep is, approximately, an order of magnitude lower than the range of V_P sweep. The number of peaks in a single sweep decreases with a size of the dot, as seen in the inset in Fig. 3.5(a). The reason for this is an increase in the charging energy e^2/C as a result

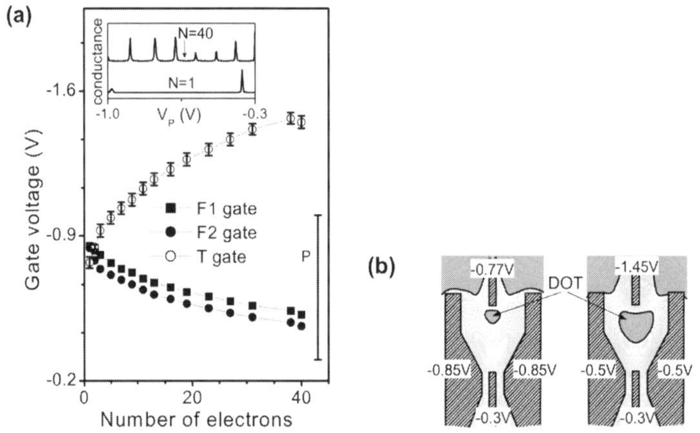

Fig. 3.5. The procedure of emptying the quantum dot shown in Fig. 4(a). (a) Voltage applied to the gates as a function of electron number N. Inset shows single plunger voltage scans at $B = 0$ for the case when dot contains one and around 40 electrons. (b) The schematic representation of the gate potential on the size of the dot in the regime of one (left panel) and 40 (right panel) electrons in the dot.

of a decrease in the total capacitance of the dot C. The triangular shape of the dot is used to maintain the dot located in the proximity of the barrier region throughout the procedure, as is shown in Fig. 3.5(b).

There are many procedures that can be used to confirm the successful emptying of a dot. Some are based on the spectroscopic features. One important feature that confirms the emptying of the dot is the presence of an open Coulomb diamond in a $V_P - V_{SD}$ spectrum of a dot containing no electrons. Such a structure is shown in Fig. 3.6. Generally, a Coulomb diamond closes when the edges of two neighboring Coulomb blockade peaks intersect, as is shown in the inset in Fig. 3.6 for the eight-electron dot. It happens for such a bias voltage V_{SD}, that $eV_{SD} \approx e^2/C$. Clearly, a peak related to adding the first electron to a dot does not have a neighboring peak on the side of higher negative values of V_P because there are no more electrons in the dot. The complete diamond cannot be formed then, on that side of the first peak, even for large source-drain bias V_{SD} applied across the dot. Recently, Sprinzak and co-workers [23] introduced the use of non-local charge detectors into similar devices. Amongst other benefits these provide an alternative signature for the removal of the final electron *since* they operate even when the exit and entrance barriers are pinched off.

We close this section by introducing a simple, but useful, spectroscopic way of visualizing CB peaks. A Coulomb blockade peak provides information about two ground states differing by one electron (e.g. N and $N + 1$), as seen from Eq. (9). The incoming $N + 1$ electron should be considered as a probe of the state of the N-electron dot. More accurately, the amplitude of the Coulomb blockade peak will depend on the relationship between the electronic configurations of N and $N + 1$ electron systems. This will be important when we introduce spin blockade spectroscopy.

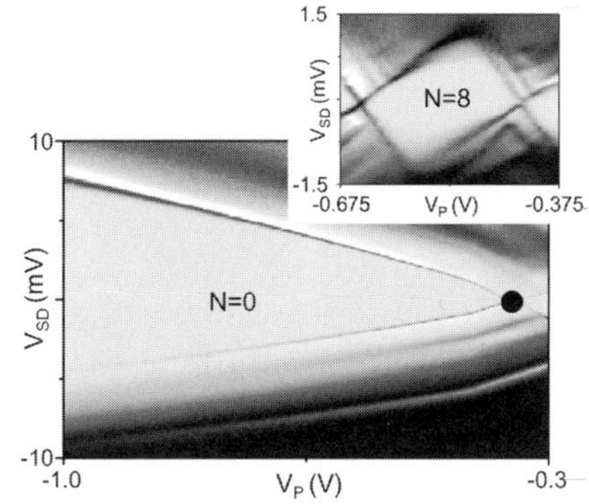

Fig. 3.6. Non-closing Coulomb diamond in the (V_P, V_{SD}) plane indicating an empty dot. Black dot marks the position of the first CB peak for $V_{SD} = 0$. Inset shows a typical full diamond (for $N = 8$) resulting from the merging of two neighboring peaks with increasing V_{SD}.

4. Spin-Polarized Injection and Detection

Developing techniques for spin injection into semiconductors [24–26] is one of the most active areas of research in the field of semiconductor spintronics. In this work we make use of a well-known 2DEG property to achieve spin-polarized injection. Figure 3.7 illustrates an energy profile through a 2DEG in the presence of a perpendicular magnetic field. The quantization of the cyclotron orbit in the bulk 2DEG leads to the formation of Landau levels and a resultant periodic modulation in the density of states characterized by the cyclotron energy. At the edges of a 2DEG this energy separation ΔE of Landau levels becomes a spatial separation Δx of magnetic edge states [27]. The important question for this work is whether spin resolved Landau levels in the bulk of a sample can be reflected in spatially separated spin resolved magnetic edge states at the edge. The answer to this question depends on the detailed properties of these magnetic edge states that are very dependent on the density profile at the edge. An edge for this purpose can be at the boundary of any depleted region, specifically that created by the gates defining our quantum dot. Figure 3.8(a) shows where the magnetic edge states are present near the quantum dot. Experimentally it is found (cf. below) that the density profile outside our dot forms spatially separated magnetic edge states at magnetic fields of around 0.5 T. Above this field, therefore, the entrance and exit leads of our dot are the outermost spin resolved magnetic edge states, as illustrated in the cartoon shown in Fig. 3.8(b). Spin-up electrons, coming from the edge states located further away from the edge of the sample, have effectively thicker tunneling barrier (drawn with a dashed line in the figure) than spin-down electrons. The tunneling rate for spin-down electrons entering/leaving the dot is then much higher than that for spin-up electrons. In this way spin-polarized injection and detection is achieved in our dots [13,28], leading to various spin blockade effects [22,29]. Experimentally, we find the magnitude of the spin selectivity varies not only with magnetic field but also the gate voltage settings.

We are therefore able to achieve quantum dots into which we are able to insert any *known* number of electrons from 0 to 50 and to probe these electrons with spin-polarized

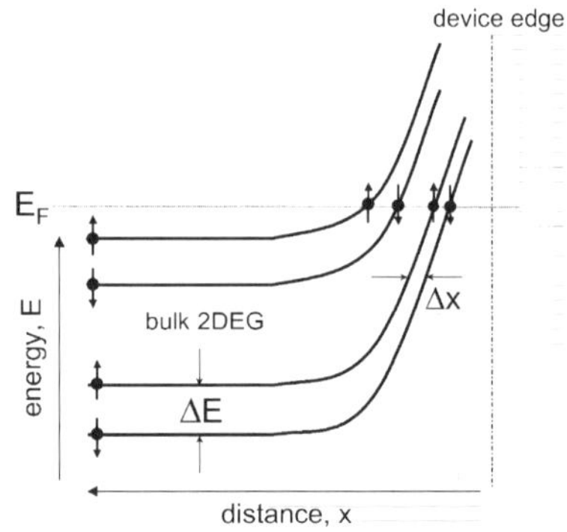

Fig. 3.7. An energy profile of a 2DEG close to the edge of the device (gates) in the presence of a perpendicular magnetic field. Arrows indicate spin polarization of different Landau levels. Energy separation ΔE between Landau levels in the bulk 2DEG is converted to a spatial separation Δx between magnetic edge states close to the edge of the device.

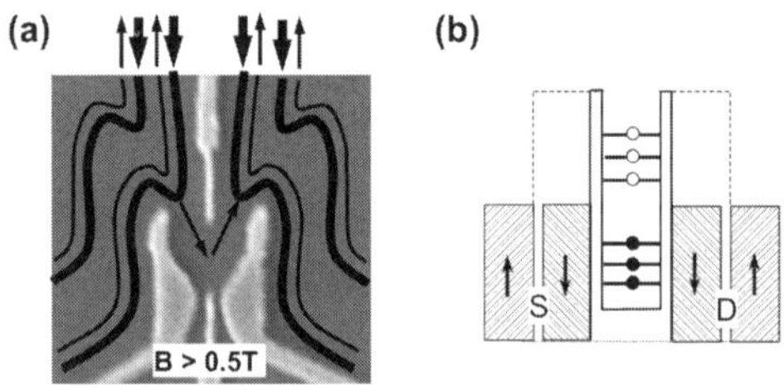

Fig. 3.8. Spin-polarized edge states as spin-polarized leads in lateral quantum dot devices. (a) Schematic representation of edge states close to the edge of the experimental device. Thick and thin lines represent, respectively, spin-down and spin-up edge states. (b) Schematic representation of the dot with spin-polarized source (S) and drain (D) leads. Dashed lines indicate effectively thicker tunneling barriers for spin-up electrons to enter/exit the dot.

leads. In the following sections we will apply this powerful combination to study the spin properties of a quantum dot.

5. Coulomb and Spin Blockade Spectrum

Figure 3.9 plots the magnetic field dependence of first 45 Coulomb blockade peaks of a typical one of our lateral dots. The peaks have been manually grouped together, which approximates removing the field independent charging energy. The plot is approximately the addition spectrum of the dot. When compared to the non-interacting Fock–Darwin spectrum for a parabolic quantum dot (see section 2) there are both similarities and differences. To extract the zero field peak spacing quantitatively (plotted in the inset) it is necessary to take into account the α parameter (the conversion factor between applied plunger gate voltage and energy) which was found to be dependent on the electron occupation number [30]. Unlike the results from transport measurements

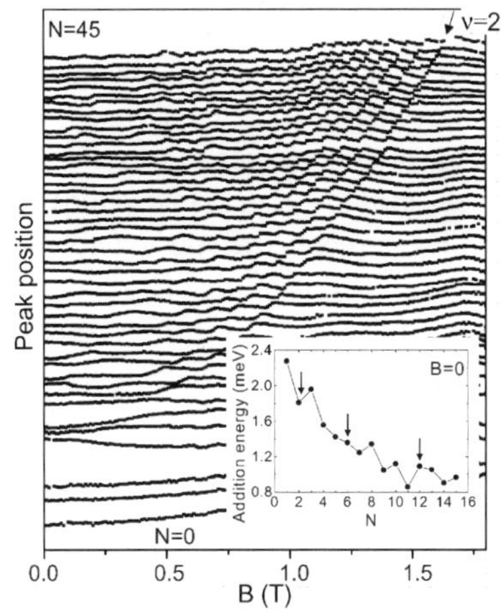

Fig. 3.9. An experimental addition spectrum of the quantum dot containing up to 45 electrons with a charging energy between peaks manually removed. Inset: addition energy vs. the number of electrons for the first 15 electrons entering the dot. Arrows mark electron numbers for which the shell-related features are observed for dots with a parabolic confinement potential.

on few electron vertical quantum dots [31], although similar to capacitance spectroscopy results on larger vertical dots [32], there is no obvious signature of shell structure visible in the peak spacing (i.e. a larger spacing at 2, 6, 12, 20, ... electrons). This is true for several reasons. Firstly, the confinement potential changes sharply with electron occupation [8], secondly, interaction effects might be expected to be bigger in our dots due to reduced screening related to the layout of the gates and contacts, and thirdly, our dot is not circular. The $\nu = 2$ line and its extrapolation to $B = 0$ as the number of electrons is reduced is clearly visible in the figure.

The CB peaks occur when the electrochemical potential of the dot and leads are matched. The chemical potential of the 2DEG oscillates with magnetic field as successive Landau levels become depopulated. This is observed very clearly on the first Coulomb blockade peak ($N = 0$ to $N = 1$ electron dot), seen in Fig. 3.10, i.e. in the regime where the quantum dot spectrum is featureless. These 2DEG features are easy to differentiate from quantum dot effects since they occur at the same field position on all of the Coulomb blockade peaks and their inverse magnetic field behavior is consistent with the bulk 2DEG density.

Figure 3.11 plots the amplitude of the peaks from Fig. 3.10. Since many of the features in Fig. 3.11 are related to spin blockade effects, we call this the spin blockade spectrum. Remarkably, the plot reveals a very similar looking spectrum to Coulomb blockade spectrum. On closer inspection, however, significant differences emerge. An example of this is the detailed behavior close to the $\nu = 2$ line which marks the low field boundary of the $\nu = 2$ singlet state. Whereas the Coulomb blockade spectrum consists of a series of upward steps in the chemical potential, the spin blockade spectrum consists of alternating increases and reductions in the peak amplitude on entering the $\nu = 2$ state [28]. This will be discussed in detail in section 7.

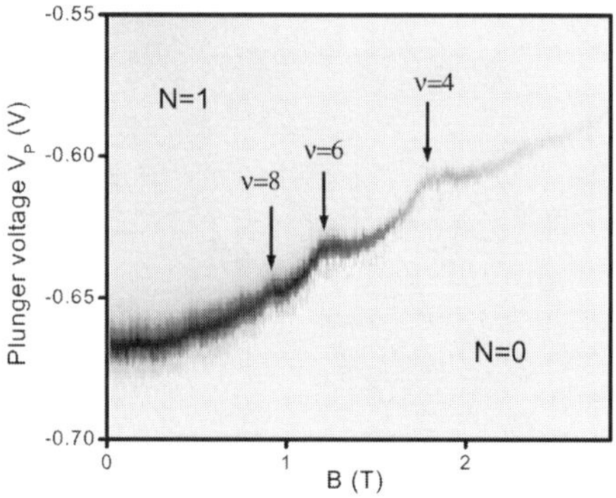

Fig. 3.10. Magnetic field evolution of the Coulomb blockade peak related to adding the first electron to the empty dot. Arrows indicate features related to the oscillations in the chemical potential of 2DEG resulting from depopulation of Landau levels. Adapted from [22].

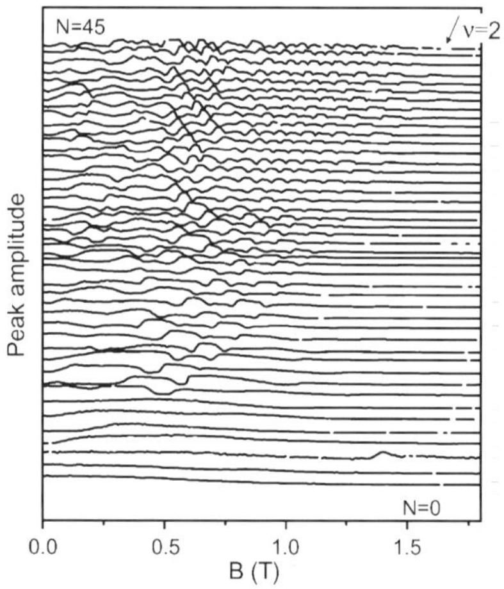

Fig. 3.11. Amplitude spectrum of the same Coulomb blockade peaks as in Fig. 3.9.

6. The First Few Electrons

In this section we consider the simplest possible quantum system consisting of a quantum dot containing 1, 2 or 3 electrons. A quantum dot containing a single electron has been considered by several groups as a possible spin (or charge) qubit for quantum information applications [33,34]. Spin properties of such a system are then on a subject of extensive studies [35]. One reason why a low number of electrons per qubit is required, is to avoid complex many body effects within the qubit, which would complicate the theoretical description. In this regards, the achievement of a purely electrostatically isolated electron is a very significant development. It is important to note, that the spin blockade technique, when applied to the single electron, provides immediately a readout function, i.e. a conversion of spin information to an electrical

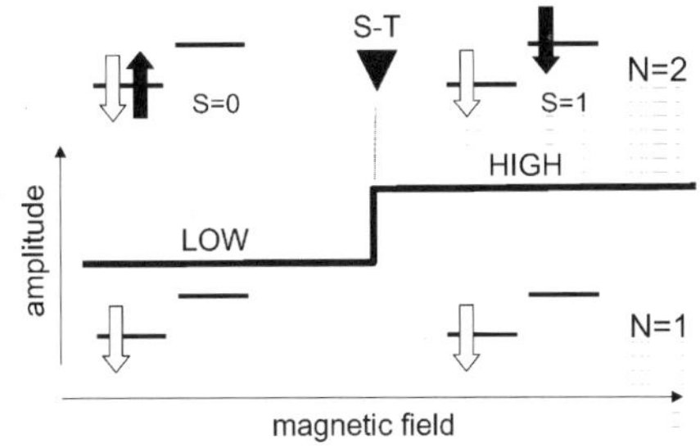

Fig. 3.12. The scheme for current readout of a singlet–triplet (S–T) transition for a two-electron dot. Black arrows indicate the spin of an electron to be added to a one-electron dot to form a two-electron dot. Current of spin-down electrons is preferred due to spin polarization of the leads. For details see text.

signal identifying the sign of the electron spin—the 0 or 1 of the qubit [34]. In order to demonstrate this quantum computer requirement non-trivially, one would ideally combine this with a read-in capability, i.e. with an ability to tune the spin of a single electron. The sophisticated techniques being pursued to achieve the read-in function are beyond the scope of this article. In the absence of this capability we are able to make use of the singlet–triplet transition [32,36,37] to demonstrate indirectly the read-out function [38].

Figure 3.12 illustrates schematically how spin blockade can be utilized for a readout. Consider a single electron in the quantum dot to which one adds a second electron at the Coulomb blockade peak. The first electron in the dot is characterized by a spin down, as depicted by empty arrows in Fig. 3.12. Electronic configurations of ground states of a two-electron dot, depending on a magnetic field regime, are also shown in Fig. 3.12. At low magnetic fields the ground state of a two-electron system is a singlet (total spin $S = 0$) state whereas at higher magnetic fields a triplet state ($S = 1$) is a ground state. The spin of the electron which constitutes the difference in configurations between the two- and one-electron systems is depicted by solid black arrows. As we can see, this is a spin-up (spin-down) electron in the singlet (triplet) regime. The second, "probe," electron entering the dot from the leads is statistically a spin-down electron once the spin polarization of the leads is set up (i.e. at $B > 0.5$ T). Therefore spin polarized leads should allow one to "read out" the singlet–triplet (S–T) transition through the observation of an increase in current amplitude from the "low" to "high" state at a magnetic field value corresponding to the S–T transition. In a quantum computer one could, by analogy, read out the state of a single electron by observing whether the addition of the second, spin-down, due to the spin polarization of the leads, electron at a Coulomb blockade peak resulted in a big or small current.

Figure 3.13 shows the experimental demonstration of the current read-out of the singlet–triplet transition for a two-electron dot. The magnetic field position of this transition can be detected without any ambiguity in non-linear Coulomb blockade measurements [39], i.e. with high V_{SD} bias across the sample. In order to locate the magnetic field position of the S–T transition independently for our demonstration, we perform a non-linear measurement of our dot with a bias voltage $V_{SD} = 350\,\mu$V. The resulting current stripe related to adding one electron to the $N = 1$ electron dot

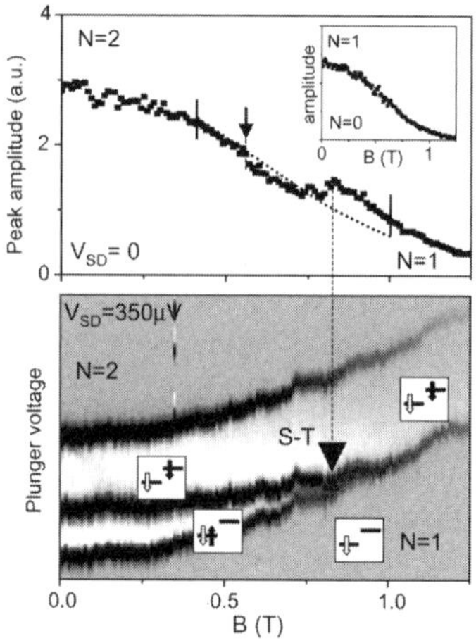

Fig. 3.13. Experimental realization of the current readout of a singlet–triplet (S–T) transition for a two-electron dot. Lower panel: differential conductance of the dot during the addition of the second electron to the one-electron dot, measured with a non-zero dc bias V_{SD} across the device. Singlet–triplet transition in the ground state of the two-electron dot is marked with a black triangle. Upper panel: the amplitude of the CB peak related to adding the second electron to the one-electron dot in a linear regime (zero dc bias). Dotted line represents the magnetic field dependence of the tunneling barriers as extracted from the amplitude of the first-electron peak (shown in the inset). Clearly visible is the increase in the peak amplitude at the magnetic field corresponding to the S–T transition.

is shown in the bottom panel of Fig. 3.13. The triplet state can clearly be observed at zero field as an excited state. The magnetic field at which the dot undergoes a singlet–triplet transition and a triplet state becomes a ground state is clear and is marked in the figure. Current read-out of the transition is shown in the top panel. The amplitude of the first Coulomb blockade peak, measured in a linear regime (i.e. purely ground state transport), is plotted as an inset. This peak shows a monotonic decrease in current amplitude. Assuming that this behavior is fully driven by the well-known magnetic field dependence of the tunneling barriers, we plot it in the main figure (dotted line) to compare with the dependence of the second peak's amplitude. The latter shows two additional features. At a magnetic field just above 0.5 T the reduction in the current accelerates (feature marked with an arrow), while at the singlet–triplet transition the current increases again. The interpretation of the data is totally consistent with our read-out scheme from Fig. 3.12. As the magnetic field is first increased, the magnetic edge states in the 2DEG leads become spin-polarized injecting predominately spin-down electrons. Since the ground state of a two-electron system is a singlet state and the incoming spin-down electron has the incorrect spin, to form the singlet state with the spin-down electron already being in the dot, the current is reduced ("low" state of the amplitude). At the singlet–triplet transition, a triplet state becomes the ground state, requiring a spin-down electron to be added. Since spin-down electrons are predominately injected the current increases, reaching the "high" state of the amplitude. The current then decreases due to the magnetic field dependence of the barriers, as described above.

We can summarize the results of this experiment in the following way: we have confirmed that the magnitude of the current through the quantum dot reveals the spin state of the two electrons, i.e. whether they form a singlet or triplet state. If one has spin-polarized leads one knows the sign of the spin of the probe electron and therefore the sign of the other, "probed" electron, can be deduced. It is interesting to note, that Fig. 3.13 reveals a smaller spin selectivity that is observed for higher electron occupations, as will be shown in the next section. This may be related to different electron density profiles at the barriers in the 2DEG leads for different occupation numbers. Further experiments are required to clarify this problem.

The current stripe in Fig. 3.13 provides more information relevant for the design of spin-based qubits. The spacing between the excited and ground state is precisely the Heisenberg direct exchange term J, which drives the two-qubit quantum gate action for spin-based qubits [34]. It has been extracted and is plotted in Fig. 3.14. A second curve of the same parameter is also plotted, obtained from the current stripe of the third peak [8]. This involved a hole excited state, corresponding to the situation when the probe electron exits the dot leaving it in the excited, triplet, state of a two-electron system (see the inset). Two features of the curves in Fig. 3.14 are rather surprising. Firstly, there is a flat region close to zero field revealing a magnetic field independent exchange splitting between the singlet–triplet states. Secondly, the two experimental measures of the exchange energy differ in by a factor of around two.

Calculations confirm that the origin of the flat region is related to the non-circular geometry of our device [8]. Similarly, the factor of two difference in J is related to the change in the quantum dot parameters (potential profile and shape) as one changes the voltages defining the quantum dot, while switching from the second to the third

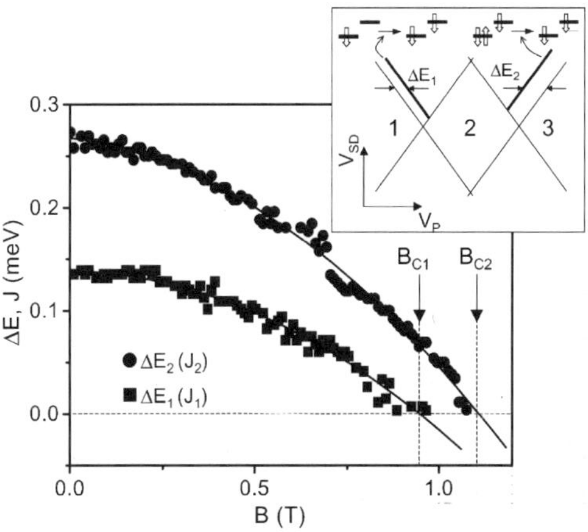

Fig. 3.14. Singlet–triplet energy gap ΔE, equivalent to the Heisenberg exchange constant J, for a two-electron system measured on adding the second electron to the one-electron dot (ΔE_1) and on removing the third electron from the two-electron dot (ΔE_2). Critical magnetic field values at which S–T transition (the change of the sign of J) is observed for both cases are marked by arrows and labeled B_{C1} and B_{C2}. The solid lines are guide for eyes. The inset shows how, at constant magnetic field, the gap ΔE can be tuned by means of the plunger gate voltage V_P.

Coulomb blockade peak. These two features raise some interesting issues. While in this case the flattening of J is fortuitous, one can speculate whether one could take advantage of this property. To operate a quantum gate one needs to switch on and off the interaction between neighboring qubits with great precision (J must be controlled to 1 part in 10^4 for error correction schemes to be effective [40]). The exchange interaction between qubits can be efficiently switched off by separating the quantum dots electrostatically. As we have demonstrated above, however, small gate voltage variations are reflected in changes in the confining potential that can change J by a factor of 2. It may be then possible to design geometries to "engineer" J to be independent of gate voltages in the same way that J is independent of B at low magnetic fields in our experiments. A second fascinating feature of Fig. 3.14 is related to the observation that the singlet–triplet transition, i.e. when J changes sign, occurs at different magnetic fields, B_{C1} and B_{C2}, for two gate voltage settings. This implies that if we place the quantum dot at an intermediate magnetic field B_0, such that $B_{C1} < B_0 < B_{C2}$, we are able to change the sign of J, i.e. to flip the spin of one of the two electrons, by purely electrostatic means—changing the value of V_P, as shown in the inset of Fig. 3.14. This functionality is fundamental for achieving a quantum gate, although, to our knowledge, this is the first demonstration of tuning J by purely electrostatic means in qubits (without varying an applied magnetic field).

There exists another important regime in the phase diagram of a quantum dot that is rather similar to that of a quantum dot containing just one or two electrons. This is the filling factor in the dot $\nu = 2$ phase. In that case there exists a sea of electrons that is in a well characterized singlet state. In the next sections we discuss the stability of that phase against spin-flip transitions occurring when the magnetic field is either lowered or raised.

7. The $\nu = 2$ Regime

The formation of $\nu = 2$ phase was described theoretically in section 2, when discussing the Fock–Darwin Spectrum. The phase develops as the magnetic field is increased. Degeneracies associated with symmetries related to the confining potential are lifted as states with a positive (negative) angular momentum increase (decrease) in energy. States with the same n quantum number gradually approach one another forming consecutive Landau levels. In general, the wave functions describing the single particle states are complex, involving several radial nodes. The $\nu = 2$ singlet state and $\nu = 1$ spin polarized states are exceptions involving the occupation of simple ring orbitals with the ring radius increasing with the energy of the orbital (see Fig. 3.1(c) in section 2). The highest occupied orbital lies at the edge of the dot.

Figure 3.15 illustrates the occupation of orbitals for odd and even electron numbers at $\nu = 2$ phase. The energetically lower orbitals are occupied in pairs. In a simple picture, these play the role of "spectator" background states and the electrons occupying them, depicted in Fig. 3.15 by empty arrows, may be treated as spectator electrons. Electrons occupying the highest orbital are depicted by full black arrows. For an odd number of electrons the highest orbital contains a single spin-down electron at the edge, which determines the total spin of the dot as $S = \frac{1}{2}$. For even number of electrons this orbital

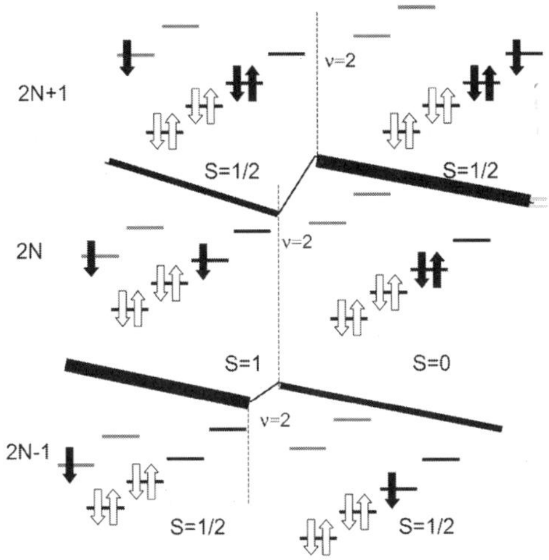

Fig. 3.15. Electronic configurations of ground states in the close vicinity of the $\nu = 2$ phase in quantum dots containing an odd and even number of electrons. Empty arrows indicate "spectator" electrons occupying the energetically lowest orbitals. Shown are the schematics of two adjacent Coulomb blockade peaks. The thickness of the lines in the schematic indicates the expected amplitude of particular CB peak (thicker lines—higher amplitude).

contains two electrons with different spin, and total spin of the dot is $S = 0$—the dot is in a singlet state. We can then see that increasing the number of electrons in a dot in the $\nu = 2$ phase makes the total spin of the dot oscillate between 0 and $\frac{1}{2}$ [38].

Also shown in Fig. 3.15 are the ground states for the first transition as the magnetic field is lowered. This involves the transfer of the electron to the lowest orbital of the second Landau level $|m = 0, n = 1\rangle$. It is important to note, that this orbital is located near the center of the dot (see section 2). Because of that we refer to configurations in which that orbital is occupied as "center" configurations, as opposed to $\nu = 2$ "edge" configurations. For an even number of electrons transfer of the electron from the edge to the center involves a spin flip and as result triplet ($S = 1$) configurations are formed with a triplet state formed by a pair of electrons occupying orbitals belonging to different Landau levels.

The relative simplicity of electronic configurations of ground states in close vicinity of $\nu = 2$ makes possible to compare the experimental results directly with theoretical predictions. This makes this region very attractive for study of spin effects in quantum dots. Typical Coulomb blockade (CB) spectroscopy data taken in this regime, showing four CB peaks, is presented in Fig. 3.16. Oscillations are clearly seen, both in position of CB peaks as well as in the conductance. Using ground state configurations from Fig. 3.15 and evaluating their energy using a Hartree–Fock approximation, we can now predict what we should observe as a function of magnetic field in both Coulomb and spin blockade spectroscopy and compare it with experimental results. The chemical potential step at the onset of the $\nu = 2$ phase boundary is the difference in energy between the ground states for two successive electron occupation numbers calculated

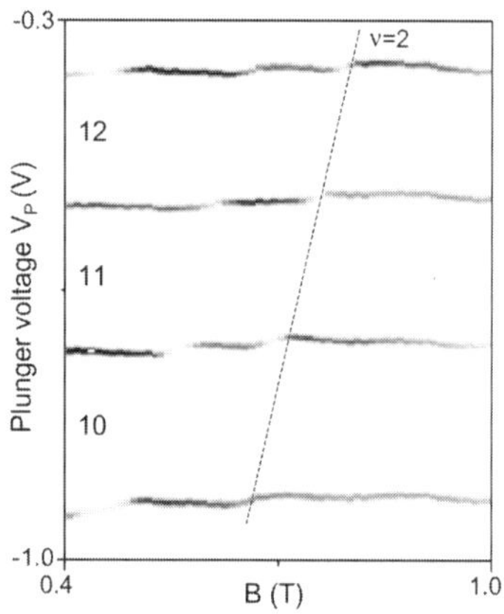

Fig. 3.16. The typical Coulomb blockade spectrum of a lateral quantum dot with a small number of electrons (shown in the picture) in the vicinity of the $\nu = 2$ phase. Plotted is a conductance of the dot. Dark (light) color indicates (high) low amplitude.

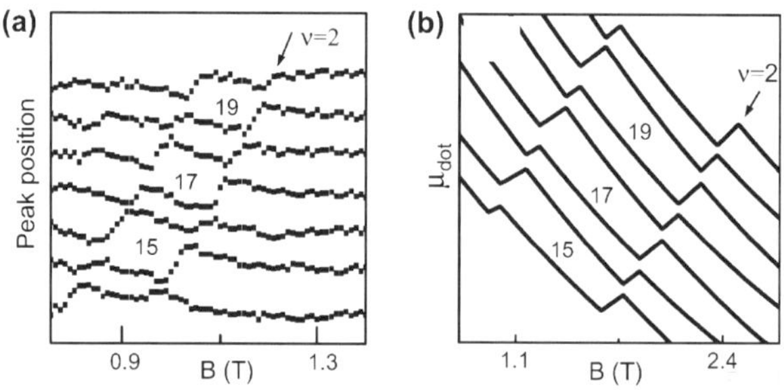

Fig. 3.17. The measured (a) and calculated (b) addition spectrum in the vicinity of $\nu = 2$ phase of a quantum dot containing between 14 and 20 electrons. Adapted from [38].

at fields before and after the boundary. This results in an odd–even effect in the height of that step observed in Coulomb blockade spectroscopy (i.e. peak position) as seen in results of calculations and experiment in Fig. 3.17. The effect has been analyzed in detail for both vertical [41] and lateral quantum dots [28,38]. It also, however, leads to a dramatic odd–even effect in spin blockade (i.e. peak amplitude) [28,38]. At $\nu = 2$, e.g. a comparison of the ground states in Fig. 3.15 indicates that electronic configurations for two successive electron numbers differ by a spin-down (up) electron at the edge of the dot when going from an even (odd) to an odd (even) number of electrons. One would then expect a large (small) current amplitude at a given CB peak, as indicated by thick (thin) lines in Fig. 3.15. A similar analysis can be done for center configurations and the expected behavior is also indicated in Fig. 3.15.

Along the vertical step at the $\nu = 2$ boundary, ground states for two successive number of electrons always differ by an electron on the center orbital. Since these are lateral quantum dots this always leads to a low current amplitude due to a small overlap

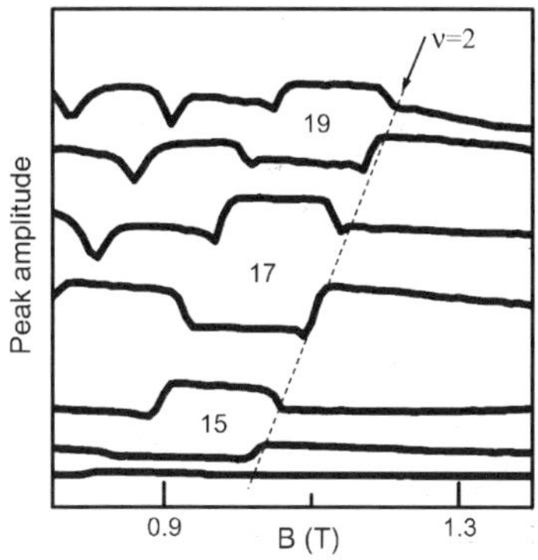

Fig. 3.18. The amplitude spectrum of the same peaks as in Fig. 3.17(a). Adapted from [38].

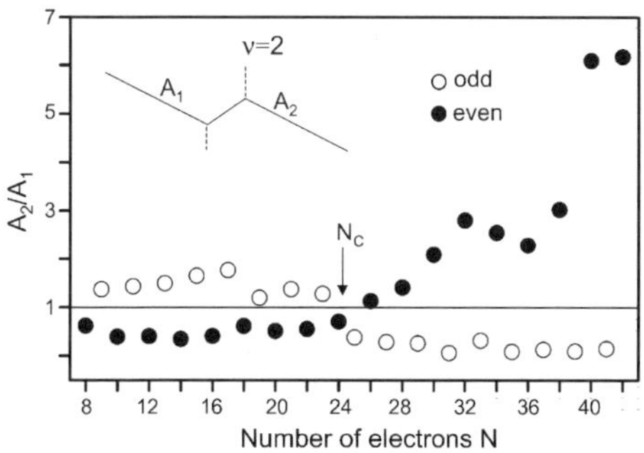

Fig. 3.19. Current readout of the collapse of the $\nu = 2$ spin singlet phase above the critical number of electrons N_C for an even number of electrons. Plotted is the ratio of the Coulomb blockade peak amplitudes on the right (A_2) and the left (A_1) side of the $\nu = 2$ boundary. Adapted from [42].

between the quantum dot and lead states (spatial blockade). Figure 3.18 confirms the expected pattern of amplitudes near the $\nu = 2$ phase boundary.

It is found, however, using spin blockade spectroscopy, that the behavior discussed above breaks down above a critical electron occupation number [42]. Figure 3.19 plots the ratio of peak amplitude before and after entering the $\nu = 2$ phase. Basing on our previous discussion we expect this ratio to oscillate between values smaller and bigger than one as a function of electron number N. While this rule is obeyed for low occupation numbers, it can be seen that at a critical number of electrons N_c (which differs slightly from dot to dot) there is a phase change in the plot. The origin of this phase change was found to be an interaction driven quenching of the $\nu = 2$ spin singlet phase [42].

Figure 3.20 illustrates the new ground states above N_c, found in calculations based on Local Spin Density Appoximation (LSDA) theory [11]. A comparison of the new ground states with those in Fig. 3.15 reveals that the new states are spatially more spread out, which lowers the Coulomb energy of the system. One can notice immediately that there is no longer a $\nu = 2$ spin singlet ($S = 0$) ground state for even electron

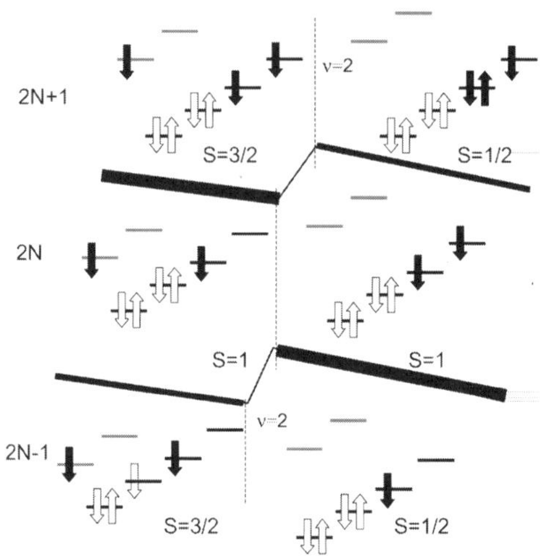

Fig. 3.20. Electronic configurations of ground states in the close vicinity of the $\nu = 2$ phase in quantum dots containing an odd and even number of electrons, larger than a critical value N_C. Comparing to Fig. 3.15, note the lack of the spin singlet phase ($S = 0$) for an even number of electrons $2N$ and an appearance of center configurations with spin-polarized edge ($S = \frac{3}{2}$) for an odd number of electrons.

numbers. The center triplet configuration is replaced as a ground state by the edge triplet configuration when the magnetic field is increased. Please note that this edge triplet configuration, with a triplet formed by a pair of electrons in the lowest Landau level, is the high electron number analogue of the triplet state for two-electron system discussed in section 6. One can see also the presence of new center configurations for an odd-electron dot, characterized by a spin-polarized edge ($S = \frac{3}{2}$). Comparing the relevant ground states in the same way we did for those from Fig. 3.15 gives us the expected amplitude pattern in spin blockade spectroscopy, as indicated in Fig. 3.20 and shown in experimental data in Fig. 3.19.

Figure 3.21 illustrates a simple way of viewing the quenching of the $\nu = 2$ state. Figure 3.21(a) plots the simple Fock–Darwin spectrum, but with an exaggerated spin splitting comparing to that derived from the Zeeman term. This is justified by the experimental observation that spin flips are governed by the Coulomb interactions (otherwise they would occur at much higher magnetic fields). The first spin flip for odd and even electron occupation numbers, that is the first crossing between orbitals within the lowest Landau level, is marked in Fig. 3.21(a). It is important to note that the first spin flip line for even electron number crosses the $\nu = 2$ line above a critical occupation number. Figure 3.21(b) plots the amplitude of the Coulomb blockade peaks, with the amplitude of the Nth (for $N > 1$) peak normalized by dividing by the amplitude of the first peak. The first spin flips for both an even and odd electron occupation numbers are identified as an increase in the amplitude of the CB peak [13], as shown in the inset in Fig. 3.21(b) (for a more detailed description see section 8). Their magnetic field positions for different N are marked in Fig. 3.21(b) and reveal a remarkable similarity to Fig. 3.21(a) with the first spin flip line for even N clearly crossing the $\nu = 2$ line

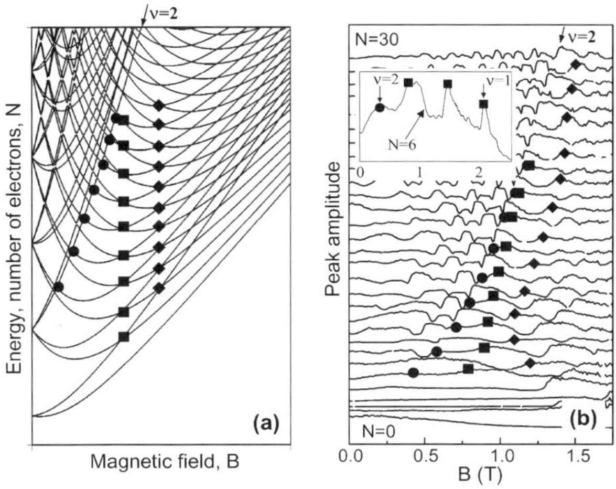

Fig. 3.21. Collapse of the $\nu = 2$ spin singlet phase as seen in the Fock–Darwin spectrum with an exaggerated spin splitting (a) and in the amplitude spectrum of CB peaks (b). Circles mark the beginning of the $\nu = 2$ spin singlet phase, squares and diamonds mark the first spin flip for an even and odd number of electrons, respectively. The amplitude plotted in (b) was normalized for $N > 1$ by dividing the amplitude of Nth peak by the amplitude of $N = 1$ peak to take into account the effect of magnetic field on tunneling barriers. The spin flips were identified as an increase in the amplitude, as shown in the inset for the $N = 6$ peak.

above a critical electron occupation number. This critical electron number is similar to that in Fig. 3.19 providing a simple explanation for the quenching of the $\nu = 2$ state. The simple arguments for the finite stability of the spin singlet phase presented above are supported by a number of more sophisticated calculations described in detail in [11].

Consider one such calculation of the $\nu = 2$ ground state, that involves the Hartree–Fock (HF) approximation including the mixing of two Landau levels [11,50]. From the calculated HF $\nu = 2$ state the spin flip excitations at the edge (positive angular momentum) and to the center (negative angular momentum inter-Landau level transitions) of the dot are calculated. An example of results of such calculations for $N = 12$ electrons, $\omega = 6$ meV and $B = 3$ T are shown in Fig. 3.22. Black dots represent the energy of HF quasiparticles, both occupied states of the lowest Landau level (below the Fermi level) and empty states of the first and second Landau levels (levels above the Fermi level). The bars represent the energy of collective spin flip excitations. For each angular momentum ΔL there are many possible excitations. They mix and interact with each other and the final result is shown in Fig. 3.22. These excitations represent proper excitations of the HF ground state and their energy is significantly lowered in comparison with HF quasiparticle excitations. When the energy of the excitation at the edge is negative, the HF spin singlet droplet is unstable against a spin flip at the edge. When the energy of the spin flip excitation to the center is negative, the droplet is unstable against transfer of electron to the second Landau level orbital in the center of the dot, accompanied by a spin flip. These two transitions determine the stability of the spin singlet droplet.

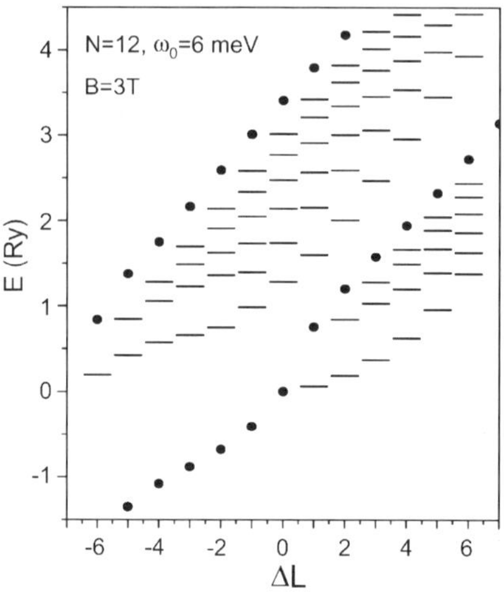

Fig. 3.22. The results of Hartree–Fock (HF) calculations of the $\nu = 2$ ground state, with the mixing of two Landau levels, for $N = 12$ electrons. Black dots represent the energy of HF quasiparticles and bars represent the energy of collective spin flip excitations. For details see text.

We can then use this computational tool and the experimentally determined magnetic field values at which the $\nu = 2$ droplet becomes unstable, to extract useful characteristic parameters of the dot. In particular, spin blockade spectroscopy provides the values of the magnetic field at which the first spin flip occurs at the edge. The addition of the first electron [22,42] gives us the confining energy of the empty dot, and one finds $\omega(N = 1) = 1$ meV. Spin blockade spectroscopy allows us to determine precisely the value of the magnetic field B_{2N} at which the first spin flip in the even electron droplet occurs. By assuming that the confining energy for two electrons has not changed from the one-electron value, i.e. $\omega(N = 2) = 1$ meV, and fitting the calculated value of B_{2N} of the first spin flip (singlet–triplet transition) for $N = 2$ to the experimental value, one can extract the strength of Coulomb interactions $\alpha = 0.22$ (we introduce an effective Coulomb interaction $\alpha e^2/\varepsilon r$). Keeping $\alpha = 0.22$ we extract $\omega(N)$ by fitting the experimental values B_{2N} of the first spin flip at the edge. The experimental and calculated values of the magnetic field corresponding to the spin flip at the edge B_{2N}, and the experimental and calculated values corresponding to the spin flip to the center orbital (the $\nu = 2$ line) are shown in Fig. 3.23. We see that there is a finite stability range for the $\nu = 2$ droplet, which decreases with increasing electron number. At a critical electron number $Nc \sim 24$, $\nu = 2$ droplet becomes unstable. This critical number is not computed well by our theoretical approach. This is because while the spin flip at the edge was fitted successfully, the $\nu = 2$ line was not. In principle, we should be able to extract both the confining energy and strength of Coulomb interactions. Pure Coulomb interactions do not allow us, however, to do so and different forms of interactions might be more successful in the future. It is still useful, however, to examine the extracted dependence of ω on N. The values of $\omega(N)$ as a function of the number of electrons N is shown in the inset. It starts at $\omega(1) = 1.0$ meV for

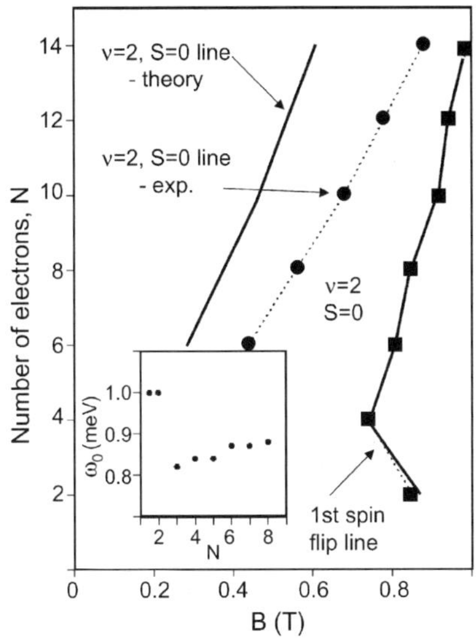

Fig. 3.23. The experimental (symbols) and calculated (solid lines) values of magnetic field corresponding to the spin flip to the "center" of the dot ($\nu = 2$ boundary) and the first spin flip at the edge of the dot. The inset plots the dependence of confining potential ω_0 on number of confined electrons N as calculated basing on the experimental results. For detailed discussion see text.

$N = 12$, drops to $\omega(4) = 0.80$ meV for $N = 4$, and then continues increasing with the number of electrons.

The above discussion concerns the finite stability of the spin singlet state, that is confirmed both experimentally and by a variety of computational methods. It is clear from the results that more work is needed to develop a *quantitative* microscopic understanding of electrons in quantum dots.

In the next section we discuss in more details the high magnetic field breakdown of the $\nu = 2$ state and its evolution toward the spin-polarized $\nu = 1$ state.

8. The Spin Flip Regime

The spin flip regime brings about the spin polarization of the quantum dot extending from $\nu = 2$ to $\nu = 1$, as a function of magnetic field. This regime has also been studied in vertical quantum dots [43]. At $\nu = 2$ the total spin S of the quantum dot is zero for even N, while at $\nu = 1$, also referred to as the maximum density droplet (MDD) regime [44], the total spin of the spin-polarized quantum dot containing N electrons is $N/2$. At still higher fields further higher angular momentum states involving both charge and spin rearrangements (reconstructions) come into play [44]. These are beyond the scope of this article and indeed much work still remains to be done to understand them.

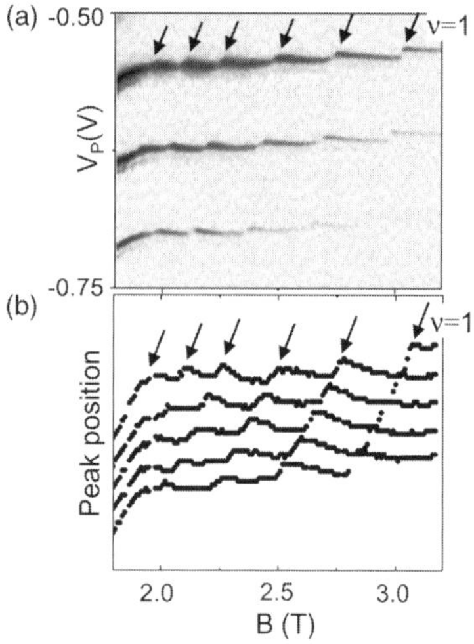

Fig. 3.24. (a) Successive CB peaks as a function of the magnetic field between $\nu = 2$ and 1; (b) position of the CB peaks with the charging energy removed. Arrows mark spin flips. Adapted from [45].

Figure 3.24(a) shows typical spin flip data close to the $\nu = 1$ regime. The spin flips are identified by a series of steps. The amplitude of the Coulomb blockade peaks drops during the upward part of the step and then increases again at magnetic field value corresponding to a spin flip transition at N-electron dot ground state. Figure 3.24(b) confirms that the features are continuous from one Coulomb blockade peak to the next—upward cusps on one peak become downward cusps on the peak below. Such a pattern was first observed by McEuen et al. [19] for quantum dots containing many electrons. While the steps themselves become more difficult to observe close to the $\nu = 2$ line, the amplitude signature may be used to count all of the steps between $\nu = 2$ and $\nu = 1$. For a dot containing less than 20 electrons we find that for a $2N$-electron dot there are N spin flips (see Fig. 3.21(b) in section 7). While this seems very intuitive, detailed calculations and measurements reveal a much richer process that occurs at spin flips. Figure 3.25 plots the results of calculations for a dot containing 8 electrons for $\omega = 6$ meV [12,13]. The plot in the top panel shows the number of spin-up electrons in the system, changing when magnetic field is increased from the $\nu = 2$ to $\nu = 1$ regime, from Hartree–Fock (solid line) and exact diagonalization (dotted line) calculations. Also plotted, in the bottom panel, is the angular momentum of the dot. The Hartree–Fock calculation is consistent qualitatively with the simple picture discussed above. At each spin flip event a single electron is moved to a higher empty orbital, characterized by a higher angular momentum, and its spin is flipped. This occurs first for ground state with the lower number of electrons (each CB peak involves two ground states) and at a slightly higher magnetic field for the ground state of the higher electron number ground state. The step in the chemical potential occurs between these two transitions. The origin of the small amplitude during the step is due to both spatial and spin blockade. The temperature dependence of this amplitude [13], which reveals that it is a very low temperature effect, indicates, however, that this simple picture must be refined.

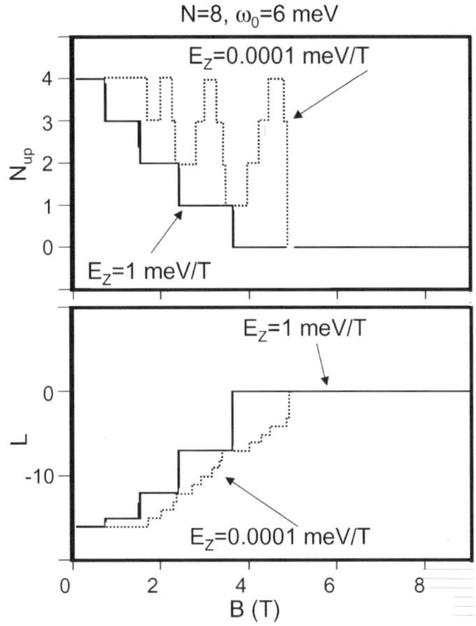

Fig. 3.25. The results of Hartree–Fock calculations (solid line) and exact diagonalization calculations with correlation effects included (dotted line) for the eight-electron dot. Plotted is the number of spin-up electrons in the dot (top) and total angular momentum L of the dot (bottom) against magnetic field B.

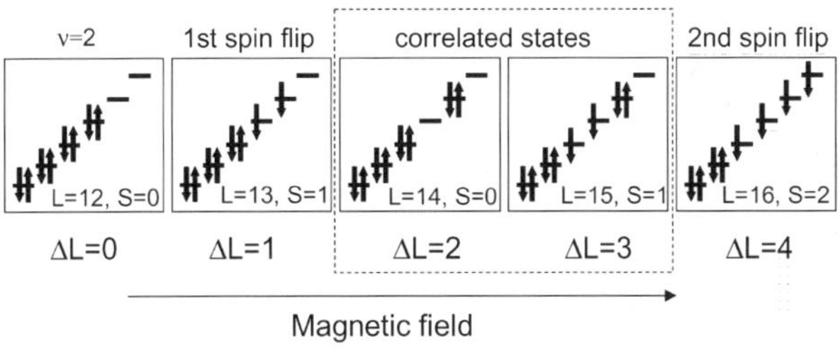

Fig. 3.26. Electronic configurations of the eight-electron dot ground states between the $\nu = 2$ phase and the second spin flip. The states are characterized by the total angular momentum L, total spin S and the change in the angular momentum ΔL. For details see text.

Exact diagonalization calculations, with included correlations, reveal a remarkable spin depolarization and polarization sequence at each spin flip [45,46]. To understand this further consider Fig. 3.26 which plots the configurations at the first two spin flips for the dot with $N = 8$ electrons. The spin polarization of the quantum dot involves a monotonic increase in the angular momentum L, when we move from the $\nu = 2$ ($L = 12, S = 0$) configuration by increasing the magnetic field. As can be seen, within the simple scenario, the first spin flip transition to the ($L = 13, S = 1$) configuration involves an increase in an angular momentum $\Delta L = 1$, when compared to the $\nu = 2$ configuration. The second spin flip transition, to ($L = 16, S = 2$)

state, involves $\Delta L = 4$. Note that no $\Delta L = 2$ or $\Delta L = 3$ transitions occurs within this picture. Exact diagonalization calculations, however, reveal that two other states briefly become ground states. These are $(L = 14, S = 0)$ and $(L = 15, S = 1)$ states with, respectively, $\Delta L = 2$ and $\Delta L = 3$. Figure 3.27(a) shows an experimental current stripe covering the first two spin flip events. At the first spin flip region we observe one transition, as expected. At the second spin flip, however, two "extra" states can be observed, as seen in the Figs 3.27(b), (c) and as predicted by the exact diagonalization calculations discussed above. The strength of a feature (ground and excited state) in the Coulomb blockade stripe or Coulomb blockade diamond is related to the overlap of the final and initial states. When one is dealing with complicated correlated states, such as the $(L = 15, S = 1)$ state discussed above, this becomes non-trivial and leads to a complicated interplay between ground and excited states. The $(L = 15, S = 1)$ state, for example, becomes visible only when the first excited state of the $N = 7$ electron quantum dot enters the tunneling window, since its overlap with this state is large while the overlap with the $N = 7$ ground state is small. Such effects can lead to complicated current stripes, Coulomb diamonds and the observation of negative differential resistance regions (in section 9 we will describe how we can use spin to generate negative differential resistance "on demand"). Figure 3.28(a) shows a similar current stripe for 16 electrons at the last spin flip before the maximum density droplet,

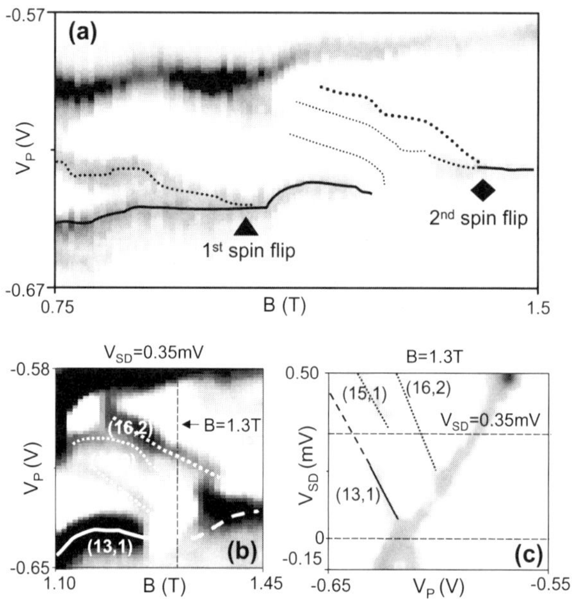

Fig. 3.27. (a) Experimental current stripe for an eight-electron dot in the region of the first two spin flips. Plotted is a differential conductance dI/dV_{SD} through the dot. Dotted lines indicate excited states, with two faint lines representing the correlated states whose configurations are shown in Fig. 3.26; (b) zoom of (a) in the vicinity of the second spin flip; (c) Coulomb diamond at a magnetic field between the first and the second spin flip transition. Numbers in brackets (L, S) represent a total angular momentum L and total spin S.

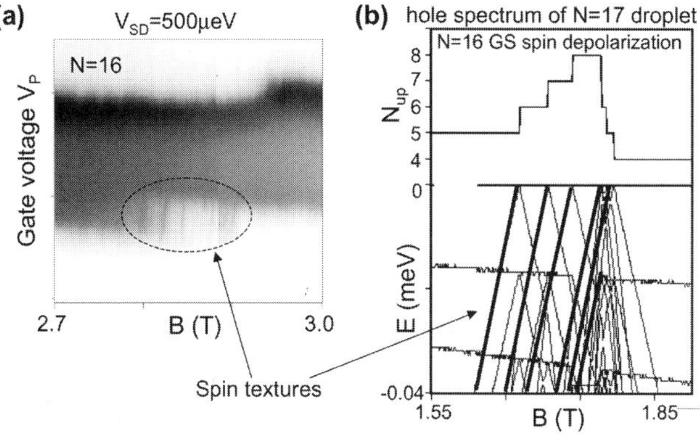

Fig. 3.28. (a) Experimental current stripe at the last spin flip before the maximum density droplet for $N = 16$ electrons in the dot. Plotted is the total current through the dot; (b) the magnetic field dependence of the number of spin-up electrons in the ground state of the $N = 16$ electron dot; (c) all possible energy levels of a particle removed from the $N = 17$ droplet. Adapted from [45].

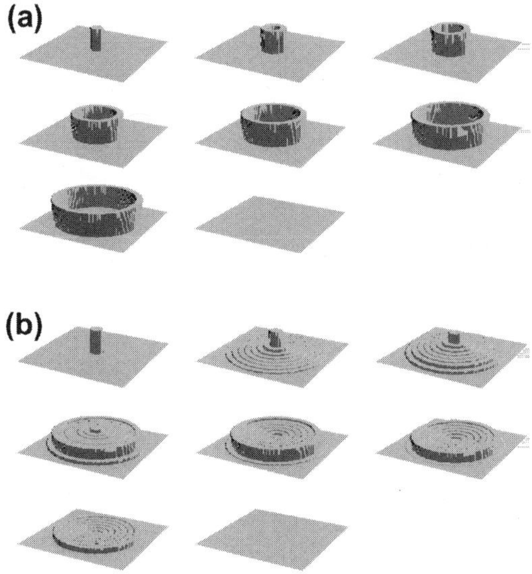

Fig. 3.29. The distribution function for spin-up electrons with increasing angular momentum at the last spin flip transition for non-interacting (a) and interacting (b) electrons. For details see text.

where even more excited states related to the spin depolarization, polarization features (see Fig. 3.28(b)) are visible.

The correlated states described above are sometimes referred to as spin textures. To visualize this it is useful to consider the last spin flip in more detail. The last spin flip corresponds to moving the electron from the center to the edge of the dot. This movement of a spin-up electron from center to edge is illustrated in Fig. 3.29

which shows the distribution function for spin-up electrons with increasing angular momentum L. For non-interacting electrons, these are rings with increasing radius. When the radius corresponds to the first unoccupied orbital, the spin-up electron can flip its spin and vanishes from this picture. The situation is drastically different for interacting electrons. When the spin-up electron is in-between the center and the edge of the droplet, it causes three additional electrons to flip their spin and the ground state is completely spin depolarized. For an insight into the wavefunctions of these spin textures see Oaknin *et al.* [47] and references therein. Experimentally, spin textures lead to the collapse of the Zeeman gap and anomalies in linear transport through quantum dots [13], as well as additional features in high source and drain experiments [45,46].

9. Negative Differential Resistance Achieved by Spin Blockade

This chapter has revealed many spin properties of few electron quantum dots. Can these properties be exploited? Potential applications as spin qubits have already been mentioned (see section 6). In this section we describe a second potential application—the use of spin to generate negative differential resistance (NDR) [48] "on demand." As described in section 3, Coulomb blockade peak observed in linear measurements, with small source-drain bias, involve electrons repeatedly entering and leaving the dot—one electron at a time. As the voltage across the quantum dot is increased excited states enter the transport source-drain "window" eV_{SD} and the non-linear regime of measurements is reached. Usually an increase in a density of states leads to an increase in total current through the system. In the Coulomb blockade driven process, however, there is an important complication—no new electron can enter the quantum dot until the previous one has left. If one adds an excited state into the transport window which has (i) a finite probability of being occupied but (ii) which has a small overlap with the exit lead, the current will drop and not increase. Spin-polarized leads are a good way to achieve condition (ii) and to get a control over the appearance of NDR in the non-linear CB spectra of quantum dots [29].

Figure 3.30 shows the configuration of interest. The spin-polarized leads provide not only spin-polarized injection but, more importantly for this section, they also result in spin-polarized detection. The spin-polarized leads are not 100% efficient but can be modeled as having tunneling barriers with different widths for the two spin species (see section 4). Consider the situation of Fig. 3.30(a). Figure shows the schematic of our dot device with spin resolved source (S) and drain (D) leads. The spin-down (up) electrons tunnel from source to quantum dot and then to drain via thin (thick) tunneling barriers. The discrete quantum dot levels are assigned spin, depicted by horizontal solid (spin down) and dotted (spin up) lines. The top panel shows the tunneling process at a low source-drain bias—a spin-down electron in the source can tunnel into the unoccupied spin-down state of the dot. The spin-down electron can now leave the dot to the spin-down polarized drain, resulting in a high current, depicted in the picture with thick solid arrows indicating a tunneling process. The bottom panel shows an additional tunneling process available at high V_{SD}. When V_{SD} is high enough, an excited, spin-up state, becomes available for tunneling. There exists a small, but finite, probability that the incoming electron will occupy this state, entering the dot through

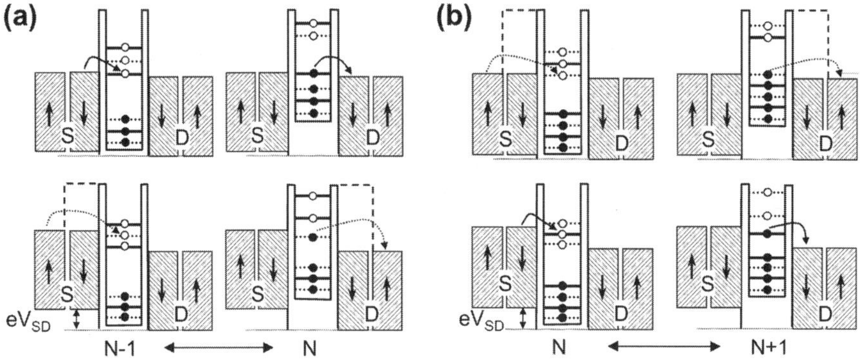

Fig. 3.30. (a) Schematics of the transport of the Nth electron tunneling through the quantum dot containing $N - 1$ electrons. Top panel: the non-blockaded process in the linear regime; bottom panel: spin-blockaded NDR mechanism in the non-linear regime. (b) Schematics of the transport of the $N+1$ electron tunneling through the quantum dot containing N electrons. Top panel: spin-blockaded process in the linear regime; bottom panel: non-blockaded PDR mechanism in the non-linear regime. Horizontal lines indicate occupied (•) and unoccupied (∘) spin-down (solid lines) and spin-up (dotted lines) levels; thick solid and thin dotted arrows show, respectively, high and low tunneling rates of the tunneling process; thick dashed lines indicate thick barrier for spin-up electrons entering/leaving the dot. Adapted from [29].

a thicker barrier (shown by dashed line). To exit the dot this electron must then flip its spin or tunnel out to the spin-up exit leads, again through a thick tunneling barrier. If the spin relaxation time in the dot is sufficiently long, the spin-up electron remains in the dot and blocks the spin-down electron channel. Increasing the bias voltage should then result in a decrease of current through the dot, i.e. in observation of NDR. The effect obviously depends on the arrangement of spin levels in the dot, that can be modified by adding an extra electron. Fig. 3.30(b) shows the schematics of the same tunneling process for $N + 1$ electron flowing through the dot already containing N electrons. Now, at low V_{SD}, the current has to flow through a spin-up channel, so its amplitude is low. Increasing V_{SD} allows spin-down electrons to tunnel through a new spin-down level of the dot. This leads to increase of the current, and positive differential resistance (PDR).

As a result we are able [29] to find a direct correlation between the ground and excited states of a dot and the presence of NDR and PDR regions. This is illustrated in Fig. 3.31. Figure 3.31(a) shows typical experimental data taken in the linear regime, illustrating the magnetic field evolution of a CB peak in the representative magnetic field range $B = 0.9$–1.3 T. The dark (light) color represents high (low) peak amplitudes. The figure shows details of the very clear amplitude modulation extending up to the boundary of $\nu = 2$ phase in the dot, discussed in detail in the previous sections. Based on our previous arguments (see section 7) this amplitude modulation can be related to spin effects occurring in quantum dots and can be explained in terms of our spin injection/detection picture.

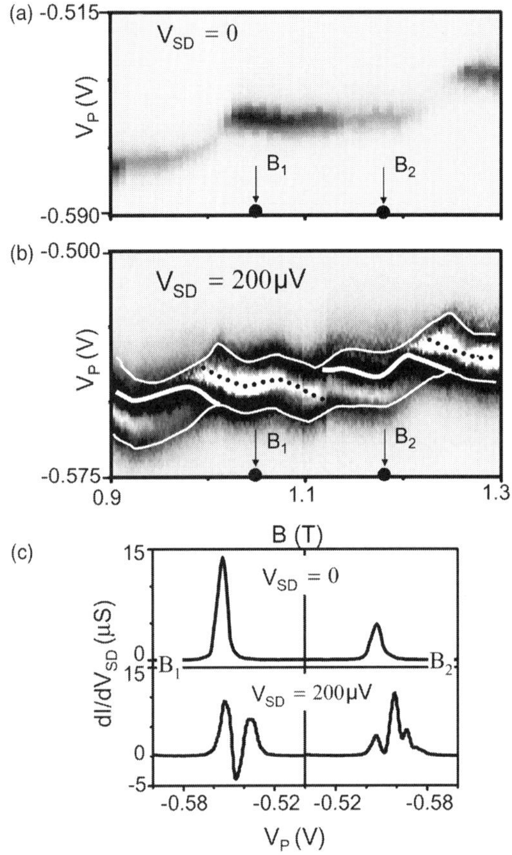

Fig. 3.31. (a) Inverted grayscale of dI/dV_{SD} showing amplitude modulation of the CB peak as a function of the plunger voltage V_P and the magnetic field B from linear transport measurements; (b) the same peak as in (a) but in the non-linear regime. The white color in the grayscale and dotted lines in the schematics indicate regions of negative differential resistance (NDR); (c) plunger voltage sweeps at two different values of B. Adapted from [29].

Figure 3.31(b) shows the current stripe resulting from the non-linear measurement of the same CB peak. A source-drain voltage $V_{SD} = 200\,\mu V$ was applied to include the first excited state in the transport window eV_{SD}. The white color in the grayscale represents the region where current decreased with increasing voltage, i.e. the NDR region. The schematic drawing is superimposed on the grayscale to show more clearly the structure of the current stripe. Thick lines seen between both edges represent the evolution of the first excited state of N electron system, with dotted lines showing the NDR regions. We can clearly see that NDR regions are observed in the field range where the high amplitude of CB peak is measured in the linear regime. This behavior is in very good agreement with spin polarized/injection detection picture and confirms that the spin blockade mechanism responsible for amplitude modulation in the linear regime also leads to NDR behavior in non-linear measurements.

One can extend this by moving to positions in the phase diagram of a few electron quantum dot where the addition of a single electron leads in a periodical way to accompanying reverses in the sign of differential resistance, i.e. three terminal NDR–PDR (positive differential resistance) devices [29]. The regime close to $\nu = 2$ fulfils these criteria. By combining the two electron's degrees of freedom, i.e. spin and charge it is hoped that spintronics will lead to new functionalities and make novel paradigms feasible. While the current device requires low temperatures and an applied magnetic field the three terminal device described here is an indication of the type of unconventional device that nano-spintronics makes possible.

10. Conclusions

The device described in this chapter is quite unique: a few electron quantum dot with spin-polarized leads. The current through the dot is dependent on the spin state of the quantum dot. The spin state of the dot can be tuned in three ways (i) varying the magnetic field at fixed electron number and confining potential (section 6) (ii) adding single electrons to the quantum dot at fixed magnetic field and confining potential (e.g. section 7) or (iii) by varying the confining potential at fixed electron number and magnetic field (section 6). In this way the device behaves as a prototype spin transistor. While many material issues need to be solved to generate similar devices which would work at room temperature and zero magnetic field, this chapter illustrates the first application of such a spin transistor—as a powerful probe for studying the spin properties of interacting few-electron systems. When combined with the traditional tools of Coulomb blockade spectroscopy and detailed theoretical studies this has led to the observation and explanation of the spin properties of few electron quantum dots and has illustrated the potential of nano-spintronics for future applications in quantum information and quantum electronics.

At even higher magnetic fields other novel spin related phenomena, not discussed in this chapter, are predicted to occur such as spin and charge reconstructions. In our devices amplitude modulation and chemical potential shifts occur all the way up to our highest available magnetic fields $B = 18$T. While some preliminary results are reported [44,49] the detailed role of reconstructions and the fractional quantum Hall effect remain to be studied.

Acknowledgments

This work summarizes the work over several years of many experimentalists and theoreticians working in our group. On the experimental side we would like to acknowledge the contributions of Charles Gould, Michel Pioro-Ladriere, Piotr Zawadzki, Yan Feng, Jean Lapointe and Zbigniew Wasilewski. Valuable theoretical contributions by Marek Korkusiński, Andreas Wensauer and Jordan Kyriakidis are also acknowledged. We would also like to acknowledge funding from National Research Council of Canada, the National Scientific and Engineering Research Council of Canada and the Canadian Institute for Advanced Research and a NATO fellowship.

References

1. S. Wolf, D.D. Awschalom, R.A. Buhrman, J.M. Daughton, S. von Molnár, M.L. Roukes, A.Y. Chytelkanova, and D.M. Treger, Spintronics: A spin-based electronics vision for the future, *Science* **294**, 1488–1495 (2001).
2. M.F. Crommie, C.P. Lutz, and D.M. Eigler, Imaging standing waves in a two-dimensional electron gas, *Nature* **363**, 524–527 (1993).
3. P.M. Petroff and S.P. DenBaars, MBE and MOCVD growth and properties of self-assembling quantum dot arrays in III-V semiconductor structures, *Superlatt. Microstruct.* **15**(1), 15–21 (1994).
4. J. Lefebvre, P.J. Poole, J. Fraser, G.C. Aers, D. Chithrani, and R.L. Williams, Self-assembled InAs quantum dots on InP nano-templates, *J. Cryst. Growth* **234**(2–3), 391–398 (2002).
5. T.J. Thorton, M. Pepper, H. Ahmed, D. Andrews and G.J. Davies, One-dimensional conduction in the 2D electron gas of a GaAs-AlGaAs heterojunction, *Phys. Rev. Lett.* **56**(11), 1198–1201 (1986).
6. P. Hawrylak, Single-electron capacitance spectroscopy of few-electron artificial atoms in a magnetic field: Theory and experiment, *Phys. Rev. Lett.* **71**(20), 3347–3350 (1993).
7. L. Jacak, P. Hawrylak, and A. Wojs, *Quantum Dots*. Springer-Verlag, Berlin (1997).
8. J. Kyriakidis, M. Pioro-Ladriere, M. Ciorga, A.S. Sachrajda, and P. Hawrylak, Voltage-tunable singlet–triplet transition in lateral quanum dots, *Phys. Rev.* B **66**(3), 35320–35327 (2002).
9. D.J. Lockwood, P. Hawrylak, P.D. Wang, C.M. Sotomayor Torres, A. Pinczuk, and B.S. Dennis, Shell structure and electronic excitations of quantum dots in a magnetic field probed by inelastic light scattering, *Phys. Rev. Lett.* **77**(2), 354–357 (1996).
10. P. Hawrylak, A. Wojs, and J.A. Brum, Magnetoexcitons and correlated electrons in quantum dots in a magnetic field, *Phys. Rev.* B **54**(16), 11397–11409 (1996).
11. A. Wensauer, M. Korkusiński, and P. Hawrylak, Theory of the spin singlet filling factor $\nu = 2$ quantum Hall droplet, *Phys. Rev.* B **67**, 035325 (2003)
12. A. Wojs and P. Hawrylak, Spectral functions of quantum dots in the integer and fractional quantum Hall regime, *Phys. Rev.* B, **56**(20), 13227–13234 (1997).
13. P. Hawrylak, C. Gould, A.S. Sachrajda, Y. Feng, and Z. Wasilewski, Collapse of the Zeeman gap in quantum dots due to electronic correlations, *Phys. Rev.* B **59**(4), 2801–2806 (1999).
14. B.J. van Wees, L.P. Kouwenhoven, C.J.P.M. Harmans, J.G. Williamson, C.E. Timmering, M.E.I. Broekaart, C.T. Foxon, and J.J. Harris, Observation of zero-dimensional states in a one-dimensional electron interferometer, *Phys. Rev. Lett.* **62**(21), 2523–2526 (1989); R.P. Taylor, A.S. Sachrajda, P. Zawadzki, P.T. Coleridge, and J.A. Adams, Aharonov-Bohm oscillations in the Coulomb blockade regime, *Phys. Rev. Lett.* **69**(13), 1989–1992 (1992).
15. G. Kirczenow, A.S. Sachrajda, Y. Feng, R.P. Taylor, L. Henning, J. Wang, P. Zawadzki, and P.T. Coleridge, Artifical impurties in quantum wires: From classical to quantum behaviour, *Phys. Rev. Lett.* **72**(13), 2069–2072 (1994); C.J.B. Ford, P.J. Simpson, I. Zailer, D.R. Mace, M. Yosefin, M. Pepper, D.A. Ritchie, J.E.F. Frost, M.P. Grimshaw, and G.A.C. Jones, Charging and double-frequency Aharonov-Bohm effects in an open system, *Phys. Rev.* B **49**(24), 17456–17459 (1994).
16. D.A. Wharam, T.J. Thornton, R. Newbury, M. Pepper, H. Ahmed, J.E.F. Frost, D.G. Hasko, D.C. Peacock, D.A. Ritchie, and G.A.C. Jones, One-dimensional transport and the quantisation of the ballistic resistance, *J. Phys.* C **21**(8), L209–L214 (1988); B.J. van Wees, H. van Houten, C.W.J. Beenakker, J.G. Williamson, L.P. Kouwenhoven, D. van der Marel, and C.T. Foxon, Quantized conductance of point contacts in a two-dimensional electron gas, *Phys. Rev. Lett.* **60**(9), 848–850 (1988).

17. J.H.F. Scott-Thomas, S.B. Field, M.A. Kastner, H.I. Smith, and D.A. Antoniadis, Conductance oscillations periodic in the density of a one-dimensional electron gas, *Phys. Rev. Lett.* **62**(5), 583–586 (1989).
18. D. Gammon, E.S. Snow, B.V. Shanabrook, D.S. Katzer, and D. Park, Homogenous linewidths in the optical spectrum of a single gallium arsenide quantum dot, *Science* **273**, 87–90 (1996); A. Kuther, M. Bayer, A. Forchel, A. Gorbumov, V.B. Timofeev, F. Schafer, and J.P. Reithhmeier, Zeeman splitting of excitons and biexcitons in single $In_{0.60}Ga_{0.40}As/GaAs$ self assembled quantum dots, *Phys. Rev.* B **58**(12), R7508–R7511 (1998).
19. P.L. McEuen, E.B. Foxman, U. Meirav, M.A. Kastner, Y. Meir, and N.S. Wingreen, Transport spectroscopy of a Coulomb island in the quantum Hall regime, *Phys. Rev. Lett.* **66**(14), 1926–1929 (1991); P.L. McEuen, E.B. Foxman, J. Kinaret, U. Meirav, M.A. Kastner, N.S. Wingreen, and S.J. Wind, Self-consistent addition spectrum of a Coulomb island in the quantum Hall regime, *Phys. Rev.* B **45**(19), 11419–11422 (1992).
20. L.P. Kouwenhoven, C.M. Marcus, P.L. McEuen, S. Tarucha, R.M. Westervelt, and N.S. Wingreen, Electron transport in quantum dots, in *Mesoscopic electron transport, NATO ASI Conference Proceedings* (eds. L.L. Sohn, L.P. Kouwenhoven, and G. Schön), Kluwer Academic, Dordrecht, (1997) (Series E 345).
21. L.P. Kouwenhoven, D.G. Austing, and S. Tarucha, Few-electron quantum dots, *Rep. Prog. Phys* **64**(6), 701–731 (2001).
22. M. Ciorga, A.S. Sachrajda, P. Hawrylak, C. Gould, P. Zawadzki, S. Jullian, Y. Feng, and Z. Wasilewski, Addition spectrum of a lateral dot from Coulomb and spin-blockade spectroscopy, *Phys. Rev.* B **61**(24), R16315–R16318 (2000).
23. D. Sprinzak, Y. Ji, M. Heiblum, D. Mahalu, and H. Shtrikman, Charge distribution in a Kondo-correlated quantum dot, *Phys. Rev. Lett.* **88**(17), 176805–176808 (2002).
24. R. Fiederling, M. Keim, G. Reuscher, W. Ossau, G. Schnidt, A. Waag, and L.W. Molenkamp, Injection and detection of a spin-polarized current in a light emitting diode, *Nature* **402**, 787–790 (1999)
25. Y. Ohno, D.K. Young, F. Matsukura, H. Ohno, and D.D. Awschalom, Electrical spin injection in a ferromagnetic semiconductor heterostructure, *Nature* **402**, 790–792 (1999).
26. P.R. Hammar and M. Johnson, Detection of spin-polarized electrons injected into a two-dimensional electron gas, *Phys. Rev. Lett.* **88**(6), 066806–066809 (2002).
27. D.B. Chklovskii, B.I. Shklovskii, and L.I. Glazman, Electrostatics of edge channels, *Phys. Rev.* B **46**(7), 4026–4034 (1992); R. Haug, Edge-state transport and its experimental consequences in high magnetic fields, *Semicon. Sci. Tech.* **8**, 131–153 (1993).
28. A.S. Sachrajda, P. Hawrylak, M. Ciorga, C. Gould, and P. Zawadzki, Spin polarized injection into a quantum dot by means of the spatial separation of spins, *Physica* E **10**, 493–498 (2001).
29. M. Ciorga, M. Pioro-Ladriere, P. Zawadzki, P. Hawrylak, and A.S. Sachrajda, Tunable negative differential resistance controlled by spin blockade in single-electron transistors, *Appl. Phys. Lett.* **80**(12), 2177–2179 (2002).
30. M. Pioro-Ladriere, Master Thesis, *Étude par spectroscopie de Coulomb d'un boîte quantique latérale contenant de 1 à 12 électrons*, University of Sherbrooke, Quebec, Canada (2002).
31. S. Tarucha, D.G. Austing, T. Honda, R.J. van der Hage, and L.P. Kouwenhoven, Shell filling and spin effects in a few electron quantum dot, *Phys. Rev. Lett.* **77**(17), 3613–3616 (1996).
32. R.C. Ashoori, H.L. Stormer, J.S. Weiner, L.N. Pfeiffer, K.W. Baldwin, and K.W. West, N-electron ground states of a quantum dot in magnetic field, *Phys. Rev. Lett.* **71**(4), 613–616 (1993).
33. J.A. Brum and P. Hawrylak, Coupled quantum dots as quantum exclusive-OR gate, *Superlatt. Microstruct.* **22**(3), 431–436 (1997).

34. D. Loss and D.P. DiVincenzo, Quantum computation with quantum dots, *Phys. Rev.* A **57**(1), 120–126 (1998).
35. T. Fujisawa, D.G. Austing, Y. Tokura, Y. Hirayama, and S. Tarucha, Allowed and forbidden transitions in artificial hydrogen and helium atoms, *Nature* **419**, 278–281 (2002).
36. T. Schmidt, M. Tewordt, R.H. Blick, R.J. Haug, D. Pfannkuche, K. v. Klitzing, A. Förster, and H. Lüth, Quantum-dot ground states in a magnetic field studied by single-electron tunneling spectroscopy on double-barrier heterostructures, *Phys. Rev.* B **51**(8), 5570–5573 (1995); B. Su, V.J. Goldman, and J.E. Cunningham, Single-electron tunneling in nanometer-scale double-barrier heterostructure devices, *Phys. Rev.* B **46**, 7644–7655 (1992).
37. W.G. van der Wiel, T.H. Oosterkamp, J.W. Janssen, L.P. Kouwenhoven, D.G. Austing, T. Honda, and S. Tarucha, Singlet-triplet transitions in a few electron quantum dot, *Physica* B **256–258**, 173–177 (1998).
38. M. Ciorga, A.S. Sachrajda, P. Hawrylak, C. Gould, P. Zawadzki, Y. Feng, and Z.Wasilewski, Readout of a single electron spin based quantum bit by current detection, *Physica* E **11** 35–40 (2001).
39. L.P. Kouwenhoven, T.H. Oosterkamp, M.W.S. Danoesastro, M. Eto, D.G. Austing, T. Honda, and S. Tarucha, Excitation spectra of circular, few-electron quantum dots, *Science* **278**, 1788–1792 (1997).
40. G. Burkard, H.A. Engel, and D. Loss, Spintronics and quantum dots for quantum computing and quantum communication, *Fortschritte der Physik* **48** (Special Issue on Experimental Proposals for Quantum Computation), 965–886 (2000).
41. S. Tarucha, D.G. Austing, Y. Tokura, W.G. van der Wiel, and L.P. Kouwenhoven, Direct Coulomb and exchange interaction in artificial atoms, *Phys. Rev. Lett.* **84**(11), 2485–2488 (2000).
42. M. Ciorga, A. Wensauer, M. Pioro-Ladriere, M. Korkusinski, J. Kyriakidis, A.S. Sachrajda, and P. Hawrylak, Collapse of the spin-singlet phase in quantum dots, *Phys. Rev. Lett.* **88**(25), 256804–256807 (2002).
43. D.G. Austing, Y. Tokura, T. Honda, S. Tarucha, M. Danoestastro, J. Janssen, T. Oosterkamp, and L. Kouwenhoven, Several- and many-electron artificial-atoms at filling factors between 2 and 1, *Jpn. J. Appl. Phys.* **38**(1), 372–375 (1999).
44. T.H. Oosterkamp, J.W. Janssen, L.P. Kouwenhoven, D.G. Austing, T. Honda, and S. Tarucha, Maximum-density droplet and charge redistributions in quantum dots at high magnetic fields, *Phys. Rev. Lett.* **82**(14), 2931–2934 (1999).
45. C. Gould, P. Hawrylak, A.S. Sachrajda, Y. Feng, P. Zawadzki, and Z. Wasilewski, Correlations effects in few-electron quantum dots between $\nu = 2$ and 1, *Physica* E **6**, 461–465 (2000).
46. M. Ciorga, M. Korkusinski, M. Pioro-Ladriere, P. Zawadzki, P. Hawrylak, and A.S. Sachrajda, Simple spin textures in a quantum dot, *Phys. Status Solidi*, (in press).
47. J.H. Oaknin, L. Martín–Moreno, and C. Tejedor, Skyrmions and edge-spin excitations in quantum Hall droplets, *Phys. Rev.* B **54**(23), 16850–16859 (1996).
48. J. Weis, R.J. Haug, K. v. Klitzing, and K. Ploog, Competing channels in single-electron tunneling through a quantum dot, *Phys. Rev. Lett.* **71**(24), 4019–4022.
49. M. Ciorga, M. Pioro-Ladriere, P. Zawadzki, P. Hawrylak, and A.S. Sachrajda, The break-up of a lateral quantum dot into multiple dots in high magnetic fields, in *Proceedings of the 26th International Conference on the Physics of Semiconductors*, Edinburgh (2002), CD-ROM edition.
50. P. Hawrylak, Spin effects in quantum Hall droplets, in *Proceedings of NATO ARW-Recent Trends in Theory of Physical Phenomena*, I. Vagner, P, Wyder (eds.) (2002).

Chapter 4

Novel Phenomena in Small Individual and Coupled Quantum Dots

A. M. Chang[1]*, H. Jeong[1], and M. R. Melloch[2]

[1] *Department of Physics, Purdue University, West Lafayette, IN 47907-1396*
[2] *School of Electrical Engineering, Purdue University, West Lafayette, IN 47907*

1. Introduction

The field of semiconductor quantum dot [1] single electron transistor [2–4] has witnessed tremendous advances in the past 10 years. Whether from the perspective of fundamental physics or potential technological applications, the quantum dot has generated a great amount of excitement as a result of its tunability and controllability. This tunability has enabled the realization of a remarkable variety of physical phenomena, associated with the charge and spin degrees of freedom of the electron(s) occupying the "upper" most quantum levels, as well as potential device applications due to the extraordinary sensitive to charge and electric fields [9,10]. To date, an impressive variety of phenomena has been observed. These include: atomic shell filling in regularly shaped QD, signatures of quantum chaotic dynamics in irregularly shaped QD manifest in the wavefunction and energy level statistics, evolution of levels in a perpendicular magnetic field (with or without interaction effects), molecular state formation in double and triple QD's, coherent molecular states, covalent and ionic states, higher-order effects such as cotunneling, under strong coupling leads to spin effects such as Kondo physics or mixed valence physics associated with the Anderson model, and interference effects such as Fano resonance line shape due to the presence of resonant and non-resonant channels. These results represent the contributions of numerous workers in the field and excellent reviews are available in the literature [1,5–8]. On the application side, several very interesting proposals, such as single-electron transistor (SET) based logic, memory [10], quantum-dot electron-spin based qubits for quantum computation [11], are currently being investigated in laboratories around the world. Furthermore, notable and useful devices have been invented, such as the

*Effective August 1, 2003, please contact A.M. Chang at Department of Physics, Duke University, Durham, NC 27708 USA.

scanning SET for the detection of local electric fields [12,13], and the RF-SET (radio-frequency single-electron transistor) for the detection of charge at the $10^{-5}e/\sqrt{Hz}$ level at 100 MHz bandwidth [14]. Although several of these applications/devices do not necessarily require semiconductor quantum dots, and are thus far implemented in metallic SET systems, there is no fundamental reason they could not also be workable in semiconductor, and in particular, GaAs/$Al_xGa_{1-x}As$ quantum dots.

To illustrate the richness of phenomena and potential impact on future technology, in this review we discuss three contributions in our own work on transport in small GaAs—$Al_xGa_{1-x}As$ quantum dots. These results pertain to three rather diverse aspects of QD physics and underscore the richness of the observable phenomena. The first topic deals with the highly non-Gaussian distribution of Coulomb Blockade conductance peak heights in individual quantum dots. In this phenomenon the spin degree of freedom does not play a significant role. While our work represented the first systematic study [15], results similar to ours are also obtained by Folk et al. [16]. Subsequent to the discovery of the Coulomb Blockade (CB) in semiconductor quantum dots [1,17,18], Jalabert, Stone, and Alhassid [19] recognized that the height fluctuations may be related to the well-known Porter-Thomas type distribution of resonance line widths in the elastic scattering of neutrons from complex nuclides such as U^{235}, U^{238}, and Th^{238}, etc. [20]. Using a random matrix theory (RMT) approach which neglected the influence of electron–electron interaction beyond the Hartree contribution responsible for the classical electrostatic charging energy, U, Jalabert et al. computed the distribution of peak heights in the single-level tunneling limit, where electrons tunnel through an individual, thermally resolved quantum level within the quantum dot. Results were obtained for both $B = 0$ corresponding to the Gaussian Orthogonal Ensemble (GOE), and $B \neq 0$ corresponding to the Gaussian Unitary Ensemble (GUE). In both cases, the most striking feature is the non-Gaussian nature with a prevalence of small valued peaks for g $\rightarrow$ 0 where g is the dimensionless peak conductance, while the $B \neq 0$ distribution attains a maximum near zero.

The second topic concerns the even–odd (pairing) behavior associated with the spin degree of freedom in the electron filling behavior of an irregularly-shaped individual quantum dot. The pairing behavior arises within a simple scenario appropriate for the individual quantum dot under appropriate conditions discussed below, and results from the spin degree of freedom and the Pauli exclusion principle for Fermions. Here we report distinct signatures of the even–odd effect observable in the pairing behavior of CB peak spacing, peak heights, and spin status. Furthermore, we found that the peak spacing pairing depends significantly on the r_s value characterizing the Coulomb interaction strength.

The third topic pertains to the Kondo effect in an artificial quantum dot molecule [21], arising from coupling of the quantum dot excess-spins to the spin of the electrons in the leads and a resultant screening of the dot-spins [22,23]. Since a quantum dot with an excess spin may be modeled as a magnetic impurity under appropriate conditions, coupled double-quantum-dots provide an ideal model system for studying interactions between localized impurity spins. We report on the transport properties of a series-coupled double quantum dot as electrons are added one by one onto the dots. When the many body molecular states are formed, we observe a splitting of the Kondo resonance peak in the differential conductance. This splitting reflects the energy difference between the bonding and antibonding states formed by the coherent superposition of the Kondo states of each dot. The occurrence of the Kondo resonance and its magnetic field dependence agree with a simple interpretation of the spin status of a double quantum dot. This direct observation of coherent coupling between an excess spin on each dot will have potential ramification for the implementation of the quantum-dot systems for quantum computation [11].

Lastly, we will briefly present suggestive evidence for sudden rearrangements of the double-dot energy ground state configuration, leading to sharp features in the CB peaks, as well as a functioning quantum dot with lithographic dimensions as small as 120 nm.

2. Models of Single and Double Quantum Dot Systems

To provide the basis for discussing the phenomena reviewed in this chapter, we present a summary of the simplest physical scenario characterizing the quantum dot system, including the external leads connected to the dot through which current can be injected or removed. In this simplest scenario, the physical system of an isolated single QD is viewed as a sort of artificial "atom" [1], characterized by several parameters and by simple even–odd filling of the individual quantum levels under appropriate conditions.

For this nearly-isolated individual dot case, in general two familiar ingredients control its characteristics: Coulomb blockade (CB) and the discretized zero-dimensional quantum level structure, which in the general case may include the effects of spin and exchange/correlation effects from the Coulomb repulsion between electrons. This simplest scenario arises when the effect of the Coulomb repulsion can be accounted for by the electrostatic charging energy alone, which is represented by the Hartree term with the neglect of exchange and correlation effects. Within this so-called "constant interaction" model [1,24] the charging energy, U (or E_C), of the small electron puddle on the quantum dot, in the limit of zero source–drain bias across the dot, is given by:

$$E = \frac{Q^2}{2C} - \frac{C_g}{C} Q V_g \tag{1}$$

where $C = C_L + C_R + C_g$ is the total capacitance of the dot to the environment, for a resistor–capacitor model of a single quantum dot depicted in Fig. 4.1. Note that to ensure charge is quantized on the dot, it is necessary for the tunneling resistances, R_L and R_R to be large. In the limit of very large tunneling resistances, the model reduces to the situation depicted in Fig. 4.1(b). Coulomb blockade arises when the chemical potential of the dot is tuned via V_g to favor an integral number $(n, n+1, \ldots)$ of electrons as depicted in Fig. 4.2(a) (solid curves). Since charge is quantized in units of e, in this case the energy can be rewritten in the form:

$$E = \frac{(-N|e| + C_g V_g)^2}{2C} - \frac{C_g^2}{2C} V_g^2 \tag{2}$$

where we have $Q = -N|e|$, i.e. charge is quantized in units of $-|e|$, yielding a Coulomb charging energy of $E_C = e^2/(2C)$ for the addition or removal of an electron. Here, the

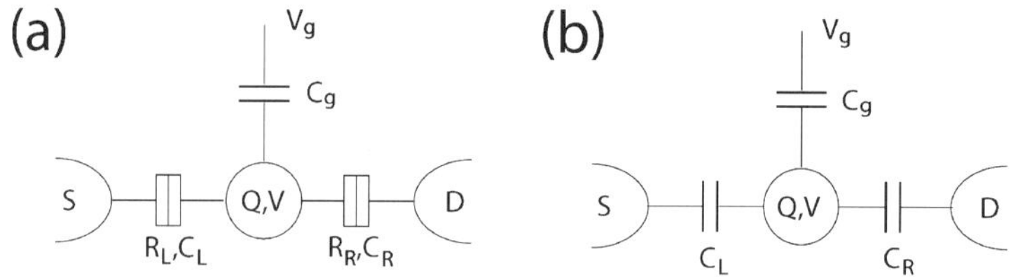

Fig. 4.1. (a) Model of a quantum dot with left (L) and right (R) tunnel junctions to the leads (source, S, and drain, D). The charge on the dot can be tuned via the gate, V_g. (b) Capactive model of the single quantum dot.

second term quadratic in V_g and independent of Q is irrelevant and suppressed in the figures, since what is important is a comparison of the system energy at different values of the charge, Q, in integral multiples of $-|e|$. At temperatures below this charging energy scale, $kT < E_C$, electron transport through the dot is suppressed. To facilitate charge transport, it is thus necessary to tune the chemical potential towards a regime where a $\frac{1}{2}$ integer number of electron is favored:

$$C_g V_g = \left(n + \tfrac{1}{2}\right)|e| \tag{3}$$

Here the N and $(N + 1)$ electron states are degenerate and charge can flow freely through the dot without the cost of a charging energy [1–3] as depicted in Fig. 4.2(a) by the dashed curve. Alternatively, we may plot the energy as a function of V_g at fixed number of electrons as shown in Fig. 4.2(b). At a V_g value where the parabolas for N electrons and $N+1$ electrons cross, Coulomb blockade is lifted and transport proceeds freely through the dot.

Within this scenario, inclusion of the effect of zero-dimensional quantum confinement introduces an additional quantum energy level structure on top of the charging energy, E_C, as depicted in Fig. 4.2(c) by the presence of the quantum level spacing, Δ. In the absence of spin degree of freedom, or for phenomena in which spin does not play a role aside from doubling the density of states, and when higher-order virtual tunneling processes are neglected the current (I) through the quantum dot in the limit of fully coherent tunneling is given by a convolution of the difference of the Fermi function at the chemical potentials in the two leads and the Breit–Wigner resonance formula [49]:

$$\begin{aligned} I(kT, eV_{\text{bias}}) = \frac{e^2}{h} \int \mathrm{d}E[f(E + eV_{\text{bias}}) - f(E)] \\ \times \frac{\Gamma_L \Gamma_R}{\Gamma^2 + (E - E_o)^2} \end{aligned} \tag{4}$$

where f is the Fermi–Dirac distribution function, $\Gamma_{L,R}$ denote the level broadening due to coupling to the left and right leads, respectively, and $\Gamma = \frac{1}{2}(\Gamma_L + \Gamma_R)$, and

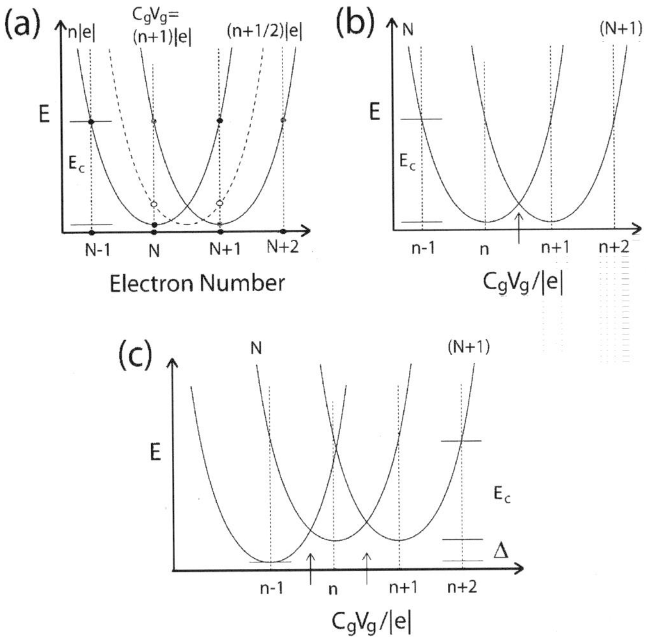

Fig. 4.2. (a) Energy parabolas of the single-electron transistor quantum dot versus the number of charge on the dot, in the absence of the contribution from quantum levels, at fixed gate voltages corresponding to $C_g V_g/|e| = n$, $(n + \frac{1}{2})$, and $(n + 1)$, respectively. A Coulomb blockade conductance peak takes when the energy for the $Q/|e| = N$ and $(N + 1)$ configurations become degenerate, as is the case for $C_g V_g/|e| = (n + \frac{1}{2})$. (b) An alternative representation of the energy parabolas plotted versus V_g, at fixed charge number, $Q/|e| = N$ and $(N + 1)$, respectively. At a gate voltage (arrow) where the two parabolas cross, conduction takes place giving rise to the Coulomb blockade peak. (c) Same as (b) but with the inclusion of quantum level spacing, Δ. Note that the gate voltage position of the energy degeneracy point is shifted for the $(N - 1)$ to N case.

E_0 the energy of the resonant level which may be tuned by the plunger gate, V_g. In this simplest scenario, if the dot shape is irregular and therefore without the special symmetries which often give rise to accidental degeneracies in the energy spectrum and shell filling effects [25], the zero-dimensional energy levels are expected to exhibit level repulsion related to the Wigner–Dyson type while at the same time the wave function fluctuation at given spatial points reflected by a resonance width, Γ, which follows the Porter–Thomas distribution. As it turns out, the resonance line-width does indeed follow the Porter–Thomas type distribution to a large extent [15,16], as will be reviewed below. However, the level spacing distribution deviates from the naive expectation of a Wigner–Dyson type due to the effect of interaction beyond the Hartree contribution including the influence of spin and exchange energies [26,27].

Even within this scenario which excludes exchange and correlation effects, spin plays a role in the filling behavior of the dot. If we fill the dot starting from empty, the first electron will enter the lowest quantum level, followed by a second electron into *the same* level, albeit with opposite spin. The two electron will form a spin singlet to ensure antisymmetrization of the total wave function including of the orbital and the spin parts. The addition of this second electron will cost a Coulomb charging energy, E_C. If we next proceed to add a third electron, it will be forced to occupy the next high quantum level (first excited level) due to Pauli exclusion. Therefore the energy required would not only involve a charging energy, E_C, but also in addition, the level spacing energy equal to the excitation energy of the first excited state. The fourth electron will again be accommodated within this same excited level and cost an energy, $\Delta E = E_C$. The electron addition energy, ΔE, therefore is expected to exhibit an oscillatory behavior of period 2, oscillating between $\Delta E = E_C$ and $\Delta E = E_C + \Delta$. This is a key signature of the so-called even–odd pairing effect associated with the spin degree of freedom.

More complex physical effects can be uncovered when higher-order processes are taken into account, particularly when the quantum dot is tuned to the Coulomb blockade valleys (integer number of electrons). Among the variety of effects which arise one of the most interesting is the Kondo resonance which results when an effective spin-interaction occurs between the quantum dot excess spin and the spin of the electron sea contained in the leads. Through virtual processes, screening of the residual spin on the quantum dot leads to the well-known Kondo effect [22,23,28,29] familiar in the case of dilute magnetic impurities in metals. Specifically, starting from the Hamiltonian for the dot-lead system:

$$
\begin{aligned}
H &= H_{\mathrm{QD}} + H_{\mathrm{R}} + H_{\mathrm{L}} + H_{\mathrm{QD\text{-}Leads}} \\
&= \sum_{i\sigma} \epsilon_i d_{i\sigma}^{\dagger} d_{i\sigma} + \sum_{i} U n_{i\uparrow} n_{i\downarrow} + \sum_{k\sigma} \epsilon_k^{\mathrm{R}} c_{\mathrm{R}k\sigma}^{\dagger} c_{\mathrm{R}k\sigma} \\
&\quad + \sum_{k\sigma} \epsilon_k^{\mathrm{L}} c_{\mathrm{L}k\sigma}^{\dagger} c_{\mathrm{L}k\sigma} + \sum_{i,k,\sigma,\sigma'} [V_{ik}^{R} d_{i\sigma}^{\dagger} c_{Rk\sigma} + V_{ik}^{\mathrm{L}} d_{i\sigma}^{\dagger} c_{\mathrm{L}k\sigma} + h.c.]
\end{aligned}
\tag{5}
$$

it can be shown that a Kondo resonance, characterized by a Kondo energy scale, $E_{\mathrm{K}} = kT_{\mathrm{K}} = \sqrt{U\Gamma \mathrm{e}^{-\pi|\epsilon_o-\mu|(U+\epsilon_o)/U\Gamma}}$, results from higher order processes when a single spin resides in the highest, occupied level within the quantum dot [22,23]. Loosely speaking, an electron from the left lead biased at a slightly higher chemical potential can tunnel into this highest, singly-occupied level, while at the same time, either electron can at once tunnel out into the right lead. Since double-occupancy occurs for a short time, no penalty in terms of the charging energy is incurred. Adding up such higher order contributions, including those which the external electron from the lead can tunnel back and forth any number of times between the left lead and the dot before an electron exits into the right lead gives rise to a Kondo resonance at the Fermi level.

2.1. Double-Quantum-Dots

When two quantum-dots are coupled together, the electrostatics naturally become more complicated as a result of the mutual capacitance between each pair of conductor, including the quantum dot metallic puddles and the various pincher and plunger gates for the formation and control of the dots. Here we also summarize the simplest scenario based on even–odd filling of each of the two dots. To clearly illustrate the role of the electrostatics, we initially consider a situation of weak coupling to the leads ($\Gamma \ll \Delta$).

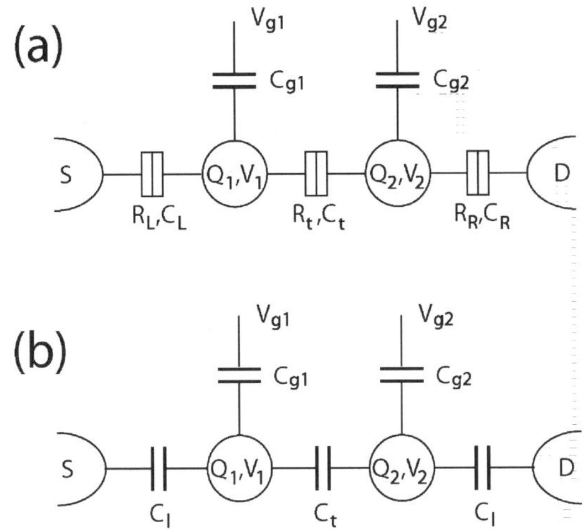

Fig. 4.3. (a) Model of a series-coupled double quantum dot system with left (L) and right (R) tunnel junctions to the leads (source, S, and drain, D), and a dot to dot tunnel junction (t). The charge on each dot can be separately tuned via the gates V_{g1} and V_{g2}. (b) Capactive model of the fully symmetric, series-coupled double quantum dot.

This means we will be able to for the time being neglect Kondo spin effects. For now we will also neglect the existence of quantum level spacing as well as higher-order effects from inter-dot coupling, which leads to an effective anti-ferromagnetic coupling between the excess spins on the two dots of $J \sim 4t_c^2/U$. These additional ingredients and their effects will be discussed subsequently.

This energy diagram based again on classical electrostatics can readily be calculated by generalizing the capactive–resistive model of a single dot (Fig. 4.1) to the double dot situation as is depicted in Fig. 4.3 [5,32–34]. It is most convenient to characterize the system using the capacitance matrix. Starting from the basic charge–voltage difference relationship on the ith conduction, the total charge Q_i, is given by:

$$Q_i = \sum_j q_{ij} = \sum_j c_{ij}(V_i - V_j) \tag{6}$$

Cast this into vector form with $\mathbf{Q} = (Q_1, \ldots, Q_i, \ldots, Q_N)^{\mathrm{T}}$, $\mathbf{V} = (V_1, \ldots, V_i, \ldots, V_N)^{\mathrm{T}}$, and defining the capacitance matrix C, where $C_{ij} = \left(\sum_k c_{jk}\right)\delta_{ij} - c_{ij}$, yield:

$$\mathbf{Q} = C\mathbf{V} \tag{7}$$

and an electrostatic energy for the system, E, of:

$$E = \tfrac{1}{2}\mathbf{V}^{\mathrm{T}}C\mathbf{V} = \tfrac{1}{2}\mathbf{Q}^{\mathrm{T}}C^{-1}\mathbf{Q} \tag{8}$$

Inclusion of the charge, $\mathbf{Q_v}$, and voltage, $\mathbf{V_v}$, of voltage sources, which can be modeled as batteries with large capacitances, the voltage, $\mathbf{V_c}$, on the conductors may conveniently be related to its charge, $\mathbf{Q_c}$, and the voltage settings on the sources, and the capacitance sub-matrices between the conductors, C_{cc}, and between the conductors and the voltage sources, C_{cv} [5,32–34]:

$$\mathbf{V_c} = C_{cc}^{-1}(\mathbf{Q_c} - C_{cv}\mathbf{V_v}) \tag{9}$$

For the series-coupled double-quantum-dot depicted in Fig. 4.3(b) such consideration leads to an electrostatics energy at zero bias of:

$$\begin{aligned} E(N_1, N_2) = {} & \frac{1}{2}N_1^2 E_{C1} + \frac{1}{2}N_2^2 E_{C2} + N_1 N_2 E_{Ct} \\ & - \frac{1}{|e|}[C_{g1}V_{g1}(N_1 E_{C1} + N_2 E_{Ct}) + C_{g2}V_{g2}(N_2 E_{C2} + N_1 E_{Ct})] \\ & + \frac{1}{e^2}\left[\frac{1}{2}C_{g1}^2 V_{g1}^2 E_{C1} + \frac{1}{2}C_{g2}^2 V_{g2}^2 E_{C2} + C_{g1}V_{g1}C_{g2}V_{g2}E_{Ct}\right] \end{aligned} \tag{10}$$

where $E_{Ci} = (e^2/C_i)[1/(1 - C_t^2/C_1C_2)]$, $E_{Ct} = (e^2/C_t)[1/(C_1C_2/C_t^2 - 1)]$, and $C_i \equiv C_{L,R} + C_{gi} + Ct$ is the total capacitance of the i-th (L or R) dot to its surroundings. Substantial intuition may be gained by examining the experimentally relevant and simple case of fully identical dots, each with the same capacitive coupling to its respective plunger gate, i.e. $C_L = C_R = C_l$ and $C_{g1} = C_{g2} = C_g$. Equation (11) then reduces to the form:

$$\begin{aligned} E(N_1, N_2) = {} & \frac{1}{2}N_1^2 E_C + \frac{1}{2}N_2^2 E_C + N_1 N_2 E_{Ct} \\ & - \frac{1}{|e|}[C_g V_{g1}(N_1 E_C + N_2 E_{Ct}) + C_g V_{g2}(N_2 E_C + N_1 E_{Ct})] \\ & + \frac{1}{e^2}\left[\frac{1}{2}C_g^2(V_{g1}^2 + V_{g2}^2)E_C + C_g V_{g1} C_g V_{g2} E_{Ct}\right] \end{aligned} \tag{11}$$

with $C = C_1 = C_2$ and $E_{C1} = E_{C2} = E_C$. In the limit of zero interdot-coupling, for which $C_t = 0$ and $E_{Ct} = 0$, the system behaves as two isolated but identical dots, with energy:

$$E(N_1, N_2) = \frac{E_C}{2}(-N_1|e| + C_g V_{g1})^2 + \frac{E_C}{2}(-N_2|e| + C_g V_{g2})^2 \tag{12}$$

where $E_C = e^2/C_o$ and $C_o = C_l + C_g$. In the opposite limit of large interdot coupling and dominant $C_t \gg (C_l + C_g)$, so that $C \approx C_t$, we have $E_{Ct} \approx E_C = e^2/2C_o$ and:

$$E(N_1, N_2) = \frac{E_C}{2}[-(N_1 + N_2)|e| + C_g(V_{g1} + V_{g2})]^2 \tag{13}$$

and the system behaves as a single large dot, albeit with a charging energy, E_C, which is one half the isolated case.

If we were to tie V_{g1} and V_{g2} together so that $V_{g1} = V_{g2} = V_g$, and plot the electrostatic energy at fixed (N_1, N_2):

$$E(N_1, N_2) = \frac{E_C}{2}\left[\left(-N_1 + \frac{C_g V_g}{|e|}\right)^2 + \left(-N_2 + \frac{C_g V_g}{|e|}\right)^2\right] + E_{Ct}\left(-N_1 + \frac{C_g V_g}{|e|}\right)\left(-N_2 + \frac{C_g V_g}{|e|}\right) \tag{14}$$

versus V_g, for the three cases of: (a) $C_t = 0$ with $E_{Ct} = 0$, (b) $C_t < (C_l + C_g)$ with $E_{Ct} > 0$, and (c) $C_t \gg (C_l + C_g)$ with $E_{Ct} \approx E_C$, we see behaviors at the charge degeneracy point(s) corresponding to two identical isolated dots, the development of a splitting due to the finite E_{Ct}, and one large dot, as shown in Figs 4.4(a)–(c) respectively, with a concomitant doubling of the period of the Coulomb blockade conductance peaks versus gate voltage compared to the isolated-dots case ($C_t = 0$), when V_{g1} and V_{g2} are tied together. Here we have plotted the energy for fixed electron numbers (N_1, N_2), which denote the excess occupancy above up-down spin paired occupied quantum levels, with $N_i = (0,1)$ indicating the excess occupancy on dot i, in Fig. 4.1(a) the parabolas depict the energy curve for the (0,0) empty state, the (1,1) singly-occupied state on each dot, and the (0,1) and (1,0) states where one electron occupies one of the two dots. When the inter-dot coupling is turned off, $C_t \to 0$, the (0,1) and (1,0) parabolas are degenerate so that the condition for which Coulomb blockade is lifted

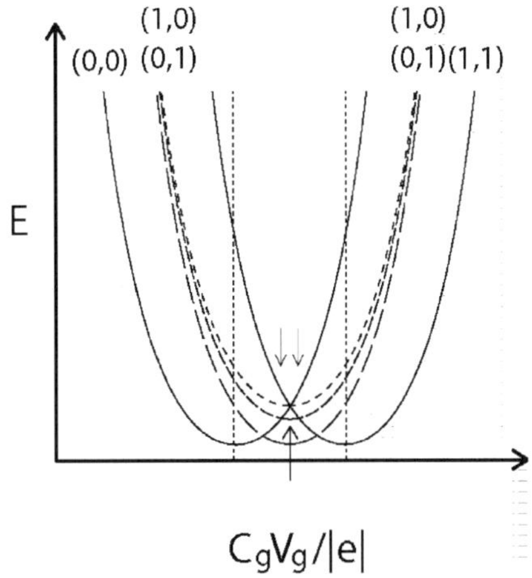

Fig. 4.4. The energy diagram for a fully symmetric series-coupled, double quantum dot, for different values of the interdot capacitance, C_t. The energy at fixed occupancy, (N_1, N_2), is plotted versus the gate voltage, $V_g = V_{g1} = V_{g2}$. For $C_t = 0$, $E_{Ct} = 0$, and the two dots are isolated (short dashed curve). For $C_t > 0$ but small compared to $(C_l + C_g)$, $E_{Ct} < E_C$ and the single degeneracy point is now split as indicated by the two short arrows (medium dashed curve). This gives rise to a splitting of the Coulomb peak when contrasted with the isolated case of $C_t = 0$. When C_t dominates, $E_{Ct} \approx E_C$, and the two dots behave as a single large dot with a doubling in the frequency of occurrence for the Coulomb peak as a function of V_g.

occurs at one point in the diagram. This is the case of two identical but isolated dots for which in each dot the Coulomb blockade is removed when it is tuned to favor a half-integer number of electrons. Note that here the parabola maybe be shifted in energy by a single particle quantum level spacing, Δ. However, such a shift does not qualitatively change the picture aside from shifting the gate voltage position where the blockade is lifted. When coupling is gradually introduced, the now non-zero inter-dot coupling, C_t, a lowering of the electrostatic energy of the (0,1) and (1,0) states. The interception of the lower curve with the (0,0) and (1,1) parabolas at two distinct points signals a splitting of the quantum dot conductance peak into two.

In the more general situation where $V_{g1} \neq V_{g2}$, it is useful to examine the stability diagram in the V_{g1} versus V_{g2} plane. Such a stability diagram can be obtained by first defining a chemical potential for each dot, μ_i:

$$\begin{aligned}\mu_i &\equiv E(N_1, N_2) - E(N_1 - \delta_{i,1}, N_2 - \delta_{i,2}) \\ &= E_C\left[\left(N_i - \frac{C_g V_{gi}}{|e|}\right) - 1\right] + E_{Ct}\left(N_j - \frac{C_g V_{gj}}{|e|}\right)\end{aligned} \tag{15}$$

and the addition energy for adding an electron on either dot, E_{add}:

$$E_{add} \equiv \mu_i(N_1 + \delta_{i,1}, N_2 + \delta_{i,2}) - \mu_i(N_1, N_2) = E_C \tag{16}$$

where $i = \{1, 2\}$ and $i \neq j$. With the definition of $\mu = 0$ for each lead at zero bias, stability for occupation (N_1,N_2) is given by the requirement of $\mu_1 < 0$ and $\mu_2 < 0$. Such stability diagram in this ideal situation is shown in Fig. 4.5 for different values of C_t and hence E_{Ct}. The three scenarios correspond to the situations depicted in the previous Fig. 4.5. The presence of the six-sided polygon for general values of E_{Ct} has given rise to the nomenclature "honeycomb" diagram. Inclusion of the quantum energy level, non-ideality in real devices such as residual gate-voltage dependence of E_C and E_{Ct}, and residual mutual capacitance between gates lead to distortions of such honeycombs such variations in the area of honeycombs and a change in the slope of the domain boundaries. Again, in the limit of very strong inter-dot coupling, the double-quantum-dot (DQD) behaves as a single large dot in accordance with expectation.

In an experiment the stable configuration (N_1,N_2) is controlled by V_{gi}. A plot in the (V_{g1}, V_{g2}) plane can be obtained and represents an extremely useful way to characterize the DQD system. The above-mentioned limits of no inter-dot coupling and strong coupling are shown in Fig. 4.5. When E_{Ct} increases from 0, within a quantum-mechanical picture where such an increase corresponds to increasing the inter-dot tunneling matrix, the non-zero inter-dot tunneling splits the energy of the symmetric and anti-symmetric orbital levels associated with the (0,1) and (1,0) states by an amount proportional to E_{Ct}, whenever the corresponding unperturbed energy levels for these states are tuned to degeneracy. Such coherent coupling is essential in the use of quantum dots as qubits for quantum computation. Experimentally, such an energy splitting due to this coherence have been demonstrated in transport as well as microwave absorption experiments [36,35]. Again this scenario takes place when coupling to the leads is negligible and the properties are single-particle in nature.

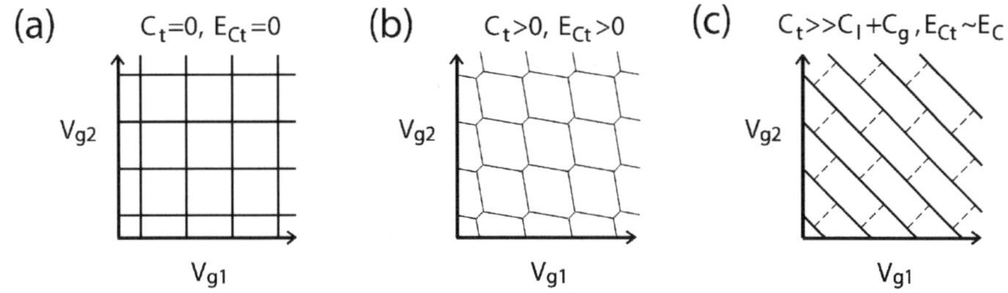

Fig. 4.5. Charge stability, honeycomb diagrams in the V_{g1} versus V_{g2} plane, for the three cases of Fig. 4.4, $C_t = 0$, $C_t > 0$ but small, and $C_t \gg (C_l + C_g)$, respectively. The situations depicted in Fig. 4.4 correspond to a cut along the diagonal in these three diagrams, respectively.

As in the case of the individual dot, inclusion of quantum level spacing simply shifts the parabolas by the respective level spacings, Δ. Going beyond single particle properties to access regimes exhibiting properties of many-body spin correlation may be accomplished by introducing coupling to the conduction electrons in the leads by increasing Γ to $\sim \Delta$. The full Hamiltonian of the coupled DQD system with leads is given by:

$$\begin{aligned} H_{\mathrm{DQD}} = & \sum_{k_\alpha \in \{L,R\},\sigma} \epsilon_{k_\alpha} c^\dagger_{k_\alpha,\sigma} c_{k_\alpha,\sigma} \\ & + \sum_{\alpha \in \{\mathrm{L,R}\},\sigma} \epsilon_{\alpha\sigma} d^\dagger_{\alpha\sigma} d_{\alpha\sigma} + V_{\mathrm{o}} \sum_{k_\alpha \in \{\mathrm{L,R}\},\sigma} (c^\dagger_{k_{\alpha,\sigma}} d_{\alpha\sigma} + h.c.) \\ & + t_{\mathrm{c}} \sum_\sigma (d^\dagger_{\mathrm{L}\sigma} d_{\mathrm{R}\sigma} + h.c.) + U(n_{\mathrm{L},\uparrow} n_{\mathrm{L},\downarrow} + n_{\mathrm{R},\uparrow} n_{\mathrm{R},\downarrow}) \end{aligned} \quad (17)$$

The introduction of the inter-dot tunneling matrix, t_c, in addition to the charging energy, $U = E_C$, level spacing, Δ, and level width characterizing coupling to the leads, $\Gamma_L = \Gamma_R \equiv \Gamma$, leads to additional and extremely interesting new physics. For weak-coupling where $\Gamma \ll \Delta$, the usual single-particle coherence between the two coupled-dot quantum levels is expected and observed [35,36]. When the coupling is strong when $\Gamma \sim \Delta$, dramatic new physics association with the many-body Kondo effect has been predicted by theory.

Compared to the single QD case, this DQD model contains one extra parameter, t_c, which parameterizes the coupling (tunneling matrix element) between the two dots. The inclusion of this energy scale gives rise to a rich variety of correlated physics beyond the single dot case. In the limit of large $U(E_C)$, it turns out to be convenient to re-parameterize the system with the following energy scale obtainable from those given in the Hamiltonian:

1. $\Gamma_{L,R} \equiv \pi V_o^2 \rho_{L,R} = \Gamma$, is the quantum level broadening in the two dots (L or R) due to their respective coupling (V_o) to the left (L) or right (R) lead. Here $\rho_{L,R}$ denotes the electronic density of states in the L- or R-lead,
2. t_c, the bare, interdot tunneling matrix element,

3. $J = 4t_c^2/U$ the effective anti-ferromagnetic coupling between a single excess (unpaired) spin on each dot, and
4. $T_K \approx \sqrt{U\Gamma} exp[-\pi|\mu - \epsilon_o|(U + \epsilon_o)/\Gamma U]$, the individual dot Kondo temperature.

Accordingly, based on Slave–Boson Mean Field (SBMF) theory [37–40], numerical Lanczos [41] and renormalization group (NRG) calculations [42], or the non-crossing approximation [43], three distinct regions of correlated spin behavior is present when an excess spin occupies each dot. When $t_c/\Gamma < 1$, the system can be mapped onto the two-impurity Kondo problem initially discussed by Jones et al. [44,45], albeit without even–odd parity symmetry [46]. In this situation, when the antiferromagnetic coupling is weak, $J \neq T_K < 2.5$, the system behaves as two separate screen Kondo spins, with screening of each dot spin by the conduction electrons in the respective leads. When J increases beyond the "critical" value where $J/T_K \approx 2.5$, a cross-over takes place to a state where the two impurity (dot) spins form a single due to the strong anti-ferromagnetic coupling. A cross-over rather than a true phase transition arises due to the relevant perturbation introduced by the hopping term, t_c, which automatically breaks the even–odd parity symmetry [42,40]. Note that in an experimental device, J can be increased by increasing t_c while keeping U relatively constant. A third region of strongly correlated behavior occurs if t_c approaches and exceeds $\Gamma(t_c/\Gamma \geq 1)$ before the antiferromagnetic transition point is reached [37–39,43]. The system forms a coherent superposition of the many-body Kondo state of each dot, leading to a strongly renormalized splitting $2\tilde{t}_c$ and a double peak in the differential conductance near zero bias, in contrast to the ordinary single peak behavior. In addition, dramatic negative differential conductance is also predicted in this regime. Further increasing t_c beyond U/4 destroys the Kondo correlation completely [40,42].

3. Non-Gaussian Distribution of Coulomb Blockade Peak Heights in Individual Quantum Dots: Porter–Thomas Distribution of Resonance Widths

For open systems, in a separate work we had experimentally demonstrated a distinction between chaotic and non-chaotic behavior [47]. In nearly isolated systems, however, non-chaotic behavior is practically unrealizable. Any residual disorder or lithographic imperfections will render the dynamics chaotic on the long trapping time-scale before eventual escape into external leads/reservoirs takes place. There are non-trivial predictions on statistical properties of nearly isolated cavities which can be tested in experiments in the transport through quantum dots, pertaining to the distribution of energy level spacings and the statistical properties of wave functions. In the case of the quantum dots, the presence of the Coulomb charging energy indicates that electron-electron interaction is present. However as a first approximation, it may be reasonable to assume that in some appropriate limit, e.g. high electron density where screening is effective, the interaction simply contributes a classical charging energy given by $E_C/e^2/C$ where C is the capacitance of the quantum dot to its surroundings (see Background Section), and does not seriously affect the RMT distributions. The assumption proves to be largely correct in the GaAs/Al_xGa_{1-x}As quantum dots studied. However, there is suggestive evidence that deviations from the RMT universal distributions maybe observable in our small ($\leq$0.25 μm) quantum dots.

The specific predictions of theory which we tested pertain to the distribution of these Coulomb blockade peak heights at $B = 0$, corresponding to the RMT GOE, and at $B \neq 0$, corresponding to the GUE. In the latter case, the magnetic field, B, is required to exceed some correlation field, B_c, which characterizes the transition from GOE to GUE statistics. According to theory [19,48], in the single level tunneling limit of thermally broadened conductance peaks, the $B = 0$ distribution is given by:

$$P_{(B=0)} = \sqrt{2/\pi\alpha}\, e^{-2\alpha} \tag{18}$$

with a square-root singularity near zero. Here, α is related to the Coulomb blockade peak conductance, $G_{\max}$, is given by [49]:

$$G_{\max} = \frac{e^2}{h}\frac{\pi}{2kT}\frac{\Gamma_L\Gamma_R}{\Gamma_L + \Gamma_R} \equiv \frac{e^2}{h}\frac{\pi\overline{\Gamma}}{2kT}\alpha \tag{19}$$

where $\Gamma_L(\Gamma_R)$ is the partial decay width into the left (right) lead. This expression is obtainable from Eq. (4) by taking the voltage derivative, $\mathrm{d}I/\mathrm{d}V$, in the limits of zero V_{bias} and low temperature $\Gamma \ll kT \ll \Delta$, where single level tunneling dominates so that electrons tunnel through a single quantum level within the quantum dot, and the conductance peak is thermally broadened beyond the natural width Γ. In a magnetic field greater than the correlation field, the breaking of time-reversal symmetry reduces the number of nearly zero values of $G_{\max}$. Nevertheless, the distribution [19,50] is still non-Gaussian and peaked near zero:

$$P_{(B\neq 0)} = 4\alpha[K_0(2\alpha) + K_1(2\alpha)]\mathrm{e}^{-2\alpha} \tag{20}$$

where K_n are the modified Bessel functions. For GaAs/Al_xGa_{1-x}As quantum dots, Γ can readily be tuned to be below 3 μV (corresponding to $T \sim 34$ mK) while typically the level spacing $\Delta \geq 200$ μV ($T \sim 230$ mK). The required conditions of thermally broadened single level tunneling limit is therefore accessible at dilution refrigerator temperatures ~70 mK. In contrast, in a typical metallic dot of size 50 nm, $\Delta \sim 0.1$ μV (~1.1 mK) and many levels are accessed. This leads to a convolution of independent single level distributions resulting in a Gaussian distribution of peak heights in accordance with the central limit theorem.

From an experimental perspective, the challenges are to fully access the single level tunneling limit, and to identify GaAs/Al_xGa_{1-x}As starting material sufficiently free of background traps to allow temporal stability of fabricated quantum dots. To fully access the single level limit, $\Delta \leq 5\,kT$ is invariably needed. The unavoidable decoupling of the electrons from the lattice at low temperatures renders it difficult to reduce the electron temperature below the 50–100 mK range [18,51] even after strongly filtering any noise (thermal or pickup) from the measurement system. For GaAs/Al_xGa_{1-x}As to satisfy this condition, small quantum dots of order 0.25 μm or smaller in its largest dimension must be fabricated. Superior electron beam lithography is requisite to produce

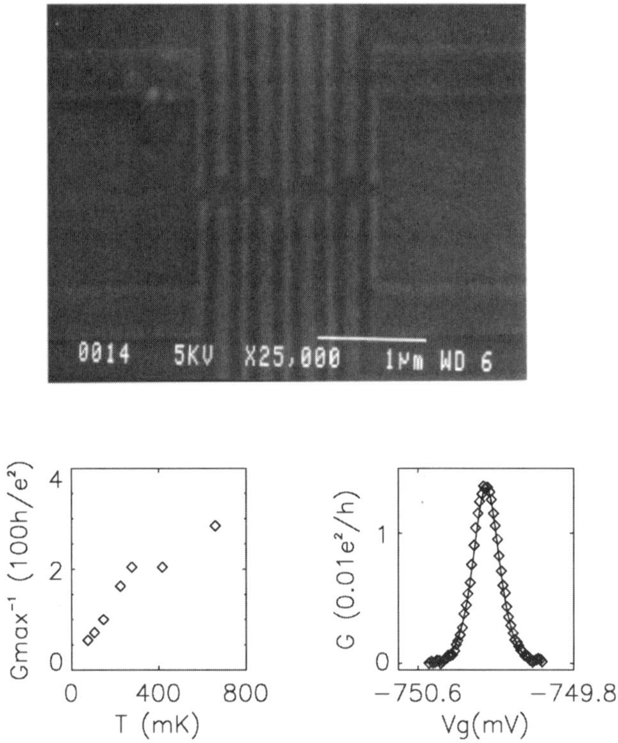

Fig. 4.6. Top: Electron micrograph of the gates defining the quantum dots. Four dots are available on each sample. Bottom-left: G_{max}^{-1} versus T for a representative peak at $B = 0$. The roughly linear behavior below ~300 mK indicates this is a single level tunneling peak. Bottom-right: A fit of the convolution of $-\partial f/\partial \epsilon$ with the Breit–Wigner resonant tunneling formula to the peak in the left panel at $T = 108$ mK.

the multiple metal gate electrodes used to form the weak-coupling, tunnel barriers to external leads (otherwise known as pinchers) and to form the central quantum dot. In Fig. 4.6 top panel we present an electron micrograph of the metal gate pattern for four dots in series used in our experiment. Each dot is roughly 0.3 μm × 0.35 μm in lithographic size but is reduced to below 0.25 μm × 0.25 μm after gating. In our experiment, individual dots were separately measured rather than the whole series of four dots. Regarding temporal stability, by nature the Coulomb blockade is sensitive to a single electron charge. Any motion of background charges in near by trap sites will strongly affect the transport both via a change in electrostatic energy, and via a distortion of the dot shape. It is found that crystals grown under varying conditions in different laboratories yield quantum dots which range from temporally quite stable to extremely unstable. The fabrication and sorting out of sufficiently stable dots to ensure reproducible Coulomb blockade peak height values has made our work and the work of Folk et al. possible. In our devices, approximately 50% of individual dots meet the stability criterion.

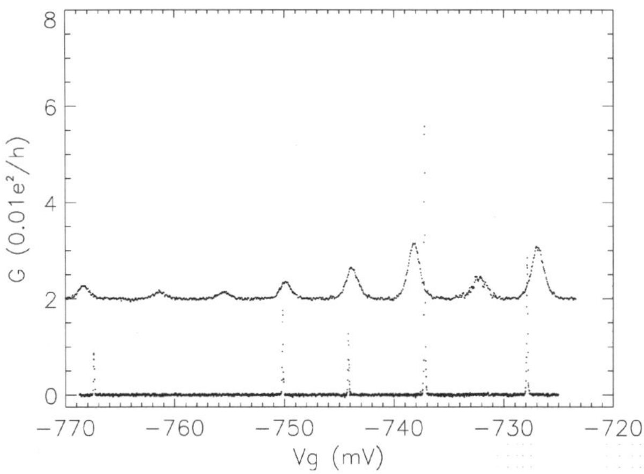

Fig. 4.7. A typical trace showing successive Coulomb blockade conductance peaks versus the center gate voltage, V_g. $B = 0$ and $T = 75\,\text{mK}$ (lower trace) or $T = 660\,\text{mK}$ (upper trace, displaced by 2 units). Note that three peaks are missing out of seven, but they emerge at higher temperature. The slight shifting in peak positions is discussed in the text.

In Fig. 4.7 we show a representative trace of Coulomb blockade peaks at $B = 0$ for $T = 75\,\text{mK}$ (bottom curve) and 660 mK (top curve) temperatures. Note the missing peaks at the gate voltages −733, −753, and −762 mV in the 75 mK curve which are observable at the higher temperature of 660 mK. *The large difference in height of adjacent peaks and the many small peaks are our primary experimental observation, and qualitatively demonstrate the outstanding feature of a prevalence of small peaks predicted by RMT in Eq. (18).* To demonstrate we are fully in the single level tunneling limit of temperature broadened peaks, in Fig. 4.6 lower-left and right panels, we plot G_{max}^{-1} max versus T and the line shape of a representative peak fitted to the theoretical $\cosh^{-2}[(E_0 - \gamma e V_g)/2kT]$, respectively. In Fig. 4.6 bottom-left the roughly linear dependence of G_{max}^{-1} on T at low temperatures indicates we are clearly in the single level regime (see Eq. (19)). In fact we find this behavior in all of the eight peaks we studied in detail.

For all peaks, we are able to follow the evolution with magnetic field. In Fig. 4.8 we show magnetic field traces of G_{max} for four representative peaks. From the fastest variation of we estimate a correlation field, B_c, of the order of 500 Gauss, somewhat larger then the theoretical value [19,50] of 200 Gauss. Panels (c) and (d) depict two peaks which are nearly zero at $B = 0$. The behavior in (d) where the peak remains small for large stretches of B and only occasionally increases to a large value is observed in roughly $\frac{1}{3}$ of peaks which start near zero. The type of behavior is not expected in the RMT picture where fluctuations should occur on the scale of B_c. The fact the height remains small for large regions in B suggests an enhancement of small peak probability above the RMT prediction (Eq. (20)) for $B \neq 0$.

The traces in Fig. 4.7 are obtained by keeping the pinchers (tunnel barriers to external leads) gate voltages constant while sweeping the central quantum dot gate voltage,

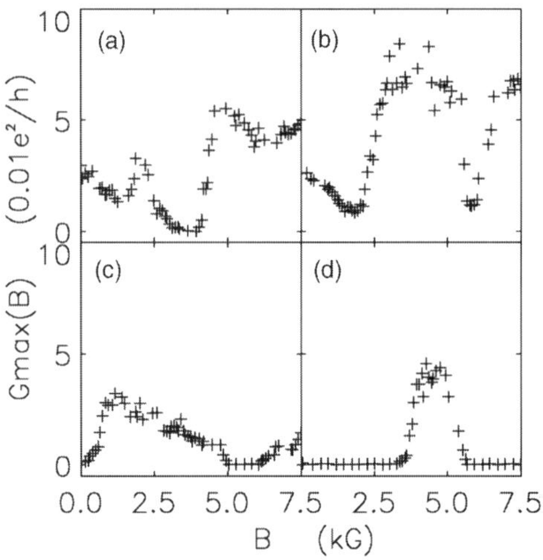

Fig. 4.8. Magnetic field sweep of four peaks at $T = 100\,\text{mK}$. The field range for ~100% change in G_{max} is ~500 Gauss. (c) and (d) show the two types of behavior of peaks which are very small at $B = 0$.

V_g. Sweeping V_g more negative, however, also affects the pinchers by gradually closing them off, thereby reducing the conductance value. When too many peaks produced by sweeping V_g alone while keeping the pinchers fixed are included, skewing of the distribution can occur. Therefore, we impose a window of acceptance corresponding to roughly five peaks per pincher setting as valid data. The details of how the window is set can be found in Chang et al. [15]. The distribution we obtain will subsequently be compared to the theoretical distributions deduced by averaging Eqs (1) and (3) over the pincher window of approximately a factor of 3.5 in the transmission probability. By collecting data at different pincher settings where care is taken to ensure the Coulomb blockade peaks are uncorrelated from previous settings, we are able to collect data from 72 independent peaks. This procedure yielded 72 peak height values for $B = 0$, and 216 values for $B \neq 0$. In the latter case height data are taken at three different magnetic fields well separated by several B_c's to triple the data set. The resulting distributions are presented as histograms in Fig. 4.9. The distributions are normalized to unit area as for a probability density. Both the $B = 0$ (a) and $B \neq 0$ (b) distributions are strongly non-Gaussian, and clearly peaks toward zero values. In the $B = 0$ case, nearly $\frac{1}{3}$ of the peaks fall in the lowest bin: 23 out of 72 peaks are less than $0.005e^2/h$ compared to a mean of $\sim 0.024e^2/h$. In contrast, for $B \neq 0$ only 43 out of 216 peaks are this small. Fig. 4.9 indicates that there is a difference between the two distributions for low values as is born out by the Kolmogorov–Smirnov statistical test.

The mean decay width needed for comparing to theory is not measured experimentally and is therefore a fitting parameter. This width should be nearly independent of B; thus we introduce a single scale parameter and fit simultaneously to the $B = 0$ and $B \neq 0$ data. Figure 4.9 shows a fit to the data using both the theory for constant pincher transmission [Eqs (1) and (3) (solid)] and this theory averaged over a variation

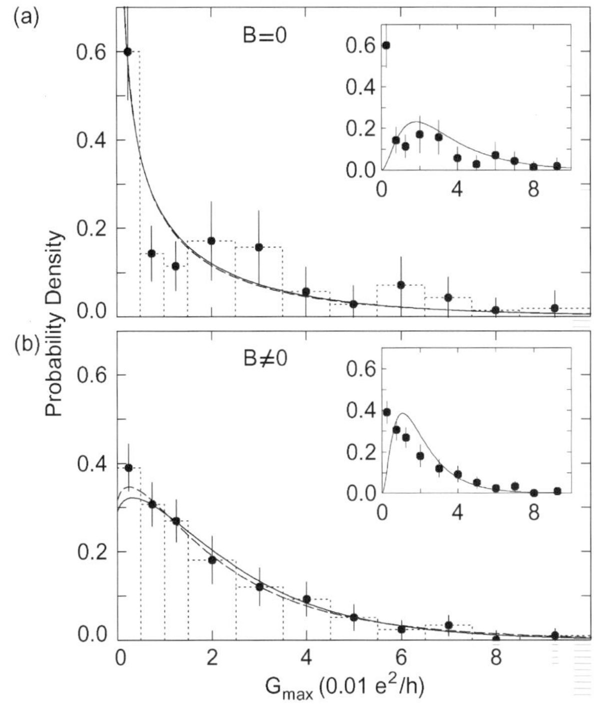

Fig. 4.9. Histograms of conductance peak heights for (a) $B = 0$ and (b) $B \neq 0$. Data are scaled to unit area; there are 72 peaks for $B = 0$ and 216 peaks for $B \neq 0$; the statistical error bars are generated by bootstrap re-sampling. Note the non-Gaussian shape of both distributions and the strong spike near zero in the $B = 0$ distribution. Fits to the data using both the fixed pincher theory (solid) and the theory averaged over pincher variation (dashed) are excellent. The insets show fits to $\chi_6^2(\alpha)$—more Gaussian distribution—averaged over the pincher variation; the fit is extremely poor.

of the pincher transmission by a factor of 3.5 (dashed). The similarity of the two curves shows that the variation in our pincher transmission can be neglected.

The experimental distribution in Fig. 4.9(b) for $B \neq 0$ appears slightly higher for the smallest height data point compared to the theoretical dashed curve. Even though the difference is within error bar, it is suggestive. Intrigued by the possibility of deviation from the RMT theoretical result, we further split this lowest bin into two, producing the histogram in Fig. 4.10. The probability of small peaks continues to increase for height values approaching zero, in stark contrast to the RMT result of a maximum at $\sim 0.025e^2/h$! This excess of small peaks is related to the discussion of Fig. 4.8(d) where certain peaks which are small when $B = 0$ remain small for large regions of $B > 0$. In fact, recent theoretical calculations aimed at accounting for electron–electron interactions appear to show exactly this trend [52]. The inuence of interaction certainly deserves further investigation. Improved statistics should further elucidate the role of electron–electron interaction in chaotic systems.

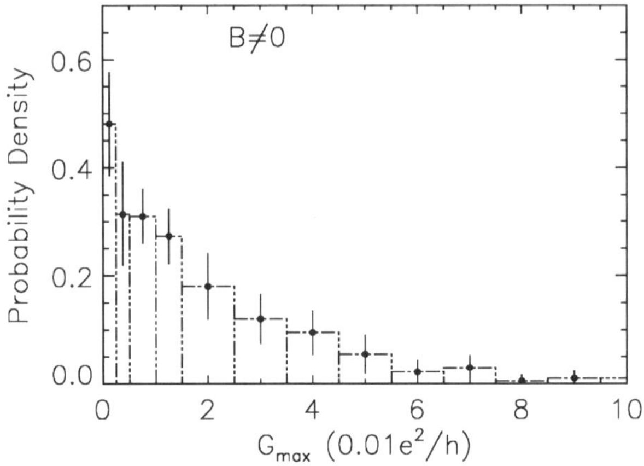

Fig. 4.10. The $B \neq 0$ distribution of Fig. 4.9(b) replotted with the lowest bin further split into two. The resulting histogram shows that the probability for small peaks continues to increase toward zero, in contrast to RMT predictions. This trend is likely a result of electron–electron interaction.

4. Spin and Pairing Effects in Ultra-Small Dots

The difference between an even and odd-numbered finite Fermion system, known as the even–odd parity effect, is a distinct feature reflecting the unique behavior of fermionic particles *in the presence of both orbital and spin degrees of freedom* [53–56]. This parity effect is expected to appear in artificially fabricated semiconductor quantum dots. The unprecedented control over experimental conditions in lateral quantum dots allows one to fill electrons one by one as manifested in the phenomenon of Coulomb Blockade [3] in transport. The peak spacing fluctuation in CB peaks provides unique information about single particle energy level, many-body interaction effect and the parity of electron numbers. In irregularly shaped dots without special symmetries the CB peaks are expected to be paired with smaller spacings in the odd electron valleys and larger spacings in the even valleys, reecting the dot spin status which is $1/2\,\hbar$ for an odd number of electrons and paired to zero for an even number (Fig. 4.11 and discussion above). The importance of unambiguous and differing effects reflecting the spin status of the dot pertains directly to the desire to take the next logical step and couple dots together for the sake of both fundamental physics (competition between the Kondo effect and indirect exchange interaction in the two impurity model [44,45,57,58]) and for technological reasons in the implementation of the double dot system as prototype quantum qu-bits in quantum computation [11]. Because these potentially important future developments depend on the success in achieving well-defined and well-controlled spin status on a dot, the clear observation of the pairing effect as well as the elucidation of the necessary conditions are of central importance.

In this section, we describe clear pairing features attributable to the even–odd effect in small $GaAs/Al_xGa_{1-x}As$ lateral quantum dots some with the smallest r_s value measured to date. These signatures are observed in the CB peak spacing, peak

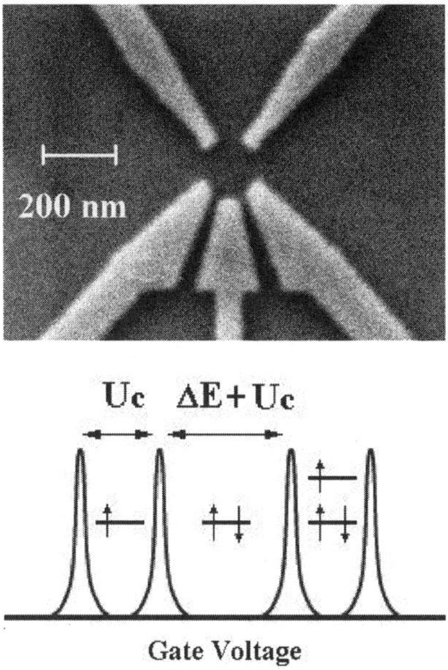

Fig. 4.11. Scanning electron micrograph of device 1 (top). Schematic diagram of the peak positions as a function of gate voltage (bottom). The narrow period corresponds to the change from odd to even numbers by adding an electron with the opposite spin into the same spin degenerate state, and the broad period to the change from even to odd numbers, occupying different dot energy levels.

height as well as spin. We find a quantum dot with smaller r_s shows more pronounced peak pairing than high r_s dots. Furthermore, qualitatively different behaviors in peak spacings are observable depending on the coupling strength between the dots and leads.

Three devices are fabricated in different r_s regimes. Device 1 containing the number of electrons, $N \approx 10$ and device 2 containing $N \approx 40$ are made from a crystal of density, $n = 3.5 \times 10^{11}\text{cm}^{-2}(r_s = 0.93)$ while device 3 with $N \approx 10$–20 from a high density crystal of $n = 9 \times 10^{11}\text{cm}^{-2}(r_s = 0.58)$. The geometry and size of the devices 1 and 2 (lithographic diameters of 160 and 230 nm, respectively) were chosen carefully to maximize functionality in spite of their small sizes (Fig. 4.11 (top)). The geometry of device 3 with a lithographic diameter of 170 nm enables us to control N between 10 and 20 (inset in Fig. 4.14(a)) by the application of six independent gate voltage settings. The minimum size of small quantum dot is to a large extent limited by the depth of two-dimensional electron gas (2DEG). In devices 1 and 2 all length scales, pincher gap, dot size, gate widths and their positions were optimized for a relatively deep 2DEG, 95 nm below the surface. While the lithographic size is 160 nm, device 1 is estimated to be 60 nm in diameter after accounting for depletion. The main advantage of our small quantum dot with multiple independent gates is that by setting different gate voltages, different realizations of a quantum dot can be implemented. Although it is often difficult to precisely determine how many electrons reside on lateral quantum dots, to check whether the dot region is completely depleted tunneling barriers are reduced to compensate increased depletion from the plunger gate while it is swept

[59]. In this way, we observed a maximum of eight CB peaks in device 1, close to the estimated electron number based on the 60 nm size.

In a standard constant interaction model [1], the peak spacing in plunger gate voltage is expressed as a combination of single particle energy level splitting and charging energy term:

$$\Delta V_{\mathrm{g}} = \frac{C}{eC_{\mathrm{g}}}\left\{\Delta + \frac{e^2}{C}\right\} \tag{21}$$

Here, Δ is zero if N = odd and greater than or equal to zero for N = even, reflecting the spin degeneracy on each energy level (Fig. 4.11 (bottom)), C is the total capacitance of the dot and C_{g} is the capacitance between the dot and plunger gate. A direct consequence is that the spacing is smaller between adjacent CB peaks separated by a valley with an odd number of electrons than that with an even number. When the dot size is large, the average level spacing Δ is much smaller than the charging energy $E_{\mathrm{C}} = e^2/C$ and the above formula implies relatively small fluctuations in peak spacing. The spacing appears nearly uniform in this case. On the other hand, a small quantum dot can be expected to give more pronounced peak spacing pairing from the increased Δ. Furthermore, in the small-number electron system, the addition or extraction of one electron can change the entire energy spectrum due to the strong Coulomb interaction, known as Koopman's theorem, resulting in discrepancy between constant interaction model and experiment.

In our experiment, clear signature of an even–odd pairing can be observed in the CB peak spacing when the coupling between the quantum dot and leads is strong in all devices. On the other hand, as the dot–lead coupling is reduced a marked difference emerges between the high r_{s} devices 1 and 2 and the low r_{s} device 3. In device 1 when the dot is nearly open, two broad peaks are observed as shown in Fig. 4.12(a). A decrease of the dot–lead coupling resolves more peaks and pairing is clearly present in Fig. 4.12(b). When the coupling is further reduced, the pairing is again no longer visible (Fig. 4.12(c)). Even with this small size dot, there is no clear signature of even–odd effect in weak tunneling regime. Figure 4.13 demonstrates similar results in device 2. The spacing between peaks straddling a valley (2, 4, 6) with an even number of electrons (paired spins) is distinctly larger than that of the odd valleys (1, 3, 5) (inset to Fig. 4.13(b)), while at the same time the valleys 1, 3, and 5 exhibit the Kondo resonance as a signature of unpaired single electron spin. The pairing gradually disappears when the dot–lead coupling is reduced for valleys 7–10 in Fig. 4.13(a) and all valleys in Fig. 4.13(b). The disappearance of pairing in the weak coupling regime for these two devices indicates that the constant interaction model is beginning to break down and explains the difficulty in observing peak pairing over a large number of peaks in previous works [26,28–31]. In stark contrast, in the small r_{s} high density device 3 *peak pairing is preserved in all regimes of the dot–lead coupling* as evidenced by the ubiquitous pairing behavior in Fig. 4.14. In fact pairing is observable under quite different gating configurations. For example, the dot contains more electrons with the configuration in Fig. 4.14(c) than Fig. 4.14(a) or (b). Nevertheless, *pairing is observable for at least 10 peaks in succession!* The relative ease of observing pairing in this device is likely a direct consequence of the low value of $r_{\mathrm{s}} = 0.58$, lower than all previous devices reported in the literature.

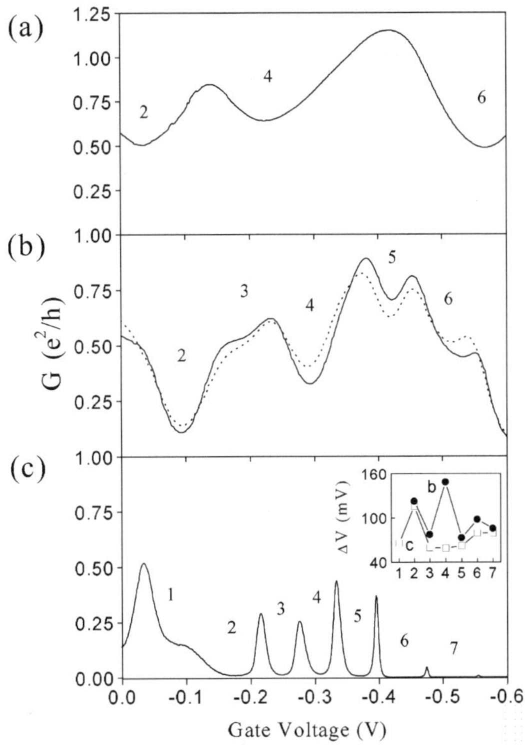

Fig. 4.12. Evolution of the peak structure for different dot–lead coupling in device 1 measured at $T = 300\,\text{mK}$. From (a) to (c), the coupling is decreased gradually. In (b), dotted line is $T = 900\,\text{mK}$ trace, illustrating temperature dependence of Kondo and non-Kondo valleys. By closing the dot, the peak spacing pairing is destroyed even with this small size dot. Inset in (c): Comparison of peak spacing in (b) (filled circle) and (c) (unfilled square) in each valley (x-axis).

We point out that the results in devices 1 and 2 indicate that the controlling parameter for observability of the peak spacing pairing is not simply the ratio Δ/E_C. By controlling the transmission coefficient of tunnel barriers, confinement strength of the dot is adjustable, which at the same time modifies Δ and E_C rendering it possible to set different values for this ratio. In the strongly coupled regime we estimate this ratio to be $\sim$1 in device 1 compared to 0.5 in device 2 which is larger in size. On the other hand in the weak coupling regime, we expect an increase in both E_C [28] and Δ, yielding a ratio between 0.5 and 0.6 for device 1. Even though still larger than device 2 in the open regime, the even–odd pairing has disappeared. Furthermore, the pairing we observe can be attributed mainly to the large Δ in these small dots and not to a spacing shift caused by the Kondo resonance, based both on a recent theoretical estimate indicating that the Kondo shift is at most 20% of Δ [60] and the fact that our results often show large even–odd spacing differences as large as $E_C \sim 1$ meV (Fig. 4.12(b)).

We turn our attention to the different behaviors of the low and high r_s dots. The observed behavior when the dots are closed (weak coupling) depends on r_s. When

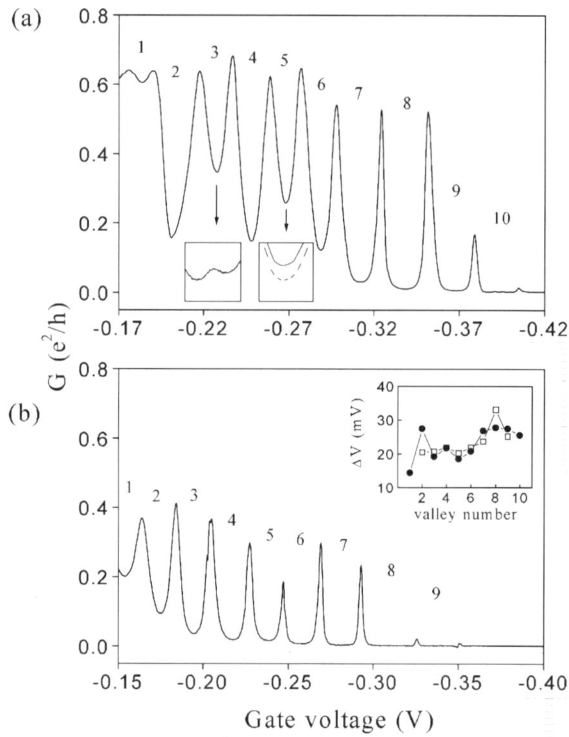

Fig. 4.13. Fluctuation of peak spacing in device 2 at $T = 75\,\text{mK}$. This device has about 40 electrons inside the dot, fabricated on a low density crystal ($r_s \sim 1$). Insets in (a) show several features of Kondo effect. Left inset: zero bias maximum (ZBM) of valley 3 in $-0.1\,\text{mV} < V_{\text{DS}} < 0.1\,\text{mV}$ range in differential conductance. Kondo valleys in this device show Zeeman splitting of ZBM under magnetic field, about $\pm 21\,\mu\text{V/Tesla}$. Right inset: expanded view ($\times 6$) of temperature dependence of the valley 5. Dotted line is at $T = 150\,\text{mK}$. Inset in (b): Filled circle symbol represents stronger coupling(a) than unfilled square symbol (b).

open, peak pairing is observable in all three devices as shown in Figs 4.12(b), 4.13(a), and 4.14(a). However when closed, in contrast to the high r_s devices 1 and 2 pairing is still preserved in the low r_s device 3. This is particularly striking in view of the fact that device 3 and device 1 contain a similar number of electrons. The difference in behavior is believed to be a consequence of the modification in the electron–electron interaction due to a reduced $r_s = 0.58$ in device 3 compared to an $r_s = 0.93$ in devices 1 and 2, and suggests that strong Coulomb interaction plays an important role in deciding peak spacing fluctuation, washing out the peak pairing in the low density high r_s regime while preserving it in a robust way in the high density low r_s regime. We propose that when the Fermi energy E_F is high enough, it offers a better chance of observing pairing effects since electrons with higher kinetic energy, $\hbar^2 k_F^2 = 2m^*$, experience relatively less Coulombic interaction and the constant interaction model may be expected to have better validity. In fact, recent theoretical [61–63] and experimental works [64] indicate exactly this observed trend resulting from the contribution of an energy fluctuation term

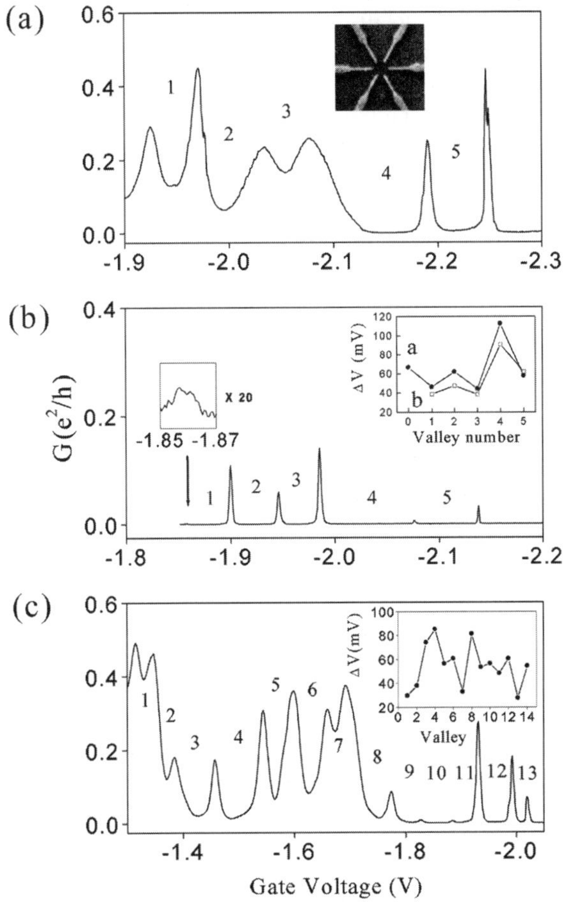

Fig. 4.14. CB peaks spacing fluctuations at $T = 300$ mK from device 3. The lithographic size of the dot is 170 nm (inset in (a)). This dot shows clear even–odd effect in the peak spacing for both (a) strong, and (b) weak coupling to the leads, as well as (c) an entirely different gating configuration where up to seven pairs (14 peaks) in succession are observable. Left inset in (b): magnified view of the first peak. Right inset in (b): peak spacing change of (a) (filled circle) and (b) (unfilled square).

proportional to r_s [61–63]. On the other hand the relative ease of observing pairing in all devices when the dot is open likely results from the presence of spectral rigidity in this regime analogous to the situation in open billiards or disordered conductors where universal conductance fluctuations are present.

Aside from the peak spacing, peak height pairing is also visible over several peaks in all devices, particularly in the weak tunneling regime. In each pair, one is small and the other is large and this alternating sequence is present over all the peaks in Figs 4.12(c), 4.13(b), and 4.14(b). However, devices 1, 2, and 3 show reversed peak height sequence. Current theories as yet do not satisfactorily explains this behavior, although promising theories have been advanced to address peak height fluctuations based on quantum chaos ideas [19] and selection rules from spin and geometry with electron correlation [65]. At present the relation between these mechanisms and the pairing effect in peak height variation is not clear.

An independent determination of the net spin for a given CB valley can be obtained from the Kondo effect, in which the coupling between unpaired localized electron spin and conduction electron spin results in enhanced density of states around Fermi level [22,23]. If the coupling between the dot and the leads are strong enough, this Kondo correlation for an unpaired spin enhances the zero-bias differential conductance as was demonstrated recently [28,29]. Typical temperature dependence of odd and even numbered electron valley conductance is illustrated in Figs 4.12 and 4.13. In each device, odd-electron number valleys showed conductance enhancement at low temperatures, zero-bias maximum (ZBM) in nonlinear conductance, and Zeeman splitting under magnetic field as evidences of Kondo effect. On the other hand, for even-number electrons in the dot the ground state is a singlet and there is no net spin to couple to the conduction electrons and the Kondo effect is suppressed.

As final points, we wish to distinguish our results from anomalous pairing effects observed by capacitive spectroscopy in vertical quantum dots [6] and shell effects [25]. In large vertical dots there exist pairing effects in the addition spectrum when electrons are successively added with some pairs occurring dramatically right on top of each other (at the same energy). Recent works [66,67] indicate these pairing arises from the spatial separation of a relatively large, nominal single dot into effectively two dots when disorder is present. The pairing corresponds to the addition of one electron each to the two effective dots, a situation completely different from our case of a single very small dot. Our results are also distinct from InGaAs-based circular vertical dots in which shell effects are observable since shell effects are present only in regularly shaped dots (notably circular dots) possessing extra degeneracies in the energy spectrum. In contrast in our GaAs/AlGaAs dots, the gating configuration employed to define the dot is completely non-circular, yielding a grossly elongated dot. Our claim that shell effect is not present is borned out by the data of Fig. 4.14(c) where up to seven pairs are observable. The shell effect predicts special positions at 2, 6, 12, and 20 electrons completely at variance with our data. Our work also differs from a recent work by Ciorga et al. [59] in which complete emptying of a lateral dot was achieved. In this work, they explicitly pointed out the absence of an even–odd effect in their data due to conflicting signatures.

5. Coupling between Two Dots and Leads–Coherent Many-Body Kondo States

[Excerpted with permission from "The Kondo effect in an artificial Quantum-Dot Molecule", H. Jeong, A.M. Chang, and M.R. Melloch, *Science* **293**, 2221 (2001). Copyright, 2001, American Association for the Advancement of Science.]

The recent discovery of the Kondo effect in artificial quantum dot systems [28,29,68–72] has raised much interest in the area of quantum impurity and the associated physics of electron spin and strong correlation. By now, the Kondo problem in a quantum dot [22,23] is well understood through experimental and theoretical studies.

Further interesting physical situations can be found when we consider the inter-dot interaction in multiple quantum dots.

The tunability of artificial quantum dots has brought unprecedented control to the investigation of the single impurity level Kondo effect. It is a natural next step to inquire about the physics of interaction between multiple localized magnetic spin moments and conduction electrons in a double quantum dot system. Here, the impurities interact via an effective antiferromagnetic coupling, $J = 4t^2/U$, where t is the tunable interdot coupling and U the intradot charging energy. These two new energy scales, t and J, are expected to introduce qualitatively new behavior. The rich physics of a double quantum dot Kondo system as a tunable, two quantum impurity system has been studied extensively by theorists [37–43]. More intriguingly, the double dot has recently been proposed as a feasible two-qubit system for quantum computation [11]. Here, we review the observation of the coherent Kondo effect in a double quantum dot which is a direct experimental realization of a two impurity Kondo model [21].

Our device is fabricated on a $\mathrm{GaAs}/\mathrm{Al}_x Ga_{1-x}\mathrm{As}$ heterostructure (Fig. 4.15(a)). The sample is mounted in the mixing chamber of a dilution refrigerator parallel to the axis of a super-conducting magnet. The lattice base temperature is 15 mK while the electronic temperature is estimated around 40 mK. It is possible to set each quantum dot to show similar characteristics by tunning gate voltages properly. Each dot exhibited two Kondo valleys represented by the pronounced ZBM (Fig. 4.15(b)). We observe the Kondo ZBM only in the odd-electron Coulomb blockade (CB) valleys between CB peaks with a spacing smaller than the adjacent even-electron valleys. These Kondo valleys exhibit an increase in the conductance with decreasing temperature. Δ and $E_C(U)$ were determined in the closed dot regime by the standard method of measuring the lever arm, α, and the differential conductance at finite source–drain bias to observe excited states. For the upper (lower dot), $\Delta = 0.45$ meV (0.37 meV) and charging energy, $E_C = 1.92$ meV (1.73 meV). These are comparable values reported previously by other groups for dots of similar size [28,68,71]. To accomplish the delicate settings for forming the double dot, the gating information of each dot is utilized since all the gates are needed. For example, by using V1, V2, and V3, the upper dot is first formed. Next V4 is biased up to the proper value and finally V5 is energized. Due to mutual capacitive coupling, the center tunneling barrier becomes higher after energizing V4 and V2 then needs to be reduced to maintain the optimum conduction characteristics through the center barrier.

Assuming simple, even–odd electron filling in each quantum dot, we can expect three unique spin configurations in a double dot. It is necessary to consider only the uppermost energy level in each dot since it is the most relevant for Kondo physics (Figs 15(c)–(e)) [73]. When both dots contain a single electron on their uppermost level, a coherent Kondo resonance can occur, and is the case which has attracted the most theoretical attention. If one or both of the dots contain a spin-singlet, pair of electrons (Fig. 4.15(d) or (e)), a coherent Kondo resonance cannot occur since Kondo coupling cannot be achieved throughout the double-dot system. In a series coupled double dot, we can readily distinguish cases (c) and (d) as even if a Kondo resonance forms in (d) in one dot, transport is greatly impaired by the presence of the paired electrons in the remaining dot. Within this scenario we expect a periodic occurrence

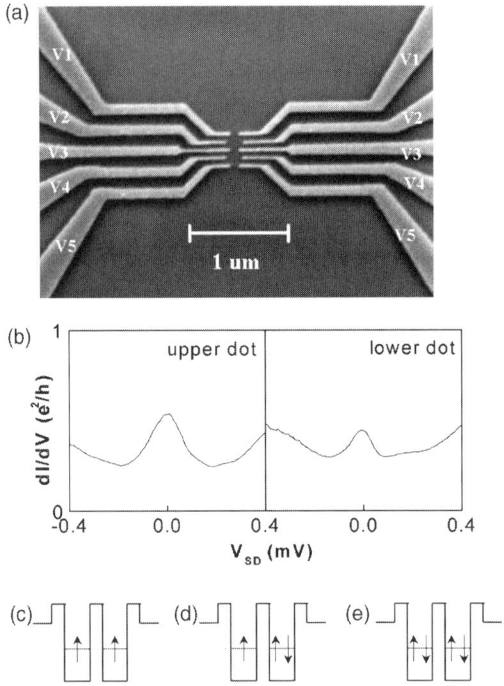

Fig. 4.15. (a) The dots are defined by 10 independently tunable gates on a GaAs/AlGaAs heterostructure containing a 2DEG located 80 nm below the surface. The low temperature sheet electron density and mobility are $n = 3.8 \times 10^{11}$ cm^{-2} and $\mu = 9 \times 10^5$ cm^2/Vs, respectively. Lithographic dot size is 180 nm in diameter and each dot contains about 40 electrons inside. To reduce unnecessary degree of freedom in controlling the double dot, gates sitting on the opposite side are connected together, giving total five pairs of controllable gates. Gate pair V1 and V5 are used to set tunneling barriers, while the V3 sets the inter-dot tunnel coupling between the dots. V2 and V4 control the number of electrons and energy levels in each dot separately. (b) Typical traces of Kondo resonance peaks when each dot is working as single dots. Upper dot shows larger Kondo resonance than lower one. (c)–(e) Three unique cases of spin states of a double dot, based on a simple electron filling with spin degeneracy. (c) is the case which has been considered to show split Kondo resonance [37,39–43]. (d) and (e) contain singlet electrons in one of the two dots and the overall Kondo resonance is not allowed.

of Kondo resonance (circled region of Fig. 4.16(a)) as electrons are added one by one.

By first carefully characterizing individual dots and subsequently the double dot systems, we were able to obtain the honeycomb charging diagram shown in Figs 4.16 and 4.17 at different coupling values, Γ, to the leads. In the measured double dot conductance as a function of plunger gate voltages V2 and V4 (Fig. 4.16(b)), the center gate V3 is set such that it gives barely enough honeycomb structure for easy identification of the valley regions and also enough cotunneling conductance for the Kondo resonance. From the splitting of the Coulomb blockade peaks, we estimate an inter-dot conductance of $\sim 0.8(2e^2/h)$ [74–77]. In this strong tunneling regime,

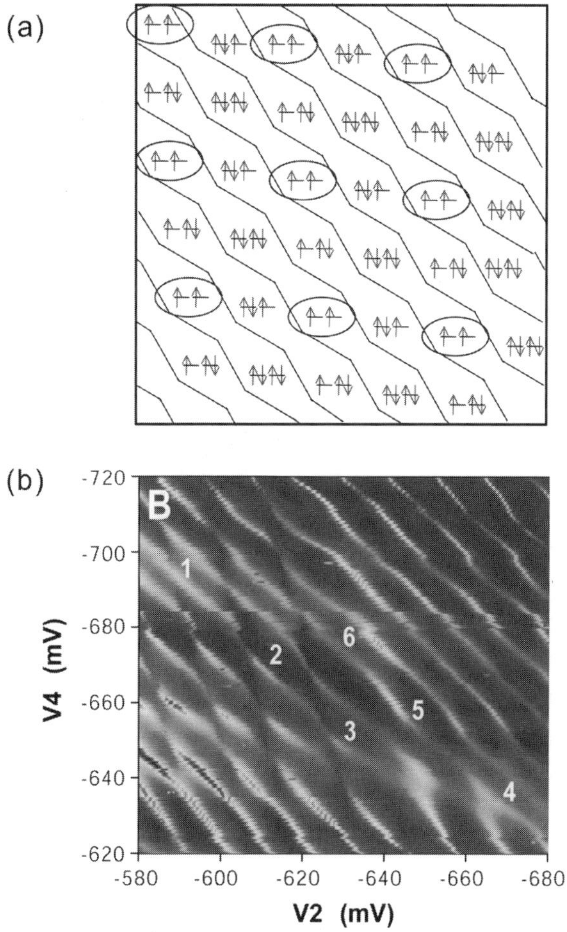

Fig. 4.16. (a) Schematic of a periodic structure of the electron spin configuration as a function of two gate voltages controlling the electron numbers in each dot in a double dot. Lines represent positions of Coulomb blockade peaks. This zig-zag pattern changes depending on the inter dot coupling strength controlled by V3. For simplicity, only the spins of last electronic levels are shown. Circled regions contain possible double Kondo impurity spin status corresponding to the region 1, 3, 4, and 6 in (b). (b) Grey [color] scale plot of the measured conductance of a double dot as a function of gate voltages V2 and V4. The center gate voltage is set $V3 = -860\,\mathrm{mV}$. Brighter (darker) [red (blue)] color signifies higher (lower) conductance. The numbered valley regions, 1, 3, 4, and 6 show ZBM (see Fig. 4.19). Note that (a) is meant as an illustration for the comparison to (b) only. In a real device, the double dot characteristics gradually change as the plunger gate voltages are swept, and the honeycomb pattern inevitably appears distorted from the ideal situation.

however, it is not straightforward to convert the conductance into a reliable estimate of t. To find out which valleys are Kondo valleys containing a single electron on each dot, the differential conductance, $\mathrm{d}I/\mathrm{d}V$, versus voltage bias across the double dot is measured in a total of 32 valley regions four of which show ZBM peaks.

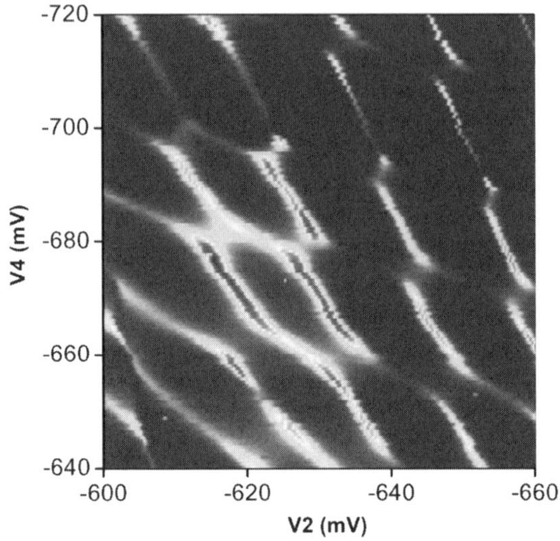

Fig. 4.17. Honeycomb pattern similar to Fig. 4.16(b), but for weaker inter-dot coupling (smaller inter-dot tunneling).

For example, in the six numbered regions in Fig. 4.16(b), we see the appearance and disappearance of ZBM Kondo resonance peaks roughly matched to Fig. 4.16(a) (Fig. 4.18).

Compared to the single dot case, the striking feature of the Kondo resonance peaks is its splitting into two peaks in zero magnetic field. Several theoretical papers have predicted that when the many-body molecular bonding and antibonding states are formed, the Kondo resonance shows a double peak structure in a coupled quantum dot [38–43]. When the split peaks are symmetric (Fig. 4.18, valley 4), they are centered about zero bias and for the non-symmetric cases (Fig. 4.18, valleys 1, 3, 6), the larger peak is closer to zero bias. We believe the prevalence of the asymmetric situation comes from the difficulty in achieving the same condition on both dots. This is supported by the fact that for the symmetric trace 4, when either dot–lead tunnel barrier is changed through V1 or V5, we obtain a similar asymmetric double peak. Attempts to make asymmetrical peaks symmetric were less successful due to slight differences in the characteristics of the two dots and the mutual capacitance and interdependence of the gates. In small devices such as ours, the gates are pushed closer together and experience stronger interdependence. The observed splitting $\delta \approx 45\,\mu\text{eV}$ is comparable to the molecular bonding-antibonding splitting of 10–120 μeV previously reported in double quantum dots [35,36]. It is important to emphasize that our result represents a clear observation of the formation of many-body bonding-antibonding Kondo states. Previous experiments could not determine the spin status of the double dots and were often configured to suppress Kondo correlation (small Γ) yielding a splitting single particle in nature. The gap of double peaks can be adjusted to a certain extent (Fig. 4.19(a)), depending on the gate voltage settings, e.g. by changing the coupling of two dots or coupling of the dots to

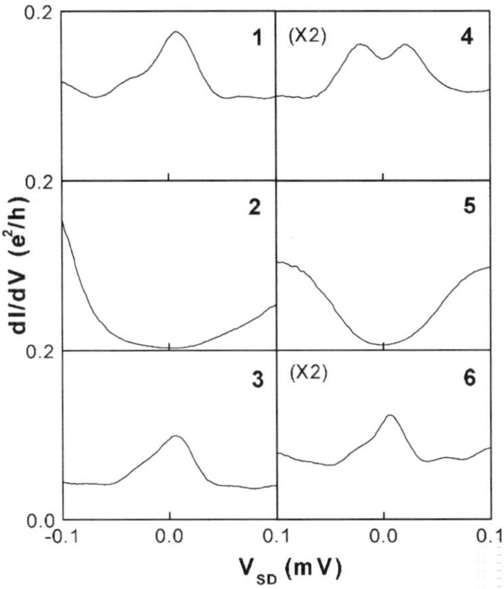

Fig. 4.18. Differential conductance traces from 1 to 6 in Fig. 4.16(b). Trace 4 and 6 are magnified by factor 2. Occurrence of Kondo resonance peaks is well contrasted. The periodicity is consistent with the diagram Fig. 4.16(a). A unique feature of the Kondo resonance peaks is their splitting compared to the single peaks from single dots (e.g. Fig. 4.15(b)).

the leads. However, we were not able to observe single peaks in the regime that the dot–dot coupling is smaller than the dot–lead coupling mentioned in theories. In case the coupling of two dots is reduced, the overall conductance is decreased too much in this series-coupled configuration and a clear signature of ZBM was no longer observable.

All four split Kondo peaks showed qualitatively similar results in magnetic and temperature dependence. The parallel magnetic field dependence (Fig. 4.19(b)) of the symmetric peak trace 4 in Fig. 4.18 shows that as the magnetic field increases, the two split peaks approach each other, merge and then split again. When the magnetic field is applied, the Zeeman effect splits the two many-body molecular states formed around the Fermi levels, giving a total of four peaks. Two of the four peaks closest to the mid point of the left and right Fermi levels overlap when the source–drain bias, V_{SD}, is applied, cross and split again. The contribution from the other two outside peaks should in principle be present but are not observed possibly due to spin decoherence at larger bias. Similar behavior is also present in the single particle, two level Kondo system [69,72] where only two out of four peaks are clearly visible. We can estimate a rough value for the electron magnetic moment g factor based on the Zeeman energy. We find a value between 0.3 and 0.6 compared to the known magnitude of 0.44. As the temperature increases, both of the split peaks have a tendency to decrease and finally disappear (Fig. 4.19(c)). In this temperature dependence, the zero bias conductance of the symmetrically split peaks increases slightly at first and decreases as the temperature goes up

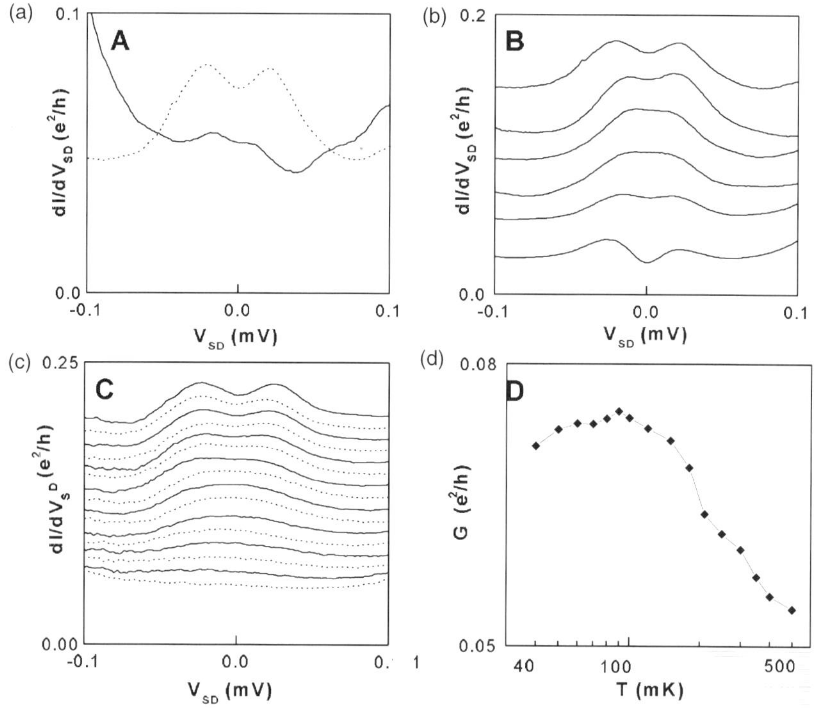

Fig. 4.19. (a) Peak splitting changes depending on the coupling of the two dots. Dotted (solid) line is for V3 = −860 mV (−870 mV). Solid line is magnified by factor 2. For each case, the splitting is 42 and 26 μV. (b) Magnetic field dependence of the symmetric Kondo resonance peak 4. Traces are for $B = 0, 0.25, 0.5, 0.75, 1.0$, and $1.25\ T$. The curves are offset by $0.02e^2/h$ for clarity. Other peaks also show qualitatively similar behavior. The Zeeman splitting from two split peaks enhance the conductance at zero bias as the field increases because of the overlap of the density of states from two peaks. For higher fields, they are going apart further like a single resonance peak case. (c) Temperature dependence of differential conductance. From top to bottom, $T = 40, 50, 60, 70, 80, 90, 100, 120, 150, 180, 210, 250, 300, 350, 400$, and 500 mK, offset by $0.01e^2/h$ for each line. Overall conductance structure decreases as temperature goes up. (d) Conductance at zero bias in log (T) scale. By increasing temperature T, the conductance increases initially and goes down.

(Fig. 4.19(d)). Even the double peak structure disappears, the overall broad peak as a single one is maintained in higher temperature range. Based on the saturation temperature of the zero bias peak height, the Kondo temperature is approximately 500 mK.

From our study, we find that the spin status of multiple dots is consistent with an interpretation based on electron spin filling in a double quantum dot. The Kondo resonance peaks in this system showed clear splitting as an indication of the Kondo effect in a quantum dot molecule. A more quantitative analysis of the competition of Kondo singlet energy versus antiferromagnetic coupling energy in a tunable manner with the advancement of quantum dot device technology will elucidate diverse physical phenomena in multiple quantum dot systems.

6. Other Ultra-Small Devices and Phenomena

In addition to the coherent Kondo effect in multiple quantum dots, other interesting phenomena are observable, such as Fano resonances as well as sharp conductance peaks likely associated with the sudden reorganization of the electronic ground state in this system. Fano-resonance type CB peaks have been observed in the small devices that exhibited the Kondo effect [78]. The asymmetric resonance peak shape was explained to result from an interference between a resonant and a nonresonant path in the system. However, this nonresonant path was not clearly identified. In our small double-quantum-dots that exhibited the coherent Kondo effect, we have also observed Fano line shapes under appropriate conditions. Asymmetric Fano line shapes were observed when the DQD was operated either as seperate single dots or a coupled double-dot. A typical trace from a double dot configuration is presented in Fig. 4.20(a). The degree of asymmetry in the peak shape varies for different cool-downs. However, proper gate voltage settings always produced Fano shapes as well as Kondo resonances within the same cool-down session. It was also possible to change the direction of asymmetry of peaks continuously by changing gate voltages. Furthermore, other interesting peak shapes shown in Fig. 4.20(b) are also believed to originate from Fano-type resonances. The first peak pair exhibits the smooth line shape normally observed in multiple quantum dots. In contrast, the second and third pairs have sharp asymmetric shapes. The condition for formation of the double-dots was different in Figs 4.20(a) and (b). This type of tunability may prove helpful to further sort out the origin of the Fano resonance in quantum dots. One unusual feature which was not previously reported, but was theoretically predicted, is the sudden reconfiguration of electrons ground state due to spin-related effects [79]. The resultant unusual peak shape observed in our device and presented in Fig. 4.20(c) had previously been obtained in the theoretical calculations [79]. These diverse and intriguing behaviors were clearly observed in our versatile, fully tunable and controllable double-quantum-dot device, and serve as testaments to the rich variety of physical phenomena avaible in this system.

As a final indication of the possibilities offered by small, lateral quantum dots defined by electron beam lithograph, in Fig. 4.21 we present the smallest lithographic double QD written in GaAs/Al$_x$Ga$_{1-x}$As to date (of which we are aware). It is generally believed very shallow 2DEG is necessary to make small devices since the depletion length around the biased gates on the surface of the semiconductor material is comparable to the depth of the 2DEG. For example, with the 2DEG at 100 nm below the surface, the device size should be larger than the twice of 2DEG depth, 200 nm. We find, however, that depends on the density of the features in the structure.

The quantum dot in the Fig. 4.21(a) is defined on the surface of a GaAs/Al$_x$Ga$_{1-x}$As heterostructure which is grown by molecular beam epitaxy. The 2DEG is about 60 nm below the surface with a carrier concentration of 3.7×10^{11} cm^{-2} at 4.2 K. While the lithographic dot size is 120 nm, the actual dimension estimated is about 40 nm by electrostatic depletion after accounting for the pinchers as well as the plunger gate. To prevent serious proximity effect, the thick lines for fan-out are half micron away from the dot area. Figure 4.21(b) shows the CB oscillations of the dot as a function of

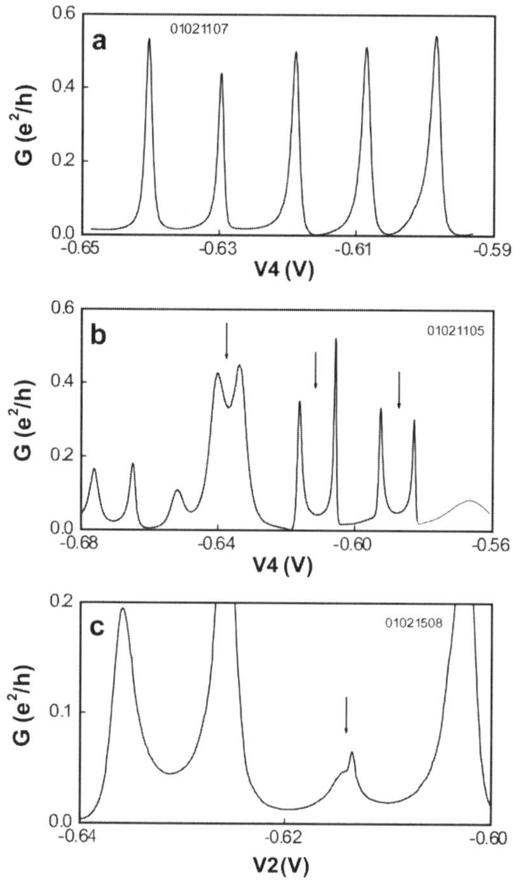

Fig. 4.20. Some of the Coulomb blockade peak traces from a double quantum dot. (a) Asymmetric Fano line shape when the two single dots are strongly coupled. (b) Paired peaks when the two single dots are weakly coupled. (c) A sharp feature observed in the intermediate coupling regime between (a) and (b).

the center plunger gates. The two plunger gates on the left and right side were swept simultaneously. The measurement was performed at 4.2 and 77 K and shows residual CB oscillations at 77 K.

From the oscillation period of CB peaks, the capacitance between the dot and a pair of plunger gates can be found , by the following formula, $C_G = e/\Delta V_G$. This dot to center (plunger) gate capacitance, C_G, is about 620 zF (10^{-21} F) in several devices. The exact number of electrons in the dot cannot be known from the measurement, but rough estimation shows the dot contains 3–5 electrons. By the fact that the CB oscillations still exist at 77 K, the charging energy, E_C, is almost comparable to the thermal energy corresponding to 77 K, which is 6.6 meV. This value is the highest one reported upto now in controllable lateral quantum dots made in GaAs materials.

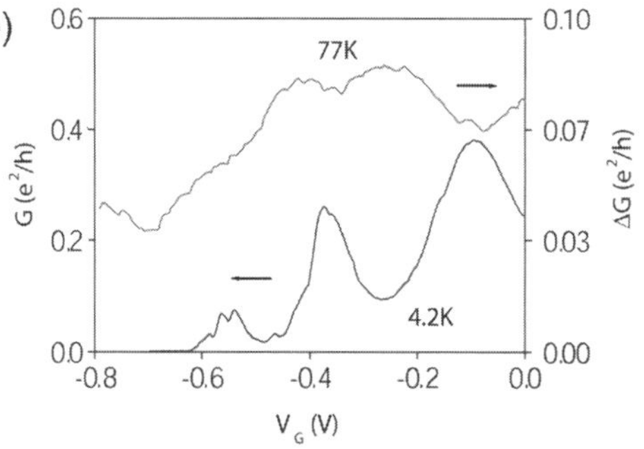

Fig. 4.21. (a) Scanning electron micrograph of a SET with 120 nm single dot size as measured from the inside diameter for the dot. (b) CB oscillations at 4.2 K (lower trace) and 77 K (upper trace). Bias gate voltages are slightly different for optimum CB peaks in the two traces. Offset in y-axis is changed and a portion of linear background conductance is subtracted for 77 K trace.

Acknowledgments

We wish to acknowledge the indispensable contributions of our colleagues Harold Baranger and T.Y. Chang. We also acknowledge helpful discussions with L. Glazman, A. Kaminski. This work was supported in part by NSF grant No. DMR-9801760.

References

1. M.A. Kastner, *Rev. Mod. Phys.* **64**, 849 (1992).
2. D.V. Averin and K.K. Likharev, *J. Low Temp. Phys.* **62**, 345 (1986).
3. T.A. Fulton and G.J. Dolan, *Phys. Rev. Lett.* **59**, 109 (1987).
4. H. Grabert and M.H. Devoret (eds.), *Single Charge Tunneling: Coulomb Blockade in Nanostructures*. Plenum Press and NATA Scientific Affairs Division, New York/London (1992).
5. W.G. van der Wiel, S. De Franseschi, J.M. Elzerman, T. Fujisawa, S. Tarucha, and L.P. Kouwenhoven, cond-mat/0205350.
6. R.C. Ashoori, *Nature* **379**, 413 (1996).
7. See articles in *Proceedings of the NATO Advanced Study Institute on Mesoscopic Electron Transport* (eds. L.L. Sohn, L.P. Kouwenhoven, and G. Schöon), Kluwer Series E345 (1997).

8. S. Tarucha et al., see Chapter 1 in this book.
9. D.V. Averin and K.K. Likharev, in *Mesoscopic Phenomena in Solids* (eds. B.L. Altshuler, P.A. Lee, and R.A. Webb), Elsevier, Amsterdam, 67 (1991).
10. K.K. Likharev and T. Claeson, *Sci. Am.* **266**, 80 (1992).
11. D. Loss and D.P. DiVincenzo, *Phys. Rev.* A **57**, 120 (1998).
12. M.J. Yoo, T.A. Fulton, H.F. Hess, R.L. Willet, L.N. Dunkleberger, R.J. Chichester, L.N. Pfeiffer, and K.W. West, *Science* **276**, 579 (1997).
13. S.H. Tessmer, P.I. Glicofridis, R.C. Ashoori, L.S. Levitov, and M.R. Melloch, *Nature* **392**, 51 (1998).
14. R.J. Schoelkopf, P. Wahlgren, A.A. Kozhevnikov, P. Delsing, and D.E. Prober, *Science* **280**, 1238 (1998).
15. A.M. Chang, H.U. Baranger, L.N. Pfeiffer, K.W. West, and T.Y. Chang, *Phys. Rev. Lett.* **76**, 1695 (1996).
16. J.A. Folk, S.R. Patel, S.F. Godijn, A.G. Huibers, S.M. Cronenwett, C.M. Marcus, K. Campman, and A.C. Gossard, *Phys. Rev. Lett.* **76**, 1699 (1996).
17. J.H.F. Scott-Thomas, S.B. Field, M.A. Kastner, D.A. Antoniadis, and H.I. Smith, *Phys. Rev. Lett.* **62**, 583 (1989).
18. U. Meirav, M.A. Kastner, and S.J. Wind, *Phys. Rev. Lett.* **65**, 771 (1990).
19. R.A. Jalabert, A.D. Stone, and Y. Alhassid, *Phys. Rev. Lett.* **68**, 3468 (1992).
20. C.E. Porter and R.G. Thomas, *Phys. Rev.* **104**, 483 (1956).
21. H. Jeong, A.M. Chang, and M.R. Melloch, *Science* **293**, 2221 (2001).
22. T.K. Ng and P.A. Lee, *Phys. Rev. Lett.* **61**, 1768 (1988).
23. L.I. Glazman and M.E. Raikh, *JETP Lett.* **47**, 452 (1988).
24. K.A. Matveev, L.I. Glazman, and H.U. Baranger, *Phys. Rev.* B **54**, 5637 (1996).
25. S. Tarucha, D.G. Austing, T. Honda, R.J. van der Hage, and L.P. Kouwenhoven, *Phys. Rev. Lett.* **77**, 3613 (1996).
26. S.R. Patel, S.M. Cronenwett, D.R. Stewart, A.G. Huibers, C.M. Marcus, C.I. Duruoz, J.S. Harris, K. Campman, and A.C. Gossard, *Phys. Rev. Lett.* **80**, 4522 (1998).
27. G. Usaj and H.U. Barange, *Phys. Rev.* B **64**, Art. No. 201319 (2001).
28. D. Goldhaber-Gordon, H. Shtrikman, D. Mahalu, D. Abusch-Magder, U. Meirav, and M.A. Kastner, *Nature* (London) **391**, 156 (1998).
29. S.M. Cronenwett, T.H. Oosterkamp, and L.P. Kouwenhoven, *Science* **281**, 540 (1998).
30. U. Sivan, R. Berkovits, Y. Aloni, O. Prus, A. Auerbach, and G. BenYoseph, *Phys. Rev. Lett.* **77**, 1123 (1996).
31. F. Simmel, T. Heinzel, and D.A. Wharam, *Europhys. Lett.* **38**, 123 (1997).
32. I. Ruzin, V. Chandrasekhar, E.I. Levn, and L.I. Glazman, *Phys. Rev.* B **45**, 13469 (1992).
33. H. Pothier, P. Lafarge, C. Urbina, D. Esteve, and M.H. Devoret, *Europhys. Lett.* **17**, 249 (1992).
34. D.C. Dixon, PhD Thesis, University of California, Berkeley (1998).
35. T.H. Oosterkamp, T. Fujisawa, W.G. van der Wiel, K. Ishibashi, R.V. Hijman, S. Tarucha, and L.P. Kouwenhoven, *Nature* **395**, 873 (1998).
36. R.H. Blick, D. Pfannkuche, R.J. Haug, K.V. Klitzing, and K. Eberl, *Phys. Rev. Lett.* **80**, 4032 (1998).
37. A. Georges and Y. Meir, *Phys. Rev. Lett.* **82**, 3508 (1999).
38. R. Aguado and D.C. Langreth, *Phys. Rev. Lett.* **85**, 1946 (2000).
39. T. Aono and M. Eto, *Phys. Rev.* B **63**, 125327 (2001).
40. B. Dong and X.L. Lei, *Phys. Rev.* B **65**, Art. No. 241304 (2002).
41. C.A. Büsser, E.V. Anda, A.L. Lima, M.A. Davidovich, and G. Chiappe, *Phys. Rev.* B **62**, 9907 (2000).
42. W. Izumida and O. Sakai, *Phys. Rev.* B **62**, 10260 (2000).
43. R. Aguado and D.C. Langreth, cond-mat/0207283.

44. B.A. Jones and C.M. Varma, *Phys. Rev. Lett.* **58**, 843 (1987).
45. B.A. Jones, C.M. Varma, and J.W. Wilkins, *Phys. Rev. Lett.* **61**, 125 (1988).
46. I. Affleck and A.W.W. Ludwig, *Phys. Rev. Lett.* **68**, 1046 (1992).
47. A.M. Chang, H.U. Baranger, L.N. Pfeiffer, and K.W. West, *Phys. Rev. Lett.* **73**, 2111 (1994).
48. V.N. Prigodin, K.B. Efetov, and S. Iida, *Phys. Rev. Lett.* **71**, 1230 (1993).
49. C.W.J. Beenakker, *Phys. Rev.* B **44**, 646 (1991).
50. H. Bruus and A.D. Stone, *Phys. Rev.* B **50**, 18275 (1994).
51. E.B. Foxman, U. Meirav, P.L. McEuen, M.A. Kastner, O. Klein, P.A. Belk, D.M. Abusch, and S.J. Wind, *Phys. Rev.* B **50**, 14193 (1994).
52. R. Berkovits and B.I. Altshuler, unpublished.
53. W. Heisenberg, *Z. Phys.* **78**, 156 (1932).
54. A. Bohr, B.R. Mottelson, and D. Pines, *Phys. Rev.* **110**, 936 (1958).
55. M.T. Tuominen, J.M. Hergenrother, T.S. Tighe, and M. Tinkham, *Phys. Rev. Lett.* **69**, 1997 (1992).
56. T.M. Eiles, J.M. Martinis, and M.H. Devoret, *Phys. Rev. Lett.* **70**, 1862 (1993).
57. C. Jayaprakash, R. Krishnamurthy, and J.W. Wilkins, *Phys. Rev. Lett.* **47**, 737 (1981).
58. R.M. Fye, J.E. Hirsch, and D.J. Scalapino, *Phys. Rev.* B **35**, 4901 (1987).
59. M. Ciorga, A.S. Sachrajda, P. Hawrylak, C. Gould, P. Zawadzki, S. Jullian, Y. Feng, and Z. Wasilewski, *Phys. Rev.* B **61**, R16315 (2000).
60. A. Kaminski and L.I. Glazman, *Phys. Rev.* B **61**, 15927 (2000).
61. Y.M. Blanter, A.D. Mirlin, and B.A. Muzykantskii, *Phys. Rev. Lett.* **78**, 2449 (1997).
62. Harold U. Baranger, Denis Ullmo, and Leonid I. Glazman, *Phys. Rev.* B **61**, R2425 (2000)
63. B.I. Halperin, unpublished.
64. S. Luscher, T. Heinzel, K. Ensslin, W. Wegscheider, and M. Bichler, *Phys. Rev. Lett.* **86**, 2118 (2001).
65. D. Weinmann, W. Husler, and B. Kramer, *Phys. Rev. Lett.* **74**, 984–987.(1995).
66. C.M. Canali, *Phys. Rev. Lett.* **84**, 3934 (2000).
67. M. Brodsky, N.B. Zhitenev, R.C. Ashoori, L.N. Pfeiffer, and K.W. West, *Phys. Rev. Lett.* **85**, 2356 (2000).
68. J. Schmid, J. Weis, K. Eberl, and K.V. Klitzing, *Physica* B **256–258**, 182 (1998).
69. S. Sasaki, S. De Franceschi, J.M. Elzerman, W.G. van der Wiel, M. Eto, S. Tarucha, and L.P. Kouwenhoven, *Nature* **405**, 764 (2000).
70. W.G. van der Wiel, S. De Franceschi, T. Fujisawa, J.M. Elzerman, S. Tarucha, and L.P. Kouwenhoven, *Science* **289**, 2105 (2000).
71. Y. Ji, M. Heiblum, D. Sprinzak, D. Mahalu, and H. Shtrikman, *Science* **290**, 779 (2000).
72. J. Nygard, D.H. Cobden, and P.E. Lindelof, *Nature* **408**, 342 (2000) and references therein.
73. Note that in these Figs 15 and 16, a single up arrow denotes an unpaired electron. It is not intended to represent the actual direction of spin alignment. In fact in a complex many-body state, the dot spin most likely contains projections from both up and down spin states.
74. F.R. Waugh, M.J. Berry, D.J. Mar, R.M. Westervelt, K.L. Campman, and A.C. Gossard, *Phys. Rev. Lett.* **75**, 705 (1995).
75. C. Livermore, C.H. Crouch, R.M. Westervelt, K.L. Campman, and A.C. Gossard, *Science* **274**, 1332 (1996).
76. K.A. Matveev, L.I. Glazman, and H.U. Baranger, *Phys. Rev.* B **53**, 1034 (1996).
77. J.M. Golden, B.I. Halperin, *Phys. Rev.* B **53**, 3893 (1996).
78. J. Gres, D. Goldhaber-Gordon, S. Heemeyer, M.A. Kastner, H. Shtrikman, D. Mahalu, and U. Meirav, *Phys. Rev.* B **62**, 2188–2194 (2000).
79. P.G. Silvestrov and Y. Imry, *Phys. Rev.* B **65**, 035309 (2002).

Chapter 5

Classical and Quantum Transport in Antidot Arrays

D. Weiss, K. Richter, and J. Eroms

Universität Regensburg, D-93040 Regensburg, Germany

1. Introduction

The story of solid state physics is the story of electrons in periodic potentials caused by the periodic arrangement of atoms in crystalline solids. It was early recognized that an additionally imposed periodic potential can significantly modify the solid's properties [1]. A major breakthrough in this respect was the concept of bandstructure engineering introduced by L. Esaki and R. Tsu [2,3]. The advent of molecular beam epitaxy with the possibility to grow semiconductor crystals atom by atom (see, e.g. [4]) allowed one not only to fabricate such one-dimensional superlattices superimposed upon the three-dimensional crystallographic lattice but also to realize two-dimensional electron systems (2DES) of superior quality [5]. Two-dimensional electron systems offered a wealth of new phenomena and are still the subject of current research. Prominent examples of effects based on the reduction of dimensionality for conduction electrons (or holes), are the quantum Hall effect (QHE) [6] and the fractional quantum Hall effect (FQHE) [7]. New phenomena were also expected for a one- or two-dimensional (2D) periodic potential imposed upon a two-dimensional electron system [8,9]. By using nanolithographic techniques it is possible these days to impress periodic potentials of different strength, period and shape upon high mobility two-dimensional electron systems such that the electron mean free path l_e and phase coherence lenght l_ϕ is much longer than the period of the periodic potentials. Two different types of potential landscapes are displayed in Fig. 5.1 showing weak and strong 2D-periodic potentials imposed upon a two-dimensional electron system. In a weakly density modulated system the modulation of the conduction band edge at the interface of a semiconductor heterojunction is much smaller than the Fermi energy in the 2DES. In contrast, in a strongly modulated system the periodic potential breaks through the Fermi energy and defines regions of vanishing electron density in the plane of the two-dimensional electron system. Although a periodic potential opens gaps in the free electron dispersion of a 2DES such band structure effects were only seen recently in weakly modulated

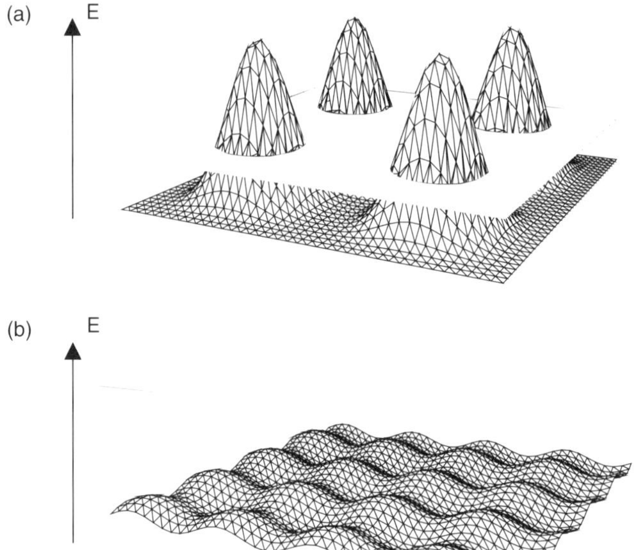

Fig. 5.1. Examples of strong (a) and weak periodic potential (b) imposed upon a two- dimensional electron system. Shown is the conduction band edge in the x-y-plane of the heterointerface and the Fermi energy (gray plane). In a weakly density modulated system the variation of the carrier density (conduction band edge) is much smaller than the (average) carrier density (Fermi energy). In the case of a strong modulation the conduction band edge breaks through the Fermi energy and defines regions of vanishing carrier density.

2DES [10,11] and demand, besides electron systems with extremely high mobility, periods in the 100 nm range (or below) and temperatures in the millikelvin regime. Hence, other effects usually dominate the resistance of density modulated systems with larger periods and at temperatures in the Kelvin range.

If a magnetic field is applied perpendicularly to a density modulated two-dimensional electron gas pronounced oscillations appear in the resistance. These oscillations reflect the interplay between the two characteristic length scales of the system, the cyclotron radius R_c and the period a of the artificial potential. Since their discovery [12] these oscillations were intensively investigated both experimentally and theoretically. The oscillations can be explained either in a picture of Landau band formation due to the periodic potential [13,14] or, classically, within the "guiding-center-drift" resonance picture [15]. Both pictures rely on a nearly unperturbed cyclotron motion in the presence of a periodic potential. While valid for low modulation amplitudes this is no longer the case if the amplitude of the modulation increases. For a large potential modulation, as shown in Fig. 5.1(b), the cyclotron motion of electrons with Fermi energy E_F and with cyclotron diameters $2R_c$ comparable or larger than the lattice period a becomes strongly distorted. The motion of the electrons in phase space turns classically chaotic. Hence such antidot lattices, as they were called [16], constitute a model system to study Hamiltonian chaos in a solid state system [17].

The physics of transport in antidot lattices is surprisingly rich and stretches from classical mathematical physics—an example is the derivation of Ohm's law in a so called Lorentz gas [18,19]—to an experimental verification of the composite fermion concept in the fractional quantum Hall regime [20,21]. Below some aspects of dc-transport in antidot arrays will be reviewed.

2. Antidot Arrays

Antidot arrays consist of a periodic lattice of nanometer-sized holes etched into semiconductor sandwich structures containing a two-dimensional electron gas (2DEG) at the interface of the heterojunction. This procedure results in a periodic potential landscape for the conduction electrons, similar to the one shown in Fig. 5.1(a). These electrons move at constant energy, the Fermi energy E_F, in between the periodically arranged potential posts. Such systems with lattice constants of usually a few hundred nanometer allow one to investigate electrical transport in an interesting regime: While the electron mean free path l_e is large (up to 20 μm) compared to the period a of the lateral superlattice, the Fermi wavelength $l_F = (2\pi/n_s)^{1/2} \sim 50\,\text{nm}$ in typical heterojunctions is smaller than a. Here, n_s is the carrier density of the two-dimensional electron gas. An antidot array can thus be considered as an artificial two-dimensional crystal with semi-classical electron dynamics. The electron dynamics in such a system is chaotic and related to the motion of a particle in a so called Sinai billiard, a system used to investigate classical chaotic dynamics and its manifestations in quantum spectra (see, e.g. [22]) This is in contrast to work carried out on systems with superimposed weak periodic potentials where the electron motion is regular [12–14].

The starting point for the fabrication of antidot lattices are very often GaAs/AlGaAs sandwich structures which contain a high-mobility 2DES at the interface between GaAs and AlGaAs. But also other material systems can be used as will be shown below. A typical layer sequence of a heterojunction together with the band structure is displayed in Fig. 5.2. The electrons are confined in a potential well of the conduction band (see Fig. 5.2(b)) and the electron motion is quantized in z-direction, giving rise to discrete energy levels E_i as is shown in Fig. 5.2(c) (for a review see [23]). In the samples described below only the lowest subband E_0 is occupied. While electron motion in z-direction is prohibited, the motion along the interface (in x- and y-direction) is free. The kinetic energy of an electron at zero magnetic field is given by $E = E_0 + \hbar^2 k_x^2/2m^* + \hbar^2 k_y^2/2m^*$ equal to the energy (apart from the subband energy E_0) of a free two-dimensional electron gas with the free electron mass replaced by the effective mass m^*. The advantage of heterojunctions is the high mobility of the electrons in the 2DEG which is due to the spatial separation of the ionized donors (located in the Si-doped AlGaAs layer, see Fig. 5.2(a)) and the electrons [5]. The other reason for the high electron mobility is the perfect crystalline heterojunction interface. The high mobility is an essential ingredient which assures that the electron mean free path, l_e, is much longer than the period a of the superimposed periodic potential. Typical mean free paths range from 5 to 20 μm before patterning.

Using electron beam lithography and dry etching techniques a repulsive antidot potential can be imposed upon the two-dimensional electron gas [24]. The fabrication

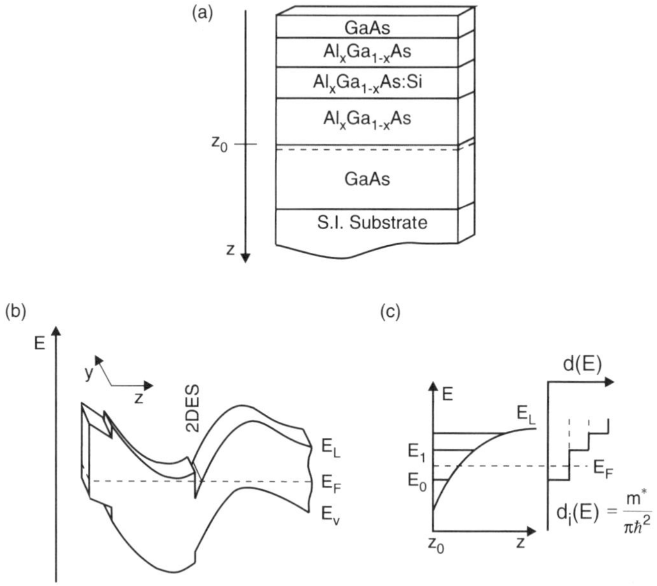

Fig. 5.2. (a) Typical layer sequence of a GaAs/AlGaAs heterojunction. The 2DES forms at the GaAs/AlGaAs interface (at position z_0). (b) Sketch of conduction and valence band showing the nearly triangular confinement potential at the interface in z-direction. (c) Confining potential and corresponding subbands. At $T = 0$ all states are occupied up to E_F. The corresponding density of states $d(E)$ is shown on the right hand side. $d(E)$ is constant within a subband. In the experiment only the lowest subband is occupied. There $d(E) = m^*/\pi\hbar^2$.

is sketched in Fig. 5.3. Typically, a periodic array of antidots is written into the PMMA-resist layer by using the highly focused electron beam of a scanning electron microscope (Fig. 5.3(a)). After a developing step, the holes in the PMMA are transferred into the heterojunction by reactive ion etching using $SiCl_4$, sketched in Fig. 5.3(b). A potential for the electrons similar to the one depicted in Fig. 5.1(a) results. Due to negatively charged surface states the conduction band is bent up at the antidot positions and defines the repulsive posts of the potential. The effective antidot diameter d consists of the lithographic diameter plus the surrounding depletion region which can be reduced by brief illumination [25,26]. Typical antidot arrays have periods between 100 and 1000 nm. For transport measurements the antidot array is incorporated in a conventional Hall bar geometry sketched in the inset of Fig. 5.4(b).

Other techniques for fabricating antidot lattices, not discussed here, involve, e.g. ion beam implantation [16,27,28], patterned gates [29–32], AFM-lithography [33,34] or local doping [35] to define a strong density modulation in the 2DES.

3. Early Experiments and Pinball Model

The periodic potential modifies the low-field magnetoresistance in a characteristic manner. To probe the magnetotransport coefficients like resistivity ρ_{xx} and the Hall

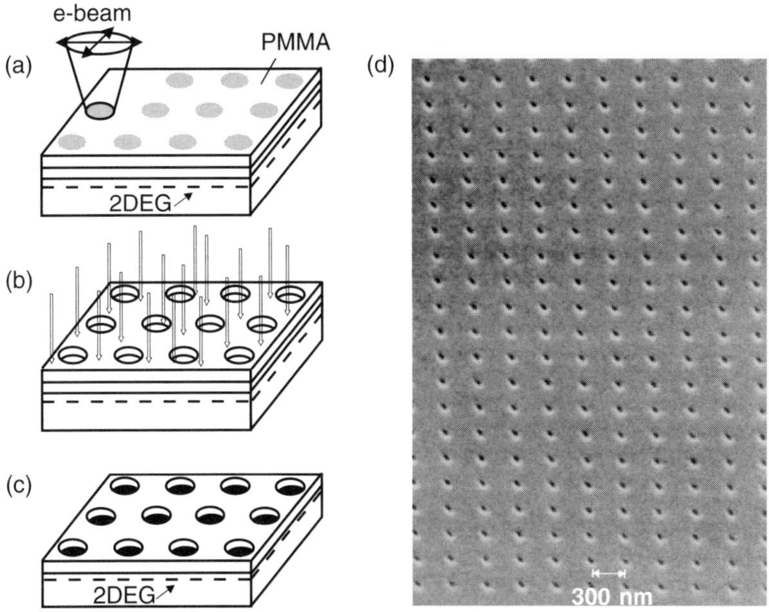

Fig. 5.3. (a) Sketch of the fabrication of an antidot array by e-beam lithography and (b) subsequent etching of the holes through the 2DEG by using reactive ion etching. (c) After etching the polymethylmethacrylate (PMMA) resist mask is removed in a solvent and the array fabrication is completed (d) An electron micrograph of a finished antidot array (top view).

resistivity ρ_{xy} a low-frequency ac-current is applied between the current probes of the Hall bar. The voltage drops V_x along or V_y across the device are measured by using standard Lock-in techniques. This experimental setup is sketched in the inset of Fig. 5.4(b). In Fig. 5.4 $\rho_{xx} = V_x w/Il$ with w, the width of the Hall bar, l the distance between potential probes and I the current through the Hall bar, and $\rho_{xy} = V_y/I$ of patterned and unpatterned areas of the device are compared. Pronounced low-field anomalies appear in the ρ_{xx} and ρ_{xy} trace of the antidot array at low magnetic fields, $B < 1$ T [25]. In this regime the transport coefficients of the unpatterned device can be approximated by the Drude expressions $\rho_{xx} = m^*/e^2 n_s \tau$, and $\rho_{xy} = B/en_s$. Here, τ is the electron scattering time. A double peak structure in ρ_{xx}, emerging at B values where the classical cyclotron radius at the Fermi energy, $R_c = \hbar(2\pi n_s)^{1/2}/eB$, equals $\sim 0.5\,a$ and $\sim 1.5\,a$, is strikingly different from the simple Drude picture. The ρ_{xx} peaks are accompanied by non-quantized steps in the Hall resistance displayed in Fig. 5.4(b). At higher B values where the cyclotron diameter becomes smaller than the period a, ρ_{xx} drops quickly, Shubnikov-de Haas (SdH) oscillations commence, and ρ_{xy} begins to display quantized Hall plateaus. In this high field regime the traces of patterned and unpatterned segments become essentially identical indicating that the intrinsic high mobility is preserved between the antidots. The differences in the position of the SdH minima indicate a slightly different carrier density in patterned and unpatterned regions. At zero field, however, the mobility is limited by scattering off the antidots

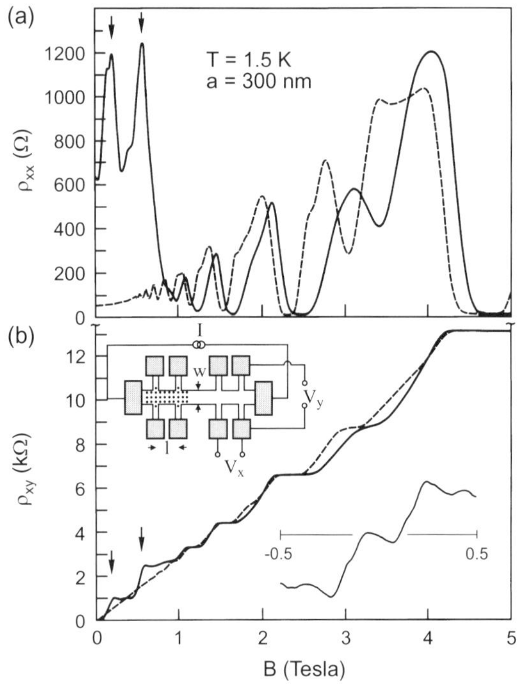

Fig. 5.4. (a) Magnetoresistance and (b) Hall resistance in patterned (solid line) and unpatterned (dashed line) sample segments at 1.5 K after brief illumination. The arrows mark magnetic field positions where $R_c/a \sim 0.5$ and ~ 1.5. The inset in (b) sketches the sample geometry with patterned (left) and unpatterned segments (right) of the two-dimensional electron gas and illustrates the measurement of the magnetotransport coefficients. Bottom inset: Magnification of the quench in ρ_{xy} about $B = 0$. After [25].

and the corresponding mean free path becomes comparable to the antidot spacing a. An additional anomaly in the ρ_{xy} trace is the negative Hall resistance close to $B = 0$; this low B regime with the quenched and negative Hall effect is displayed in the bottom inset of Fig. 5.4(b). Here, negative Hall resistance means that the electrons move on average opposite to the Lorentz force.

The number of peaks and steps resolved at low magnetic fields critically depends on the effective diameter d and the period a of the antidots. The larger the ratio d/a the less peaks are observable in the resistivity. In Fig. 5.5 the magnetoresistance traces of three samples with different periods and effective antidot diameters are compared. For curves with smaller zero-field resistance a progressively greater number of peaks becomes resolved. While the trace of device A ($a = 200$ nm, $d \sim 100$ nm) displays only one pronounced peak at a magnetic field position where the cyclotron diameter is equal to the period, $2R_c = a$, sample C ($a = 300$ nm, $d \sim 100$ nm) exhibits the largest sequence of ρ_{xx} peaks and ρ_{xy} plateaus. At each peak, R_c can be associated with a commensurate (circular) orbit encircling a specific number of antidots sketched in the upper right inset of Fig. 5.5. The success of such a simple circular orbit analysis

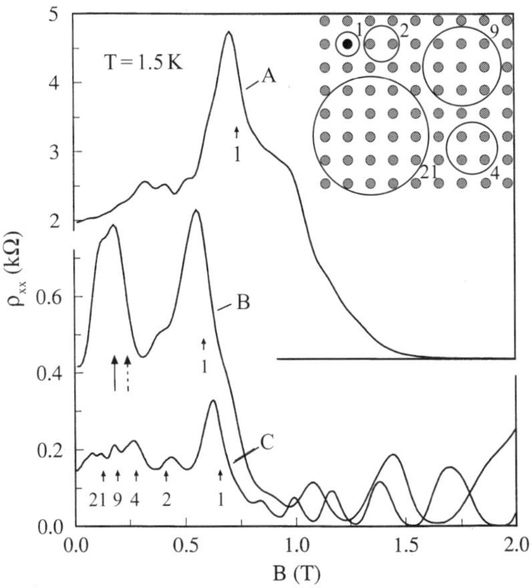

Fig. 5.5. Low-B anomalies from three different samples. For smaller d/a more structure in ρ_{xx} evolves. All peaks in trace C can be ascribed to commensurate orbits around 1, 2, 4, 9, and 21 antidots, as is sketched in the inset. Corresponding R_c/a values, marked by arrows, are 0.5, 0.8, 1.14, 1.7, and 2.53, respectively. The dashed arrow for trace B marks the position of an unperturbed cyclotron orbit around four antidots ($R_c/a = 1.14$). The shift of the corresponding resistance peak towards lower B indicates the deformation of the cyclotron orbit in a "soft" potential [17]. After [25].

led to an intuitive model ("pinball model") based on a Drude type of picture [25]. The orbits involved in this picture are shown in the inset of Fig. 5.6. At low magnetic fields where the cyclotron diameter is larger than the antidot spacing scattered (s) orbits and pinned orbits (p) coexist. With increasing magnetic field drifting orbits (d) start to dominate phase space. A magnetic field dependence of the phase space fractions of scattered, pinned and drifting orbits is displayed in Fig. 5.6. Pinned orbits, within this simple picture, remain localized about their orbit centers and cannot contribute to transport: hence the resistivity of the pinned orbits $\rho_p = \infty$. Nonetheless, these orbits play a central role in the story since they remove a fraction $f_p(B)$ of carriers from the transport process. As in a pinball game, the scattered orbits constitute the collection of paths leading between two pairs of antidots. This fraction of orbits $f_s(B)$ scatters with an effective relaxation time $\tau' = (\tau^{-1} + \tau_{ex}^{-1})^{-1}$. Here, τ and τ_{ex} are the intrinsic (of the unpatterned material) and the extrinsic scattering time, respectively. The latter is of order $\tau_{ex} \sim a/v_F$ with the Fermi velocity v_F. The resistivity for scattered orbits is thus $\rho_s = \rho_0\tau/f_s\tau'$ with $\rho_0 = m^*/e^2 n_s \tau$ the resistivity of the unpatterned material. At higher magnetic field, however, it is drifting orbits involving the fraction $f_d(B)$ of carriers which dominate transport. These behave as if in an unpatterned sample and $\rho_d = \rho_0/f_d$. In an antidot array each contingent contributes to the total resistivity, which is obtained from the inverted sum of the individual conductivity tensors (for

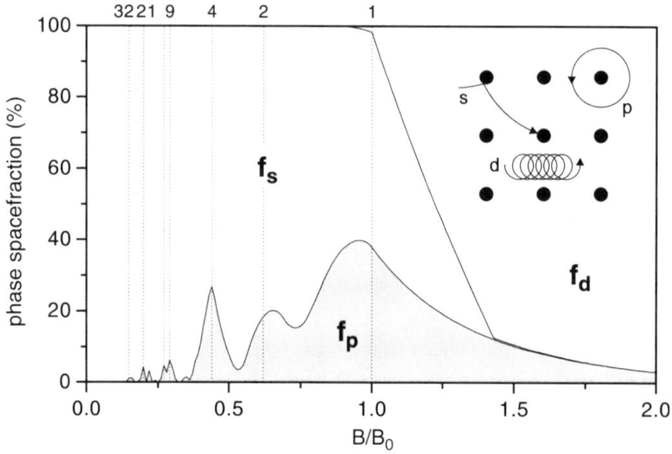

Fig. 5.6. Magnetic field dependence of the phase space fractions of scattered (f_s), pinned (f_p) and drifting (f_d) orbits for $d/a = 1/3$. At low B only scattered and pinned orbits exist while at high B the drifting ones dominate. From [102].

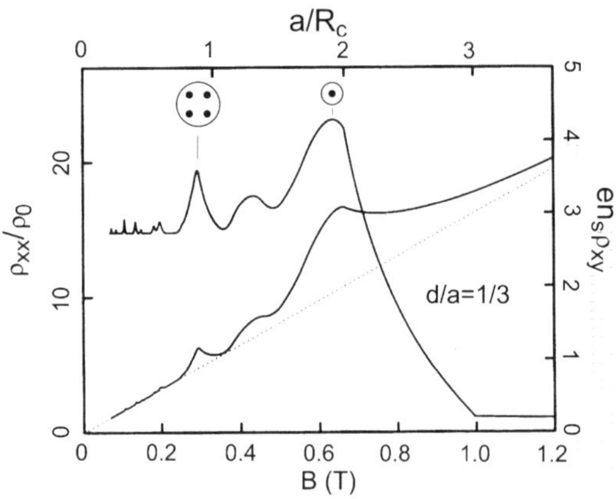

Fig. 5.7. Calculated resistivities within the "pinball-model". Features attributed to pinned orbits around $n = 1$ and 4 antidots are denoted. After [25].

more details see [25]). A calculation based on this simple pinball model is shown in Fig. 5.7. The magnetic field dependent phase space fractions are obtained numerically and are shown in Fig. 5.6. At cyclotron radii $2R_c > a$ the resistance level is high and dominated by scattering off the antidots with scattering rate $\sim 1/\tau'$. Superimposed upon this elevated resistance level are resistance peaks which stem from the fraction of pinned orbits f_p. They appear in a square lattice of antidots with sufficiently small diameter when $R_c/a = 0.5, 0.8, 1.14, 1.7, 2.53, \ldots$ holds. For increasing magnetic field the fraction of pinned and scattered orbits drops and the resistivity of the unpatterned 2DES is recovered once $f_d \to 1$. The calculated ρ_{xx} trace in Fig. 5.7 displays peaks which closely correspond to the ones observed in experiment. Also shown is the calculated Hall resistance modeled along the same lines as described above [25]. The calculation reproduces the characteristic step structure which, within this model, is due to a reduction of current carrying electrons at commensurate magnetic fields. Although the experimentally observed low-field maxima in Fig. 5.7 (equivalent to trace C of Fig. 5.5) are well described by unperturbed circular cyclotron orbits, the simple orbit analysis fails to explain the prominent low B peak in curve B of Fig. 5.5: the R_c/a value of ~ 1.5 deduced from the peak position at ~ 0.18 T is not commensurate with the lattice. Also the negative Hall resistance can not be explained within such a modified Drude picture. This deviations with respect to the pinball model are due to the real "soft" antidot potential sketched in Fig. 5.1(b). The circular orbit approximation is reasonable only as long as the Fermi energy (with respect to the bottom of the potential landscape) is constant between the antidots: this only holds in the limit of small antidots and large periods or in the limit of "hard wall" antidot potential posts. In general, the equations of motion in a realistic antidot model potential have to be solved in order to get information about the classical electron dynamics. Corresponding model calculations, sketched below, show that the distinct low field peak in trace B of Fig. 5.5 can be associated with electrons on chaotic trajectories trapped on paths around four antidots [17]. Compared to the free electron case the "orbit-around-four-antidots-peak" is shifted towards lower B and provided a first experimental signature of the complex electron dynamics.

The quenching and even negative value of the Hall resistance shown in the bottom inset of Fig. 5.4(b) also was a surprising result. While such quenching has been observed in small, mesoscopic junctions where the electron motion between the sample boundaries is ballistic [36], the antidot array itself is macroscopic; its dimension is large compared to l_e. The quenching of the Hall resistance is just another example which underlines the importance of the classical chaotic electron dynamics pointed out in the next section.

The results presented above were obtained from square lattices. A variety of other geometrical arrangements and also material systems have been investigated up to now and an ample amount of literature exists. Triangular antidot lattices [37–41] as well as rectangular ones [42–46] have been investigated by a number of groups. In addition Penrose lattices [47], disordered arrays, and antidot lattices with anisotropic antidot positions were explored [45,48–51]. Also the dependence of the commensurability features on the shape of the antidots was investigated [52–54]. Different material systems were used to fabricate (semiconductor) antidot lattices and involve SiGe heterostructures [55–57] and InAs based devices [58–60].

4. Chaotic Dynamics in Antidot Lattices

While the simple pinball model introduced in the previous section can describe some basic features observed in experiment some important ingredients are missing. Of course, quantum mechanical effects are not included in this description based on the Drude model and will be discussed later. But also on this classical level, introduced above, deficiencies exist. It turns out (see below) that the phase space fraction of pinned orbits is much to small to explain the pronounced commensurability features. On the other hand, the role of orbits was ignored which can "hop" from one antidot to the next at commensurate magnetic fields and increase the conductivity in the hopping direction (run-away trajectories [61]). Depending on the aspect ratio d/a or the geometry of the lattice these trajectories can become important for the commensurability features observed in experiment [46,62–64]. Apart from the neglect of run-away trajectories also the assumption of a hard, pillar-like antidot potential does not correctly reflect the experimental situation. Instead the potential landscape looks more like depicted in Fig. 5.1(a). A theoretical picture including all but the pinned orbits in a soft potential was put forward by Fleischmann, Geisel and Ketzmerick (FGK) [17]. The pinned orbits on stable periodic and quasi-periodic orbits, which where at the heart of the pinball model, can be neglected as they constitute only a small fraction in phase space and can not explain the pronounced features observed in experiment [17]. Calculated trajectories (see below) also show that the classical motion of an electron "injected" into an antidot potential with constant energy E_F is chaotic. The chaotic motion is manifested by an extreme sensitivity of the electron trajectories on the initial conditions. Fleischmann, Geisel and Ketzmerick [17,65] were able to explain the magnetotransport anomalies observed by accounting for the non-linear classical electron dynamics in antidot lattices with soft periodic potentials: Using $U(x, y) = U_0[\sin(\pi x/a)\sin(\pi y/a)]^\beta$ as a model potential, FGK studied the phase space structure of the classical Hamiltonian,

$$H = E_F = \frac{1}{2m^*}(\mathbf{p} - e\mathbf{A})^2 + U(x, y) \tag{1}$$

by integrating the Hamiltonian equations of motion. Here, **p** is the electron momentum and **A** the vector potential. Model potentials with $\beta = 4$ and $\beta = 64$ are shown in Fig. 5.8 together with calculated trajectories. As shown in Fig. 5.8, the phase space is generally divided into regions with regular cyclotron-like motion (with phase space

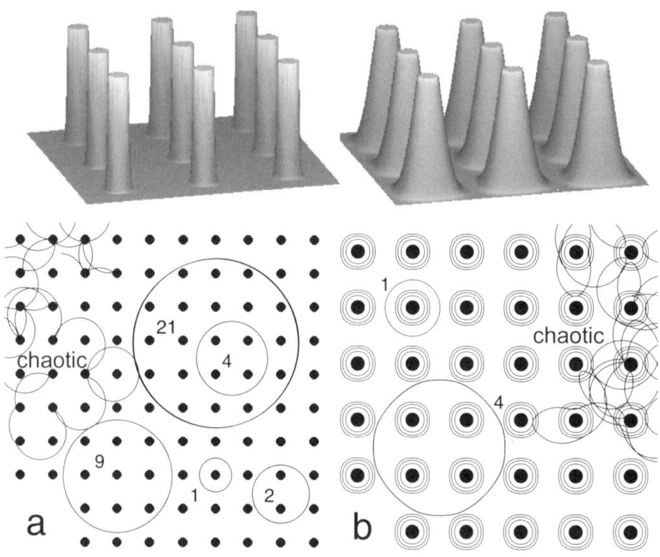

Fig. 5.8. Model antidot potential for $\beta = 64$ (upper left) and $\beta = 4$ (upper right) with corresponding trajectories. Courtesy of R. Fleischmann.

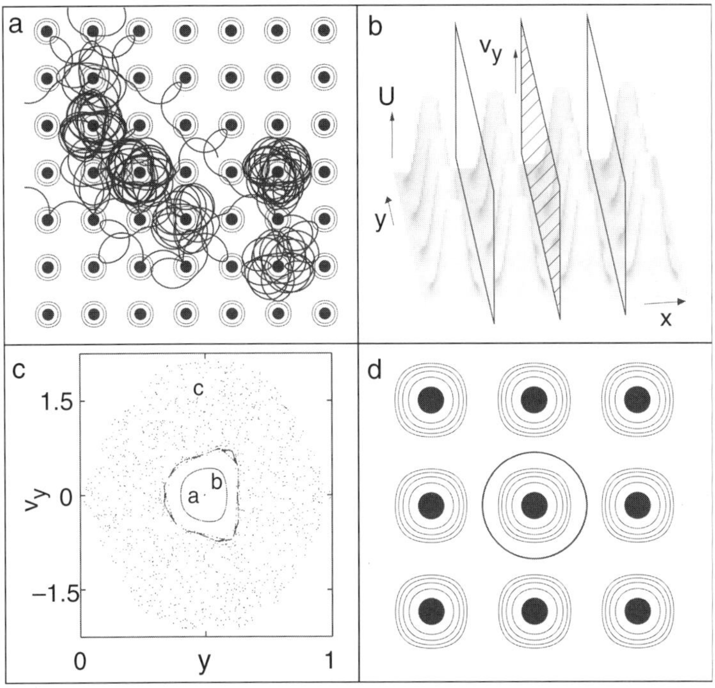

Fig. 5.9. (a) Calculated trajectory for $2R_c = a$ illustrating their chaotic nature. (b) Definition of Poincaré surfaces of section in an antidot array. (c) Typical Poincaré surface of section for $2R_c \approx a$. Areas of regular motion with periodic and quasi-periodic orbits (labeled a and b) coexist with chaotic ones (c). (d) Periodic orbit in the center of the stable island. The contour lines of the soft antidot potential are shown.

fraction p_r) and chaotic motion (phase space volume p_c with $p_c + p_r = 1$). An example of a chaotic trajectory, calculated for a magnetic field where the cyclotron diameter is equal to the period, $2R_c = a$, is displayed in Fig. 5.9(a). To visualize chaotic and regular motion in such a mixed phase space usually Poincaré surfaces of section are used. The definition of such sections is depicted in Fig. 5.9(b): plotted are, e.g. the y- and v_y-coordinates of the points of intersections of the trajectories with the dashed plane and with all the other parallel planes [$x = 0(\text{mod}a)$]. A typical Poincaré surface of section is shown in Fig. 5.9(c). The central point, labeled a, corresponds to the periodic orbit displayed in Fig. 5.9(d) while the dotted region reflects chaotic motion. The dots stem from one chaotic electron trajectory which is scattered throughout the antidot lattice. The closed lines, denoted by b, describe quasi-periodic motion on tori. The electrons within the stable island behave like the pinned electrons of the simple pinball model: they do not contribute to transport [17] (if scattering out of the regular islands is neglected). Interestingly, however, Fleischmann et al. found that it is not primarily pinned regular orbits but structure in the velocity–velocity correlation function which is responsible for the ρ_{xx} peaks at commensurate B's. In the FKG approach, only electrons which are part of the chaotic phase space contribute to the conductivity σ.

Within a classical Kubo-type of model FGK find for the conductivity

$$\sigma_{ij}^{cl} = (1 - p_r)\frac{n_s e^2}{k_B T}\int_0^\infty e^{-t/\tau}\langle v_i(t)v_j(0)\rangle \, dt \tag{2}$$

where the velocity correlation function solely includes unperturbed chaotic orbits obtained numerically. The indices i, j stand for x- or y-direction and the bracket denotes averaging over phase space. As will be shown below such a formula naturally arises as the smooth part of a semi-classical trace formula for magnetoconductivity. For low temperatures the prefactor $n_s e^2/k_B T$ has to be replaced by $e^2 \bar{d}$ where $\bar{d}$ is the density of states at the Fermi energy [66,67]. The fraction p_r of the phase space volume of the regular motion is determined numerically. In recent work pinned orbits which contribute due to scattering (scattering out of periodic orbits) were included in the classical Kubo formula [68]. Note that the unperturbed chaotic motion is cut off by the intrinsic scattering time τ. Evaluating Eq. (2) gives the classical longitudinal conductivity σ_{xx} and Hall conductivity σ_{xy}; to obtain the corresponding resistivities the conductivity tensor has to be inverted. For an isotropic system one obtains:

$$\rho_{xx} = \frac{\sigma_{xx}}{\sigma_{xx}^2 + \sigma_{xy}^2} \qquad \rho_{xy} = \frac{\sigma_{xy}}{\sigma_{xx}^2 + \sigma_{xy}^2} \tag{3}$$

The calculated resistivities, based on the evaluation of Eqs (2) and (3), are displayed in Fig. 5.10(b) and show good agreement with the experimental traces displayed in Fig. 5.10(a). The two peaks at $B/B_0 \sim 1$ and at $B/B_0 \sim \frac{1}{3}$ (where $B_0 = 2\hbar k_F/ea$, is the field where $2R_c = a$) as well as the corresponding steps in the Hall resistance ρ_{xy} can be ascribed to structure in the correlation function of chaotic trajectories moving through the antidot lattice as depicted, for example, in Fig. 5.9(a). It is not only temporarily trapped orbits (see, e.g. the trajectories orbiting around a single antidot in Fig. 5.9(a)) which contribute to the structure in the transport coefficients but also trajectories which jump from one antidot to the neighboring ones and enhance the electron diffusion in the lattice [69]. These orbits which come into existence by a magnetic focusing effect for commensurate magnetic fields (an electron hitting an antidot is focused onto a neighboring one in the next "hop") crucially contribute to the magnetotransport coefficients [69]. A systematic study of the contributions of different types of trajectories, also based within the framework of the classical Kubo formula, was carried out Ishizaka [68] and is in good agreement with experimental data.

The classical Kubo formula also correctly predicts the shift of the "orbit-around-four" antidots peaks towards lower magnetic fields as compared to the circular cyclotron motion around four antidots, sketched in the inset of Fig. 5.5. This shift reflects "squeezed" orbits, shown for a periodic orbit in Fig. 5.11.

Also the negative Hall resistance which appears at very low B in experiment (see Fig. 5.4(b) and Fig. 5.10(a)) can be understood in the picture of classical motion. "Channeling" trajectories confined over many antidot unit cells between two rows of antidots can cause a suppression and even a negative Hall voltage [65]. Such trajectories which exist only at sufficiently low magnetic field are displayed in Fig. 5.12. A slightly modified Eq. (2) can also be used to describe the quenching of the Hall effect [65]. A factor

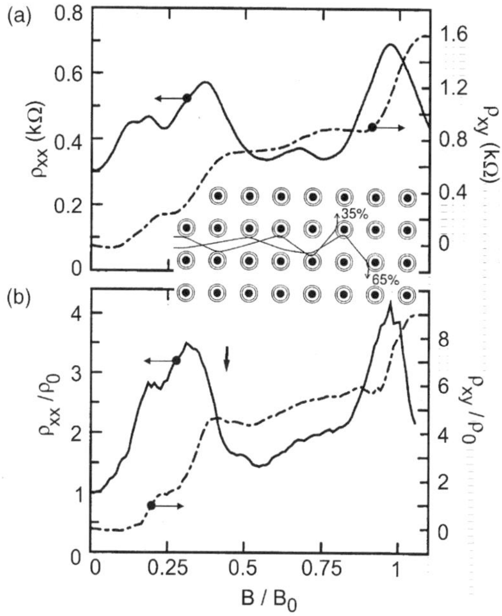

Fig. 5.10. (a) Experimental and (b) calculated (normalized) values of ρ_{xx} and ρ_{xy}. B_0 is the field where, for an unpatterned 2DEG, $2R_c = a$ holds. The magnetic field position of a circular orbit (free cyclotron motion) around four antidots is marked by an arrow. The inset illustrates the mechanism leading to a negative Hall effect: long chaotic trajectories following the Lorentz force are shifted to the upper row of antidots and are hence predominantly scattered in the direction *opposite* to the Lorentz force. Calculations are taken from [65]. After [63].

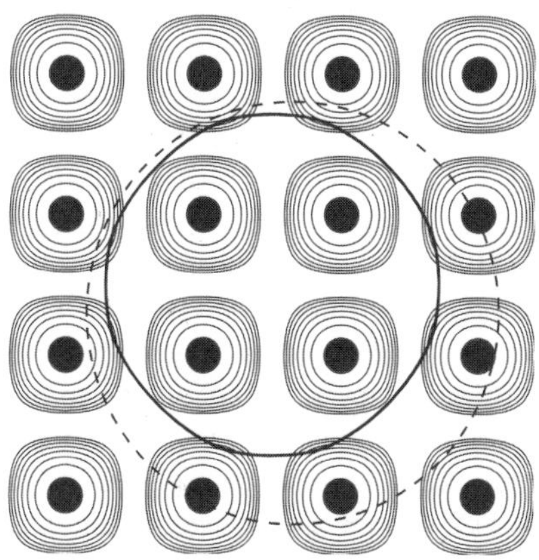

Fig. 5.11. Squeezed periodic orbit around four antidots in a soft antidot potential with $\beta = 4$, responsible for the shift of the "orbit-around-four" peak [17]. Courtesy of R. Fleischmann.

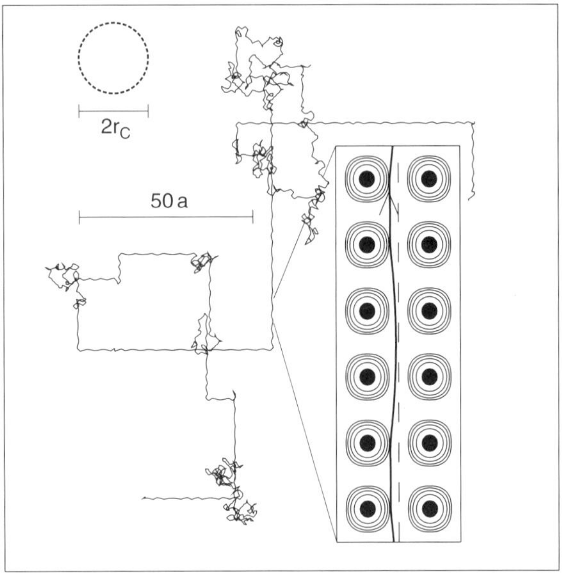

Fig. 5.12. Chaotic trajectory at low B. The magnified inset shows a chaotic trajectory trapped for a while in between two rows of antidots. These trajectories are responsible for the suppression of the Hall effect at low B [65]. Courtesy of R. Fleischmann.

$p_r \sigma_{ij}^r$ has to be added to Eq. (2) which includes the conductivities of non-chaotic trajectories "channeling" forever (in an infinite antidot array) between two rows of antidots at very low magnetic fields. For them $\sigma_{xx}^r = \sigma_{yy}^r = p_r n_s \tau e^2/m^*$ and $\sigma_{xy}^r = \sigma_{yx}^r = 0$ holds. Here, the superscript r denotes contributions of electrons along non-chaotic channeling trajectories. The trajectories which cause the negative Hall resistance are chaotic ones which follow the regular motion over many lattice constants. Shifted by the Lorentz force towards one row of antidots they scatter predominantly opposite to the direction of the Lorentz force (see inset of Fig. 5.10). The scattering against the Lorentz force is the origin of the negative Hall resistance at sufficiently low magnetic fields where these channeling trajectories exist [65]. The negative Hall effect at very low B is nicely reproduced in the model calculation displayed in Fig. 5.10(b).

So far only the resistivities were discussed but it is also instructive to look at the conductivities. While not directly accessible to experiment (on Hall bars) the conductivities σ_{xx} and σ_{xy} can be extracted from the experimentally determined resistivities using expressions similar to Eq. (3). In Fig. 5.13 σ_{xx} and σ_{xy} taken from three samples with different d/a values are displayed [63]. For the sample with the largest period ($a = 400$ nm) a maximum in σ_{xx} emerges at $B/B_0 \sim 1$ (see Fig. 5.13(a)). For samples with larger values of d/a these conductivity peak positions are shifted towards lower magnetic fields, away from the commensurate position $B/B_0 = 1$. On the other hand, all three samples (and all others investigated) show a minimum in σ_{xy} around $B/B_0 = 1$ (Fig. 5.13(b)) as is expected for trapped trajectories. It is the minima in σ_{xy} at $B/B_0 \sim 1$ which in most cases dominate the commensurability peaks in ρ_{xx} after inversion of the conductivity tensor (see Eq. (3)). Similar results were found by Schuster et al. [62].

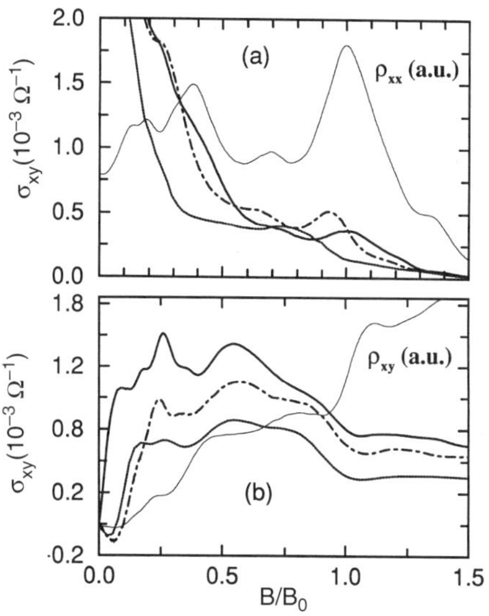

Fig. 5.13. Low-temperature conductivities σ_{xx} (a) and σ_{xy} (b) for three samples with differing d/a values. Also displayed is the longitudinal resistivity ρ_{xx} (a) and the Hall resistance ρ_{xy} (b) taken from the same sample as the dash-dotted conductivities. Only for the sample with the smallest d/a value (solid trace) the maximum in σ_{xx} coincides with the maximum position of ρ_{xx}. After [63].

The characteristic features of the experimental data displayed in Fig. 5.13—a peak in σ_{xx} which shifts, starting from $B/B_0 = 1$, to lower B with increasing aspect ratio d/a and a dip in σ_{xy} at $B/B_0 = 1$—were nicely reproduced theoretically [68]. While the peak in σ_{xx} can be ascribed to run-away trajectories in an open antidot lattice, the dip in σ_{xy} is due to orbits which change direction [68] (i.e. spend more time in the vicinity of an antidot).

5. Quantum Effects in Antidot Arrays

5.1. Semi-Classical Picture: Quantized Periodic Orbits

The oscillations in the resistivity observed at magnetic fields above 1 T in Fig. 5.4(a) and the corresponding quantized Hall steps in Fig. 5.4(b) are the consequence of Landau quantization. The density of states of a two-dimensional electron gas splits in a perpendicular magnetic field into a series of discrete Landau levels with occupancy $n_L = eB/2\pi\hbar$. The levels are separated by the cyclotron energy $\hbar\omega_c$ (see, e.g. [23]). At high magnetic fields where $2R_c < a - d$ holds the Landau band structure is relatively simple despite the presence of the antidots. It consists of the usual Landau levels in between the antidots and of quantized edge states surrounding the antidots. For lower B, however, where a the cyclotron orbit is larger than the lattice period a, the Landau levels mix and a complicated Landau band structure evolves. The magnetotransport experiments described in the previous sections were carried out at temperatures where quantum effects in the regime of commensurability anomalies were negligible. In fact the transport anomalies reported so far are observable up to high temperatures $T \sim 50\,\mathrm{K}$ supporting the classical description given above. In this section, transport is explored in a temperature regime where the quantization of chaotic electron motion comes into play [70,71]. Measurements of ρ_{xx} at $T \sim 0.4\,\mathrm{K}$ display quantum oscillations superimposed upon the low B resistance anomalies. Corresponding data for sample A (see Fig. 5.5) with $d/a \sim 0.5$ are shown in Fig. 5.14(a) where ρ_{xx} from both patterned and unpatterned sample segments are compared. In the unpatterned part, the $1/B$-periodic Shubnikov-de Haas oscillations reflect the Landau energy spectrum. The quantum oscillations in the antidot segment reveal quite different behavior. The oscillations are periodic in B with a period $\Delta B \sim 0.105\,\mathrm{T} \sim h/ea^2$ corresponding to the addition of approximately one flux-quantum h/e through the antidot unit cell. At 4.7 K, the quantum oscillations are smeared out while the characteristic ρ_{xx} peak at $2R_c = a$ persists. The oscillations periodic in B dominate only the low B regime ($2R_c > a - d$); at high B, the sample behaves as if unpatterned, and ρ_{xx} displays minima which are $1/B$-periodic and reflect quantization of essentially unperturbed cyclotron orbits (Fig. 5.14(a), inset). In Fig. 5.14(b) the oscillation index η for both the high and low field regime is plotted versus inverse magnetic field positions of the ρ_{xx} minima. Clear deviation from a straight line, characteristic for $1/B$ periodic SdH oscillations, is evident. At high B, η is identical with the filling factor $\upsilon = n_s h/eB$ counting the number of occupied (spin-split) Landau-levels. At low B, however, the filling factor looses its physical meaning since the antidot potential strongly mixes different Landau levels.

The temperature dependence of the ρ_{xx}-oscillations is very similar to the one of SdH oscillations. This suggests that the B-periodic oscillations, like the SdH oscillations, reflect the structure of the density of states (DOS) at the Fermi energy which will be explained below in more detail. In Fig. 5.15(a) the energy spectrum for different directions in the magnetic Brillouin zone and the corresponding DOS (Fig. 5.15(b)) for a model antidot potential is shown [72–74]. In comparison, the unperturbed Landau level spectrum of the unpatterned electron gas is also shown in Fig. 5.15(a). The insertion of the antidots into the 2DES leads to a complex magnetic bandstructure: the corresponding DOS at $T = 0$ (thin solid line in Fig. 5.15(b)) which, due to the complex arrangement of bands, has a lot of structure which is not observed in experiment. Finite temperatures (thick solid line in Fig. 5.15(b)) and in particular scattering processes lead to a smoothing of the DOS and only the most prominent features in the spectrum survive in experiment. In fact, the quantum oscillations observed (Fig. 5.14) can be explained by the quantization of few periodic orbits as is pointed out below.

A smoothed DOS can be described semi-classically. For the antidot lattice where the underlying classical electron motion is essentially chaotic the Gutzwiller trace formula is the appropriate semi-classical tool which expresses the DOS $\mathrm{d}(E, B)$ as a sum over the classical periodic orbits of the system [75]: Each periodic orbit gives rise to an oscillatory (in E or B) contribution to the DOS.

Similar semi-classical expressions can be derived for the magneto-conductivity tensor [66,76,77]. A brief discussion of the diagonal conductivity will be sketched below [66,78]; the semi-classical derivation of the Hall conductivity follows similar lines.

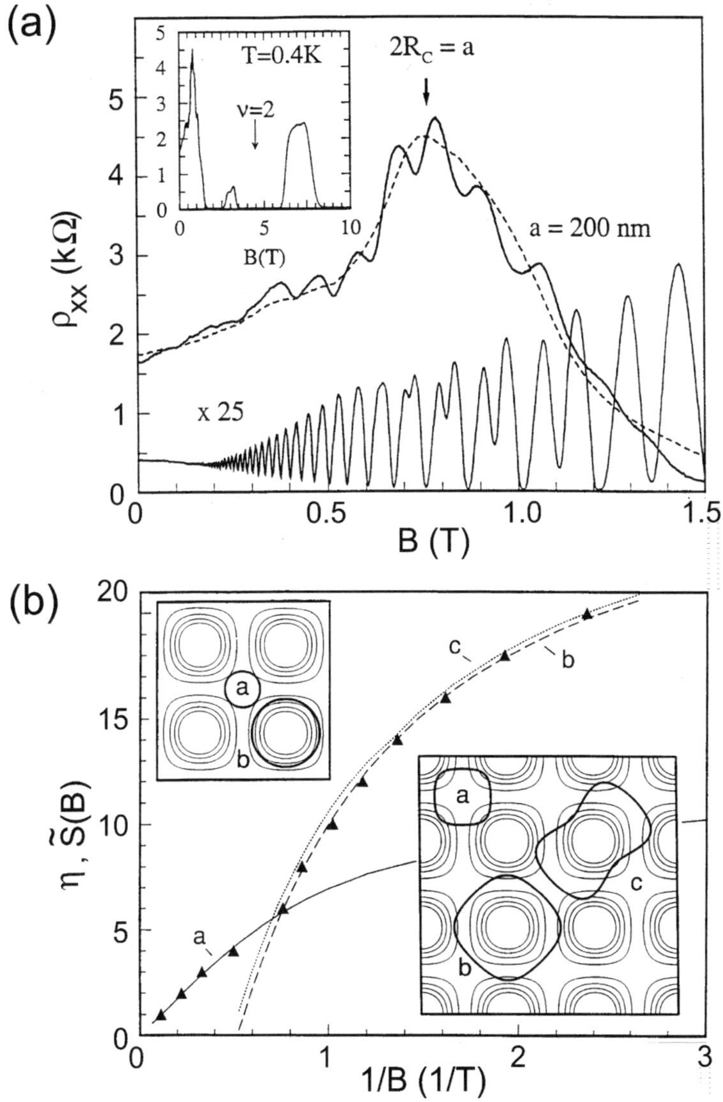

Fig. 5.14. (a) ρ_{xx} measured in patterned (top traces) and unpatterned (bottom trace) segments of the same sample at $T = 0.4$ K (solid lines) and 4.7 K (dashed line). The inset displays the ρ_{xx} trace from the patterned segment up to 10 T; the filling factor $\upsilon = 2$ is marked. At high B, the emergence of SdH-oscillations reflects the quantization of essentially unperturbed orbits. (b) The triangles mark all (up to 10 T) $1/B$ positions of the ρ_{xx} minima. At high B the resistance minima lie equidistant on the $1/B$ scale; at low B the spacing becomes periodic in B. Solid, dashed and dotted lines are calculated reduced actions $\tilde{S}(1/B)$ of orbit (a), (b), and (c), respectively. These orbits are shown for $1/B = 0.6\,\mathrm{T}^{-1}$ (top) and $1/B = 2.7\,\mathrm{T}^{-1}$ (bottom inset). The potential steepness in the calculation was $\beta = 2$. After [70].

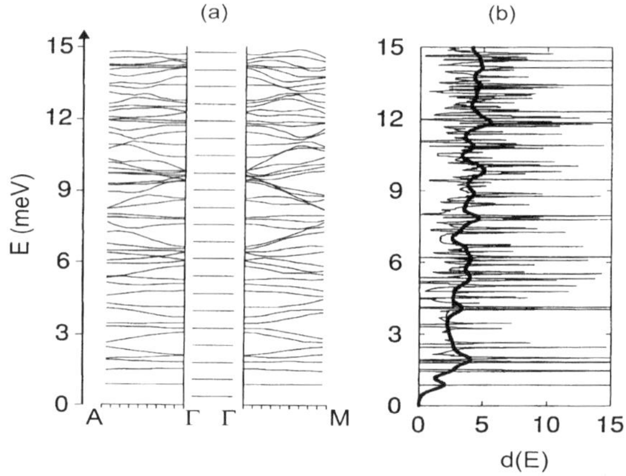

Fig. 5.15. Energy spectrum (a) and corresponding density of states (b) of electrons in an antidot lattice calculated for $B = 0.41$ T, $a = 200$ nm and $\beta = 2$ by H. Silberbauer [72,73]. The Landau level structure with equidistant level spacing is also displayed in (a). The density of states is shown for $T = 0$ (thin solid line) and for $T = 1.5$ K (thick solid line). After [51,63,73].

Within the Kubo formalism which is appropriate to describe transport in macroscopic antidot lattices the static conductivity at temperature T is given by [63]

$$\sigma_{xx}(E_{\mathrm{F}}, B; T) = g_{\mathrm{s}} \int \mathrm{d}E \left(-\frac{\mathrm{d}f}{\mathrm{d}E} \right) \langle \sigma_{xx}(E, B; T = 0) \rangle \tag{4}$$

where g_{s} accounts for the spin degeneracy and $f(E) = 1/[1+\exp((E - E_{\mathrm{F}})/k_{\mathrm{B}}T)]$ is the Fermi function. $\langle \ldots \rangle$ denotes an average over weak disorder in the antidot system. The conductivity is semi-classically evaluated within the Kubo formalism by approximating the Green functions involved by sums over classical paths. As a result the impurity averaged conductivity can be semi-classically expressed as

$$\sigma_{xx}(E_{\mathrm{F}}, B; T) = \sigma_{xx}^{cl}(E_{\mathrm{F}}, B) + \sigma_{xx}^{\mathrm{osc}}(E_{\mathrm{F}}, B; T) \tag{5}$$

Here, the smooth classical part, which depends only weakly on temperature, is

$$\sigma_{xx}^{cl}(E_{\mathrm{F}}, B) = e^2 \bar{d} \int_0^\infty \mathrm{d}t \langle v_x(t) v_x(0) \rangle_{\mathbf{pq}} \mathrm{e}^{-t/\tau} \tag{6}$$

where $\langle \ldots \rangle_{\mathbf{pq}}$ denotes a classical phase space average. Equation (6) corresponds to expression (2) which was used to explain the pronounced classical resistance peaks discussed in section 4.

Here, the focus lies on the additional oscillatory part $\sigma_{xx}^{\mathrm{osc}}$ accounting for quantum interference effects at low temperatures. It is semi-classically expressed in terms of a sum over contributions from classical periodic orbits (po) (and their higher repetitions j)

of electrons in the antidot lattice [66]:

$$\sigma_{xx}^{\rm osc}(E_{\rm F}, B; T)$$
$$= \frac{2g_s}{V}\frac{e^2}{h}\sum_{\rm po}\zeta_{xx}^{\rm po}\sum_{j=1}^{\infty}\frac{R(jT^{\rm po})\exp(-jT^{\rm po}/2\tau)}{\left|\det\left(\mathbf{M}_j^{\rm po}-\mathbf{1}\right)\right|^{1/2}}\cos\left[j\left(\frac{S^{\rm po}}{\hbar}-\alpha^{\rm po}\frac{\pi}{2}\right)\right] \quad (7)$$

Here, $S^{\rm po}(E, B) = \int_{\rm po} \mathbf{p}d\mathbf{q}$ is the classical action, $T^{\rm po}$ is the period and $\alpha^{\rm po}$ the Morse index of a periodic orbit. The monodromy matrix $\mathbf{M}_j^{\rm po}$ measures its stability [75]. The weights

$$\zeta_{xx}^{\rm po} = \int_0^{\infty} dt e^{-t/\tau}\int_0^{T^{\rm po}} dt' v_x(t')v_x(t+t') \quad (8)$$

are velocity-correlation functions along the periodic orbits. In addition to the exponential damping due to weak disorder (τ is the elastic scattering time) temperature effects are accounted for by the damping function $R(t) = (t/\tau_\beta)/\sinh\,(t/\tau_\beta)$ with $\tau_\beta = \hbar/\pi k_{\rm B}T$. Apart from $\zeta_{xx}^{\rm po}$ Eq. (7) for $\sigma_{xx}^{\rm osc}$ is essentially the same as Gutzwiller's trace formula for the density of states. The quantum oscillations are semi-classically related to interference (due to the phase differences $\sim jS^{\rm po}/\hbar$ between pairs) of paths along a periodic orbit differing by j in their number of full traversals.

In order to understand the measured B-periodic resistance oscillations in view of Eq. (7) we first focus on the periodicity which is essentially the same for conductivity and density of states. It is given by the actions of the periodic orbits involved. Finite temperature and impurity broadening leads to a smoothing of the observed quantum oscillations related to a cut off in the lengths of the periodic orbits involved. To understand the basic results of the experiment one therefore needs to consider only the shortest periodic orbits depicted in the insets of Fig. 5.14(b). The periodic orbits cause a modulation of $\sigma_{xx}(E_{\rm F}, B)$ with maxima given by their quantized B-dependent actions

$$S(B) = \oint(m^*\mathbf{v} + e\mathbf{A})d\mathbf{r} = m^*\oint \mathbf{v}d\mathbf{r} - eB\mathcal{A}(B) = 2\pi\hbar N \quad (9)$$

where $N = n + \gamma/2 + \alpha/4$ contains the quantum number n, the Morse index α and, in the case of a stable orbit, the winding number γ. In Eq. (9) $B\mathcal{A}(B)$ is the enclosed flux through a periodic orbit and $\mathbf{v}$ is the electron velocity given by Eq. (1). To compare with experiment the reduced action $\tilde{S}(B)$

$$\tilde{S}(B) \equiv 2\frac{S(B)}{h} - \gamma(B) - \frac{\alpha}{2} - 1 = 2n, \quad n = 1, 2, \ldots, \quad (10)$$

is calculated where $2n$ now labels minima in $\rho_{xx}(E, B)$. The y-axis of Fig. 5.14(b) represents $\tilde{S}$, where even η is equal to $2n$, described by Eq. (10). At high B, $2n$ corresponds to even filling factors υ. The calculated traces of $\tilde{S}(B)$ are shown in Fig. 5.14(b). The three periodic orbits, displayed in the insets of Fig. 5.14(b), perfectly explain the minima positions of ρ_{xx}. These are (i) an orbit between four antidots, denoted as (a), (ii) an orbit around one antidot, (b), and (iii) orbit (c) emerging from a bifurcation of orbit (b). They are sufficient to explain the periodicity of the quantum oscillations.

The $\tilde{S}(B)$ curves in Fig. 5.14(b) are calculated for a particular steepness of the model potential $U(x, y)$ (see Eq. (1)), namely $\beta = 2$. Using steeper potentials like $\beta = 4$ one finds less deviations from the usual $1/B$ periodic Shubnikov–de Haas oscillations of an unpatterned 2DEG. This behavior is observed experimentally [70]: data taken from a superlattice with smaller $d/a \sim 0.4$ deviate less from the usual $1/B$ periodicity of the Subnikov–de Haas oscillations. The crossover from B-periodic to $1/B$-periodic oscillations, dependent on the normalized cross section d/a, together with the fact that the antidot array is much larger than the phäse breaking length in a 2DEG [79], suggest that the B-periodic oscillations observed in experiment are not due to the Aharanov–Bohm interference mechanisms discussed in previous work [71,80]. An approach based on Eq. (7) requires only phase coherence on the length scale of a periodic orbit, i.e. a few lattice constants, and not throughout the entire system. In the calculations described above β is the only free parameter since d/a-values are taken from the experiment [25].

The magnetic field dependence of $\tilde{S}(B)$ in Fig. 5.14(b) can be qualitatively explained by a simplified approach evaluating the B dependence of the enclosed area $\mathcal{A}$. For unperturbed cyclotron motion $S(B) = eB\mathcal{A}$ holds, $\mathcal{A}(B) = \pi R_c^2$ scales with $1/B^2$, and $1/B$-periodic resistance oscillations result. At high B, orbit (a) is essentially unperturbed and in this realm oscillations periodic in $1/B$ are prominent in Fig. 5.14(b). Deviations from $\mathcal{A}(B) \propto 1/B^2$ destroy the $1/B$-periodicity: smaller action is caused by impeding the expansion of a cyclotron orbit. B-periodic oscillations result when $\mathcal{A}$ is essentially independent of B. This condition is closely fulfilled by orbit (b) calculated for $\beta = 2, d/a = 0.5$ and shown in the bottom inset of Fig. 5.14(b). This trajectory encloses an area $\sim a^2$ causing the B-periodic oscillations with $\Delta B \approx h/ea^2$ displayed in Fig. 5.14(a).

So far we used the analogy to periodic orbit theory for the DOS to analyze the measured quantum oscillations in Fig. 5.14. In Fig. 5.16 we compare our results from a direct calculation of the conductivity using Eq. (7) with the experimental traces. The experimental data for σ_{xx} (upper panel) are obtained by inverting the resistivity tensor and by subtracting the nonoscillatory part of the conductivity, taken at higher temperatures. Under the experimental conditions of finite temperature ($T = 0.4$ K) and weak disorder only the six shortest periodic orbits had to be taken into account in the calculation [66]. Interference effects due to the summation over different periodic orbits, resulting in an amplitude modulation of the σ_{xx}-oscillations, are clearly manifested in both, experimental (Fig. 5.16, upper panel) and calculated traces (middle panel) for an antidot array. Also the overall magnitude and temperature dependence is in accordance with experiment. On the other hand, the response of a homogeneous 2DEG (lower panel in Fig. 5.16), for which the semiclassical conductivity is given in terms of cyclotron orbits, is $1/B$ periodic.

As the oscillations for the antidot lattice are, for sufficiently large aspect ratio, nearly periodic in B they are often referred to as Aharanov–Bohm type of oscillations [71,81]. The picture given above involving the quantization of periodic orbits suggest that these quantum oscillations are closely related to Shubnikov–de Haas oscillations. In recent work where the shape of the antidots was systematically changed a clear correlation between the existence of the fundamental periodic orbits and the quantum oscillations was found [82]. Triangular lattices investigated by Nakamura et al. [81], however,

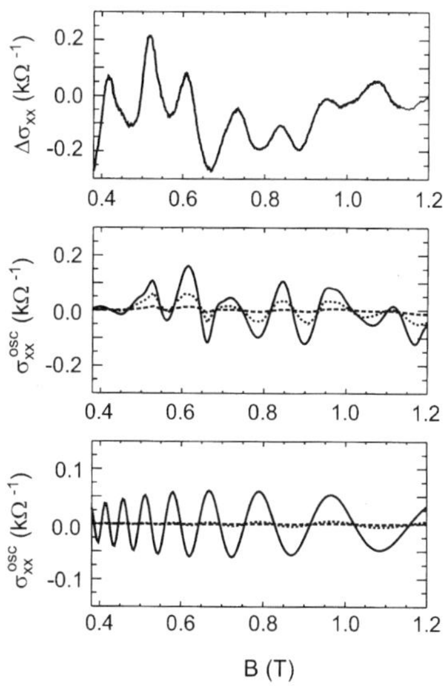

Fig. 5.16. Top panel: oscillatory part $\delta\sigma_{xx} = \sigma_{xx}(0.4\,\mathrm{K}) - \sigma_{xx}(4.7\,\mathrm{K})$ of the experimental conductivity $\sigma_{xx} = \rho_{xx}/(\rho_{xx}^2 + \rho_{xy}^2)$ of an antidot array. The corresponding resistivities are shown in Fig. 5.14(a). Intermediate panel: semi-classically calculated oscillating part of σ_{xx} for $T = 0.4\,\mathrm{K}$ (solid line), 2.5 K (dotted line) and 4.7 K (dashed line). Bottom panel: Semiclassical Shubnikov–de Haas oscillations of an unpatterned 2DEG under the same conditions as for an antidot lattice. After [66].

display Al'tshuler-Aronov-Spivak (AAS) oscillations [83] with $\Delta B \approx h/2ea^2$ (interference of time-reversed path in a ring) in the limit where the antidot lattice consists rather of weakly coupled rings than of an open lattice. This situation was addressed theoretically, e.g. by Uryu et al. [84,85].

5.2. *Quantum Mechanical Aspects*

While semi-classical methods were used above to describe the quantum oscillations in antidot arrays, the oscillatory behavior of the magnetotransport coefficients can also be reproduced by numerical quantum mechanical calculations. Such calculations are rather involved due to the strong coupling between the lattice potential and the magnetic field in the presence of disorder. Early quantum mechanical calculations on antidot arrays were carried out by Silberbauer et al. [72,73]. The calculated magnetoresistance traces were in good agreement with experiment [73,86]. Besides this technique based on appropriately chosen eigenstates of the magneto-translation group (used in [73,74,86]) also other approaches have been opted: the self-consistent Born-approximation on the basis of Green functions [87,88], the recursive Green function technique [89,90] or a S-matrix formalism [91–93]. The quantum mechanical density of states has been shown to correlate well with the semi-classical energy levels [87,94] calculated by periodic orbit theory, introduced in the last section. A comparison between quantum mechanically and semi-classically calculated resistivities and conductivities can be found in [95] and are in line with the experimental results. A quantum mechanical description of transport in rectangular antidot lattices can be found in [96,97]. The field of quantum transport in antidot structures was recently reviewed by Ando [69,98] and Suhrke and Rotter [99]. Instead of reviewing in more detail theoretical work on quantum phenomena in antidot arrays, we describe an experiment in which a prominent quantum mechanical effect, tunneling, is involved.

It is well known that patterned gates provide a way to impose a periodic potential on a two-dimensional electron gas [100,101]. By tuning the carrier density and/or the antidot potential the transition between an antidot array and an array of weakly coupled quantum dots can be investigated [31]. Such a transition is depicted in Fig. 5.17. In the experiment a patterned gate, sketched in Fig. 5.18(a), depletes the electrons underneath the metallic posts and allows one to tune the antidot lattice similar to the situation sketched in Fig. 5.17. At sufficiently large negative gate bias transport through the antidot lattice takes place via quantum mechanical tunneling and the conductivity of the array drops. Corresponding temperature dependent data are shown in Fig. 5.18(c) [31,102]. The data are compared with calculated conductivities which are in good agreement with experiment. In the model tunneling between two dots connected by

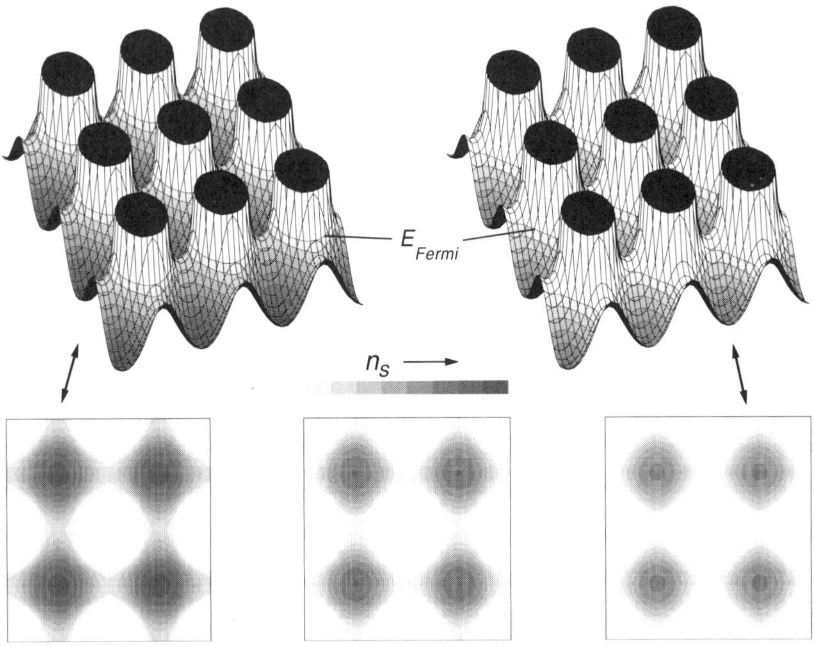

Fig. 5.17. Sketch of confinement potential tuning using a patterned gate which consists of an array of connected metallic posts. Underneath the posts the electron density is depleted, corresponding to the maxima of the potential landscape shown in the upper graph. With increasing negative bias the Fermi energy is lowered such that it lies below the saddle points. The lower three graphs show the transition from an open antidot lattice (left; vanishing carrier density corresponds to the white areas) to an array of dots. From [102].

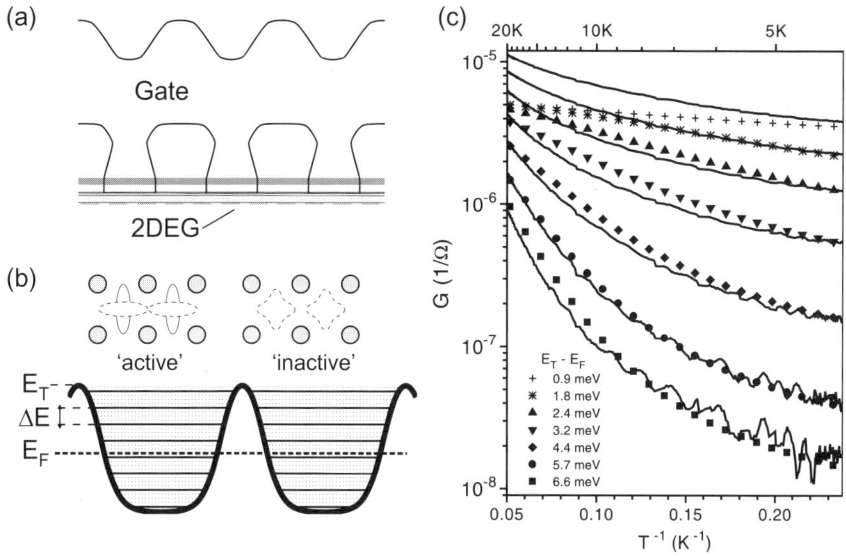

Fig. 5.18. (a) Schematic design of the device. The metallic gate posts are immersed into the top layers of the heterojunction. (b) Model used to explain tunneling between neighboring quantum dots. The dots form, at sufficiently negative gate bias, in the square confined by four antidots. ΔE is the energy separation between active modes, sketched in the inset. (c) Calculated temperature dependence of the conductance (symbols) in comparison to the experimental data (solid lines). For modelling $\alpha = 1.1\,\text{meV}^{-1}$ and $\Delta E = 0.9\,\text{meV}$ was used. The gate voltage was varied from -0.21 to -0.24 V in 5 mV steps. From [31,102].

a barrier was assumed (see Fig. 5.18(b)). It was further assumed that the conductance is due to tunneling from occupied to unoccupied single electron states in the dot on either side of the barrier where only active states (those dot states with wave functions of sufficient overlap, sketched in the inset of Fig. 5.18(b)) were taken into account [31,103]

$$\sigma = \frac{e^2}{h} \sum_{i \to j} f_i(1 - f_j) \exp[-\alpha(E_\mathrm{T} - E_i)] \tag{11}$$

Here, $(E_\mathrm{T} - E_i)$ is the relative barrier height in a dot (see Fig. 5.18(b)). The ratio of active states/number of electrons per dot is small. Their level spacing is energy independent, and is deduced from the dot geometry [103]. Three-dimensional quantum mechanical calculations carried out with the SCRAPS code of F. Stern, A. Kumar and S. E. Laux confirm the picture of active states [102]. The temperature dependence arises from the $f(1 - f)$ Fermi factor. We assumed a simple but realistic exponential decay of the tunnelling rate with respect to the energy of the state below the barrier top $(E_\mathrm{T} - E_i)$. The constant α is highly constrained by requiring the barrier width and the ratio barrier height/dot depth at cut-off to take physical values [103]. The remaining free parameter is the Fermi energy in respect to the barrier $(E_\mathrm{T} - E_\mathrm{F})$. This depends on gate voltage and is thus fitted (see legend of Fig. 5.18(c)).

Related work on metal-insulator transition and localization in antidot arrays was addressed in several investigations. Examples involve both experimental [84,104–108] and theoretical work [84].

6. Random Antidot Arrays

So far regularly ordered antidot arrays were the subject of the investigations, i.e. square antidot arrays, rectangular ones or triangular lattices. Due to irregularities in the fabrication process and "natural" scattering centers in the 2DES disorder is present which affects the transport properties [98]. While such type of disorder can hardly be influenced, the order of the array can be varied. In a series of experiments and corresponding calculations the actual antidot sites were displaced by a certain amount from their regular positions and an increasing suppression of commensurability features with increasing disorder was found [48–50,109,110]. A different situation arises in completely random antidot lattices where the order of the antidot positions is completely destroyed [102]. This is achieved by defining both the x- and y-coordinate of an antidot location by random numbers in the pattern generator of the e-beam system. Resulting electron micrographs of antidot arrays are displayed in Fig. 5.19. The random antidot lattices are characterized by their average period $\bar{a}$ which is given by the total number of antidots in the patterned area. As the electron mean free path l_e is much longer than the average spacing of the antidots one expects, at first glance, that the situation is very similar to the usual Drude picture with a magnetic field independent magnetoresistance. The

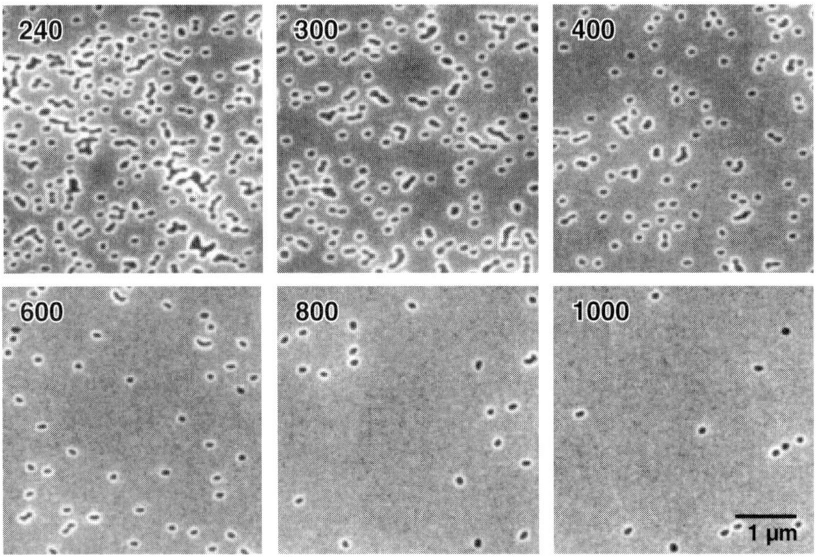

Fig. 5.19. Electron micrographs of completely random antidot arrays with different average period $\bar{a}$. The diameter of individual antidots was ~90 nm. The electron mean free path of the electrons in the unpatterned 2DES is about 13 μm. From [102].

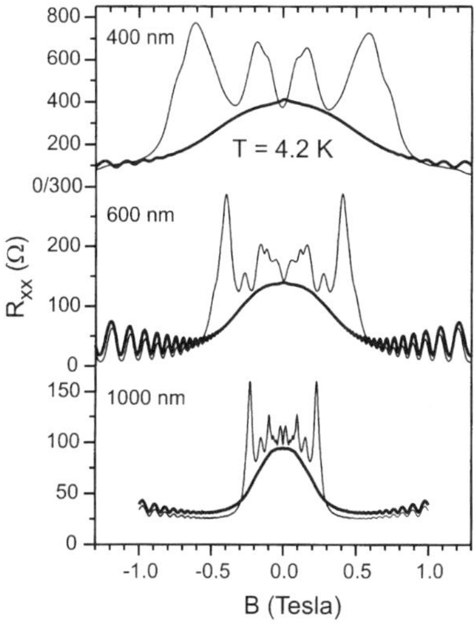

Fig. 5.20. Magnetoresistance of periodic (thin solid line) and random (thick solid lines) antidot arrays with the same (average) period. The data were taken after brief illumination with a LED. The carrier density of the random/periodic antidot lattices was 4.6/4.4, 4.7/4.6 and 4.9/4.9 $\times 10^{15}\ \mathrm{m}^{-2}$. The high field peak is the fundamental commensurability maximum at which $2R_c = a$ holds. From [102].

experiment tells a different story [102]. In Fig. 5.20 the magnetoresistance of periodic square lattices is compared to the one of random arrays with the same (average) period. The periodic arrays show pronounced commensurability features with a fundamental peak which shifts according to the condition $2R_c = a$. Commensurability features are completely absent in the magnetoresistance of the random antidot lattices. Also the Hall-resistance (not shown) displays no quenching and no additional steps and grows nearly linearly with B, in contrast to periodic arrays (see, i.e. Fig. 5.4(b)). The zero-field resistance of periodic and random antidot lattices is almost the same and elucidates that the antidots act as additional scattering centers. The mean free path is dominated by the antidots and increases quadratically with the period $\bar{a}$, both for random and periodic arrays. For $\bar{a} > 600$ nm l_e saturates. The random arrays display a negative magnetoresistance with a saturation field which is clearly connected to the (average) period. The dependence of the magnetoresistance saturation field on $\bar{a}$ can also clearly be seen in Fig. 5.21 where normalized R_{xx}-traces of all investigated random arrays are compared. The saturation field B_F, extracted by linear extrapolation of the traces at the inflection point to $R_{xx} = 0$ is plotted in the inset of Fig. 5.21 and is nearly the same for periodic and random arrays. A simplified view on the origin of the negative magnetoresistance is to assume that backscattering of electron trajectories gets suppressed for increasing B (decreasing R_c) and drifting orbits, in the picture of the pinball model given above, start to dominate phase-space and hence R_{xx}. A detailed analysis of the magnetoresistance in a system with two types of random potential—randomly placed hard disks that scatter electrons and a smooth random potential due to the remote dopants—is rather involved and shows a wealth of different regimes [111,112].

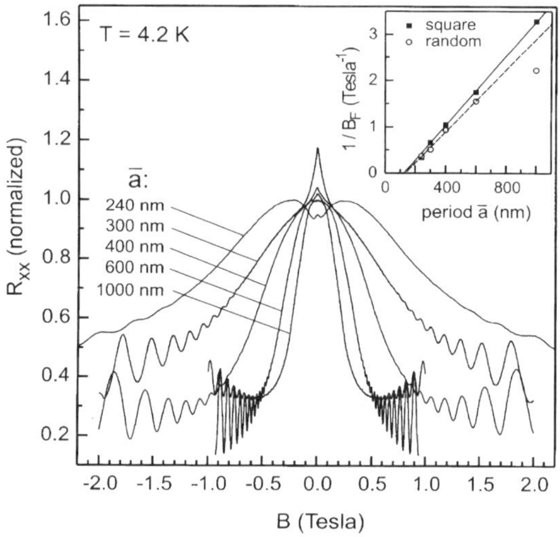

Fig. 5.21. Dependence of the magnetoresistance of random arrays on the density of antidots (expressed as average period). Data were taken after illumination and normalized to the high-temperature zero field resistance. Samples with smaller densities (larger $\bar{a}$) display a Lorentzian shaped magnetoresistance which is ascribed to suppression of backscattering with increasing magnetic field. Except for the $\bar{a} = 1000$ nm trace a narrow peak is found around $B = 0$. For the smallest average period two maxima appear at about ± 0.22 T. The inset depicts the saturation field B_F for random and periodic arrays. From [102].

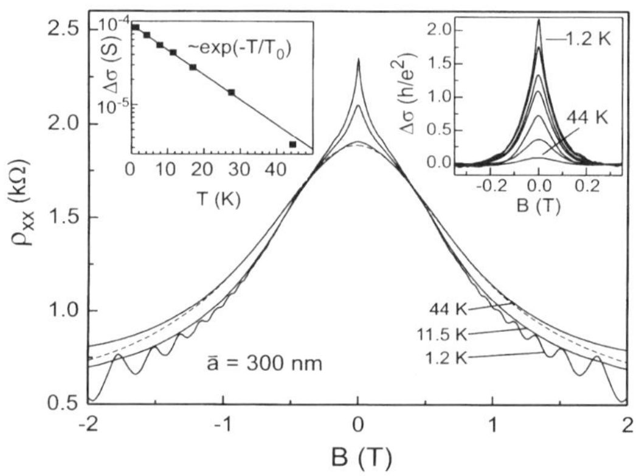

Fig. 5.22. Temperature dependence of ρ_{xx} of a sample with $\bar{a} = 300$ nm. The dashed line is the Lorentzian fitted to the high temperature data and used to extract $\Delta\sigma$, plotted in the upper right inset. The left inset shows the temperature dependence of the peak height. From [102].

Besides the negative magnetoresistance a pronounced narrow peak close to $B = 0$ can be found for the smaller periods in Fig. 5.21. These structures have a stronger temperature dependence than the Lorentzian shaped background. This is shown in Fig. 5.22 for the sample with $\bar{a} = 300$ nm. While the broad ρ_{xx}-peak is only weakly temperature dependent between 1 and 45 K (apart from the smearing of the SdH-oscillations at higher B) the zero-field spike grows with decreasing temperature. After converting the resistivities into conductivities using the Hall-resistivity, the background (a Lorentzian fitted to the resistance shown in Fig. 5.22 as dashed line which was also inverted) was subtracted. The difference $\Delta\sigma$ is plotted in the upper right insert for temperatures between 1.2 K and 44 K. The temperature dependence of the peak value drops approximately exponentially $\sim \exp(-T/T_0)$ with $T_0 = 14.5\,\mathrm{K}$ and is shown in the upper left inset of Fig. 5.22. This quantum correction to the conductivity can be interpreted as a weak localization effect. However, the observed temperature dependence is not compatible with results from usual weak localization theory for diffusive systems involving quantum impurity scattering. For the antidot ensembles, coherent backscattering arises from reflections at the (random) antidots acting as classical scatterers. Hence the weak localization effect in ballistic antidot structures has called for a weak localization theory beyond the diffusion approximation in order to account for subtle correlations in the chaotic classical dynamics. This requires a refined semi-classical theory for quantum transport [113] beyond the so-called diagonal approximation. In weak localization for the antidot systems another time scale comes into play [114]: the Ehrenfest time t_E for the spreading of a wave packet of initial size λ_F over a distance of the size d of the antidots: $t_E = (1/\lambda)\ln(d/\lambda_F)$. Here λ is the mean Lyapunov exponent of the related classical system. The Ehrenfest time sets the scale for the observed decay $\Delta\sigma \sim \exp(-T/T_0)$ (upper left inset of Fig. 5.22) [51].

The $\bar{a} = 240\,\mathrm{nm}$ trace displays in addition to the zero-field peak maxima at finite magnetic field. This is reminiscent of magnetotransport in ballistic wires [115] and suggests that random antidot arrays with a high density of random scatterers can be viewed as a network of narrow wires.

While the investigation of magnetotransport in random antidot arrays is interesting on its own such arrays can also be used to investigate the breakdown of the quantum Hall effect [116–118] or as a reference to prove commensurability features around filling factor $v = \frac{1}{2}$ in the fractional quantum Hall regime [21].

7. Finite Antidot Lattices

The experiments described above were carried out on antidot arrays with dimensions much larger than electron mean free path l_e and phase coherence length l_ϕ. Novel effects are expected for arrays smaller than l_ϕ. A corresponding experiment in which the antidots were periodically arranged in a micron sized two-dimensional "cavity" [62], displayed resistance fluctuations very similar to the ones observed by Marcus et al. [119]. The statistics of these fluctuations can, in principle, be understood in a semi-classical framework pioneered by Baranger, Jalabert, and Stone [120]. According to theory, the power spectrum of the aperiodic fluctuations depends on whether the underlying classical dynamics of the electrons is chaotic or regular, in agreement with experiment [119]. This is essentially due to the different distribution of areas enclosed by chaotic and regular trajectories in an appropriately shaped cavity. Semi-classical theory also predicted that regular and chaotic scattering should result in different line shapes of the ballistic weak localization (WL) peak around zero magnetic

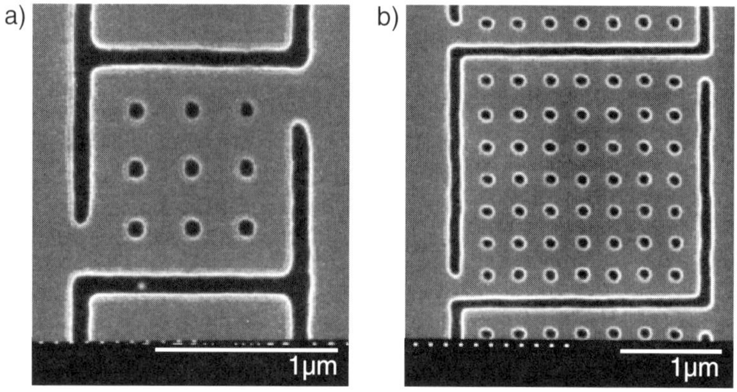

Fig. 5.23. Electron micrography of mesoscopic antidot arrays. The squares with entrances in the lower left and upper right corner were defined by e-beam and etched with reactive ion etching. The left structure (a) is an individual cavity containing 9 antidots while the right one (b) is part of a row of antidot filled cavities. From [102].

field [121,122], which is directly accessible to experiment without the need for statistical analysis of the resistance fluctuations. Indeed, a Lorentzian line shape for chaotic stadia, and a more triangular line shape for regular circular billiards have been reported [123]. In this section work on ballistic weak localization in square cavities filled with antidots is described [102,124]. The antidots in the cavities were expected to convert the integrable electron dynamics of the squares into a chaotic one.

An electron micrograph of finite antidot arrays, fabricated by electron beam lithography and dry etching techniques, is displayed in Fig. 5.23. The transport mean free path l_e is $\sim 16\,\mu$m at 4.2 K which is significantly longer than the length of the squares, $L = 2.3\,\mu$m and $L = 1.2\,\mu$m. The resistance measured at $T = 30$ mK across a single cavity displays reproducible characteristic fluctuations (not shown). Antidot peaks corresponding to chaotic motion of electrons around one and four antidots are superimposed upon a negative differential magnetoresistance (on a larger B-scale of several kG compared to the one relevant for weak localization) associated with the small openings of the cavity [102]. In order to suppress the conductance fluctuations and to enable the unperturbed observation of the weak localization peak, we did ensemble averaging by employing a technique used by Chang et al. [123]. While the averaging procedure suppresses conduction fluctuations and allows to investigate the negative magnetoresistance at very low B, it introduces additional structure due to electron focusing into neighboring cavities [124,125]. The strong focusing features observed in experiment can be used to probe "composite fermion" transport in the fractional quantum Hall regime [126–128] but shall not be discussed here.

The left hand side of Fig. 5.24 displays characteristic resistance peaks at small fields, centered at $B = 0$. The data are obtained from cavities filled with 49 antidots (see right hand side of Fig. 5.23) averaged over an arrangement of 52 cavities placed in 4 rows of 13 on a single Hall bar [124]. The phase coherence length was decreased from top to bottom for the traces shown by the successive increase of the bias current from 4 nA

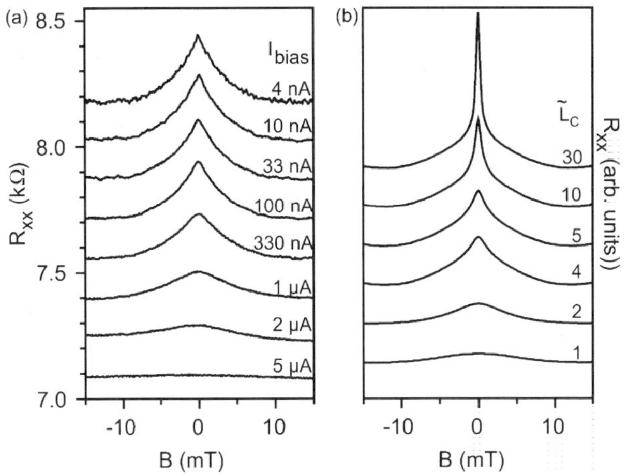

Fig. 5.24. (a) Measured magnetoresistance of the ensemble of square cavities containing 7×7 antidots as a function of bias current (corresponding to temperature). An offset was added for clarity. (b) Calculated quantum corrections to the resistance in dependence of the cut-off parameter $\tilde{L}_c$. After [124].

to 5 mA which leads to current heating of the electrons while the dilution refrigerator was kept at a base temperature of 30 mK. The line shape of the WL peak evolves by the increase of the phase breaking length from a Lorentzian profile to a cusp-like peak at lowest current (temperature). This strong deviation from a Lorentzian line shape is rather unexpected for a chaotic ballistic cavity and seems, at first glance, to be in contrast to the result of Chang et al. [123]. To understand this surprising experimental outcome semi-classical transport theory, following [121,122] was applied. The magnetic field dependence of the quantum corrections to the reflection coefficient R, which is linked to the average resistance by the Landauer transport theory, can be approximated by [113,121]

$$\langle \delta R(B) \rangle \propto \int_{-\infty}^{\infty} \mathrm{d}\theta \, N(\theta) \cos\left(\frac{2\theta B}{\phi_0}\right) \tag{12}$$

This equation represents the cosine transform of the area distribution $N(\theta)$ of backscattered paths which enclose an effective area θ, with $\phi_0 = h/e$ denoting the flux quantum. The area distribution of 10^6 electrons injected in a cavity filled with antidots was evaluated numerically [78] and serves as input to obtain the ballistic WL correction given by Eq. (12). Only areas enclosed by trajectories with a length smaller than or comparable to the phase breaking length l_ϕ can contribute to $N(\theta)$: hence a cut-off length L_c was used in the calculations which defines the maximum length of paths contributing to $N(\theta)$. The right hand side of Fig. 5.24 shows the evaluation of the computed peak profile by decreasing the normalized cut-off length $\tilde{L}_c = L_c/L$ from 30 to 1. Clearly this results in a crossover from a cusp-like peak to a smooth, flat profile very similar to that found in experiment. By relating the simulations to the experimental traces (for the lowest current of 4 nA) an estimate for the phase coherence length of $l_\phi \approx 5\tilde{L}_c = 12\,\mu$m was obtained. The deviation from a Lorentzian WL peak expected

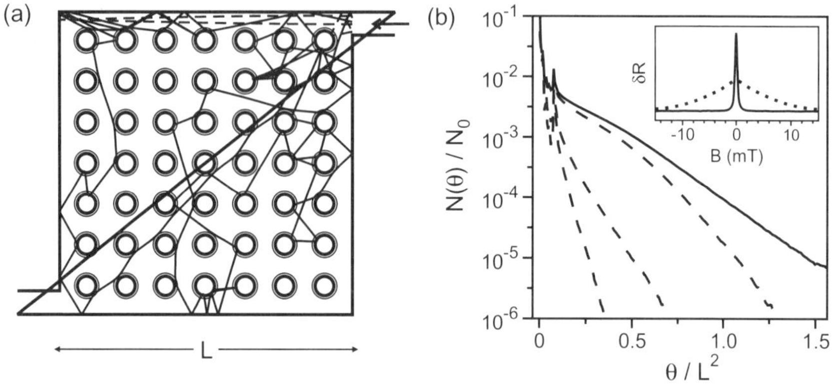

Fig. 5.25. Model of a 7 × 7 antidot cavity. Classical electrons injected from the right lead are either directly backscattered by nearby antidots or the opposite wall (dashed lines), or follow chaotic diffusive paths through the array of scatterers (solid line). (b) Classical distribution $N(\theta)/N_0$ of the effective areas θ enclosed by backscattered paths for the antidot billiard (solid line). The nearly linear part at larger θ, representing the diffusive particles, is suppressed by a decrease of the cut-off length $\tilde{L}_c$ as shown for the dashed lines for $\tilde{L}_c \approx 30, 5, 2$. Inset: Decomposition of the WL peak into the contribution from long chaotic orbits (Lorentzian like spike, solid line) and orbits shorter than $3L$ (dotted line). After [63,124].

for classically chaotic systems is due to the coexistence of two types of trajectories displayed in Fig. 5.25(a): short paths of electrons directly reflected by the antidots near to the cavity entrance or by the opposite cavity wall (dashed trajectories), and longer trajectories where the electrons move diffusively through the array of scatterers before leaving the cavity (solid lines). Directly backscattered paths enclose small areas and thus contribute to the area distribution $N(\theta)$ as a peak at small θ which translates into a broad weak localization peak shown in Fig. 5.25(b). On the other hand the long chaotic trajectories show an exponential decrease of $N(\theta)$ in accordance with chaotic scattering theory giving rise to the narrow resistance cusp, strongly dependent on l_ϕ. From this experiment we concluded that in a typical chaotic system quantum transport phenomena usually reflect the coexistence of short time dynamics of non-universal paths and chaotic behavior on longer time scales. Therefore the generic chaotic feature of a pure Lorentzian weak localization peak appears to be quite exceptional while in general the resistance profile is more complex and not unique.

8. InAs Based Arrays

Metal-InAs contacts are characterized by the absence of a Schottky barrier between metal and semiconductor. This property stems from Fermi level pinning above the conduction band edge at surfaces and interfaces [129]. Hence InAs is the material of choice for superconductor/semiconductor hybrid structures. The absence of a barrier supports the exchange of carriers between semiconductor and superconductor. This property will be important for the fabrication of "superconducting" antidots discussed in section 8.2. The Fermi level pinning in the conduction band and hence the absence of depletion regions leads to a different potential landscape which supported the observation of some novel features in the magnetoresistance of antidot arrays. This is the topic of the next section.

8.1. Transport in InAs Antidot Lattices

While antidot lattices are usually based on GaAs/Al_xGa_{1-x}As heterojunctions, we used here the system InAs/(Al)GaSb to fabricate antidot arrays. As stated above, the depletion regions around the antidots are smaller than in GaAs based systems. Due to the resulting steeper potential posts very short period antidot arrays or, as is the case here, large antidots with small constrictions b between neighboring ones can be fabricated. A central requirement for studying commensurability effects are two-dimensional electron systems with sufficiently long mean free path. Mean free path of orders 6 μm can be achieved in InAs/(Al)GaSb heterojunctions where the two-dimensional electron gas is confined in a 15 nm wide InAs quantum well. More information about the material system can be found in [130,131].

For large diameter antidots an unusually broad peak structure in the region between 1.5 and 3.5 T was found in this material system [59]. The extra structure can be ascribed to electron trajectories skipping around one antidot on a rosette-shaped trajectory and is a consequence of the steep antidot potential feasible in InAs based materials. This type of orbit also proved to be relevant in a focusing type of experiment [132].

Figure 5.26 shows the dependence of the magnetoresistance traces on the carrier density (changed by illumination). While the commensurability peaks observed at low B are similar to those discussed above a broad maximum, marked in Fig. 5.26, emerges at higher B. The unusual peak structure shifts with varying carrier density similar to the main commensurability maximum at $2R_c = a$. This suggests, together with the weak temperature dependence of the peaks [59], a classical origin of the phenomenon. Associating the magnetic-field value at which the broad maxima start to appear (marked by arrows) with the matching condition $R_c = b$ we found a value of $b = 110$ nm. This value b corresponds to the distance between neighboring antidots in the device investigated. This finding is consistent with the picture that the broad maximum is associated with rosette-shaped orbits skipping around one antidot. These trajectories become important for $R_c \sim b$.

Since the broad maximum appears at relatively high magnetic-field values, most of the phase space is occupied by regular non-chaotic trajectories. Therefore both regular and chaotic trajectories were taken into account in model calculations, again based on the classical Kubo formula, Eq. (6). By inverting the conductivity tensor the calculated resistance trace $\rho_{xx}(B)$ can be compared to the experimental one. The result of a simulation modeling the antidot potential by perfect hard walls (depicted in the inset) is shown in Fig. 5.27 [59]. The calculated traces are remarkably close to the experimental results: the additional structure between 1.5 and 3.5 T was nicely reproduced. To check whether rosette type of orbits are also present if soft wall effects are taken into account Poincaré surfaces of section of trajectories in softer potentials were analyzed. The existence of rosette type of orbits was also confirmed by these calculations [59]. The evaluation of the Poincaré surfaces of section also uncovered

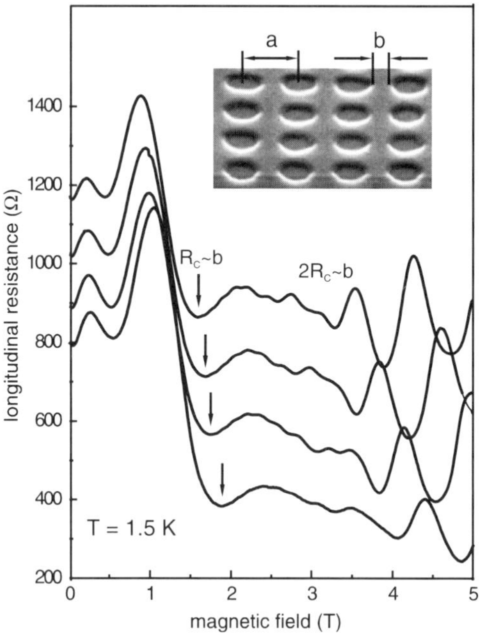

Fig. 5.26. Magnetoresistance for different carrier densities from $1.13 \times 10^{16}\,\mathrm{m}^{-2}$, $1.24 \times 10^{16}\,\mathrm{m}^{-2}$, $1.33 \times 10^{16}\,\mathrm{m}^{-2}$ to $1.42 \times 10^{16}\,\mathrm{m}^{-2}$. The carrier density was varied with a red LED. The mean free path of the electrons in the unpatterned 2DES is about 6 μm, much larger than lattice period $a = 380$ nm. Arrows highlight the onset of the broad maximum (at $R_c \approx b$), vertical line mark the onset of strong SdH-oscillations ($2R_c \approx b$). The inset shows an electron micrograph of the antidots etched into the GaSb/InAs heterojunction. The layer sequence of the heterojunction was: GaAs substrate, 1.1 μm AlGaSb, 0.5 μm GaSb buffer, a ten-period AlSb/GaSb superlattice, 50 nm GaSb barrier, a 15 nm InAs quantum well containing the 2DES and a 5 nm thin GaSb cap layer. After [59].

the origin of the minima marked by arrows in Fig. 5.27: In this magnetic field range no space for stable orbits exists. Only upon increasing the magnetic field a large fraction of phase space is occupied by periodic and quasi-periodic trajectories. These are rosette-shaped orbits which encircle one antidot and remain stationary even if an electric field is applied in the simulation, giving rise to a maximum in the magnetoresistance around 2.5 T [59].

8.2. Superconducting Antidots

In this section a system is explored where the specular reflection at the antidots is, at least partly, replaced by Andreev reflection [133]. In the Andreev reflection process, occurring at normal-superconductor interfaces at low bias, an electron can enter the superconductor only if it is paired with a second electron of opposite spin and **k**-vector. As a consequence a hole with positive charge e and with all velocity components reversed is reflected from the interface. Without applied magnetic field the retroreflected hole retraces the path of the incident electron. This constitutes an unusual scattering

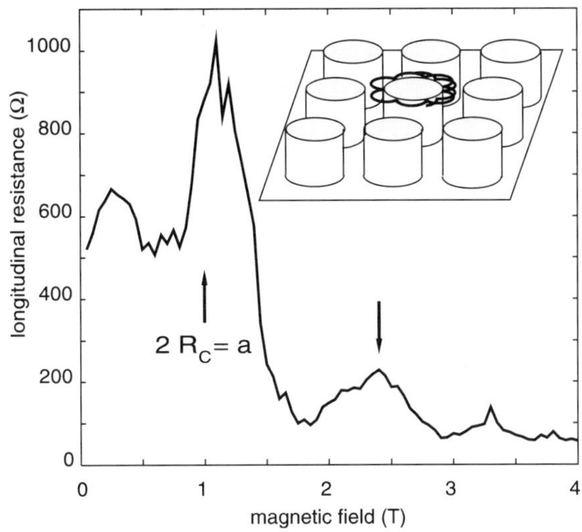

Fig. 5.27. Magnetoresistance traces obtained with a simple simulation based on the Kubo formula. The commensurability condition $2R_c = a$ is shown for $n_s = 1.3 \times 10^{16}\,\mathrm{m}^{-2}$ used in the calculation. The inset sketches the hard-wall potential and a rosette-shaped orbit encircling an antidot. Its magnetic field position is marked with an arrow in the large graph. From [59].

process which should leave its mark on the commensurability features in lateral superlattices. To aim at Andreev reflecting antidots an array of superconducting niobium disks was placed on an InAs quantum well containing a 2DES. While in a perfectly Andreev reflecting antidot array the trajectories of electrons and retro-reflected holes behave regularly at $B = 0$, their dynamics becomes again chaotic at finite B [134], as is illustrated in Fig. 5.28(d). The system can be considered as an experimental version of an Andreev billiard whose properties were first addressed theoretically [134–137]. An early experimental attempt of realizing a similar system was reported by Rahman et al. [138]. The question addressed below is: How does retroreflection, instead of specular reflection, alter the magnetoresistance features of an antidot array?

To reach an experimental version of an Andreev billiard square lattices of niobium filled antidots were fabricated from an InAs-based 2DES structure. Due to the lack of a Schottky-barrier, InAs is ideally suited for highly transparent semiconductor-superconductor contacts [139–142]. The observations described below were most clearly visible at small d/a-values [143]. The samples were fabricated from high-mobility InAs/$Al_{0.30}Ga_{0.70}Sb$ heterostructures. The patterning process is sketched in Fig. 5.29 and is described in more detail in [144]. Hall bars were prepared employing photolithography and wet chemical etching and the 2DEG was contacted with non-alloyed Cr/Au pads. The antidot lattices were defined by electron beam lithography and a selective wet chemical etch removing only the top AlGaSb layer at the antidot sites. Subsequently, the niobium disks were deposited with dc-magnetron sputtering and lift off. Both patterning steps were carried out using the same PMMA mask. In this way a complicated realignment was avoided and, more importantly, the exposure to air after etching was minimized. To compensate for the inevitable underetching of

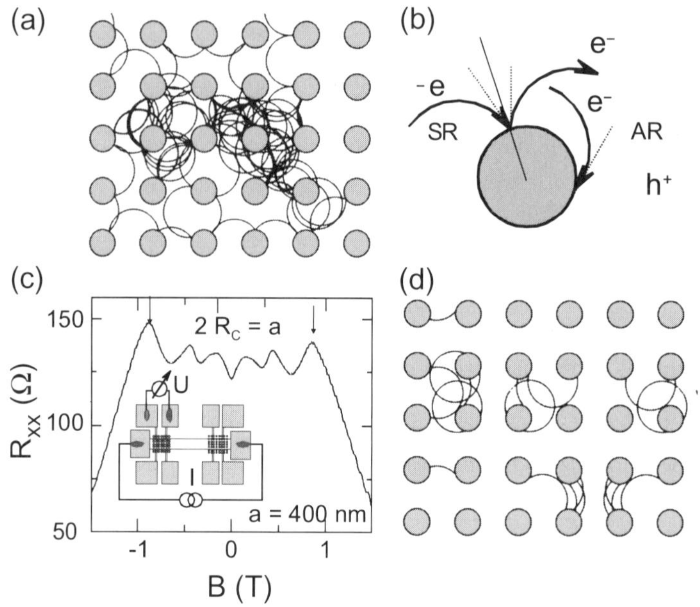

Fig. 5.28. Chaotic trajectories calculated at $2R_c = a$ in specularly reflecting (a) and Andreev reflecting (d) antidot lattices. (c): Magnetoresistance of an antidot lattice (holes etched into the AlGaSb layer) taken at $T = 1.5$ K. The inset shows the Hall bar geometry and measurement setup. (b): Trajectories of specularly (SR) and Andreev reflected (AR) particles. Black (gray) trajectory sections: electrons (holes). After [143].

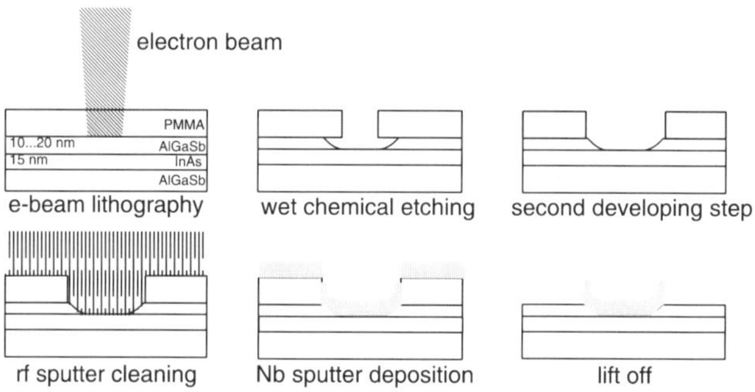

Fig. 5.29. Sketch of the fabrication procedure (see text).

the PMMA mask by the wet etching step, a modified developing scheme was used [144]. This procedure proved to be crucial for a highly transparent contact between the 2DEG and the superconductor. The samples were etched and developed immediately before loading them into the load-lock chamber of the sputtering system. Despite the short exposure to air it was necessary to remove residual contamination on the InAs surface with a low power Argon sputter etch [140]. By this technique, high quality

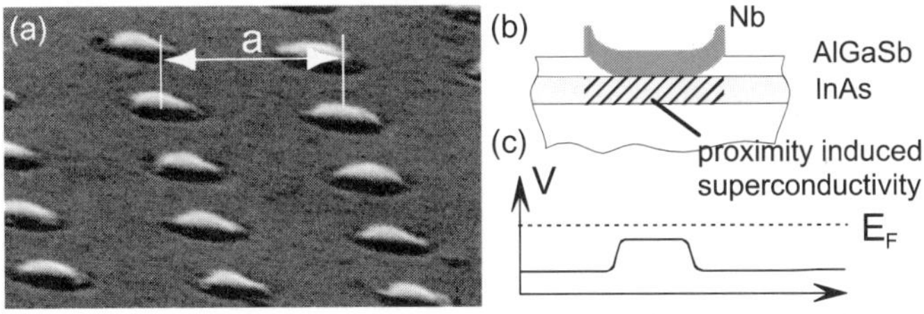

Fig. 5.30. (a) The SEM-picture taken at an angle of 60° shows Nb filled antidots with a lattice period a of 400 nm. Periods in the experiments ranged between $a = 300$ nm and $a = 1500$ nm; the ratio of antidot diameter and lattice period d/a was varied between 0.20 and 0.75, with a number of different diameters for each lattice period. (b) A cross-sectional view of the sample design. The InAs underneath the Nb dot is superconducting due to the proximity effect. The InAs-layer containing the 2DEG was 15 nm thick and the top barrier layer of $Al_{0.30}Ga_{0.70}Sb$ had a thickness of 10 nm [130]. (c) The electrostatic antidot potential is sketched below. The relatively weak antidot structures observed in experiments, where the antidots were etched into the AlGaSb top layer only, suggest that the region underneath the Nb-disks is not completely depleted. Control experiments with antidots etched through the InAs quantum well showed much stronger commensurability features. After [143].

samples could be routinely fabricated, which consistently displayed similar experimental results. Unpatterned niobium films prepared in the same system had a critical temperature T_c of about 8 K and a critical field B_c of about 2.4 T at 1.6 K. Niobium dots with a minimum diameter of 170 nm were still superconducting with a T_c of about 5 K. An electron micrograph of an array of Nb-dots immersed in InAs is displayed in Fig. 5.30(a). A cross section of the "superconducting antidot" is sketched in Fig. 5.30(b). In the following it is assumed that at sufficiently low temperature the proximity effect induces a pair potential underneath each Nb-dot [145] and the normal/superconductor interface forms in the 2DES (see Fig. 5.30(b)). At the lowest temperatures (0.3 K) the whole 2DES turned superconducting [146]. In addition to the region of proximity induced superconductivity a discontinuity of the conduction band edge is present underneath a Nb disk. Such a density modulation already exists after "drilling" the holes in the AlGaSb layer and before the deposition of the Nb. Since the respective magnetoresistance traces (shown in Fig. 5.28(c)) resemble those of antidot arrays described above we assumed a reduction of the carrier density underneath the Nb antidots. Note, however, that any discontinuity of the carrier density underneath the Nb can give rise to quantum mechanical reflection of charge carriers.

The dashed graphs in Fig. 5.31(a) and (b) show data taken at 9 K, above T_c of the niobium dots. For both samples the fundamental antidot peaks at $2R_c = a$ are clearly visible. The sample with the smaller period of $a = 400$ nm displays even more structure at lower B, ascribed to higher order commensurate orbits. From the small amplitude of the commensurability peaks we concluded that the 2DEG underneath the Nb disks is not completely devoid of electrons and that hence the repulsive potential maxima do not exceed the Fermi energy, as sketched in Fig. 5.30(c). Thus a fraction of the electrons in phase space is able to explore the 2DEG underneath the Nb disks. The traces change substantially if cooled below T_c. The thin solid lines in Fig. 5.31 were

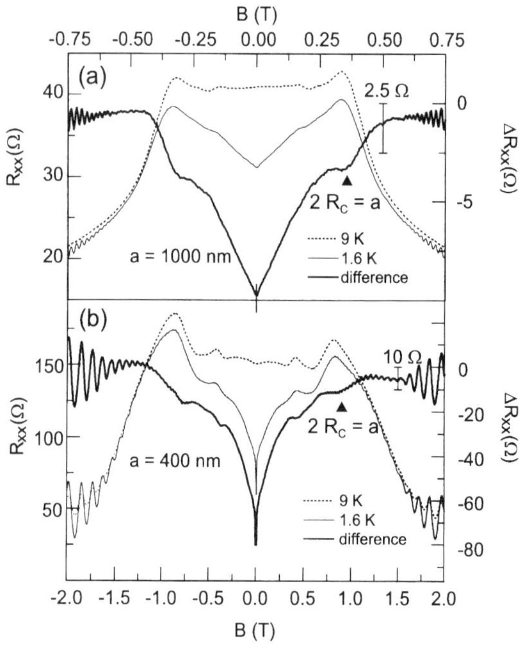

Fig. 5.31. Magnetoresistance curves for samples with two different lattice periods a. (a): $a = 1000\,\text{nm}$ and diameter of the disks $d = 200\,\text{nm}$. The difference between the curves below T_c and above T_c shows a clear dip at $2R_c = a$. (b): $a = 400\,\text{nm}$ and $d = 170\,\text{nm}$. Two dips are visible for magnetic fields at which the antidot graph ($T = 9\,\text{K}$) shows peaks. After [143].

taken at 1.6 K and show a magnetic field dependent reduction of the resistance which is especially pronounced at low B. The "V-shaped" magnetoresistance around $B = 0$ is characteristic for good Nb/2DES contacts and is not observed in devices with lower quality contacts. We ascribe this behavior to Andreev reflection taking place between the normal 2DES and a region of proximity induced superconductivity in the InAs underneath the Nb disks [145,147]. In the B-field region where the commensurability peaks are observed, reflection is mostly specular and, consequently, the antidot peaks are still present in the 1.6 K data. The extra reduction of the resistance in the millitesla range in Fig. 5.31 is a precursor of Josephson coupling between the niobium dots. At $T = 0.5\,\text{K}$, the whole array becomes superconducting up to fields of several millitesla. In this section, however, only the classical features visible at higher temperatures and magnetic fields are dealt with.

To extract the influence of Andreev reflection on the commensurability features the above-T_c-trace was subtracted from the below-T_c-trace. The result is shown as ΔR_{xx} in Fig. 5.31(a), (b). Surprisingly, ΔR_{xx} exhibits dips at commensurate magnetic fields where the original traces display maxima. In contrast, in conventional density modulated electron systems the amplitudes of the commensurability features grow with decreasing temperature due to the reduced broadening of the Fermi distribution [148,149]. This would result in peaks of ΔR_{xx} instead of dips. The dips observed in

ΔR_{xx} hence indicate a suppression of commensurability peaks. Such a suppression is expected for electrons being Andreev reflected from the antidots as will be shown below. In case of pure Andreev scattering (no specularly reflected electrons), surprisingly, all commensurability features are expected to disappear.

To demonstrate this the classical Kubo formula, used before, can be employed again. Andreev reflection is taken into account by using a time-dependent charge $q(t)$, which is $+e$ $(-e)$ on the hole (electron) part of the trajectory:

$$\sigma_{ij} = e^2 \bar{d} \int_0^\infty \mathrm{d}t \langle q(t) v_i(t) q(0) v_j(0) \rangle \mathrm{e}^{-t/\tau} \tag{13}$$

Thus, in our description Andreev reflection—which is a quantum effect that can be described using the Bogliubov–de Gennes equations—is only taken into account by the resulting classical electron and hole trajectories in the normal region of the 2DES. Transport through the metallic or superconducting dots is not included in this model. Since the mean free path in the samples is much longer than the lattice period a, the superconducting regions do not provide a transport channel with a much lower resistance than the ballistic 2DEG. As long as the Nb dots are not Josephson coupled, the current is carried by normal state electrons and holes in the 2DES.

For perfect Andreev reflection, the Kubo formula yields a surprisingly simple result: Since all the velocity components are reversed upon Andreev reflection, the velocity correlation function $\langle q(t) v_i(t) q(0) v_j(0) \rangle$ of an unpatterned 2DEG is recovered. This is illustrated in Fig. 5.32 where the velocity components v_x and v_y are compared for the case of pure specular (b) and pure Andreev reflection (c). In between reflection events

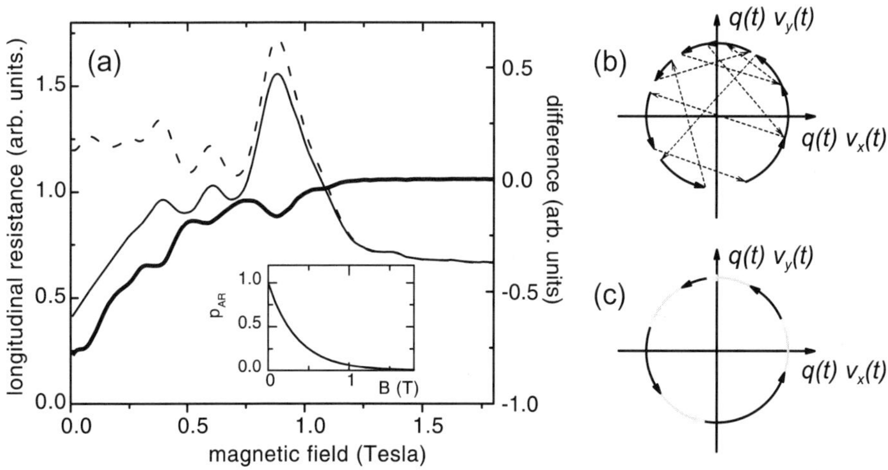

Fig. 5.32. (a) Calculated magnetoresistance curves for $p_{\mathrm{AR}} = 0$ (dashed line) and a magnetic field dependent p_{AR} (thin solid line). The difference of both graphs (thick solid line) has dips at the commensurability positions. The calculation was performed using $a = 400$ nm and $d = 170$ nm. Inset: Exponential dependence of p_{AR} on B used for the calculations. (b) Sketch of the velocity components of a trajectory in an antidot lattice with specular reflection. (c): The same, but with Andreev reflecting boundaries. Black: electrons, gray: holes. After [143].

the values of the velocity components move along a circle with the Fermi velocity v_F. In the case of perfect Andreev reflection both velocity components are reversed but $q(t)v_x$ and $q(t)v_y$ continue to move along the circle since the sign of the charge $q(t)$ is also reversed (conversion of holes into electrons and vice versa). Hence the correlation function of a perfect Andreev antidot billiard is identical to the correlation function of the free electron motion in a perpendicular magnetic field. In sharp contrast, reflection on a conventional antidot boundary gives different velocity correlations depending on the exact initial conditions of the trajectory (Fig. 5.32(b)).

In experiment scattering off the antidots will be due to a mixture of both processes. Hence both, specular and Andreev reflection has been taken into account within a simple model. To allow for specular reflection, which is due to the potential discontinuity underneath the Nb-platelets, the interface between normal and superconducting 2DES was modeled by a hard-wall potential. If the trajectory hits the interface, the charge carrier is either Andreev reflected or specularly reflected with a certain probability p_{AR}. The Andreev reflection probability p_{AR} was not assumed to depend on the angle of incidence on the boundary since this should have little effect on the transport coefficients: in the extreme case of grazing incidence, where Andreev reflection is most strongly suppressed [150], the contribution to the conductivity is the same for Andreev reflection and normal reflection. A set of magnetotransport curves was obtained by varying p_{AR} from 0 to 1 [146]. For $p_{AR} = 0$, the behavior of a conventional antidot lattice is reproduced, whereas $p_{AR} = 1$ gives a much lower resistance independent of B just as for an unpatterned 2DES. Andreev reflection therefore provides a mechanism to explain both the reduced zero-field resistivity and the suppression of commensurability features on a common footing.

To explain the magnetic field dependence of the resistance below T_c, i.e. the V-shaped magnetoresistance, we assumed a decrease of the Andreev reflection probability p_{AR} with increasing magnetic field [143]. While a priori not justified, this might be a consequence of a finite pair-amplitude in the superconducting regions of the InAs [145,147,151–153], which decreases with increasing B. Within this picture, p_{AR} is affected by the pair-amplitude inside the 2DES, which depends on temperature and magnetic field in a complicated way. No detailed explanation is presently available for this dependence.

To compare with experiment an Andreev reflection probability p_{AR} decreasing exponentially with B was introduced phenomenologically and is plotted in the inset of Fig. 5.32(a). While p_{AR} is at maximum at $B = 0$, reflection off the antidots is mostly specular at higher magnetic fields, where p_{AR} is small. This explains why the commensurability peaks are not suppressed completely, but only partly. The result of a corresponding model calculation is shown as a thin solid line in Fig. 5.32(a) [143]. The difference to the specularly reflecting antidot trace ($p_{AR} = 0$, above T_c-trace) is shown as a thick solid line and shows the same dips, characteristic for the suppression of commensurability peaks, as observed in experiment. The values $d = 170\,\text{nm}$ and $a = 400\,\text{nm}$ used in the calculation correspond to the experiment shown in Fig. 5.31(b). These findings are independent of the exact dependence of p_{AR} on B. As long as there is a small degree of Andreev reflection present, the commensurability peaks are damped. Note that the calculated curves show more structure than the experimental ones. This is due to the hard wall potential used to model the antidots and the absence of any imperfections (e.g. fluctuations in the antidot diameter or shape). Nevertheless, the simple model seems to cover the essential physics.

9. Other Experiments

Dc-transport in antidot lattices was described to some extent in the preceding sections though we surely missed some important works. Not mentioned, for example, was dc-transport in classical antidots (periodic holes in a thin semiconductor or metal film) where the superposition of local currents around the obstacles leads to interesting angular dependencies of the resistivity in this diffusive regime [154]. A large field omitted above is optical excitations in antidot arrays which started with the work of Kern et al. [155] and was the subject of many other experimental investigations [101,156–163]. Corresponding theoretical work was done by a number of authors [164–172]. The cyclotron resonance in antidot lattices was probed by photoconductivity measurements [173,174] and modeled in the picture of chaotic dynamics [173]. A recent theoretical description of cyclotron resonance in antidot lattices was given by Polyakov et al. [175]. Commensurability effects were also observed in surface acoustic wave [176] and magneto-photoluminescence experiments [177]. In the latter experiment the signature of commensurate orbits could be detected below the Fermi energy. The detailed origin of this effect is still not clear. In antidot arrays with broken symmetry a photovoltage was observed which was interpreted as a ratchet type of effect [178].

A fascinating field is transport of composite fermions (CF) in lateral superlattices which was recently reviewed by Smet [179]. Composite fermions are composed of an electron and two flux quanta [180–182] and allow to map the fractional quantum Hall effect, which results from the electron-electron interaction at high B, onto the integer quantum Hall effect. The composite fermions around filling factor $\upsilon = \frac{1}{2}$ (the lowest Landau level is half filled) behave like electrons around zero magnetic field. Hence, resistance oscillations stemming from commensurate orbits of composite fermions should also be observable around filling factor $\frac{1}{2}$. At half integer filling the effective magnetic field B_{eff} for composite fermions vanishes, $B_{\text{eff}} = B - B_{\upsilon=1/2} = 0$. Here, $B_{\upsilon=1/2}$ is the magnetic field at which the lowest Landau level is half filled. For composite fermions in antidot arrays the fundamental commensurability peak, corresponding to a periodic CF-orbit around one antidot, is expected at $B_{\text{eff}} = \hbar k_{\text{F,CF}}/ea$. The Fermi wave vector of composite fermions, $k_{\text{F,CF}} = \sqrt{4\pi n_s}$, differs by a factor $\sqrt{2}$ from the electron Fermi wave vector due to complete spin polarization of the composite fermion sea at these high magnetic fields. Such resistance oscillations were indeed found by Kang et al. in antidot arrays [20]. Similar data from Smet et al. are displayed in Fig. 5.33 [21] where commensurability oscillations due to electrons around $B = 0$ are compared to ones from composite fermions around $B_{\text{eff}} = 0$. The data show that the fundamental commensurability peaks of electrons and CF coincide if the zeros of B and B_{eff} are aligned and if the B_{eff} axis is divided by $\sqrt{2}$. Due to the much smaller mean free path of composite fermions the commensurate orbit around four antidots is, in contrast to the one of electrons, not resolved. The

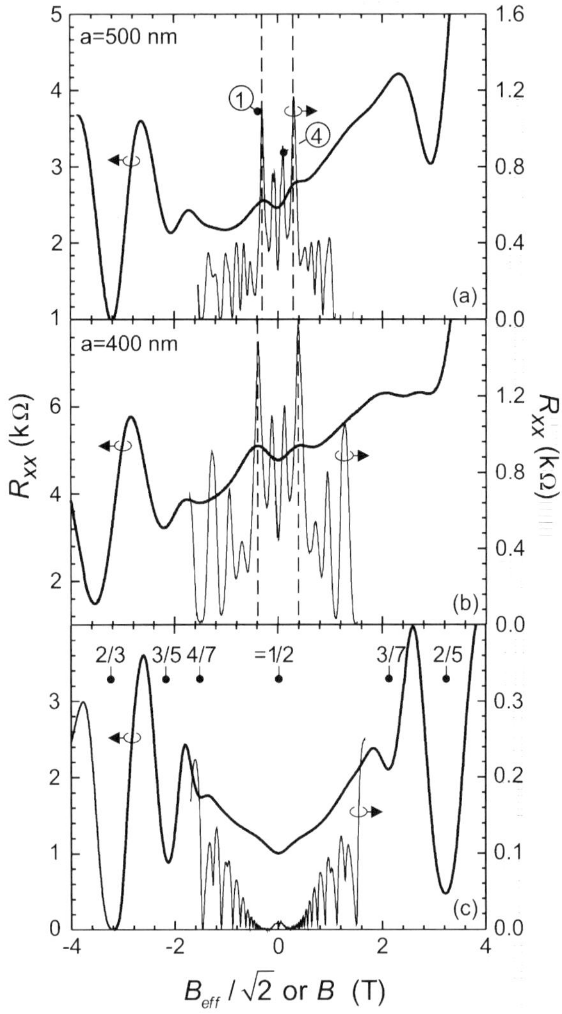

Fig. 5.33. The magnetoresistance of an antidot lattice with (a) 500 nm and (b) 400 nm near $\upsilon = \frac{1}{2}$ (left axis, thick solid lines) and near zero magnetic field (right axis, thin solid lines) measured at $T = 0.45$ K. In (c) the corresponding magnetoresistance of an unpatterned reference section of the Hall bar is shown. For ease of comparison, the magnetic field scale of the CF curves has been divided by $\sqrt{2}$ and shifted horizontally to coincide with $B = 0(B_{\text{eff}} = B - B_{\upsilon=1/2})$. After [21].

data in Fig. 5.33 show that the fundamental commensurability peak of CF at negative effective magnetic fields is more pronounced than at positive B_{eff}. This asymmetry can be explained by the classical trajectories of composite fermions in the soft potential of the antidot lattice [183]. It is thus surprising how far the concepts of classical physics reach even in the fractional quantum Hall regime where quantum effects prevail.

References

1. L.V. Keldish, Effect of ultrasonics on the electron spectrum of crystals, *Sov. Phys. Solid State* **4**, 1658–1659 (1963).
2. L. Esaki and R. Tsu, Superlattice and negative conductivity in semiconductors, *IBM Res. Note* RC-2418 (1969).
3. L. Esaki and R. Tsu, Superlattice and negative differential conductivity in semiconductors, *IBM J. Res. Dev.* **14**(1), 61–65 (1970).
4. A. Cho, *Molecular Beam Epitaxy*. AIP Press, Woodbury (1994).
5. R. Dingle, H.L. Störmer, A.C. Gossard, and W. Wiegmann, Electron mobilities in modulation-doped semiconductor heterojunction superlattices, *Appl. Phys. Lett.* **33**(7), 665–667 (1978).
6. K. von Klitzing, G. Dorda, and M. Pepper, New method for high-accuracy determination of the fine-structure constant based on quantized Hall resistance, *Phys. Rev. Lett.* **45**(6), 494–497 (1980).
7. D.C. Tsui, H.L. Störmer, and A.C. Gossard, Two-dimensional magnetotransport in the extreme quantum limit, *Phys. Rev. Lett.* **48**(22), 1559–1562 (1982).
8. H. Sakaki, K. Wagatsuma, J. Hamasaki, and S. Saito, Possible applications of surface-corrugated quantum thin films to negative-resistance devices, *Thin Solid Films* **36**(2), 497–501 (1976).
9. R.K. Reich, R.O. Grondin, D.K. Ferry, and G.J. Iafrate, Transport in surface superlattices, *Phys. Lett. A* **91A**(1), 28–30 (1982).
10. C. Albrecht, J.H. Smet, D. Weiss, K. von Klitzing, R. Hennig, M. Langenbuch, M. Suhrke, U. Rössler, V. Umansky, and H. Schweizer, Fermiology of two-dimensional lateral superlattices, *Phys. Rev. Lett.* **83**(11), 2234–2237 (1999).
11. R.A. Deutschmann, W. Wegscheider, M. Rother, M. Bichler, G. Abstreiter, C. Albrecht, and J.H. Smet, Quantum interference in artificial band structures, *Phys. Rev. Lett.* **86**(9), 1857–1860 (2001).
12. D. Weiss, K. von Klitzing, K. Ploog, and G. Weimann, Magnetoresistance oscillations in a two-dimensional electron gas induced by a submicrometer periodic potential, *Europhys. Lett.* **8**(2), 179–184 (1989).
13. R.R. Gerhardts, D. Weiss, and K. von Klitzing, Novel magnetoresistance oscillations in a periodically modulated two-dimensional electron gas, *Phys. Rev. Lett.* **62**(10), 1173–1176 (1989).
14. R.W. Winkler, J.P. Kotthaus, and K. Ploog, Landau-band conductivity in a two-dimensional electron system modulated by an artificial one-dimensional superlattice potential, *Phys. Rev. Lett.* **62**(10), 1177–1180 (1989).
15. C.W.J. Beenakker, Guiding-center-drift resonance in a periodically modulated two-dimensional electron gas, *Phys. Rev. Lett.* **62**(17), 2020–2023 (1989).
16. K. Ensslin and P.M. Petroff, Magnetotransport through an antidot lattice in GaAs-$Al_xGa_{1-x}As$ heterostructures, *Phys. Rev.* B **41**(17), 12307–12310 (1990).
17. R. Fleischmann, T. Geisel, and R. Ketzmerick, Magnetoresistance due to chaos and nonlinear resonances in lateral surface superlattices, *Phys. Rev. Lett.* **68**(9), 1367–1370 (1992).
18. N.I. Chernov, G.L. Eyink, J.L. Lebowitz, and Y.G. Sinai, Derivation of Ohm's law in a deterministic mechanical model, *Phys. Rev. Lett.* **70** (1993).
19. A. Kuzmany and H. Spohn, Magnetotransport in the two-dimensional Lorentz gas, *Phys. Rev.* E **57** (1998).
20. W. Kang, H.L. Störmer, L.N. Pfeiffer, K.W. Baldwin, and K.W. West, How real are composite fermions?, *Phys. Rev. Lett.* **71**(23), 3850–3853 (1993).

21. J.H. Smet, D. Weiss, K. von Klitzing, P.T. Coleridge, Z.W. Wasilewski, R. Bergmann, H. Schweizer, and A. Scherer, Composite fermions in periodic and random antidot lattices, *Phys. Rev.* B **56**(7), 3598–3601 (1997).
22. M.V. Berry, Quantizing a classically ergodic system: Sinai's billiard and the KKR method, *Ann. Phys.* **131**(1), 163–216 (1981).
23. T. Ando, A.B. Fowler, and F. Stern, Electronic properties of two-dimensional systems, *Rev. Mod. Phys.* **54**(2), 437–672 (1982).
24. D. Weiss, P. Grambow, K. von Klitzing, A. Menschig, and G. Weimann, Fabrication and characterization of deep mesa etched 'anti'-dot superlattices in GaAs-AlGaAs heterostructures, *Appl. Phys. Lett.* **58**(25), 2960–2962 (1991).
25. D. Weiss, M.L. Roukes, A. Menschig, P. Grambow, K. von Klitzing, and G. Weimann, Electron pinball and commensurate orbits in a periodic array of scatterers, *Phys. Rev. Lett.* **66**(21), 2790–2793 (1991).
26. G. Müller, D. Weiss, K. von Klitzing, K. Ploog, H. Nickel, W. Schlapp, and R. Lösch, Confinement-potential tuning: from nonlocal to local transport, *Phys. Rev.* B **46**(7), 4336–4339 (1992).
27. P.M. Petroff, Y.J. Li, T.J. Li, Z. Xu, W. Beinstingl, S. Sasa, and K. Ensslin, Nanostructures processing by focused ion beam implantation, *J. Vac. Sci. Technol.* B **9**(6), 3074–3078 (1991).
28. G.M. Sundaram, N.J. Bassom, R.J. Nicholas, G.J. Rees, P.J. Heard, P.D. Prewett, J.E.F. Frost, G.A.C. Jones, D.C. Peacock, and D.A. Ritchie, Magnetoconductivity in a mesoscopic antidot array, *Phys. Rev.* B **47**(12), 7348–7353 (1993).
29. A. Lorke, J.P. Kotthaus, and K. Ploog, Magnetotransport in two-dimensional lateral superlattices, *Phys. Rev.* B **44**(7), 3447–3450 (1991).
30. G. Berthold, T. Suski, J. Smoliner, R. Maschek, E. Gornik, G. Böhm, and G. Weimann, Pressure- and gate voltage-induced transitions from lateral surface superlattice to antidot systems, *Semicond. Sci. Tech.* **8**(8), 1512–1516 (1993).
31. G. Lütjering, D. Weiss, R.W. Tank, K. von Klitzing, A. Hülsmann, T. Jakobus, and K. Köhler, Metal-non-metal transition at the crossover from antidots to quantum dots, *Surf. Sci.* **361–362**, 925–929 (1996).
32. A.S. Sachrajda, Y. Feng, R.P. Taylor, G. Kirczenow, L. Henning, J. Wang, P. Zawadzki, and P.T. Coleridge, Magnetoconductance of a nanoscale antidot, *Phys. Rev.* B **50**(15), 10856–10863 (1994).
33. M. Wendel, S. Kuhn, H. Lorenz, J.P. Kotthaus, and M. Holland, Nanolithography with an atomic force microscope for integrated fabrication of quantum electronic devices, *Appl. Phys. Lett.* **65**(14), 1775–1777 (1994).
34. A. Dorn, M. Sigrist, A. Fuhrer, T. Ihn, T. Heinzel, K. Ensslin, W. Wegscheider, and M. Bichler, Electronic properties of antidot lattices fabricated by atomic force lithography, *Appl. Phys. Lett.* **80**(2), 252–254 (2002).
35. R.A. Deutschmann, C. Stocken, W. Wegscheider, M. Bichler, and G. Abstreiter, Commensurability effects in lateral surface-doped superlattices, *Appl. Phys. Lett.* **78**(15), 2175–2177 (2001).
36. M.L. Roukes, A. Scherer, S.J. Allen, Jr., H.G. Craighead, Jr., R.M. Ruthen, E.D. Beebe, and J.P. Harbison, Quenching of the Hall effect in a one-dimensional wire, *Phys. Rev. Lett.* **59**(26), 3011–3014 (1987).
37. H. Fang and P.J. Stiles, Novel magnetoresistance oscillations in a two-dimensional superlattice potential, *Phys. Rev.* B **41**(14), 10171–10174 (1990).
38. T. Yamashiro, J. Takahara, Y. Takagaki, K. Gamo, S. Namba, S. Takaoka, and K. Murase, Commensurate classical orbits on triangular lattices of anti-dots, *Solid State Commun.* **79**(11), 885–887 (1991).

39. J. Takahara, Y. Takagaki, K. Gamo, S. Namba, S. Takaoka, and K. Murase, Fabrication and transport properties of anti-dot triangle lattices, *Microelectron. Eng.* **17**(1–4), 509–512 (1992).
40. J. Takahara, K. Gamo, S. Namba, S. Takaoka, and K. Murase, Transport properties in hexagonal arrays of antidots with different carrier densities, *Jpn. J. Appl. Phys.* 1 **31**(12A), 3786–3790 (1992).
41. F. Nihey, S.W. Hwang, and K. Nakamura, Observation of large h/2e oscillations in semiconductor antidot lattices, *Phys. Rev.* B **51**(7), 4649–4652 (1995).
42. K. Ensslin, S. Sasa, T. Deruelle, and P.M. Petroff, Anisotropic electron transport through a rectangular antidot lattice, *Surf. Sci.* **263**(1–3), 319–323 (1992).
43. R. Schuster, K. Ensslin, J.P. Kotthaus, M. Holland, and C. Stanley, Selective probing of ballistic electron orbits in rectangular antidot lattices, *Phys. Rev.* B **47**(11), 6843–6846 (1993).
44. J. Takahara, A. Nomura, K. Gamo, S. Takaoka, K. Murase, and H. Ahmed, Magnetotransport in hexagonal and rectangular antidot lattices, *Jpn. J. Appl. Phys.* 1 **34**(8B), 4325–4328 (1995).
45. K. Tsukagoshi, M. Haraguchi, K. Oto, S. Takaoka, K. Murase, and K. Gamo, Current-direction-dependent commensurate oscillations in GaAs/AlGaAs antidot superlattice, *Jpn. J. Appl. Phys.* 1 **34**(8B), 4335–4337 (1995).
46. S. Lüthi, T. Vancura, K. Ensslin, R. Schuster, G. Böhm, and W. Klein, Electron trajectories in rectangular antidot superlattices, *Phys. Rev.* B **55**(19), 13088–13092 (1997).
47. G.M. Gusev, P. Basmaji, D.I. Lubyshev, L.V. Litvin, Yu.V. Nastaushev, and V.V. Preobrazhenskii, Magneto-oscillations in a two-dimensional electron gas with a Penrose lattice of artificial scatterers, *Phys. Rev.* B **47**(15), 9928–9930 (1993).
48. G.M. Gusev, P. Basmaji, Z.D. Kvon, L.V. Litvin, A.I. Toropov, and Yu.V. Nastaushev, Negative differential magnetoresistance and commensurability oscillations of two-dimensional electrons in a disordered array of antidots, *J. Phys.: Condens. Matter* **6**(1), 73–78 (1994).
49. K. Tsukagoshi, S. Wakayama, K. Oto, S. Takaoka, K. Murase, and K. Gamo, Transport properties in artificial lateral superlattice, *Superlatt. Microstruct.* **16**(3), 295–301 (1994).
50. K. Tsukagoshi, S. Wakayama, K. Oto, S. Takaoka, K. Murase, and K. Gamo, Magnetotransport through disordered and anisotropic antidot lattices in GaAs/Al_x/Ga_{1-x}As heterostructures, *Phys. Rev.* B **52**(11), 8344–8347 (1995).
51. O. Yevtushenko, G. Lütjering, D. Weiss, and K. Richter, Weak localization in antidot arrays: signature of classical chaos, *Phys. Rev. Lett.* **84**(3), 542–545 (2000).
52. G.M. Gusev, X. Kleber, U. Gennser, D.K. Maude, J.C. Portal, D.I. Lubyshev, P. Basmaji, M.De.P.A. Silva, J.C. Rossi, Y.V. Nastaushev, and M.R. Baklanov, Oscillation of the scattering time in a 2D electron system with oval antidots, Solid-State Electron. **40**(1–8), 441–446 (1996).
53. T. Azuma and T. Osada, Antidot shape dependence of the commensurability oscillation of magnetoresistance in two-dimensional antidot arrays, *Physica* B **256–258**, 397–400 (1998).
54. S. de Haan, A. Lorke, R. Hennig, M. Suhrke, W. Wegscheider, and M. Bichler, Magnetotransport properties of arrays of cross-shaped antidots, *Phys. Rev.* B **60**(12), 8845–8848 (1999).
55. D. Többen, M. Holzmann, G. Abstreiter, A. Kriele, H. Lorenz, J.P. Kotthaus, F. Schäffler, Y.H. Xie, P.J. Silverman, and D. Monroe, Antidot superlattices in two-dimensional hole gases confined in strained germanium layers, *Semicond. Sci. Tech.* **10**(10), 1413–1417 (1995).

56. M. Holzmann, D. Többen, and G. Abstreiter, Transport in silicon/germanium nanostructures, *Appl. Surf. Sci.* **102**, 230–236 (1996).
57. S. Hameau, Y. Guldner, R. Ferreira, S. Zanier, G. Faini, E. Cambril, and Y. Campidelli, Transition from an antidot to a dot array in etched Si/SiGe heterostructures: Influence of the etch depth, *Phys. Rev.* B **60**(23), 15965–15969 (1999).
58. S.-I. Osako, T. Sugihara, Y. Yamamoto, T. Maemoto, S. Sasa, M. Inoue, and C. Hamaguchi, Quantum anti-dot arrays and quantum wire transistors fabricated on InAs/$Al_{0.5}Ga_{0.5}Sb$ heterostructures, *Semicond. Sci. Tech.* **11**(4), 571–575 (1996).
59. J. Eroms, M. Zitzlsperger, D. Weiss, J.H. Smet, C. Albrecht, R. Fleischmann, M. Behet, J. De Boeck, and G. Borghs, Skipping orbits and enhanced resistivity in large-diameter InAs/GaSb antidot lattices, *Phys. Rev.* B **59**(12), R7829–7832 (1999).
60. J. Eroms, M. Zitzlsperger, D. Weiss, J.H. Smet, C. Albrecht, R. Fleischmann, M. Behet, J. De Boeck, and G. Borghs, Magnetotransport in large diameter InAs/GaSb antidot lattices, *Physica* B **256–258**, 409–412 (1998).
61. E.M. Baskin, G.M. Gusev, Z.D. Kvon, A.G. Pogosov, and M.V. Entin, Stochastic dynamics of 2D electrons in a periodic lattice of antidots, *JETP Lett.* **55**(11), 678–682 (1992).
62. R. Schuster, G. Ernst, K. Ensslin, M. Entin, M. Holland, G. Böhm, and W. Klein, Experimental characterization of electron trajectories in antidot lattices, *Phys. Rev.* B **50**(11), 8090–8093 (1994).
63. D. Weiss, G. Lütjering, and K. Richter, Chaotic electron motion in macroscopic and mesoscopic antidot lattices, *Chaos Solitons Fractals* **8**(7–8), 1337–1357 (1997).
64. K. Tsukagoshi, T. Nagao, M. Haraguchi, S. Takaoka, K. Murase, and K. Gamo, Commensurability oscillations by runaway and pinned electrons, *Superlatt. Microstruct.* **23**(2), 493–496 (1998).
65. R. Fleischmann, T. Geisel, and R. Ketzmerick, Quenched and negative Hall effect in periodic media: Application to antidot superlattices, *Europhys. Lett.* **25**(3), 219–224 (1994).
66. K. Richter, Phase coherence effects in antidot lattices: a semiclassical approach to bulk conductivity, *Europhys. Lett.* **29**(1), 7–12 (1995).
67. Wenchang Lu, Theoretical study on commensurability oscillation in anisotropic antidot lattices, *Phys. Rev.* B **54**(11), 8049–8056 (1996).
68. S. Ishizaka and T. Ando, Detailed analysis of the commensurability peak in antidot arrays with various periods, *Phys. Rev.* B **55**(24), 16331–16338 (1997).
69. T. Ando, S. Uryu, and S. Ishizaka, Chaos and quantum transport in antidot lattices, *Jpn. J. Appl. Phys.* 1 **38**(1B), 308–314 (1999).
70. D. Weiss, K. Richter, A. Menschig, R. Bergmann, H. Schweizer, K. von Klitzing, and G. Weimann, Quantized periodic orbits in large antidot arrays, *Phys. Rev. Lett.* **70**(26), 4118–4121 (1993).
71. F. Nihey and K. Nakamura, Aharonov-Bohm effect in antidot structures, *Physica* B **184**(1–4), 398–402 (1993).
72. H. Silberbauer, Magnetic minibands in lateral semiconductor superlattices, *J. Phys.: Condens. Matter* **4**(36), 7355–7364 (1992).
73. H. Silberbauer and U. Rössler, Quantum study of magnetotransport in antidot superlattices, *Phys. Rev.* B **50**(16), 11911–11914 (1994).
74. P. Rotter, M. Suhrke, and U. Rössler, Observability of the magnetic band structure of lateral superlattices, *Phys. Rev.* B **54**(7), 4452–4455 (1996).
75. M.C. Gutzwiller, *Chaos in Classical and Quantum Mechanics*. Springer, Berlin (1990).
76. G. Hackenbroich and F. von Oppen, Semiclassical theory of transport in antidot lattices, *Z. Phys.* B **97**(2), 157–170 (1995).
77. G. Hackenbroich and F. von Oppen, Periodic-orbit theory of quantum transport in antidot lattices, *Europhys. Lett.* **29**(2), 151–156 (1995).

78. K. Richter, Semiclassical theory of mesoscopic quantum systems, *Springer Tracts Mod. Phys.* **163** (2000).
79. A. Yacoby, U. Sivan, C.P. Umbach, and J.M. Hong, Interference and dephasing by electron-electron interaction on length scales shorter than the elastic mean free path, *Phys. Rev. Lett.* **66**(14), 1938–1941 (1991).
80. D.K. Ferry, Quantum magnetotransport in lateral surface superlattices, *Prog. Quant. Electron.* **16**(4), 251–317 (1992).
81. K. Nakamura, S. Ishizaka, and F. Nihey, Quantum oscillations in artificial impurity lattices, *Physica* B **197**(1–4), 144–150 (1994).
82. S. Uryu and T. Ando, Aharonov-Bohm oscillation and periodic orbits in antidot lattices, *Physica* B **256–258**, 388–391 (1998).
83. B.L. Al'tshuler, A.G. Aronov, and B.Z. Spivak, Aharonov-Bohm effect in disordered conductors, *JETP Lett.* **33**(2), 94–97 (1981).
84. S. Uryu and T. Ando, Localization oscillation in antidot lattices, *Physica* B **249–251**, 308–311 (1998).
85. S. Uryu and T. Ando, Numerical study of localization in antidot lattices, *Phys. Rev.* B **58**(16), 10583–10588 (1998).
86. H. Silberbauer, P. Rotter, M. Suhrke, and U. Rössler, Quantum transport and quantum chaos in antidot superlattices in a magnetic field, *Semicond. Sci. Tech.* **9**(11S), 1906–1911 (1994).
87. S. Ishizaka, F. Nihey, K. Nakamura, J. Sone, and T. Ando, Quantum transport in antidot arrays in magnetic fields, *Phys. Rev.* B **51**(15), 9881–9890 (1995).
88. S. Ishizaka, F. Nihey, K. Nakamura, J. Sone, and T. Ando, Numerical studies on quantum transport in antidot arrays in magnetic fields, *Jpn. J. Appl. Phys.* 1 **34**(8B), 4317–4320 (1995).
89. R.B.S. Oakeshott and A. MacKinnon, On the conductivity of lateral-surface superlattices in magnetic fields, *J. Phys.: Condens. Matter* **5**(37), 6991–6998 (1993).
90. R.B.S. Oakeshott and A. MacKinnon, On the conductivity of antidot lattices in magnetic fields, *J. Phys.: Condens. Matter* **6**(8), 1519–1528 (1994).
91. S. Uryu and T. Ando, Scattering-matrix formalism for antidot lattices, *Jpn. J. Appl. Phys.* 1 **34**(8B), 4295–4297 (1995).
92. S. Uryu and T. Ando, Electronic states in antidot lattices: scattering-matrix formalism, *Phys. Rev.* B **53**(20), 13613–13623 (1996).
93. S. Uryu and T. Ando, Aharonov-Bohm type oscillation in antidot lattices, *Solid-State Electron.* **42**(7–8), 1141–1145 (1998).
94. S. Uryu and T. Ando, Analysis of antidot lattices with periodic orbit theory, *Physica* B **227**(1–4), 138–140 (1996).
95. S. Ishizaka and T. Ando, Quantum transport in square and triangular antidot arrays with various periods, *Phys. Rev.* B **56**(23), 15195–15201 (1997).
96. O. Steffens, T. Schlösser, P. Rotter, K. Ensslin, M. Suhrke, J.P. Kotthaus, U. Rössler, and M. Holland, From the two-dimensional electron gas to antidot superlattices: magneto-resistance effects in the transition regime, *J. Phys.: Condens. Matter* **10**(17), 3859–3872 (1998).
97. R. Onderka, M. Suhrke, and U. Rössler, Anisotropic magnetotransport in a rectangular antidot superlattice: Classical and semiclassical aspects, *Phys. Rev.* B **62**(16), 10918–10922 (2000).
98. T. Ando, S. Uryu, S. Ishizaka, and T. Nakanishi, Quantum transport in antidot lattices, *Chaos Solitons Fractals* **8**(7–8), 1057–1083 (1997).
99. M. Suhrke and P. Rotter, in *Theory of transport properties of semiconductor nanostructures*, (eds. E. Schöll), Chapman and Hall, London (1997).

100. A. Lorke, J.P. Kotthaus, and K. Ploog, Localization in GaAs electron-dots and anti-dots, *Superlatt. Microstruct.* **9**(1), 103–106 (1991).
101. A. Lorke, I. Jejina, and J.P. Kotthaus, Far-infrared response of lateral superlattices in high magnetic fields, *Phys. Rev.* B **46**(19), 12845–12848 (1992).
102. G. Lütjering, Nanostrukturierte Antidot-Systeme: Vom ballistischen Billard zum Metall-Isolator-Übergang, Doctoral Thesis, Fakultät für Physik, Universität Stuttgart, 1996.
103. R.W. Tank and R.B. Stinchcombe, Lateral conductance of perfect and disordered dot lattices, *J. Phys.: Condens. Matter* **7**(5), 849–866 (1995).
104. G.M. Gusev, U. Gennser, X. Kleber, D.K. Maude, J.C. Portal, D.I. Lubyshev, P. Basmaji, M.A.P. de Silva, J.C. Rossi, and Yu.V. Nastaushev, Absence of delocalised states in a 2D electron gas in a magnetic field below $\omega_c\tau = 1$, *Solid State Commun.* **100**(4), 269–273 (1996).
105. F. Nihey, M.A. Kastner, and K. Nakamura, Insulator-to-quantum-Hall-liquid transition in an antidot lattice, *Phys. Rev.* B **55**(7), 4085–4088 (1997).
106. F. Nihey, K. Nakamura, M.A. Kastner, T. Takamasu, and G. Kido, Insulator-quantum Hall liquid transition in antidot lattices, *Physica* B **249–251**, 302–307 (1998).
107. G.M. Gusev, U. Gennser, X. Kleber, D.K. Maude, J.C. Portal, D.I. Lubyshev, P. Basmaji, M.A.P. de Silva, J.C. Rossi, and Yu.N. Nastaushev, Percolation network in a smooth artificial potential, *Phys. Rev.* B **58**(8), 4636–4643 (1998).
108. M.V. Budantsev, Z.D. Kvon, A.G. Pogosov, G.M. Gusev, J.C. Portal, D.K. Maude, N.T. Moshegov, and A.I. Toropov, 2D lattice of coupled Sinai billiards: metal or insulator at $g \ll 1$?, *Physica* B **256–258**, 595–599 (1998).
109. G.M. Gusev, Z.D. Kvon, L.V. Litvin, Yu.V. Nastaushev, A.K. Kalagin, and A.I. Toropov, Magnetoresistance of a two-dimensional electron gas in a disordered artificial array of scatterers, *Superlatt. Microstruct.* **13**(2), 263–265 (1993).
110. T. Nagao, Magnetotransport through random antidot lattices, *J. Phys. Soc. Jpn.* **65**(8), 2606–2609 (1996).
111. D.G. Polyakov, F. Evers, A.D. Mirlin, and P. Wölfle, Quasiclassical magnetotransport in a random array of antidots, *Phys. Rev.* B **64**(20), 205306 (2001).
112. F. Evers, A.D. Mirlin, D.G. Polyakov, and P. Wölfle, Magnetotransport in a random array of antidots, *Physica* E **12**(1–4), 260–263 (2002).
113. K. Richter and M. Sieber, Semiclassical theory of chaotic quantum transport, *Phys. Rev. Lett.* **89**(20), 206801 (2002).
114. I.L. Aleiner and A.I. Larkin, Divergence of classical trajectories and weak localization, *Phys. Rev.* B **54**(20), 14423–14444 (1996).
115. T.J. Thornton, M.L. Roukes, A. Scherer, and B.P. Van de Gaag, Boundary scattering in quantum wires, *Phys. Rev. Lett.* **63**(19), 2128–2131 (1989).
116. G. Nachtwei, G. Lütjering, D. Weiss, Z.H. Liu, K. von Klitzing, and C.T. Foxon, Breakdown of the quantum Hall effect in periodic and aperiodic antidot arrays, *Phys. Rev.* B **55**(11), 6731–6734 (1997).
117. G. Nachtwei, Z.H. Liu, I.I. Kaya, G. Lütjering, D. Weiss, K. von Klitzing, and K. Eberl, Nonequilibrium quantum transport in antidot arrays, *Phys. Status Solidi* B **204**(1), 329–334 (1997).
118. G. Nachtwei, Z.H. Liu, G. Lütjering, R.R. Gerhardts, D. Weiss, K. von Klitzing, and K. Eberl, Critical currents in quantum Hall conductors with antidot arrays, *Phys. Rev.* B **57**(16), 9937–9944 (1998).
119. C.M. Marcus, A.J. Rimberg, R.M. Westervelt, P.F. Hopkins, and A.C. Gossard, Conductance fluctuations and chaotic scattering in ballistic microstructures, *Phys. Rev. Lett.* **69**(3), 506–509 (1992).

120. R.A. Jalabert, H.U. Baranger, and A.D. Stone, Conductance fluctuations in the ballistic regime: a probe of quantum chaos?, *Phys. Rev. Lett.* **65**(19), 2442–2445 (1990).
121. H.U. Baranger, R.A. Jalabert, and A.D. Stone, Weak localization and integrability in ballistic cavities, *Phys. Rev. Lett.* **70**(25), 3876–3879 (1993).
122. H.U. Baranger, R.A. Jalabert, and A.D. Stone, Quantum-chaotic scattering effects in semiconductor microstructures, *Chaos* **3**(4), 665–682 (1993).
123. A.M. Chang, H.U. Baranger, L.N. Pfeiffer, and K.W. West, Weak localization in chaotic versus nonchaotic cavities: a striking difference in the line shape, *Phys. Rev. Lett.* **73**(15), 2111–2114 (1994).
124. G. Lütjering, K. Richter, D. Weiss, J. Mao, R.H. Blick, K. von Klitzing, and C.T. Foxon, Weak localisation in ballistic cavities filled with antidot arrays, *Surf. Sci.* **361–362**, 709–713 (1996).
125. P.D. Ye and S. Tarucha, Internal magnetic focusing in an array of open quantum dots, *Jpn. J. Appl. Phys.* 1 **38**(1B), 319–321 (1999).
126. J.H. Smet, D. Weiss, R.H. Blick, G. Lütjering, K. von Klitzing, R. Fleischmann, R. Ketzmerick, T. Geisel, and G. Weimann, Magnetic focusing of composite fermions through arrays of cavities, *Phys. Rev. Lett.* **77**(11), 2272–2275 (1996).
127. J.H. Smet, D. Weiss, K. von Klitzing, R. Fleischmann, R. Ketzmerick, T. Geisel, W. Wegscheider, P.T. Coleridge, Z.W. Wasilewski, and G. Weimann, Enhanced soft-wall effects for composite fermions in magnetic focusing and commensurability experiments, *Physica* E **1**(1–4), 153–159 (1997).
128. J.H. Smet, D. Weiss, K. von Klitzing, R. Fleischmann, R. Ketzmerick, T. Geisel, W. Wegscheider, P.T. Coleridge, Z.W. Wasilewski, and G. Weimann, Composite fermions in magnetic focusing and commensurability experiments, *Physica* B **249–251**, 15–22 (1998).
129. S. Bhargava, H.-R. Blank, V. Narayanamurti, and H. Kroemer, Fermi-level pinning position at the Au-InAs interface determined using ballistic electron emission microscopy, *Appl. Phys. Lett.* **70**(6), 759–761 (1997).
130. M. Behet, S. Nemeth, J. De Boeck, G. Borghs, J. Tummler, J. Woitok, and J. Geurts, Molecular beam epitaxy and characterization of $InAs/Al_{0.2}Ga_{0.8}Sb$ heterostructures for magnetic sensing applications, *Semicond. Sci. Tech.* **13**(4), 428–432 (1998).
131. M. Behet, J. Das, J. de Boeck, and G. Borghs, InAs/(Al,Ga)Sb quantum well structures for magnetic sensors, *IEEE Trans. Magn.* **34**(4), 1300–1302 (1998).
132. C. Gould, A.S. Sachrajda, Y. Feng, A. Delage, P.J. Kelly, K. Leung, and P.T. Coleridge, Phase coherence and trajectory trapping around one or two independently controllable antidots in quantum wires, *Phys. Rev.* B **51**(16), 11213–11216 (1995).
133. A.F. Andreev, The thermal conductivity of the intermediate state in superconductors, *Sov. Phys.–JETP* **19**(5), 1228–1231 (1964).
134. I. Kosztin, D.L. Maslov, and P.M. Goldbart, Chaos in Andreev billiards, *Phys. Rev. Lett.* **75**(9), 1735–1738 (1995).
135. J.A. Melsen, P.W. Brouwer, K.M. Frahm, and C.W.J. Beenakker, Induced superconductivity distinguishes chaotic from integrable billiards, *Europhys. Lett.* **35**(1), 7–12 (1996).
136. W. Ihra, M. Leadbeater, J.L. Vega, and K. Richter, Semiclassical theory of integrable and rough Andreev billiards, *Eur. Phys. J.* B **21**(3), 425–435 (2001).
137. I. Adagideli and P.M. Goldbart, Quantal Andreev billiards: Density of states oscillations and the spectrum-geometry relationship, *Phys. Rev.* B **65**(20), 201306 (2002).
138. F. Rahman, T.J. Thornton, R. Huber, L.F. Cohen, W.T. Yuen, and R.A. Stradling, Superconductor–semiconductor interaction effects in mesoscopic hybrid structures, *Phys. Rev.* B **54**(19), 14026–14031 (1996).

139. H. Takayanagi and T. Kawakami, Superconducting proximity effect in the native inversion layer on InAs, *Phys. Rev. Lett.* **54**(22), 2449–2452 (1985).
140. T. Kawakami and H. Takayanagi, Single-crystal n-InAs coupled Josephson junction, *Appl. Phys. Lett.* **46**(1), 92–94 (1985).
141. C. Nguyen, H. Kroemer, and E.L. Hu, Anomalous Andreev conductance in InAs-AlSb quantum well structures with Nb electrodes, *Phys. Rev. Lett.* **69**(19), 2847–2850 (1992).
142. A.F. Morpurgo, S. Holl, B.J. van Wees, T.M. Klapwijk, and G. Borghs, Phase conjugated Andreev backscattering in two-dimensional ballistic cavities, *Phys. Rev. Lett.* **78**(13), 2636–2639 (1997).
143. J. Eroms, M. Tolkiehn, D. Weiss, U. Rössler, J. De Boeck, and G. Borghs, Commensurability effects in Andreev antidot billiards, *Europhys. Lett.* **58**(4), 569–575 (2002).
144. J. Eroms, D. Weiss, J. De Boeck, and S. Borghs, Magnetotransport properties of periodic Nb-2DEG structures, *Physica* C **352**(1–4), 131–134 (2001).
145. A. Jacobs, R. Kümmel, and H. Plehn, Proximity effect, Andreev reflections, and charge transport in mesoscopic superconducting/semiconducting heterostructures, *Superlatt. Microstruct.* **25**(5–6), 669–681 (1999).
146. J. Eroms, M. Tolkiehn, D. Weiss, U. Rössler, J. De Boeck, and S. Borghs, Chaotic motion and suppression of commensurability effects in an Andreev antidot billiard, *Physica* E **12**(1–4), 918–921 (2002).
147. A. Chrestin, Y. Matsuyama, and U. Merkt, Evidence for a proximity-induced energy gap in Nb/InAs/Nb junctions, *Phys. Rev.* B **55**(13), 8457–8465 (1997).
148. D. Weiss, *Elektronen in künstlichen Kristallen*. Verlag Harri Deutsch, Frankfurt (1994).
149. J. Heremans, B.K. Fuller, C.M. Thrush, and V. Bayot, Temperature dependence of the magnetoresistance of $In_xGa_{1-x}As$ antidot lattices, *Phys. Rev.* B **54**(4), 2685–2690 (1996).
150. N.A. Mortensen, K. Flensberg, and A.-P. Jauho, Angle dependence of Andreev scattering at semiconductor-superconductor interfaces, *Phys. Rev.* B **59**(15), 10176–10182 (1999).
151. J. Nitta, T. Akazaki, and H. Takayanagi, Magnetic-field dependence of Andreev reflection in a clean Nb-InAs-Nb junction, *Phys. Rev.* B **49**(5), 3659–3662 (1994).
152. P.C. van Son, H. van Kempen, and P. Wyder, New method to study the proximity effect at the normal-metal-superconductor interface, *Phys. Rev. Lett.* **59**(19), 2226–2228 (1987).
153. P.C. van Son, H. van Kempen, and P. Wyder, Andreev reflection and geometrical resonance effects for a gradual variation of the pair potential near the normal-metal-superconductor interface, *Phys. Rev.* B **37**(10), 5015–5023 (1988).
154. M. Tornow, D. Weiss, K. von Klitzing, K. Eberl, D.J. Bergman, and Y.M. Strelniker, Anisotropic magnetoresistance of a classical antidot array, *Phys. Rev. Lett.* **77**(1), 147–150 (1996).
155. K. Kern, D. Heitmann, P. Grambow, Y.H. Zhang, and K. Ploog, Collective excitations in antidots, *Phys. Rev. Lett.* **66**(12), 1618–1621 (1991).
156. D. Heitmann, K. Kern, T. Demel, P. Grambow, K. Ploog, and Y.H. Zhang, Spectroscopy of quantum dots and antidots, *Surf. Sci.* **267**(1–3), 245–252 (1992).
157. Y. Zhao, D.C. Tsui, M. Santos, M. Shayegan, R.A. Ghanbari, D.A. Antoniadis, and H.I. Smith, Magneto-optical absorption in a two-dimensional electron grid, *Appl. Phys. Lett.* **60**(12), 1510–1512 (1992).
158. D. Heitmann, Far infrared spectroscopy of quantum-dots and antidot arrays, *Physica* B **212**(3), 201–206 (1995).

159. K. Bollweg, T. Kurth, D. Heitmann, E. Vasiliadou, K. Eberi, and H. Brugger, Circular polarization of far-infrared modes in antidot arrays, *Phys. Rev.* B **52**(11), 8379–8383 (1995).
160. A. Huber, I. Jejina, H. Lorenz, J.P. Kotthaus, S. Bakker, and T.M. Klapwijk, Far-infrared excitations in antidot systems on silicon MOS structures, *Semicond. Sci. Tech.* **10**(3), 365–368 (1995).
161. T. Deruelle, Y. Guldner, and B. Meurer, Collective excitations in tunable antidot arrays, *Solid State Commun.* **96**(11), 877–880 (1995).
162. M. Hochgrafe, B.P. van Zyl, C. Heyn, D. Heitmann, and E. Zaremba, Far-infrared excitations in rectangular antidot arrays, *Phys. Rev.* B **63**(3), 033316 (2001).
163. P.D. Ye, L.W. Engel, D.C. Tsui, J.A. Simmons, J.R. Wendt, G.A. Vawter, and J.L. Reno, High magnetic-field microwave conductivity of two-dimensional electrons in an array of antidots, *Phys. Rev.* B **65**(12), 121305 (2002).
164. D. Huang and G. Gumbs, Magnetoplasmon excitations in a two-dimensional square array of antidots, *Phys. Rev.* B **47**(15), 9597–9604 (1993).
165. G.Y. Wu and Y. Zhao, Magnetoplasmons and FIR response of an antidot array, *Surf. Sci.* **305**(1–3), 601–605 (1994).
166. V. Fessatidis, H.L. Cui, and O. Kuhn, Hydrodynamical theory of magnetoplasma excitations in an antidot system, *Phys. Rev.* B **47**(11), 6598–6602 (1993).
167. S.A. Mikhailov and V.A. Volkov, Theory of electromagnetic response and collective excitations in antidots, *Phys. Rev.* B **52**(24), 17260–17268 (1995).
168. A.V. Chaplik, Magnetoexcitons in quantum rings and in antidots, *JETP Lett.* **62**(11), 900–904 (1995).
169. S.A. Mikhailov, Theory of electromagnetic response and collective excitations of a square lattice of antidots, *Phys. Rev.* B **54**(20), R14293–14296 (1996).
170. V. Gudmundsson and S.I. Erlingsson, presented at the *1997 Advanced Workshop on Frontiers in Electronics, WOFE '97 Proceedings*, New York, NY, USA (1997) (unpublished).
171. V. Gudmundsson, Far-infrared-active collective modes of short-period arrays of quantum dots and antidots, *Phys. Rev.* B **57**(7), 3989–3993 (1998).
172. A. Emperador, M. Pi, M. Barranco, E. Lipparini, and L. Serra, Far-infrared excitations in an antidot at finite magnetic fields, *Jpn. J. Appl. Phys.* 1 **40**(2A), 518–524 (2001).
173. E. Vasiliadou, R. Fleischmann, D. Weiss, D. Heitmann, K. von Klitzing, T. Geisel, R. Bergmann, H. Schweizer, and C.T. Foxon, Cyclotron-resonance anomalies in an antidot array measured by microwave photoconductivity, *Phys. Rev.* B **52**(12), R8658–8661 (1995).
174. B.G.L. Jager, S. Wimmer, A. Lorke, J.P. Kotthaus, W. Wegscheider, and M. Bichler, Edge and bulk effects in Terahertz photoconductivity of an antidot superlattice, *Phys. Rev.* B **63**(4), 045315 (2001).
175. D.G. Polyakov, F. Evers, and I.V. Gornyi, Cyclotron resonance in antidot arrays, *Phys. Rev.* B **65**(12), 125326 (2002).
176. G.R. Nash, S.J. Bending, M. Riek, and K. Eberl, Commensurate and rosette-shaped electron orbits probed by surface acoustic wave attenuation, *Phys. Rev.* B **63**(11), 113316 (2001).
177. I.V. Kukushkin, D. Weiss, G. Lütjering, R. Bergmann, H. Schweizer, K. von Klitzing, K. Eberl, P. Rotter, M. Suhrke, and U. Rössler, Manifestation of commensurate orbits in the magnetoluminescence spectrum of electrons in antidot arrays, *Phys. Rev. Lett.* **79**(9), 1722–1725 (1997).

178. A. Lorke, S. Wimmer, B. Jager, J.P. Kotthaus, W. Wegscheider, and M. Bichler, Far-infrared and transport properties of antidot arrays with broken symmetry, *Physica* B **249–251**, 312–316 (1998).
179. J.H. Smet, in *Composite Fermions* (eds. O. Heinonen) World Scientific Publishing, Singapore, 443–491 (1998).
180. J.K. Jain, Composite-fermion approach for the fractional quantum Hall effect, *Phys. Rev. Lett.* **63**(2), 199–202 (1989).
181. B.I. Halperin, P.A. Lee, and N. Read, Theory of the half-filled Landau level, *Phys. Rev.* B **47**(12), 7312–7343 (1993).
182. O. Heinonen, *Composite Fermions: A Unified View of the Quantum Hall Regime*. World Scientific Publishing Co., Singapore (1998).
183. R. Fleischmann, T. Geisel, C. Holzknecht, and R. Ketzmerick, Nonlinear dynamics of composite fermions in nanostructures, *Europhys. Lett.* **36**(3), 167–172 (1996).

Chapter 6

On the Influence of Resonant States on Ballistic Transport in Open Quantum Dots: Spectroscopy and Tunneling in the Presence of Multiple Conducting Channels

R. Akis[1], J. P. Bird[1], D. Vasileska[1], D. K. Ferry[1], A. P. S. de Moura[2], and Y.-C. Lai[2]

[1] *Department of Electrical Engineering and Center for Solid State Electronics Research, Arizona State University, Tempe, AZ 85287-5706*
[2] *Department of Mathematics and SSERC, Arizona State University, Tempe, AZ 85287*

1. Introduction

A fundamental issue in quantum mechanics concerns the manner in which the discrete level spectrum of an isolated system is modified when it is coupled to some external, macroscopic measuring environment. From an experimental perspective, an ideal system for the study of this issue is provided by semiconductor quantum dots, which are quasi-zero dimensional semiconductor structures in which the flow of electrical current is confined on length scales comparable to the size of the electron itself [1]. The key components of a quantum dot are illustrated schematically in Fig. 6.1(a), in which the basic idea is that current flow between the macroscopic source and drain reservoirs is forced to occur via a central cavity.

In most situations of interest here, the dots are realized using the split-gate technique [2], the basic principles of which are illustrated schematically in Figs 6.1(b) and (c). According to this approach, metal gates with a fine-line pattern defined by electron beam lithography are first deposited on the surface of a GaAs/AlGaAs heterojunction, which has a two-dimensional electron gas (2DEG) confined near the interface between GaAs and AlGaAs layers. Application of a suitable negative bias to the gates depletes the regions of 2DEG from directly underneath them, forming a dot whose lead openings are defined by means of quantum point contacts (QPCs) [3,4]. Figure 6.1(c) displays one such confining potential, calculated by solving Poisson's equation (see section 6 for more details). What is actually pictured is the potential slice $V(x, y, z_o)$, where

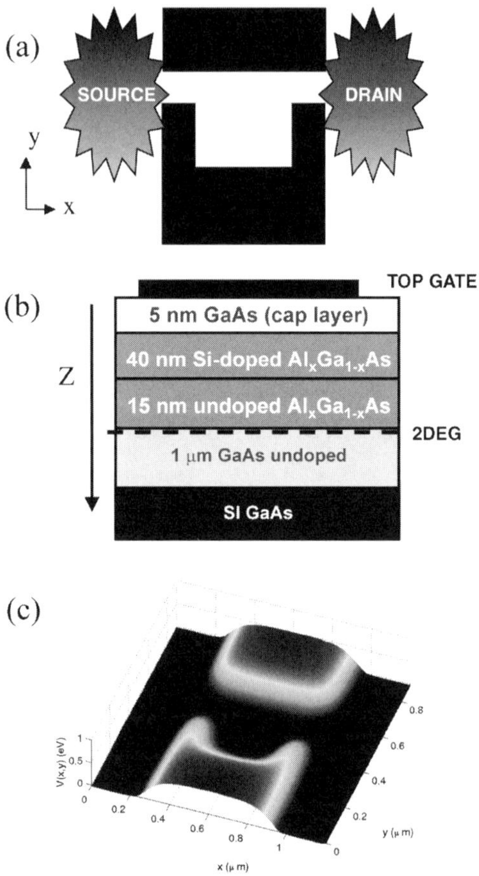

Fig. 6.1. (a) A schematic diagram illustrating the key features of a quantum dot. A central scattering region is connected to the source and drain by means of quantum point contacts (QPCs), each of which supports a number of propagating modes. (b) A dot is defined by a top gate overlying a GaAs/AlGaAs heterojunction in which conduction electrons are constrained to move in a 2 dimensional electron gas (2DEG). (c) The confining potential, $V(x, y)$ experienced by the electrons in the 2DEG when a finite bias is applied to the top gate.

z_0 is the position of the 2DEG along the z-axis, obtained assuming a top gate pattern similar to that shown in Fig. 6.1(a).

The use of electron beam lithography to define the gates allows a realization of sub-micron sized dots, so that the size is therefore comparable to the spatial extent of short range potential fluctuations in the underlying two dimensional electron gas [5]. These fluctuations are associated with the statistical distribution of donors in the AlGaAs layer and recent studies suggest that minimum-energy considerations cause their ionization to order into a quasi-lattice structure [6]. In the presence of the resulting weak disorder, electronic motion within the dots should therefore be predominantly ballistic in nature, with large-angle scattering events occurring primarily at the confining walls [7]. In this regime, open quantum dots thus represent an actual physical implementation of what is known as a billiard system.

With a sufficiently negative bias applied to the quantum point contacts (QPCs) of the dot, electron transmission through the dots may only occur by tunneling and the transport behavior is dominated by the Coulomb blockade effect [8–15]. Here, however, we focus almost exclusively on the behavior exhibited by *open* quantum dots, whose point contacts are instead biased to allow electron transmission via a number, $N > 1$, of transversely quantized modes [16–20]. In these open dots, the Coulomb blockade effect is suppressed, and the conductance, G, in units of $2e^2/h$ is typically larger than unity.

Given the close connection between quantum dots and classical billiard systems, quantum dots have generated tremendous interest as a test bed for predictions regarding quantum chaos. These predictions, generally involving the signatures of chaos that might be found in the transport, were developed by applying semiclassical methods to the problem of open cavities [16,17,19,21–32]. In this regard, a popular approach has been to make the simplifying assumption that the underlying classical scattering dynamics is completely chaotic (or hyperbolic), and then apply Random Matrix Theory (RMT) [32]. An important limitation of the semiclassical approach in this context is that it is only applicable to situations where the time scale for the electrons to be present in the device is relatively short compared to the characteristic quantum time scale. A simple equation that expresses this condition mathematically is the requirement:

$$\frac{\hbar}{\tau_0} \gg \Delta \tag{1}$$

where τ_0 is the escape time and Δ is the mean level separation. Since open dots would naturally appear to be strongly coupled to their external environment, it has been suggested that their discrete level spectrum should be obscured by lifetime broadening effects and, as such, Eq. (1) should essentially be satisfied automatically in quantum dots provided that the QPCs are sufficiently open [21].

In fact, the situation is more complicated than this naïve view. Simply opening the dots up is unsufficient to guarantee that the above condition is satisfied. Even with many modes propagating through the QPCs, features in the transport can still largely be determined by the discrete level spectrum of the dot. Moreover, rather than simply broadening energy levels, we believe that the QPCs act as a spatial filter, determining which levels will take an active role in the resulting electrical behavior. Many transport measurements on dots have actually served as an experimental probe of this filtering effect. What follows is a summary of the body of evidence that we have accumulated for this position in recent years. The main focus will be on theoretical work and simulation studies. However, some key experimental results shall also be discussed in detail. We believe that the evidence compiled here is rather compelling, and trust that the reader will concur upon reaching the end of this review.

The organization of this review is as follows. In section 2, we briefly comment on some of the details of the semiclassical theory that has been applied to quantum dot systems. In section 3, we summarize the fully quantum mechanical method we have used to simulate these systems. In section 4, we discuss conductance resonances in open quantum dots and show how they can significantly influence the conductance, in particular, producing distinct periodicities in the fluctuations that the dots exhibit. In section 5, we quantify the relationship between these open dot resonances and eigenstates of a suitably chosen (closed) system and show that the level broadening in these systems is highly nonuniform. In particular, we show that certain groups of states, those with amplitude localized in interior dot regions and typically along the paths of periodic orbits, are so robust that they survive even in situations where the QPCs are wide enough to be almost comparable in size to the entire dot. In section 6, we consider the effects of thermal broadening and ensemble averaging and demonstrate that distinct features relating to the underlying discrete dot spectrum remain present even when such averaging is done. We also discuss the zero field peaks in dot resistance that have been atttributed to weak localization and demonstrate how spectral artifacts in the conductance may render ambiguous the semiclassical prediction that the lineshape of these peaks should depend on whether or not the underlying dynamics is chaotic. In section 7, we make direct comparisons between transport experiments and simulations in which self-consistent solutions for the confining potentials were performed. Examining conductance as a function of gate voltage, we show results for two different QPC lead configurations which are consistent with the idea that the QPCs are causing selective excitation of states. In section 8, we make the case that, to be truly relevant to these systems, any semiclassical approach must take into account contributions from the quantum-mechanical tunneling of electrons through KAM islands. Finally, a summary is contained in section 9.

2. Some Comments about Semiclassical Theories and their Underlying Assumptions

In semiclassical theory, the particle trajectories are calculated using classical mechanics [31,33,34]. However, for each of these trajectories, there is an associated quantum mechanical amplitude and phase. The latter will depend on the action, S, as computed by doing a path integral over the trajectory, $\int \mathbf{p} \cdot d\mathbf{q}$, where $\mathbf{p}$ and $\mathbf{q}$ represent the appropriate momentum and coordinate variables for the problem. Quantization arises by noting that this integral has value modulo Planck's constant, h. Semiclassical theory can be thought of a simplification and approximation of the Feynman path-integral approach, in which *all* paths that connect the initial and final points are included in the calculation, not just the classically allowed ones [31,33,34].

Given the central role of classical trajectories in semiclassical theory, the nature of the dynamics of the system under study becomes of prime importance. Figure 6.2(a) shows a square billiard, a cavity in which particle motion is constrained only by specular reflections from the boundary walls. A square billiard is classically *regular*, that is, *all* trajectories result in stable periodic orbits that retrace themselves ad infinitum. The diamond orbit depicted in Fig. 6. 2(a) is a representative example. In the phase space of an appropriately chosen $\mathbf{p}$ and $\mathbf{q}$, such orbits are all constrained to move along invariant tori, which show themselves as a set of dots that lie on an ellipse when Poincaré sections of the phase space are taken [31,33,34]. Such dynamical systems are referred to as being *integrable*, i.e. if there are N coordinates, there are N independent constants of motion.

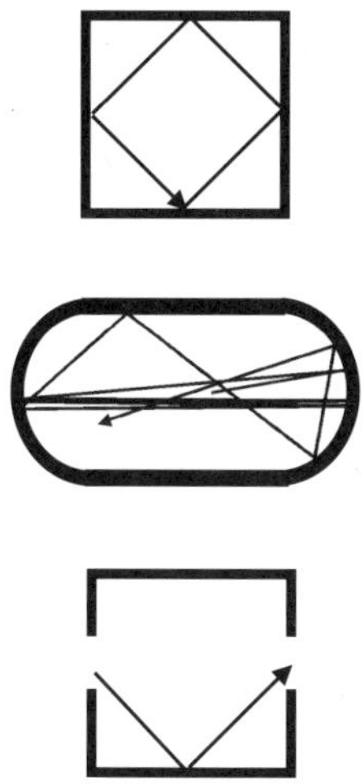

Fig. 6.2. An illustration of three different classical billiard systems. The closed square (top) exhibits regular behavior. The closed stadium (middle) exhibits chaotic behavior. The open square (bottom) allows short lived orbits that can quickly enter and exit through the openings.

In contrast, the stadium billiard shown in Fig. 6.2(b) is classically *chaotic* [31,33–35]. In such a system, the trajectories show extreme sensitivity to the initial conditions, and pairs of trajectories initially close to one another will diverge away from each other exponentially in time. While periodic orbits do exist in the stadium, they are all unstable. Moreover, these periodic orbits are imbedded in a sea of non-periodic orbits, which are vastly dominant (in comparison, the set of periodic orbits is of measure zero) and tend to fill the phase space. The non-periodic orbit depicted in Fig. 6.2(b) is one such orbit destined never to retrace itself. Chaotic systems are not integrable.

There are also situations where the phase space is *mixed*, where in a chaotic sea there also exist islands of stability, typically referred to as Kolmogorov-Arnold-Moser (KAM) islands [31,33,34]. In section 8, we shall show that the soft-walled potential shown in Fig. 6. 1(c) can generate such a mixed phase space. Moreover, a mixed phase space can also result when an initially integrable system is subjected to a perturbation, a point we shall return to in section 5.

It has been found that there is an intimate connection between classical periodic orbits, their energy levels, and the resulting density of states (DOS) of the corresponding quantum system, even in the presence of chaos. In semiclassical periodic-orbit theory, the DOS of a quantum system may be expressed in terms of trace formulae that involve summations over contributions from the periodic orbits [21,31,33,34,36–39]. For a *closed* system with energy levels, E_n, the DOS, $\rho(E)$, is expressed in terms of sum of two pieces, a smooth part which represents an average and an oscillatory contribution:

$$\rho(E) = \sum_n \delta(E - E_n) = \bar{\rho}(E) + \delta\rho(E) \tag{2}$$

The oscillatory contribution has the form

$$\delta\rho(E) = \sum_{\Gamma}\sum_{k=1}^{\infty} A_{\Gamma k}(E)\cos\left[kS_{\Gamma}(E) - \lambda_{\Gamma k}\right] \tag{3}$$

where the Γ index corresponds to a particular *periodic* orbit, and k represents the number of times that orbit is retraced. The amplitude $A_{\Gamma k}(E)$ is a quantity that depends on the period and stability of the orbit. The Maslov index $\lambda_{\Gamma k}$,is an additional phase factor that represents a winding number. The action, $S_{\Gamma}(E)$, is given by a path integral of the momentum along the periodic orbit:

$$S_{\Gamma}(E) = \frac{1}{\hbar}\oint_{\Gamma} \mathbf{p}\cdot\mathbf{dq} \tag{4}$$

With regard to how orbits contribute to the s, an orbit of period T_{Γ} plays a role in determining features in the DOS that have a spacing of $\Delta E \approx 2\pi\hbar/T_{\Gamma}$ [34]. Thus, to numerically obtain finer structure in the DOS, one must include longer orbits before truncating the summation.

The trajectories shown in Fig. 6.2(a) and (b) are constrained to the interior regions of the cavities since they are closed systems. The quantum dots we will be generally studying are *open* and are connected to the outside via QPC leads. In contrast to the closed billiards, such a system can also allow trajectories to enter and leave after only a few bounces off the boundary walls. An example of such an open orbit, directly connecting the input and output openings of an open square billiard after a single bounce, is shown in Fig. 6.2(c).

Which orbits will then take a dominant role in determining the transport properties of quantum dots- the periodic orbits, or the open ones? Naively, one might think only the open orbits are necessary, since only they can connect input and output leads in an obvious fashion. However, in tunneling dots where the coulomb blockade is prevalent, there is a direct correspondence between resonant peaks in conductance and the energy levels in the dot [8–10,12–15]. Thus, even with leads that pass current, the discrete DOS, as determined by periodic orbits, can dominate. A simpler example of a situation where there is a correspondence between resonant transmission and a discrete DOS is the one-dimensional double-barrier problem. This correspondence is eventually lost however as the potential barriers are lowered, since the resonances will broaden and eventually overlap. Might not a similar crossover occur in two dimensional quantum dots? A criterion for such a crossover in dots has been suggested [27,40] based on a variation of an argument originally applied by Thouless to the problem of diffusion-induced broadening of energy levels in disordered conductors [41].

The Thouless argument follows by noting that the energy levels in a conductor of length L should be uncertain to an amount $\Gamma \approx hD/L^2$, where D is the diffusion constant and L^2/D is the time required to diffuse across the sample. Since the average level spacing in the conductor may be written as $\Delta = 1/N_{\mathrm{E}}L^d$, where N_{E} is the density of energy states and d is the dimensionality of the conductor, the ratio Γ/Δ may be written as [40]

$$\frac{\Gamma}{\Delta} = \frac{hD}{L^2}N_{\mathrm{E}}L^d \tag{5}$$

Using the Einstein relation for a degenerate conductor ($\sigma = e^2 N_E D$) to relate the density of states to the conductivity σ, Eq. (5) may be simplified to yield

$$\frac{\Gamma}{\Delta} = \frac{h}{e^2}\sigma L^{d-2} = g \tag{6}$$

where g is a *dimensionless* conductance, measured as a fraction of e^2/h. Equation (6) is the Thouless argument and reveals that a metallic conductor ($g \geq 1$) is one in which the level broadening is always comparable to, or greater than, the average level spacing. When this argument is applied to dots, a different expression for Δ is usually invoked, $\Delta = 2\pi\hbar^2/m^* A$, where A is the dot area and m^* the effective mass [23,26,27]. In a dot whose QPCs each support N propagating modes, then the conductance may be written as $g = N$ by assuming Ohmic addition of the two point contacts. Since $N \geq 1$ is required for the dot to be open, it is therefore often argued that Eq. (6) provides a fundamental proof that the energy levels of open dots can *never* be resolved ($\Gamma \geq \Delta$ for $N \geq 1$) and that under such circumstances, only the open orbits connecting the leads need be considered in determining the transport. Unfortunately, this naïve argument ignores the actual physics of the dots.

Based on the above assumption, approximate semiclassical methods for open dot transport have been developed and employed [19,21,22, 26–30, 42]. These typically involve performing summations not unlike that of the DOS trace formula, Eq. (3), however with the crucial difference being that the relevant summations are over open orbits instead of closed, periodic ones. In such calculations, the open orbits that are typically included are short and show a clear and relatively simple connection with the input and output QPCs. A cut-off length, expressed in terms of the number of bounces an orbit makes, must be imposed in order to allow the calculations to finish in a reasonable amount of time (fifteen is a typical number) [Wirtz L., Private communication, 1998]. As such, trying to examine resonant behavior is automatically out of the question using such methods, since the necessary orbits corresponding to long dwell times cannot be included in the calculation. General predictions with regard to what should be expected for a chaotic, as compared to a regular, cavity have been made, based on the expected statistical differences in the average areas enclosed by open orbits in these two limits [16,17,21–27]. One such prediction that has gained particular notoriety involves the expected lineshapes for the so-called weak localization peak in the magnetoresistance of quantum dots [17,26,27], a matter which we shall discuss in great detail in section 6.

In this work, we completely avoid the short-time limitations inherent in such semi-classical approaches by employing a fully quantum mechanical treatment to obtain the conductance for such structures, actually solving the Schrödinger wave equation rather than using some approximate sum over orbits. In doing such simulations, we have discovered that there is actually a very crucial problem with applying the Thouless argument to open dots. The assumption of *uniform* level broadening that underlies the derivation of Eq. (6) in fact does not hold. In reality, the level broadening is highly nonuniform, and selected levels can still be resolved even in instances when $N > 10$. Moreover, this surviving resonance structure can have a very significant effect on the transport. Even though the fully quantum mechanical approach does not involve the direct use of classical orbits in the calculation, it can still yield indirect evidence regarding which periodic orbits in fact contribute to this resonance structure, through *wave function scarring* and related effects.

3. The Method of Calculation Used Primarily in this Work: A Fully Quantum Mechanical Treatment

3.1. The Transmission Problem

Within a GaAs-AlGaAs heterostructure the conducting electrons are confined in a 2DEG, so that the z-direction generally need not be considered explicitly. The electrons behave as free electrons, however they bare the effective mass of GaAs, $m^* = 0.067$.

The simulation results shown in subsequent sections are all performed on a square finite-difference lattice with lattice constant, a. Position can thus be specified as $x = ia$ and $y = ja$, where i and j are integers. Keeping only the lowest order terms in the approximations of the second derivatives, the 2D Schrödinger equation becomes

$$-t(\psi_{i+1,j} + \psi_{i-1,j} + \psi_{i,j+1} + \psi_{i,j-1}) + (V_{i,j} + 4t)\psi_{i,j} = E\psi_{i,j} \tag{7}$$

where $t = \hbar^2/2m^*a^2$ is kinetic energy term for "hopping" between lattice sites, $V_{i,j}$ represents the potential at site i, j and E is the energy. Since we are interested in current flow, the typical situation which we consider is one in which the device is enclosed inside an ideal quantum wire, which e,.tends outward to $\pm\infty$ along the x-axis. Figure 6.3(a) is meant to represent a quantum dot partially enclosed by square well barriers, thus portions of the picture that are shaded black represent regions where $V_{i,j}$ is made to be a large number, and $V_{i,j} = 0$ in the unshaded regions. More general cases where $V_{i,j}$ is an arbitrary varying function can easily be treated as well.

Along the top and bottom boundaries we use Dirichlet boundary conditions, so for a wire M lattice spacings high,

$$\psi_{i,j=0} = \psi_{i,j=M+1} = 0 \tag{8}$$

Given this, the wavefunction along a particular slice i on the x-axis can be specified by a M-dimensional vector $\vec{\psi}_i$. Defining the diagonal matrix $\boldsymbol{t} = t\boldsymbol{I}$, Eq. (7) can be rewritten as a matrix equation:

$$\mathbf{H}_{0i}\vec{\psi}_i - \mathbf{t}\vec{\psi}_{i+1} - \mathbf{t}\vec{\psi}_{i-1} = E\mathbf{I}\vec{\psi}_i \tag{9a}$$

where

$$\mathbf{H}_{0i} = \begin{bmatrix} (V_{i,M}+4t) & -t & 0 & \cdots & & & \\ -t & (V_{i,M-1}4t) & -t & \cdots & & & \\ & & & \ddots & & & \\ & & & \cdots & -t & (V_{i,2}+4t) & -t \\ & & & \cdots & 0 & -t & (V_{i,1}+4t) \end{bmatrix} \tag{9b}$$

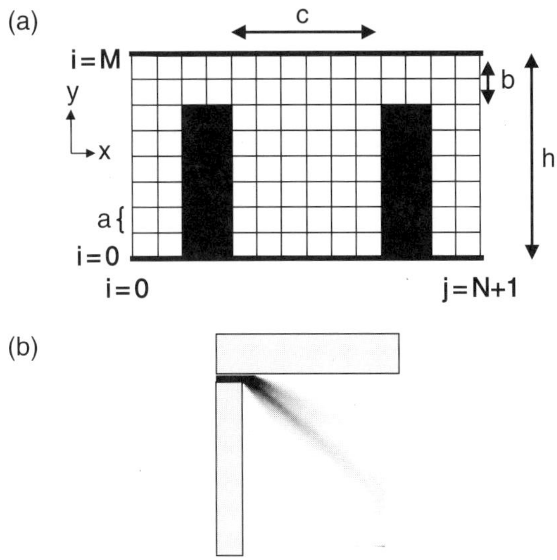

Fig. 6.3. (a) Schematic of an open quantum dot enclosed by square well potentials. The grid represents the underlying finite-difference mesh on which the calculations are performed. In practice, the grid is much finer than that shown here. The dark regions represent places where the potential at a grid point, $V_{i,j}$, is high. (b) The result of a wave function calculation showing a collimated electron wave beam exiting a QPC.

Here, the tridiagonal $\mathbf{H}_{0\mathrm{i}}$ represents the Hamiltonian for the individual, isolated slices i. The coupling terms in Eq. (9a) involving slices $i+1$ and $i-1$ can be thought of as a perturbation to this single slice Hamiltonian. Combining this with the trivial equation, $\vec{\psi}_i = \vec{\psi}_i$, one can derive a transfer-matrix equation that relates adjacent slices:

$$\begin{bmatrix} \vec{\psi}_i \\ \vec{\psi}_{i+1} \end{bmatrix} = \begin{bmatrix} 0 & \mathbf{I} \\ -\mathbf{I} & \left(\frac{\mathbf{H}_{0\mathrm{i}}-E}{t}\right) \end{bmatrix} \begin{bmatrix} \vec{\psi}_{i-1} \\ \vec{\psi}_{i1} \end{bmatrix} = \mathbf{T}_{\mathrm{i}} \begin{bmatrix} \vec{\psi}_{i-1} \\ \vec{\psi}_i \end{bmatrix} \tag{10}$$

Since the quantum wire acts as a waveguide, the actual current is carried by the propagating modes of the wire. Thus, we begin the calculation by solving the eigenvalue problem for the transfer-matrix on the first slice:

$$\mathbf{T}_1 \begin{bmatrix} \vec{\psi}_1 \\ \vec{\psi}_0 \end{bmatrix} = \begin{bmatrix} \mathbf{T}_{11} & \mathbf{T}_{12} \\ \mathbf{T}_{12} & \mathbf{T}_{22} \end{bmatrix} \begin{bmatrix} \vec{\psi}_1 \\ \vec{\psi}_0 \end{bmatrix} = \lambda \begin{bmatrix} \vec{\psi}_1 \\ \vec{\psi}_0 \end{bmatrix} \tag{11}$$

Since two adjacent slices are considered in tandem, the eigenvectors of Eq. (11) have the form

$$\begin{bmatrix} \vec{u}_m(\pm) \\ \lambda_m(\pm)\vec{u}_m(\pm) \end{bmatrix} \tag{12}$$

Since there are q propagating wave modes ($|\lambda| = 1$) and $M - q$ evanescent modes ($|\lambda| \neq 1$) the corresponding eigenvalues can be expressed as

$$\begin{aligned}\lambda_m(\pm) &= e^{\pm i k_m a}, m = 1, \ldots, q \\ \lambda_m(\pm) &= e^{\mp \kappa_m a}, m = q+1, \ldots, M\end{aligned} \tag{13}$$

The $\pm$ symbol refers to the fact that the modes actually come in pairs, those that travel to the right (+) and those to the left (−). For the transmission problem, it is useful to collect these together in a $2M \times 2M$ matrix

$$\mathbf{T}_0 = \begin{bmatrix} \mathbf{U}_+ & \mathbf{U}_- \\ \boldsymbol{\lambda}_+ \mathbf{U}_+ & \boldsymbol{\lambda}_- \mathbf{U}_- \end{bmatrix} \tag{14a}$$

where

$$\mathbf{U}_\pm = \begin{bmatrix} \vec{u}_1(\pm) & \ldots & \vec{u}_m(\pm) \end{bmatrix} \tag{14b}$$

and

$$\boldsymbol{\lambda}_\pm = diag\begin{bmatrix} \lambda_1(\pm) & \ldots & \lambda_M(\pm) \end{bmatrix} \tag{14c}$$

Multiplying by $\mathbf{T}_0$ converts the representation from the mode basis to the site basis, while multiplying by $\mathbf{T}_0^{-1}$ reverses this operation. To calculate the transmission through a device, one imposes the boundary conditions that the "+" modes are injected, each with unit amplitude, from the left side and there are no "−" modes coming from the right. For a structure N slices long, one must thus solve the transfer matrix problem:

$$\begin{bmatrix} \mathbf{t} \\ \mathbf{0} \end{bmatrix} = \mathbf{T}_0^{-1} \mathbf{T}_N \mathbf{T}_{N-1} \ldots \mathbf{T}_1 \mathbf{T}_0 \begin{bmatrix} \mathbf{I} \\ \mathbf{r} \end{bmatrix} \tag{15}$$

where $\mathbf{t}$ is a $2M$ by $2M$ matrix of transmission amplitudes of waves exiting from the right part of the structure, and $\mathbf{r}$ is the corresponding matrix of amplitudes of waves reflected back towards the left. The unit matrix, $\mathbf{I}$, and the zero matrix, $\mathbf{0}$, set the transport boundary conditions mentioned above. Given the matrix elements of $\mathbf{t}$, one can calculate the conductance, G, using the Landauer–Büttiker formula [43–46]:

$$G = \frac{2e^2}{h} \sum_{m,n} \frac{v_n}{v_m} \left| t_{n,m} \right|^2 \tag{16}$$

where $t_{n,m}$ represents the transmission amplitude of mode n to mode m and the summation is only over propagating modes. Here v_n represents the velocity in the x-direction of nth mode, which can be obtained by using the finite difference representation of the probability current in the x-direction [47] for the case where $\vec{\psi}_i = \vec{u}_n$. Exploiting the fact that $\vec{\psi}_{i+1} = e^{i k_n a} \vec{u}_n$ in the lead region and summing over the j components to get the net current, one finds that, up to a constant prefactor, $v_n = \sum_j 2t \sin(k_n a) |u_{n,j}|^2$.

Unfortunately, Eq. (15) in its current form is made numerically unstable by the exponentially growing and decaying contributions of the evanescent modes that accumulate when the product of transfer matrices is taken. Usuki et al. [48] overcame this difficulty be rewriting the transfer matrix problem in terms of an iterative scheme. Rather than using the simple relationship given by Eq. (15), slices i and $i+1$ can be related by:

$$\begin{bmatrix} \mathbf{C}_1^{i+1} & \mathbf{C}_2^{i+1} \\ \mathbf{0} & \mathbf{I} \end{bmatrix} = \mathbf{T}_i \begin{bmatrix} \mathbf{C}_1^{i} & \mathbf{C}_2^{i} \\ \mathbf{0} & \mathbf{I} \end{bmatrix} \mathbf{P}_i \tag{17a}$$

where

$$\mathbf{P}_i = \begin{bmatrix} \mathbf{I} & \mathbf{0} \\ \mathbf{P}_{i1} & \mathbf{P}_{i2} \end{bmatrix} \tag{17b}$$

$$\mathbf{P}_{i1} = -\mathbf{P}_{i2}\mathbf{T}_{i21}\mathbf{C}_1^i, \quad \mathbf{P}_{i2} = [\mathbf{T}_{i21}\mathbf{C}_2^i + \mathbf{T}_{i22}]^{-1} \tag{17c}$$

The iteration is started by the condition $\mathbf{C}_1^0 = \mathbf{I}$ and $\mathbf{C}_2^0 = \mathbf{0}$. As shown schematically, in Fig. 6.3(a), this implies a situation in which the modes start off incident (**I**) from the left with unit amplitude and there are no waves coming in from the right. What results are reflected (**R**) and transmitted (**T**) waves. At the right end of the structure, the final transmission matrix **t** obeys the relationship:

$$\mathbf{t} = -(\mathbf{U}^+\boldsymbol{\lambda}^+)^{-1}\left[\mathbf{C}_1^{N+1} - \mathbf{U}^+(\mathbf{U}^+\boldsymbol{\lambda}^+)^{-1}\right]^{-1} \tag{18}$$

The numerical stability of the Usuki et al. method in large part stems from the fact that the iteration implied by Eq. (17) involves products of matrices with *inverted* matrices. Taking such products tends to cancel out most of the troublesome exponential factors. It should be noted that the Usuki et al. method is really just a variation on the "cascading scattering matrix" method developed by Ko and Inkson [49]. Moreover, Usuki et al. recognized that their approach, as described by Eq. (17), was equivalent to using recursive Green's functions [134]. The relation given above for the $\mathbf{P}_{l2}$ matrices acts as a form of Dyson's equation, with the resulting renormalization helping the stability.

3.2. *Obtaining the Wave Functions and the Density*

Besides calculating the conductance, we can also obtain the electron density by reconstructing the electron wave functions using the $\mathbf{P}_{l1}$ and $\mathbf{P}_{l2}$ matrices. Usuki *et al.* outlined a method for doing this starting from the left and working back to the end of the structure. Unfortunately, it entails performing a calculation similar to that for obtaining the conductance, but for every single slice. As a result, while the time it takes to calculate G goes as N, the time to reconstruct the wave function instead goes as $N!$, which makes it very time consuming. Since self-consistency requires the density to be recalculated numerous times, including it thus becomes impractical for most calculations. We however have found a simple way to make the reconstruction far more efficient. Instead of going from left to right, one starts at the end of the structure and works backword. Manipulating Usuki et al.'s equations, it can be shown that for the final slice:

$$\psi_N = \mathbf{P}_{N2} \tag{19}$$

Note here that ψ_N is a matrix, the columns of which represent the separate contributions of the individual modes to the total wave function on slice N. Going towards the left, one then does the iteration:

$$\psi_i = \mathbf{P}_{i1} + \mathbf{P}_{i2}\psi_{i+1} \tag{20}$$

The **P**'s here are the same ones obtained during the G calculation and so are recalled from memory rather than being recalculated. The density at site i, j, given there are q propagating modes, becomes

$$n(x, y) = n(i, j) = \sum_{k=1}^{q} \left|\psi_{ijk}\right|^2 \tag{21}$$

Obtaining $n(x, y)$ in this modified way takes about the same amount of time as the original G calculation and can be *orders of magnitude faster* than Usuki et al.'s technique depending on the size of the structure.

3.3. *Modifications Required in the Presence of a Magnetic Field*

The method must be modified when the device is in the presence of a perpendicular magnetic field, $\vec{B} = (0, 0, B)$. The most suitable choice for obtaining such a field is the Landau gauge, $\vec{A} = (-By, 0, 0)$. The standard procedure when performing calculations on a lattice is to include the field via Peierls' phase factors, which are obtained by performing the path integral of $\vec{A}$ between adjacent lattice sites. To do these path integrals, we first specify that

$$for\ x_i \le x \le x_{i+1} \qquad y_j(x) = y_{i,j} + \frac{(y_{i+1,j} - y_{i,j})}{(x_{i+1} - x_i)}(x - x_i) \tag{22}$$

Thus, for the right and left directions on the lattice, the appropriate phase factors are

$$\begin{aligned} \theta_{R,i,j} &= \frac{2\pi e}{h}\int \vec{A} \bullet d\vec{l} = \frac{2\pi e}{h}\int_{x_i}^{x_{i+1}} -By_j(x)dx \\ &= -\frac{\pi e B}{h}\left[(y_{i+1,j} + y_{i,j})(x_{i+1} - x_i)\right] \end{aligned} \tag{23a}$$

and

$$\theta_{L,i,j} = \frac{2\pi e}{h}\int_{x_i}^{x_{i-1}} -By_j(x)dx = -\frac{\pi e B}{h}\left[(y_{i,j} + y_{i-1,j})(x_{i-1} - x_i)\right] \tag{23b}$$

Since we chose the Landau gauge, the phase factors for the upper and lower directions are zero, $\theta_{U,i,j} = \theta_{D,i,j} = 0$. For the case of a uniform grid, which we have chosen to use here, $\theta_{L,i,j} = \theta_{R,i,j} = -2\pi e B a^2/h$.

Given these factors, the 2D Schrödinger matrix equation, our Eq. (9a), must be modified

$$\mathbf{H_{0i}}\vec{\psi}_i - \tilde{\mathbf{t}}_{R,i}\vec{\psi}_{i+1} - \tilde{\mathbf{t}}_{L,i}\vec{\psi}_{i-1} = E\mathbf{I}\vec{\psi}_i \tag{24}$$

where $\tilde{t}_{R,i}$ and $\tilde{\mathbf{t}}_{L,i}$ are diagonal matrices with elements $\tilde{\mathbf{t}}_{R,i,j} = e^{i\theta_{R,i,j}} t\delta_{i,j}$ and $\tilde{t}_{L,i,j} = e^{i\theta_{L,i,j}} t\delta_{i,j}$.

In addition, the eigenvalues of the wave modes also become shifted:

$$\lambda_m(\pm) = e^{\pm ik_m a + i\theta_{R,0,j'}},\ m = 1, \ldots, q$$
$$\lambda_m(\pm) = e^{\mp\kappa_m a + i\theta_{R,0,j'}},\ m = q+1, \ldots, M \tag{25}$$

Here, j' is the index for which $y_{0,j'} = y_{max}/2$. The expression used for determining the mode velocities now takes the form $v_n = \sum_j 2t_{R,0,j} \sin(k_n a + \theta_{R,0,j})|u_{n,j}|^2$.

3.4. *Obtaining the Eigenstates of a Corresponding Closed Dot System*

In subsequent sections, we will be decomposing open dot wave functions in terms of closed dot eigenstates as part of our analysis. Assuming Dirichlet boundary conditions on all four sides ($\psi = 0$ along all boundaries), the eigenvalue equation for a closed dot that is N lattice spacings long along the x-direction and M spacings high becomes:

$$\mathbf{H}_{dot}\vec{\psi} = \begin{bmatrix} \mathbf{H}_{01} & -\tilde{\mathbf{t}}_{R,1} & \mathbf{0} & \cdots & & & \\ -\tilde{\mathbf{t}}_{L,2} & \mathbf{H}_{02} & -\tilde{\mathbf{t}}_{R,2} & \cdots & & & \\ & & & \ddots & & & \\ & & & \cdots & -\tilde{\mathbf{t}}_{L,M-1} & \mathbf{H}_{0M-1} & -\tilde{\mathbf{t}}_{R,M-1} \\ & & & \cdots & \mathbf{0} & -\tilde{\mathbf{t}}_{L,M} & \mathbf{H}_{0M} \end{bmatrix} \begin{bmatrix} \vec{\psi}_1 \\ \vec{\psi}_2 \\ \vdots \\ \vec{\psi}_{M-1} \\ \vec{\psi}_M \end{bmatrix}$$
$$= E_n\vec{\psi} \tag{26}$$

Since $\mathbf{H}_{dot}$ is made up entirely of blocks that are either diagonal, tridiagonal, or zero (in the above equation, $\mathbf{0}$ denotes an M by M zero matrix) this is a sparse matrix problem. This can be efficiently solved numerically by using ARPACK fortran subroutines which are publicly available off the internet www.caam.rice.edu/software/ARPACK/index.html. These subroutines employ Lanczos/Arnoldi factorization.

The first simulation result we show (Fig. 6.3(b)) is of a single open QPC that supports a single propagating mode. Wave function amplitude is plotted, with darker shading indicating higher amplitude. A consequence of diffraction, this QPC displays beam collimation, with the electron waves coming out at a well defined angle. Given an opening of width w, the single mode has a quantized transverse wave-number, $k_y = \pi/w$ at zero field. This is what sets the exit angle, as the electrons come out with a transverse velocity, $v_y = -\hbar k_y$. Note that the electrons can also have $v_y = +\hbar k_y$, but these are reflected downward by the upper barrier. The fact QPCs can produce such strong beam collimation should obviously have an affect on what orbits are excited and participate in quantum dot transport. This orbit selection effect is perhaps most obvious and dramatic in the case where a series of quantum dots coupled are together in a chain [50,51]. That said, the more significant role of the QPCs may be in the "anti-selection" of orbits. Just what this means and its significance will be made clear in the next section.

4. Conductance Resonances in Open Dots

4.1. Nominally Square Dots in the Presence of a Magnetic Field: Periodic Fluctuations and the Diamond Orbit

We first present an illustrative example of our calculations, an example for which the confinement has been assumed to be hard wall in nature, as in Fig. 6.3. In Fig. 6.4(a), conductance *fluctuations*, δg, as a function of magnetic field, B, are plotted for a $c = h = 0.3\,\mu\text{m}$ square dot with $b = 0.04\,\mu\text{m}$ QPC openings, which allow two modes to enter and exit the dot for the given Fermi energy of 16 meV [52,53], This value of energy was chosen as it coincides with the electron density typically found in experimental dots [54–56], a value roughly $4 \times 10^{11}\,\text{cm}^{-2}$. The fluctuations are obtained from the raw conductance by doing a background subtraction so that the average value of δg approaches zero.

Instead of random aperiodic fluctuations with magnetic field, there are sets of features that show a nearly periodic variation. In particular, we point out a series of resonances at $B \sim 0.069, 0.173, 0.283$, and 0.397 T (while the feature is difficult to see at $B = 0.173$ T, it appears as a small glitch in the curve). These are marked with arrows. Note here that these features occur with virtually periodic spacing, with $\Delta B \sim 0.11$ T. The wave functions corresponding to these four resonances are also displayed below. What is shown in each case is $|\psi(x, y)| = n^{1/2}$, with the darker shading corresponding to higher wave function amplitude. Strikingly, essentially the same diamond-shaped pattern occurs in each case, with the amplitude of the wave function apparently peaking along an underlying classical orbit similar to that displayed in Fig. 6.2(a). This behavior is highly reminiscent of the wave function *scarring* effect observed in eigenstates of chaotic billiards such as the stadium [57–60]. Technically speaking, the term *scar* should only be applied to situations where the appearance of such a pattern should seem to be unlikely, specifically when the classical orbit underlying the pattern is known to be unstable [61]. That is a criterion that generally is not satisfied in our calculations, particularly in a situation such as this where a closed square is known to yield completely regular classical behavior at zero field and a mixed phase space for finite field. In such circumstances, we shall us the term "scar-like" when referring to these kinds of features.

A semiclassical scarring formula applicable to closed systems has also been derived in a manner similar to that of the DOS formula discussed in section 2 [62]. It relates the squared wave function amplitude integrated over a range of energy to a weighted summation over periodic orbits. As a result of the nature of the weighting, the more stable orbits contribute more strongly than the others and as such are reflected in the wave functions. This is the origin of the scarring effect. While this semiclassical formula requires an integration or average over energy, it does allow for the possibility of *individual* eigenstates being scarred, provided that the energy level spacing is relatively large in comparison to the window over which the average is performed.

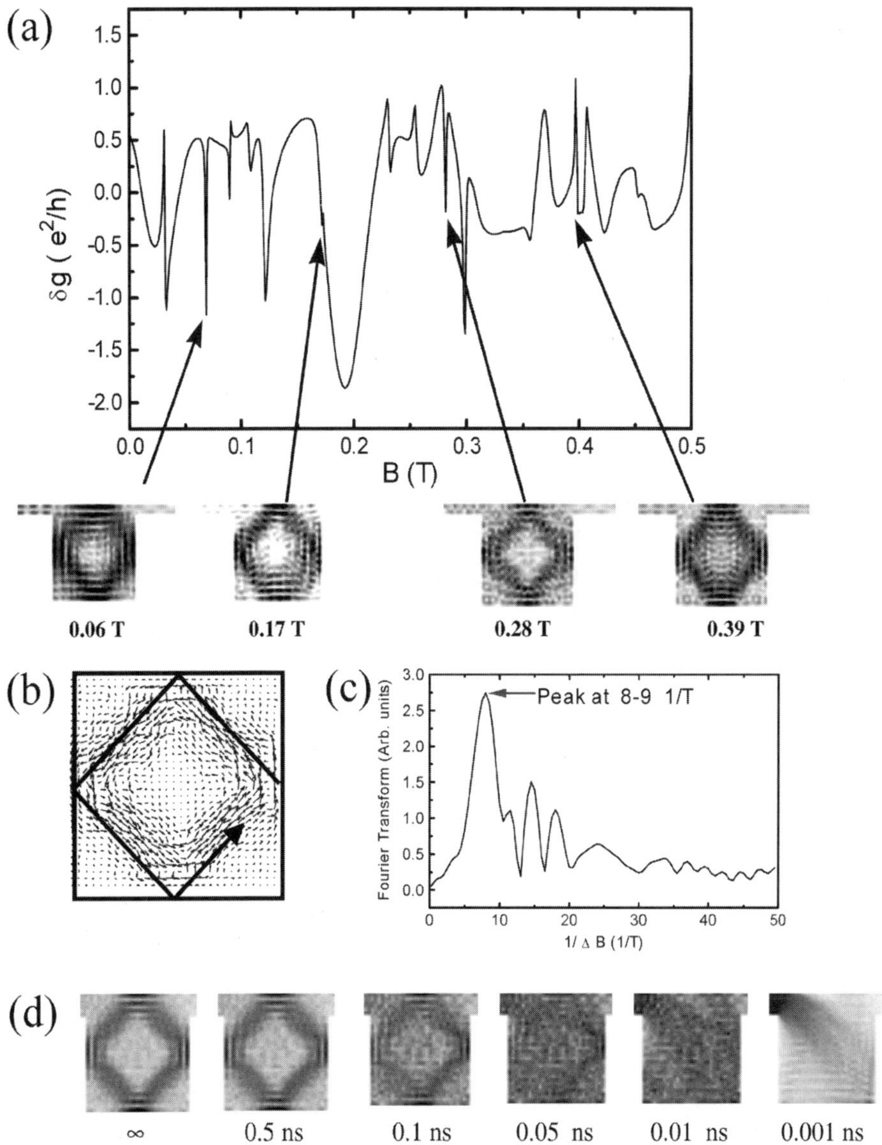

Fig. 6.4. (a) The conductance fluctuations versus magnetic field for the 0.3 μm dot discussed in the text. Four diamond resonance features that appear in the curve are also indicated, and $|\psi(x, y)|$ versus x and y in the quantum dot plotted for each of these features. Darker shading corresponds to higher amplitude. (b) The quantum mechanical current flow associated with a diamond resonance. (c) The Fourier transform of the fluctuations shown in (a). (d) The evolution of a wave function corresponding to a diamond resonance as inelastic scattering time is reduced.

To begin to approach the semiclassical limit, we must employ our fully quantum mechanical theory to examine comparatively high energy states, such that a large number of nodes are present in the wave function. Examining the ripples in the wave functions in Fig. 6.4(a), one can see about ~18 nodes along the x and y directions. Indeed, relatively large number of nodes are required in order to resolve the diamond features. In contrast, Ji and Berggren [63], who also examined what may be considered open quantum dots (actually a quantum wire with double stub side branches), only considered the very lowest part of the energy range and the first few resonant states and so did not resolve any such features. Amongst other things, they also examined the quantum mechanical current flow and saw that the resonant states yielded vortices. We note that it can be dangerous to associate quantum mechanical current flow lines with classical trajectories, particularly at low energy. For example, the lowest resonant state in a quantum wire with a cross branch has a single node trapped in the region where wire and branch intersect and is held in place by the sharp corners of the structure. The flow lines associated with the current vortex corresponding to such a state have *no* direct relationship to any classical orbit. We have found that a simple correspondence between current flow and a classical trajectory only occurs when the scarring of the resonant state is particularly obvious, such as the diamond feature shown here. As shown in Fig. 6.4(b), current circulates around and along the diamond pattern.

In Fig. 6.4(c), we plot the Fourier transform obtained from the fluctuations [52]. It is dominated by a single peak at ~8–9 T^{-1}. Meanwhile, the spacing of the scar-like features yield values of $(\Delta B)^{-1}$ of ~8.7 T^{-1}. Thus, there appears to be a correlation between the periodicity of the fluctuations and that of the scar-like features. As we shall demonstrate below, this correlation is not simply a coincidence. It should also be noted that the periodicity obtained in these simulations matches up quite well with experimental measurements done on a quantum dot with lithographic dimensions $c = h = 0.4\,\mu m$ square, for which the estimated effective size accounting for depletion was 0.3 μm [56,64]. As with the simulation, the Fourier transform of the fluctuations yielded a single dominant peak at ~8–9 T^{-1} A much more detailed discussion regarding comparisons with experiment appears in section 5.

An interesting and very important aspect of these scar-like features is that the diamond periodic orbit reflected by them is not connected to either QPC, even though the QPC is responsible for the injection of electrons into the dot. Needless to say, this runs contrary to the assumptions of the semiclassical theory described in section 2, which assumes that orbits that are connected to the QPCs that are responsible for the transport properties. One reason for the apparent disconnection from the QPCs shown by these features is the fact that they correspond to *resonant* states and, as such, the electrons can stay in the dot for a rather long time. One way of estimating the storage time associated with a resonance is to include inelastic scattering. The modification of the calculation to take into account inelastic scattering and phase randomization can be done phenomenologically by introducing an imaginary potential [65]. For an inelastic scattering time, τ_ϕ, the original potential, V, is modified to become $V{-}iV_{in}$, where $V_{in} = \hbar/(2\tau_\phi)$. In our calculations, we impose the phase breaking only in the dot itself and not in the QPC regions. One calculates the transmission and reflection coefficients as before. Current conservation is maintained in the presence of the imaginary potential since one can show that the current that is apparently lost from the propagating modes actually goes into an inelastic-scattering channel. The total inelastic scattering current is given by

$$j_{in} = (1/\tau_\phi) \iint \psi^* \psi dx dy \tag{27}$$

We assume that this inelastic channel current is equally divided between the two leads and add this to the results of the Landauer–Büttiker formula (which constitute the elastic contribution) to obtain the *total* conductance. Focusing on the resonant feature for $B = 0.28\,\text{T}$, the series of panels in Fig. 6.4(d) shows the evolution of $|\psi(x, y)|$ for progressively shorter inelastic scattering times. Note the exponential fashion in which the diamond is disrupted by increased phase breaking. In particular, reducing τ_ϕ from infinity to 0.5 ns introduces virtually no change in the pattern. In the experiments, 0.5 ns is a typical value of τ_ϕ for temperatures below 100 mK, and in this regime the periodic fluctuations have been clearly observed [56]. Reducing τ_ϕ by a further factor of five, however, causes a significant disruption of the diamond, although it is still clearly apparent. A further factor of two decrease virtually destroys what remains of the original pattern. It is only when τ_ϕ approaches 0.001 ns that the collimated beam corresponding to electron injection by the QPC clearly becomes apparent. As the diamond patterns are destroyed by the phase breaking, the amplitude of periodic magneto-resistance fluctuations is also being damped. Experimentally, the amplitude of the fluctuations typically decays in an exponential fashion by an order of magnitude on raising the temperature to a degree Kelvin, at which point τ_ϕ is estimated as being of the order 0.05 ns [29,56]. This corresponds to an inelastic path length $l_\phi = v_f \tau_\phi$ of about 15 μm, still sufficient to make nearly twenty complete circuits of the diamond loop. Simulations yield a similar exponential damping of the fluctuations as a function of τ_ϕ [53]. The decay of the fluctuations coinciding with the decay of the diamond suggests that the interference that yields the fluctuations may be dominated by the remnant of the diamond orbit.

The persistence of the diamond up to $\tau_\phi \sim 0.1$ ns gives an approximate indication of how many circuits around the orbit, over which phase coherence must be maintained, for this feature to exist. The answer being about forty. Needless to say, given that the injection from the QPCs appears only when $\tau_\phi \sim 0.001$ns, it appears the electrons are spending orders of magnitude more time in the diamond orbit than they are entering or exiting via the QPCs. Since the wave function yields the probability of finding an electron at a particular location, it is not surprising that the amplitudes in the QPC regions are comparatively small under these conditions. The conclusion that scar-like features correspond to long lived states is supported by the dwell time analysis of Zozoulenko et al. [66]. Forty orbits corresponds to one hundred and sixty bounces off the walls, much longer than the semiclassical cut-off of ~15 bounces mentioned earlier. Obviously, these diamond scar-like resonance features are inaccessible to such calculations. Moreover, the wave function amplitude associated with such relatively short paths is understandably dwarfed when resonances are apparent.

4.2. Analogous Behavior in a Stadium Dot

As mentioned above, the square corresponds to a classically regular geometry. What about a classically chaotic geometry? Does it show similar behavior? The answer

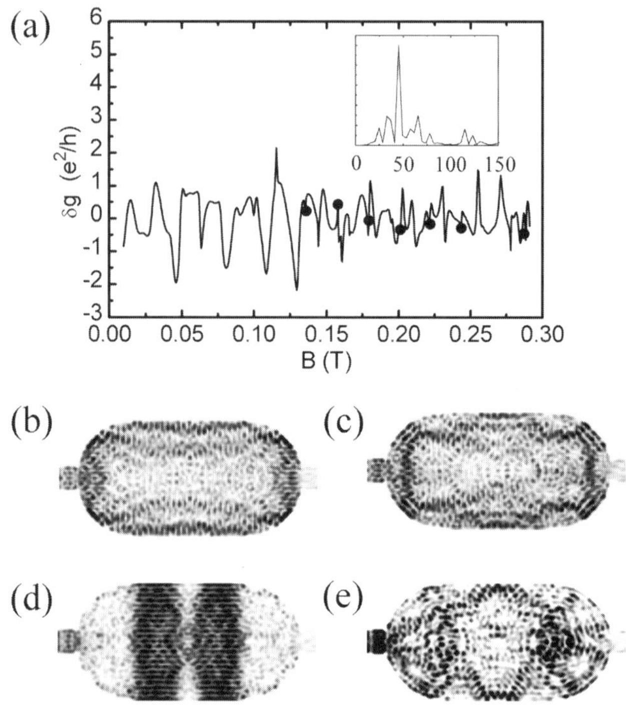

Fig. 6.5. (a) Conductance fluctuations versus magnetic field for an open stadium. The inset is the corresponding Fourier transform. $|\psi(x, y)|^2$ versus x and y is plotted for (b) $B = 0.135\,\mathrm{T}$ and (c) $B = 0.288\,\mathrm{T}$, showing a 'rectangular' scar in both cases. (d) The 'bouncing-ball' scar. (e) The 'double-V' scar.

to this question is yes. Figure 6.5 shows the results obtained for a stadium with a centrally aligned QPC [67]. Here, $R = 0.2\,\mu\mathrm{m}$ and the total length is $0.8\,\mu\mathrm{m}$, yielding an area of $0.29\,\mu\mathrm{m}^2$. The electron density in this case was set to $4 \times 10^{11}\,\mathrm{cm}^{-2}$. In Fig. 6.5(a), we plot the conductance fluctuations for this configuration and the corresponding Fourier spectrum is shown in the inset. A single peak at $47\,\mathrm{T}^{-1}$ clearly dominates. The corresponding magnetic field period is 0.021 T, in good agreement with the spacings between successive minima that appear in the fluctuations. The solid circles correspond to values of field where a rectangular scar was observed in the wave function. These appear at very nearly periodic intervals, with the period also being 0.021 T, though one period is missed (a scar is expected at $B \sim 0.26\,\mathrm{T}$). In Fig. 6.5(b) and (c), we plot $|\psi(x, y)|^2$ versus x and y for the cases corresponding to the first and last circles, $B = 0.1355\,\mathrm{T}$ and 0.288 T respectively. Both wave functions reveal similar looking but not identical scars (Fig. 6.5(c) clearly shows extra "bounces"), following trajectories running along the periphery of the stadium. This result is consistent with an observation of Heller et al. [59], who found that, in closed stadiums, there is a difficulty in making a correspondence between a scarred wave function and a *specific* orbit, as each simple orbit has a large number of similar but more complex "cousins", consisting of progressively more bounces, only matching up with themselves after several trips around the stadium. Thus, the observed periodic scarring is likely the result of a whole series of such "cousins".

These recurrent scarring features all occurred above $B = 0.1$ T. One might argue that we have crossed over into the regular regime, as a transition from chaotic to regular behavior occurs classically as the magnetic field is increased. However, using the formula $B^* = \phi_0/(\rho_{\min}\lambda_F)$, where ρ_{min} is the smallest radius of curvature of the device and λ_F is the Fermi wavelength, Marcus et al. [16,23] estimated this to occur at $B^* = 0.45$ T for the stadium dots he examined, which were comparable in size to this simulation. For the example used in Fig. 6.6, $B^* = 0.48$ T, since the dot is smaller and λ_F is larger. Thus, with this criteria, the scarring effects we have shown are at fields in which the classical scattering is expected to be chaotic.

We have also seen other scars familiar from the literature on closed stadiums [57–60, 62]. For example, as shown in Fig. 6.5(d), the classic "bouncing ball" scar appears quite strongly for a field of 0.145 T, while Fig. 6.5(e) shows the double V scar that occurred at $B = 0.13$ T (compare the latter with Fig. 6.2 of the paper of Heller and Tomsovic [60]).

4.3. Strengthening the Case for the Significance of the "Scarred" Resonances

Though the square and stadium examples we have shown so far suggest there is a correlation between the spacing of the resonant features and the periodicity of the conductance fluctuations, one could question whether there is any causal relationship. Might not the resonances simply represent an effect that is superimposed over the true fluctuations and is actually independent of them? We shall demonstrate that this is not the case.

The true significance of the resonant features is made clearer if one assembles sets of conductance plots calculated at different energies [68–71]. In Fig. 6.6(a), G, the conductance in units of $2e^2/h$, is plotted versus B, for a $c = h = 0.3$ μm square dot with $b = 0.065$ μm QPC openings. The three curves correspond to the three indicated Fermi energies (for clarity, the curves are offset by $2e^2/h$). No background subtraction has been carried out in this case. For energies ranging between 14.0 and 15.5 meV, the QPCs permit between four and five modes to propagate.

Referring to the $E = 14$ meV curve, the resonances marked "b" all correspond to the positions of where diamond scar-like features are found, such as the one illustrated in Fig. 6.6(b). Comparing this wave function to those shown in Fig. 6.4(a), it is clear that the wider QPC (4 propagating modes instead of 2) has had relatively little effect on the diamond. For the $E = 14.75$ meV and the $E = 15.5$ meV, there are also diamond resonances (again marked "b") but these occur at different fields.

The evolution of the diamond resonances as a function of both energy and field can be tracked by combining many such conductance traces and creating a shaded contour plot of $G(E, B)$ such as the one shown in Fig. 6.6(c). Here, lighter shading corresponds to larger values of G, which takes on values ranging from $\sim 2e^2/h$ to $\sim 8e^2/h$. This picture clearly shows a *regular* cross hatched pattern of linear striations running tilted in energy but parallel to each other. These striations are identified in

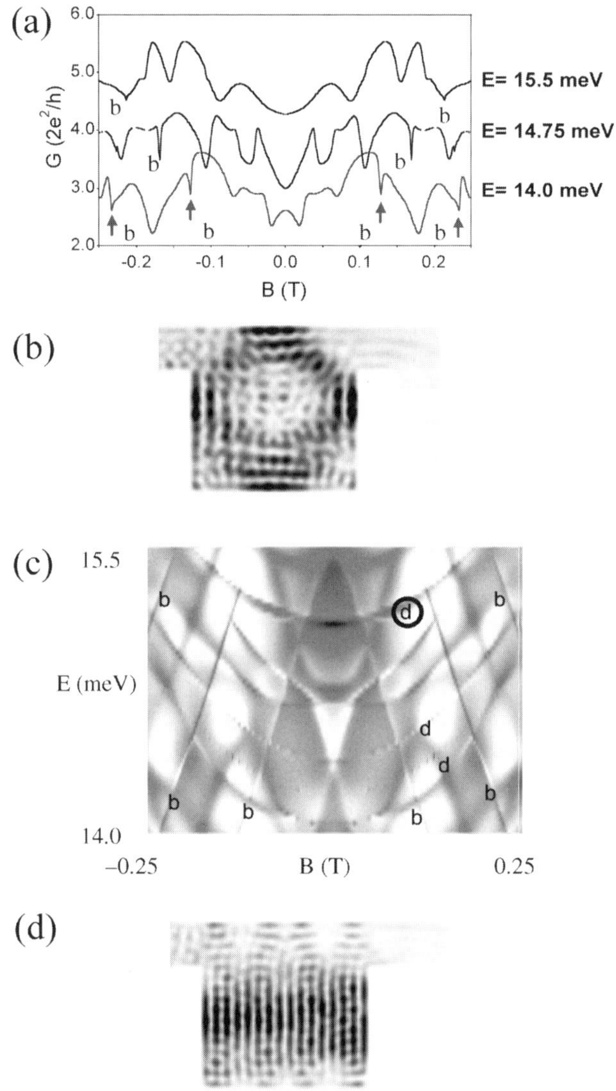

Fig. 6.6. (a) Traces of conductance, G, versus magnetic field, B, for an open 0.3 μm square dot for three different energies. (b) The diamond wave function that appears at each resonance marked "b" in (a). (c) Here $G(E, B)$, is plotted as a function of both B and energy E. The lighter regions of shading correspond to higher values of the conductance. (d) The "bouncing ball" type wave function that occurs along the resonance lines marked "d" in (c).

the figure by the "b" marks. Given this result, it is easy to understand why dots of this configuration would yield the periodic conductance fluctuations that have been observed in experiments. Significantly, the diamond scar-like feature occurs along the entire length of these resonant grooves.

Besides the linear striations, there are also resonance lines that show much more of a parabolic variation with respect to B, a few of these are marked "d" in the figure. Fig. 6.6(d) shows the wave function corresponding specifically to the position of the circled "d", for which $E = 15.25$ meV, $B = 0.112$ T. Here, the wave function indicates a standing wave pattern trapped between the left and right edges of the dot and filling most of it. As with the linear striations, the wave functions corresponding to these parabolic resonances tend to hold their form along the lines.

Where resonance lines cross, one line, typically the broader one, dominates and obscures the other. This allows us to understand some of the behavior displayed in the previous two figures. Recall in Fig. 6.4(a) that one of the resonances in the diamond family appeared only as a tiny glitch, while in Fig. 6.5(a), one of the expected periodic whispering-gallery resonances was missing. Other resonant features that happened to be coinciding with them were simply obscuring these features.

One may ask why wave functions apparently corresponding to the diamond periodic orbit lie along the linear grooves while the parabolic grooves seem to yield wave functions that show a more uniform, sine wave type excitation. In closed *circular* dots, the energy level structure follows the Darwin–Fock spectrum [72,73] which depends on radial and angular momentum quantum numbers, n and l, respectively. Importantly, $E_{n,l}$ has terms both parabolic and linear with magnetic field, with the linear term being proportional to l. Thus, states with higher angular momentum will show more linear behavior. Almost certainly something similar is happening here in the square dot. The current flow diagram in Fig. 6.4(b) clearly shows circulation along the diamond in one direction, which is consistent with it being analogous to some sort of angular momentum state. Changing the sign of the magnetic field reverses the direction of the quantum mechanical current. One may ask: what happens at the point where the linear grooves intersect at $B = 0$ T? In that case, one can obtain a wave function that looks very similar to the diamond states shown here, but which yield no net circulation. This to be expected since, at $B = 0$ T, there is no preferred direction for current flow. Such wave functions shall be examined in detail later in this section (see Fig. 6.11). With regard to the parabolic resonances, these reflect the additional confinement potential created by the magnetic field, which creates a shift in energy proportional to ω_c^2 ,which of course varies as B^2.

Figure 6.7(a) depicts a different configuration, a $c = h = 0.3\,\mu$m square dot with the input and exit QPCs are shifted to the middle. In this example, the QPCs support three modes [68]. Fig. 6.7(b), depicts a shaded contour plot of$G(E, B)$. In this case, it is apparent that many of the sharp resonance lines present in Fig. 6.6(b) are either weaker or have vanished completely, while new resonance lines have appeared in different locations. Since the QPCs are now in the middle, the diamond periodic orbit can no longer be effectively contained and so the corresponding resonance lines are now absent in the conductance. The primary orbit for generating resonances in this case is a rectangular one that indicates bounces occurring between diagonally opposite corners of the dot [68,74]. A wave function corresponding to one of these resonances is shown in Fig. 6.7(b). It displays a crossing pattern that is indicative of a pair of such orbits. As with the previous example, the wave function maintains its form along several lines of resonances, which are indicated by "c" markers in Fig. 6.7(b). Moreover, further simulation indicates that the cross resonances recur periodically in field, with a period

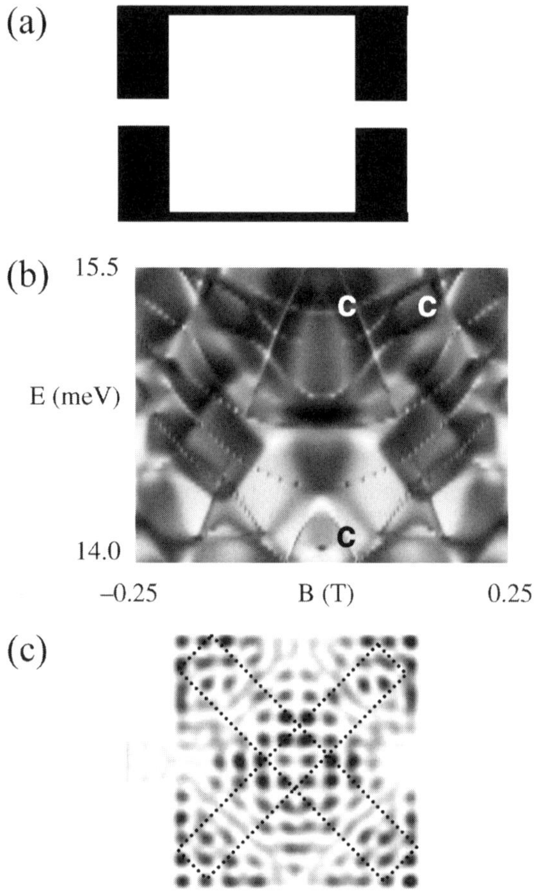

Fig. 6.7. (a) Schematic for a dot with centered QPCs . (b) A shaded contour plot of$G(E, B)$ for this dot, with lighter shading indicating higher G. (c) The wave function with a crossed pattern that occurs along the resonance lines marked "c" in (b).

of $\Delta B \sim 0.11$ T, about the same as the diamond. Thus, the behavior is quite similar, but the resonance pattern has changed since moving the QPC has caused different orbits to be automatically disallowed.

In Fig. 6.8, we return to the square dot with the QPCs aligned at the top. The results shown were obtained when we explored the effect of making the QPCs wider [75]. Fig. 6.8(a), depicts a shaded contour plot of$G(E, B)$with lighter shading corresponding to lower conductance in this instance. Here, the width of the QPCs have been adjusted so that a single mode propagates ($b = 0.02\,\mu$m) and $G < 2e^2/h$ for the entire energy range shown in this picture. As was the case in 6.6(b) and 6.7(b), $G(E, B)$ is striated by lines of resonances superimposed on the conductance. However, the pattern is much more complex in this case, as if the results of the previous two examples were superimposed on top of each other. Since the QPC is particularly narrow in this

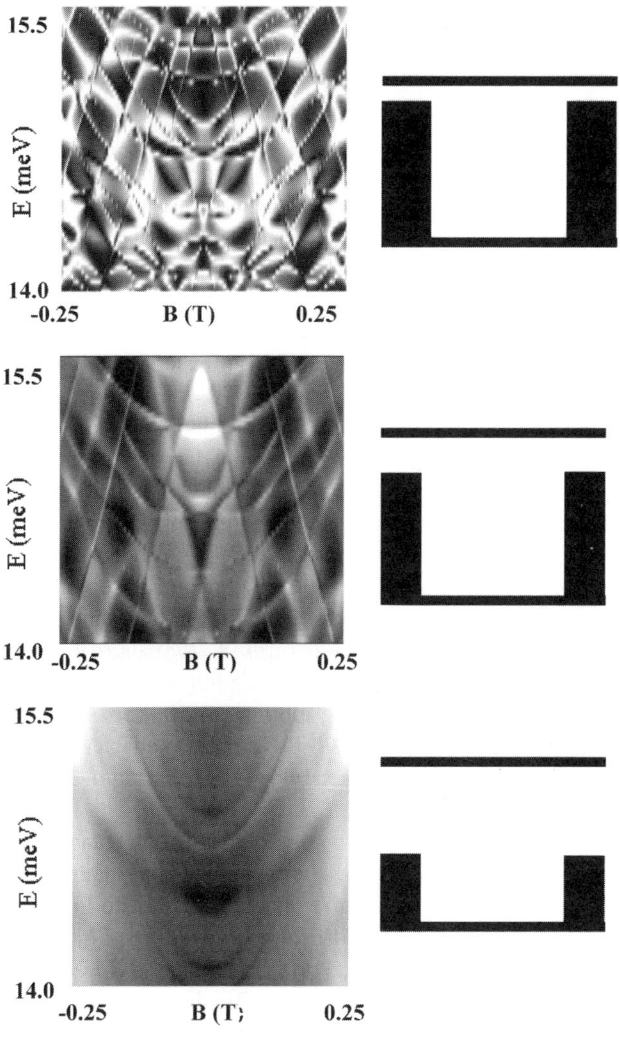

Fig. 6.8. Left panels: a series of shaded contour plots of $G(E, B)$ for dots with progressively wider QPC openings, shown schematically in the panels on the right.

instance, the crossing orbits can enter the corners without exiting the dot, thus crossing and diamond orbits both make significant contributions to the pattern. Figure 6.8(b) is a reproduction of Fig. 6.7(b), the case for which the QPCs permit between four and five modes to propagate over the given energy range ($b = 0.065\,\mu$m). Comparing Fig. 6.8(b) with 6.8(a), opening the QPCs has clearly reduced the complexity of the pattern. However, it appears to have done so in an "all-or-nothing" fashion—either the resonances are sharp, or they don't appear at all. For Fig. 6.8(c), the QPCs have been widened to $b = 0.16\,\mu$m, more than half the width of the dot. The parallel striations associated with the diamond orbit are now gone since the structure can no longer contain them. What is left are the parabolic resonance lines we previously associated with "bouncing ball" type orbits. Despite the dot being so wide open that one hesitates in referring to it as still being a dot, these orbits can still be trapped by the structure.

If the long orbits set up these periodic striation patterns, how do the short orbits, the ones normally included in semiclassical calculations, contribute to the conductance? For the most part, they set the background conductance over which this striation pattern is overlayed. This background however can have significant structure. For example, some peaks in resistance at finite field have been found to coincide with very simple backscattered trajectories (less than 10 bounces) in both triangular quantum dots [76,77] and doubly corrugated quantum wires [50]. However, these background features tend to evolve over a comparatively large magnetic field range and are readily distinguishable from the effects of the relatively low field resonant features that we focus on here.

4.4. The Fano Resonance Lineshape

We now examine the nature of the conductance resonances in greater detail. The individual traces shown in Figs 6.4(a), 6.5(a) and 6.6(a), as well as the plots of $G(E, B)$ shown in Figs 6.6, 6.7 and 6.8, reveal several different types of resonant phenomena going on. We see resonant maxima, resonant minima, and combinations thereof. The different variety of transmission line shapes possible when a cavity is attached to quantum wires has been the subject of extensive study (refer to [78,79], and the innumerable references within). A transmission minimum occurring in close conjunction with a maximum results from what is known as a Fano resonance [80]. Importantly, such asymmetric resonances are generic in situations *when a system with discrete energy levels is coupled to a system in which the levels are a continuum.* In the simple model by Shao et al. [78], which applied to single mode transmission, the Fano line shape was characterized by three parameters, the energy of the transmission zero, E_{o},the energy of the transmission pole, E_{p}, and the inverse lifetime of the quasibound state, Γ. In the limit where $E_{\mathrm{o}} = E_{\mathrm{p}}$, one obtains a *Lorentzian* shaped reflection line or antiresonance. The Shao et al. model also shows that resonance, antiresonance and more general Fano resonance behavior can all occur in a single system, which is exactly what we are seeing here. While our results are for the multi-mode case, one suspects that the physics for our system is qualitatively identical.

For fitting Fano resonances in open quantum dots with multiple propagating modes, the following functional form is appropriate [79]:

$$G = G_b + \frac{G_0(\varepsilon + q)^2}{\varepsilon^2 + 1} \tag{28}$$

where $\varepsilon = (E - E_{\mathrm{p}})/\Gamma$, $q = (E_{\mathrm{p}} - E_{\mathrm{o}})/\Gamma$, is an asymmetry parameter that depends on the *background* phase shift, G_{b} the background conductance that the resonance sits upon, and G_0 determines the magnitude of the resonance. One recovers the usual Lorentzian resonance in the limit where $q \to \infty$. When $q \to 0$, instead of there being a resonant peak in conductance, one obtains a resonant minimum, or antiresonance, representing resonant *reflection* from a quasi-bound state. This antiresonance phenomenon has been examined in detail in quantum wire T-stub quantum wire structures [78],and we will discuss it further below.

If the process of making the QPCs wider to allow more propagating modes were truly analogous to lowering the potential barriers of a double barrier resonant tunneling device, one would expect that the resulting behavior would simply be a broadening of the resonant energy levels. Figure 6.9 illustrates a clear example where this is obviously

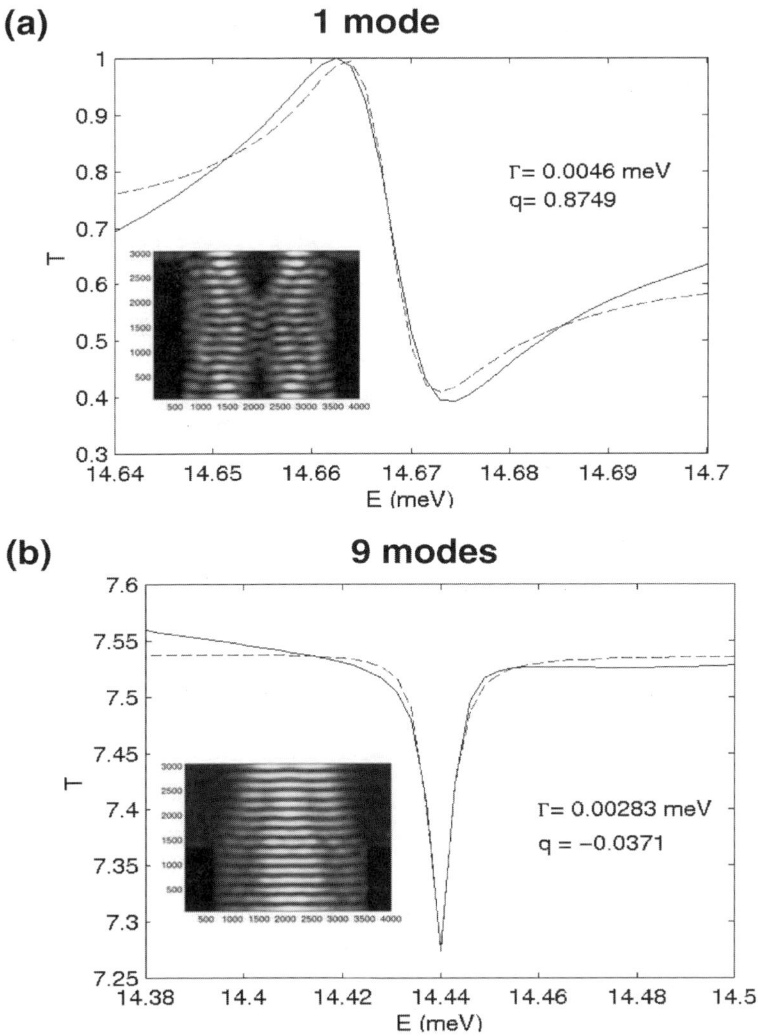

Fig. 6.9. $G(E)$ versus E for a dot with QPCs that support one propagating mode (a) and nine (b). The insets show the wave functions corresponding to the resonances displayed in the traces. The dashed lines represents the fit to the Fano formula.

not the case. In both panels, G is plotted as a function of E for a $c = h = 0.3\,\mu\text{m}$ square dot. The magnetic field is zero. In Fig. 6.9(a), the QPCs has been set to allow only $N = 1$ propagating mode and the resonance that is illustrated corresponds to the "bouncing-ball" resonance shown in the inset. In Fig. 6.9(b), $N = 9$ modes are now allowed into the dot. Here a "bouncing-ball" still occurs (see inset), but the main effect of opening the dot is to lower the energy at which the resonance occurs from 14.67 meV to 14.44 meV. By fitting to the Fano lineshape form (the dashed lines in both panels), it is apparent that the width of the resonance has actually *decreased*, from $\Gamma = 0.0046$ meV to $\Gamma = 0.0028$ meV. Note that $q = -0.0371$ for the $N = 9$ resonance and thus, not surprisingly, we are close to the antiresonance limit in that case. *Clearly, just opening the dots so that $N > 1$ is far from being a sufficient condition for broadening the resonant levels.*

5. The Correspondence Between Conductance Resonances in Open Dots and Closed Dot Eigenstates

5.1. Comparing the Spectrum of a Closed Dot with the Conductance of an Open Dot

From the previous discussion, it is clear that the broadening of resonances in two dimensional quantum dots is a more complex and subtle process than what had been assumed in deriving the "Thouless argument". In this section, we now examine in detail the connection between these resonances in open dots and the eigenstates of closed systems. As will be made evident, deciding what closed system to use for the comparison is quite crucial in establishing a direct connection between resonances and specific eigenstates.

Figure 6.10(a) shows a portion of the energy spectrum calculated as a function of field for a $c = h = 0.3\,\mu\text{m}$ square quantum dot closed to the outside environment [70,81,82]. To obtain this spectrum, the sparse matrix eigenvalue problem described at the end of section 2 was used. As can be seen, the spectrum in a square dot in the presence of a magnetic field can be quite complicated, with many crossings and anticrossings. This is in some sense a reflection of the "incompatibility" of symmetries (four-fold symmetry for the dot, radial symmetry for the magnetic field). However note the appearance of energy levels that obviously vary as B^2 on the scale displayed, while others have a more linear dispersion, which harkens back to the discussion in the previous section about comparisons with the Darwin–Fock spectrum and angular and radial momentum quantum numbers.

In Fig. 6.10 (b), a shaded contour plot of $G(E, B)$ is displayed for the case where the dot is now connected to outside reservoirs by two tunneling barriers $0.05\,\mu\text{m}$ thick, with a potential barrier height of 15.5 meV, which was sufficient to insure that most of the transmission was of a resonant nature over the given energy range [68]. Importantly, since transmission here occurs along the entire length of the barriers, there is no specially favored point or angle of entry into the dot in this case. The pattern of resonance lines in this picture resembles the energy level structure for the closed dot, though the resonances corresponding to the eigenstates with B^2 dispersion have a much higher conductance. The shading in this instance was done on a logarithmic scale, so that the

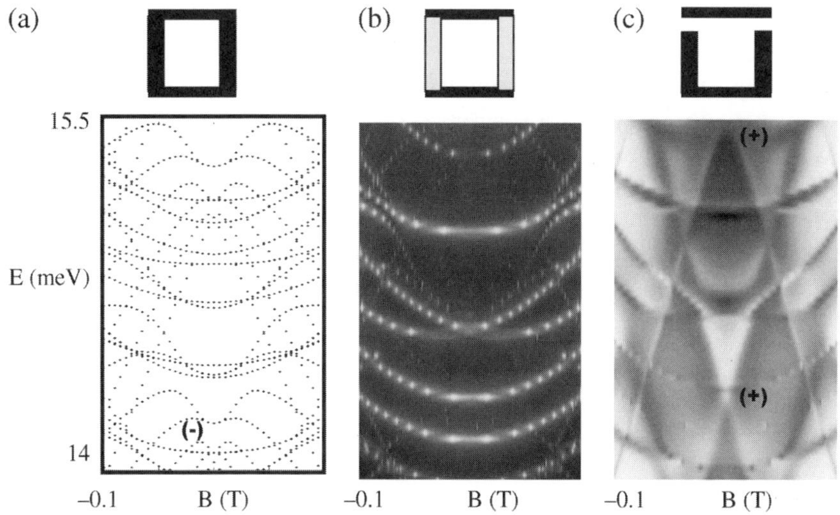

Fig. 6.10. (a) A portion of the *spectrum* is plotted as a function of E and B for a *closed* 0.3 μm square dot. The (-) represents the position of an anti-crossing. (b) A shaded contour plot of $G(E, B)$, is plotted for the 0.3 μm square dot connected to the outside via tunneling barriers. The lighter regions of shading correspond to higher values of the conductance G. (c) As in (b), but with QPCs allowing four modes propagating into the dot. The (+) represents the position where resonance lines apparently cross.

fainter resonance lines are actually several orders of magnitude lower in conductance than the darkest spots (the highest value of G here is $\sim 2e^2/h$). As noted previously for the dots with QPCs, the resonances with B^2 variation tend to have wave functions that display "bouncing ball" type patterns. These have an important property in square dots—their amplitude fairly uniformly fills the entire dot. In contrast, the "diamond" resonant wave functions are highly localized, with amplitude near the boundaries only being concentrated at the midpoints of the boundary walls. As such, "bouncing ball" resonances are much more strongly coupled to the tunneling barriers and thus display a much higher transmission probability.

Figure 6.10(c) reproduces a portion of the shaded contour plot of $G(E, B)$ from Fig. 6.6(c), which was obtained for an open dot with $b = 0.065$ μm QPC openings. Comparing this with the closed dot spectrum, it is apparent that there is some correspondence between the striation pattern for the open dot and the positions of the closed dot energy levels, particularly those with a simple B^2 dispersion. With regards to striations with linear variation, there are significantly fewer such lines in the open dot conductance than in the closed dot spectrum, which is a manifestation of the resonance "subtraction" discussed in the previous section.

There are also certain features in the conductance that *apparently* do not have a spectra analog. In particular, there are linear resonance features that actually *cross* at $B = 0$ T, indicated here by the (+) symbols. The fact that they do cross allows the regular cross-hatched noted earlier to form. In contrast, the spectrum shows levels that have linear variation for much of the range shown, but curve over in the region near $B = 0\,\mathrm{T}$. Thus, rather than crossing, they appear to form a type of *anticrossing*. Such energy level splitting effects are well understood from perturbation theory. A few of these features are indicated with the (−) symbols in Fig. 6.10(a).

5.2. Eigenstate Decomposition and the Appropriate Choice of Basis States

The relationship between open dot resonances and closed dot eigenstates can be established in a quantitative way. Since closed dot eigenstates form an orthogonal basis set, the wave functions of an open dot can be expressed as a linear combination of these states by means of projection in the dot region:

$$\psi = \sum_n c_n \phi_n^{closed} \quad \text{where } c_n = \left\langle \psi | \phi_n^{closed} \right\rangle \tag{29}$$

Figure 6.11 shows the results of such a calculation done for a nearly square dot [83]. Non-commensurate dimensions, $a = 0.3\,\mu\mathrm{m}$ by $b = 0.307\,\mu\mathrm{m}$, were chosen to insure that the levels of the rectangular cavity were not degenerate. Fig. 6.11(a) shows $G(E)$ versus E for a case where the QPCs connected to dot are $0.07\,\mu\mathrm{m}$ wide, allowing 2 modes in the energy range displayed. At the top are markers that indicate the positions of the 68th through 70th energies levels for a closed rectangular cavity with the indicated dimensions as obtained using the method described in section 2. At the bottom are markers for the 73rd and 74th eigenenergies for a T-shaped cavity (the two rectangular extensions are $0.07\,\mu\mathrm{m}$ by $0.05\,\mu\mathrm{m}$ have been added to sides of the central rectangle to make the T). The conductance over this energy range shows two major resonances, the first of which, at E = 5.988 meV, which is marked "b", lines up with both rectangular and T eigenstates. Figure 6.11(b) shows the corresponding open dot wave function (which appears to have a variation of the "bouncing ball" pattern) which happens to closely resemble the 68th rectangular state and the 73rd T state. The second resonance, at $E = 5.9915\,\mathrm{meV}$, marked "c", lines up only with a T state. Figure 6.11(c) shows the corresponding open dot wave function for that case. It appears to be quite similar to the diamond states found at finite magnetic field, however it carries no net current. The left inset shows the decomposition of this wave function in terms of rectangular states ($|C_n|^2$ is plotted versus n). Two states, $n = 69$ (Fig. 6.11(c)) and $n = 70$ (Fig. 6.11(d)), which bracket the resonance, yield the vast majority of the total amplitude (note: $|\Psi(x, y)|$ is plotted). These states have the form $sin(k\pi x/b)sin(m\pi y/a)$, with $(k, m) = (6, 8)$ and $(k, m) = (8, 6)$ respectively. These would be degenerate if $a = b$ (Note: the state in Fig. 6.11(b) corresponds to $(k, m) = (1, 10)$). The linear combination these two states produces Fig. 6.3(b), which is virtually identical to Fig. 6.11(c) in the interior region. A similar result was obtained by Zozoulenko et al. [66], who did a decomposition of a wave function with a crossing pattern in a square dot, and found that a number of square dot eigenstates contributed to create the pattern. If we stopped here, the conclusion would be that, despite the resonance, the fact that a number of eigenstates *apparently* contribute indicates that the level structure of the dot is not preserved. This however is *wrong*, because the rectangular cavity is actually not the appropriate system for comparison.

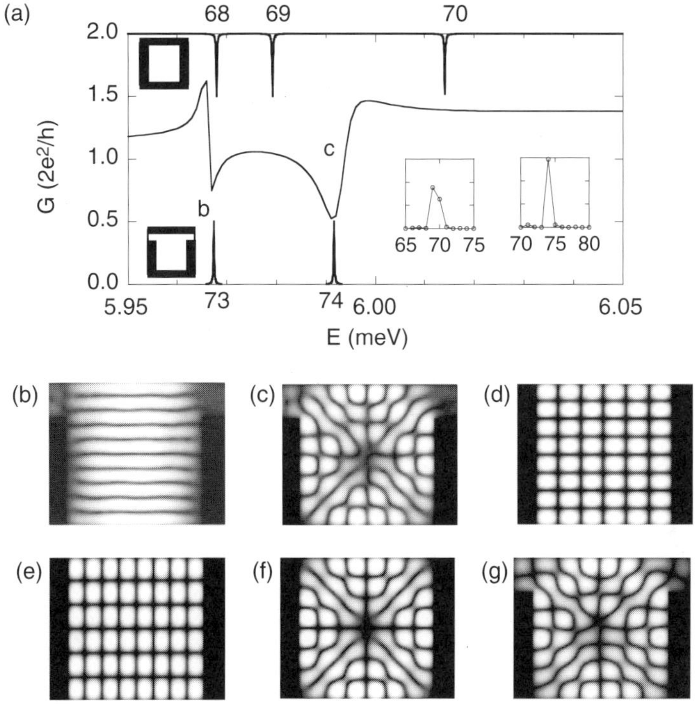

Fig. 6.11. (a) *G(E)* versus *E* for a rectangular dot. The marks at the top indicate the positions of the eigenenergies for the closed rectangle, the marks at the bottom the T-shaped cavity. Left inset: the rectangular decomposition for the wave function shown in (b). Right inset: the T decomposition. (b) $|\Psi(x, y)|$ versus x and y, the wave function in the interior region of the *open* dot at $E = 5.9915$ meV. (c) The 74th eigenstate of the T-shaped cavity (d). The 69th eigenstate of the rectangle. (e) The 69th eigenstate of the rectangle.

This is illustrated by the right inset of Fig. 6.11(a), which gives the T state decomposition of the open dot wave function at this resonance. Here, a *single state*, $n = 74$ (Fig. 6.11(g)) of the T cavity yields the vast majority of the amplitude. Comparing Fig. 6.11(g) with Fig. 6.11(b), it is difficult to pick out the open state from the closed one. The 74th T state can be viewed as a hybridized state resulting from a perturbation (the extensions added onto the sides) which has mixed primarily the 69th and 70th rectangular states together. However, while hybridized, it is still an eigenstate. Importantly, while the presence of the side extensions (or the QPCs in the open case) is crucial for the formation of the state, the amplitude of the hybrid state is highly concentrated in the central cavity. As with the diamond states discussed above, there is an apparent disconnection from the QPCs.

Classically, the square is regular or integrable. What about the T cavity? As one might expect, the presence of additional corners or vertices complicates the classical picture, making this billiard "psuedo-integrable" [84,85], with non-integrable contributions being created by the bifurcations of orbits at the vertices. Significantly, such billiards can closely mimic the behavior of truly chaotic systems [84].

5.3. The Selective Persistence of Eigenstates in Open Stadium Dots: The Relative Independence of Broadening on the Proximity of Levels

We have shown that resonances in an open rectangular dot with more than one propagating mode can correspond to single eigenstates of a closed system, provided that the closed system chosen for comparison is the appropriate one. A rectangular cavity is classically regular at zero field. What about a cavity that is classically chaotic (at least when it is closed), such as the stadium? We have found the behavior to be essentially the same [86]. To illustrate this in what follows, we examine the behavior of an open stadium and compare it to two closed systems, the standard stadium (Fig. 6.12(a)) and a perturbed stadium for which a portion of the QPCs have been included as part of the enclosed structure (Fig. 6.12(b)). The ratio of the QPC width w (indicated in Fig. 6.12(b)) to stadium height H (indicated in Fig. 6.12(a)) is a parameter we make use of below.

Figure 6.12(c) displays eigenstates 128–136 of the standard stadium, while in Fig. 6.12(d) we show eigenstates 134–142 of the perturbed stadium. The energy range

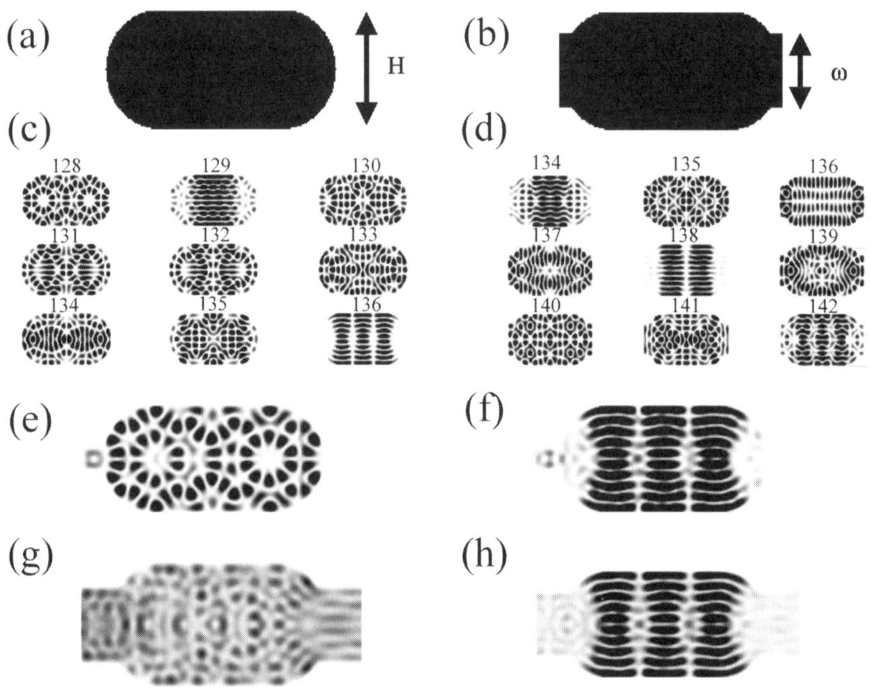

Fig. 6.12. (a) The unperturbed stadium. (b) The perturbed stadium. (c) $|\Psi(x, y)|^2$ versus x and y for eigenstates 128 through 136 of the unperturbed stadium. (d) $|\Psi(x, y)|^2$ versus x and y for eigenstates 134 through 142 of the perturbed stadium. (e) $|\Psi(x, y)|^2$ for an *open* stadium with two modes passing through the leads at the energy of the 128th stadium eigenstate. (f) As in (e), but now the energy is set to that of the 136th eigenstate. (g) As in (e), but with widened QPCs supporting 6 modes. (h) As in (f), but with 6 mode QPCs.

covered by these two groups is roughly the same. Figure 6.12(e) shows the open dot wave function for a Fermi energy equal to the 128th eigenvalue of the standard stadium. Compared to Fig. 6.12(c), it is evident that the eigenstate has maintained most of its integrity, even though leads allow two propagating modes ($w/H = 0.2$) in this case. For Fig. 6.12(g), the energy is fixed, but the QPCs have been opened further ($w/H = 0.6$), allowing six propagating modes. Now the correspondence of the open dot wave function to the eigenstate is lost. Contrastingly, Figs 6.12(f) and (h) show that the 136th state (which features a classic stadium bouncing ball scar) survives essentially unchanged in the open dot, even when the QPCs allow six propagating modes. As was the case with the rectangular cavity, it is the state that has its amplitude concentrated away from the QPC opening region that has survived.

This result is further illustrated in Fig. 6.13(a). Here for the $w/H = 0.6$ open stadium we plot conductance, G, versus E/Δ, where the average level spacing is given by the formula $\Delta = 2\pi\hbar^2/m^*A$, mentioned earlier. The squares represent the positions of eigenvalues of the standard stadium (128th and 136th eigenvalues are specifically identified). The range displayed coincides with 11 eigenvalues. Over this same range, three well defined resonances appear in G, and are labeled (i), (ii) and (iii). The circular dots superimposed on the 3 resonances represent the fits we obtained to the generic Fano lineshape discussed in the previous section. Of particular interest are the values that are obtained for the level width, Γ which features prominently in the Thouless argument. The values obtained were $\Gamma_{(i)} = 0.075\Delta$, $\Gamma_{(ii)} = 0.097\Delta$, and $\Gamma_{(iii)} = 0.104\Delta$. Thus, for these particular states $\Gamma/\Delta \ll 1$ and, despite the large number of modes in the QPCs, these states are truly being resolved.

In Fig. 6.13(b), the decomposition of the open dot wave functions is plotted as a function of the unperturbed stadium eigenstate number (n), and E/Δ. The grayscale corresponds to $|C_n|^2/|C_n|^2_{max}$, so that the peak value of each decomposition is normalized to unity. While for much of the range shown it appears that the open-dot wave functions involve a mixture of states, at energies corresponding to resonances (i) and (iii) the decompositions are clearly dominated by a *single* eigenstate of the unperturbed stadium (the 129th and 136th, respectively). In contrast, resonance (ii) corresponds to a mixture of at least two states. Moreover, none of the unperturbed stadium eigenvalues appear to line up with the position of resonance (ii). However, a logical assumption is that, like the rectangular example discussed above, the QPCs are providing a perturbation of the basic stadium structure, and this is why we have also studied the perturbed stadium shown in Fig. 6.12(b). The energy levels of this stadium are represented by the triangles in Fig. 6.13(a), from which we see that the 138th level of this structure coincides with the position of resonance (ii). The wave function decomposition in terms of this perturbed-stadium basis set is shown in Fig. 6.13(c), from which we see that all three conductance resonances are dominated by the contribution of *single* eigenstates. Referring back to Figs 6.12 (c) and (d), we see that states 134 and 142 of the perturbed stadium bear a strong resemblance to states 129 and 136 of the unperturbed stadium, and occur at similar energies. State 138 of the perturbed stadium, which coincides with resonance (ii) and which closely resembles the open-dot wave function at this energy, has no ideal-stadium analog.

In Fig. 6.14, we show the decomposition in terms of the perturbed stadium basis set again, however this time without normalization and in this case we show a three

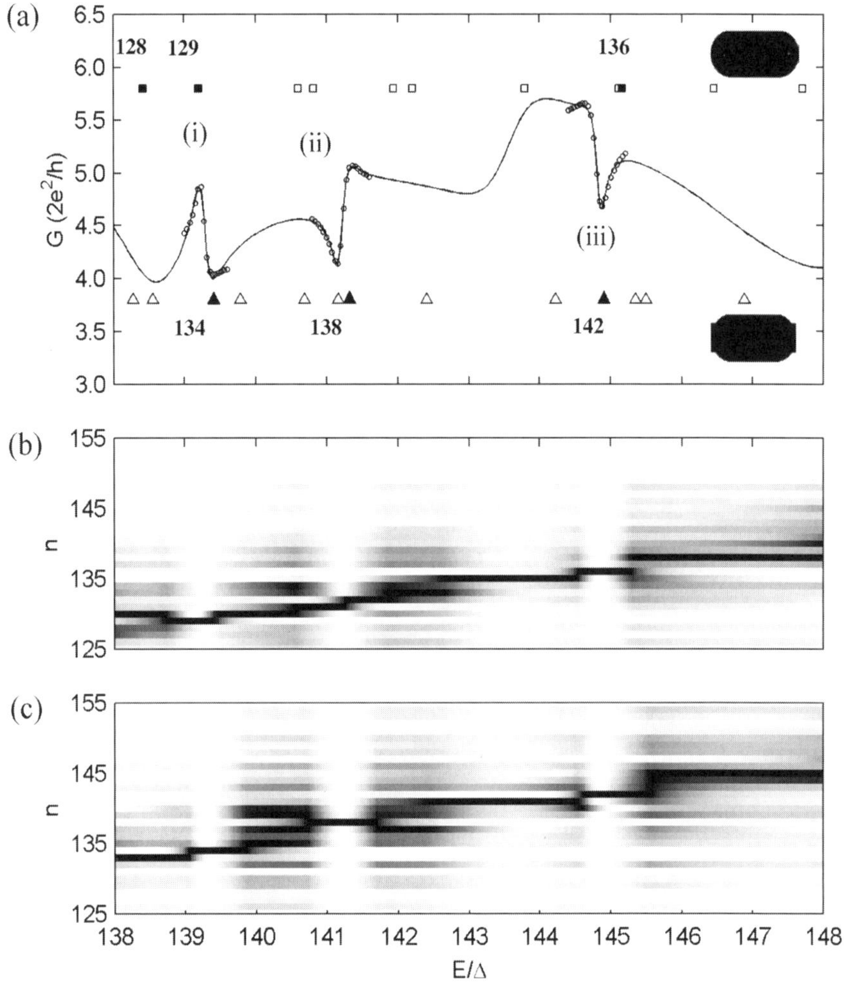

Fig. 6.13. (a) G versus E/Δ for the six-mode open stadium. The squares and triangles represent the energies of the eigenstates of the unperturbed and perturbed stadium, respectively. The circles are fits to the Fano formula. (b) The *normalized* decomposition of the open-dot wave functions into the unperturbed-stadium eigenstates. (c) As in (b), but using the perturbed-stadium eigenstates. (d) $\Sigma_n |C_n|^2$ versus E/Δ (solid line, left scale) and G versus E/Δ (dashed line, right scale) for the six-mode open stadium. (e) $\Sigma_n |C_n|^2$ indicated by the shading as a function of E/Δ and fractional lead width, w/H. (f) $\Sigma_n |C_n|^2$ versus E/Δ plotted over a larger range. The open-stadium wave functions coinciding with the numbered states are included as insets.

dimensional plot. The positions of the three resonant states are indicated, and it is evident that these states dominate the contribution, relative to all other states. Note that on the eigenvalue axis, we have plotted the energy of the perturbed stadium relative to the average level spacing (E_n/Δ), so that the spacing of successive levels is indicated on this axis. Significantly, we see that the width of the 134th state, as inferred from

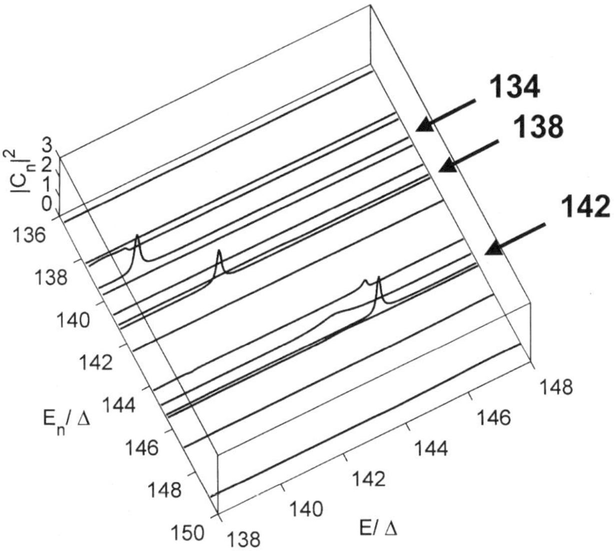

Fig. 6.14. The perturbed stadium decomposition coefficients $|C_n|^2$ versus E/Δ and E_n/Δ.

the breadth of the decomposition peak, is *larger* than the spacing between the 134th and 135th levels. Similar observations may also be made concerning the other resonant states. Nonetheless, the decompositions remain dominated by the contribution of single eigenstates. *Contrary to naive assumptions, the mere proximity of levels in energy need not be an obstacle to being able to resolve individual states.*

As shown in Fig. 6.15(a), even after performing the summation $\Sigma_n|C_n|^2$ over the total of 175 states, the contributions from states 134, 138 and 142 still dominate as peaks rising a factor of $\sim$3 above the background yielded by the total over all the other states. Importantly, this summation essentially represents the DOS of the open dot, and performing the summation with the standard (unperturbed) stadium basis set gives an identical result. To demonstrate the stability of these peaks, in Fig. 6.15(b) we plot $\Sigma_n|C_n|^2$ versus E/Δ and w/H. As indicated by the shading, the peaks persist over a wide range of lead opening. Also apparent in this plot are peak contours that survive over only a much shorter range of w, before shifting dramatically to lower energy and disappearing. The downward shift as w is increased is understandable, as the *effective* dot size is being increased. Figure 6.15(c) shows a wider range of energy. Evident here is a group of five resonant states whose basic structure is essentially repeated. This behavior is somewhat analogous to the *shell structure* of the hydrogen atom, or for that matter, the *circular* dot, which is a classically *regular* system. Moreover, these states all appear to be scarred. The conclusion one draws from these results is that opening up the stadium enhances the influence of classically regular behavior. The more open the dot, the more important the role played by periodic orbits in determining the fluctuations. Significantly, with regards to the statistical properties of the eigenvalues of stadiums, it has been found that the contributions from such states, which are associated with classically periodic behavior, must be removed in order for the results of the closed

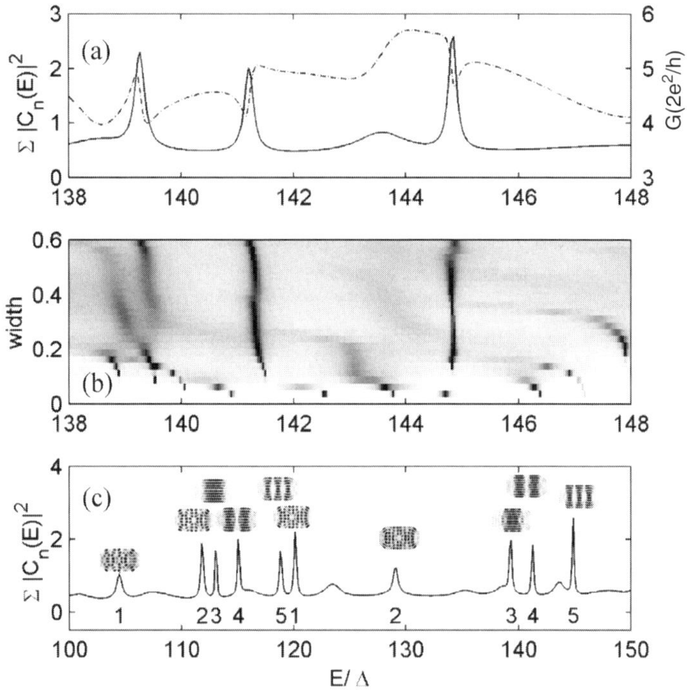

Fig. 6.15. (a) $\Sigma_n|C_n|^2$ versus E/Δ (solid line, left scale) and G versus E/Δ (dashed line, right scale) for the six mode open stadium. (b) $\Sigma_n|C_n|^2$ indicated by the shading as a function of E/Δ and fractional lead width. (c) $\Sigma_n|C_n|^2$ versus E/Δ plotted over a larger range, the numbers indicate the positions of perturbed stadium eigenstates. The open stadium wave functions coinciding with these numbered states are included as insets.

stadium to match the Gaussian orthogonal ensemble prediction for a chaotic system as obtained using random matrix theory [87].

5.4. *What Happens at Higher Energies*

One possible criticism regarding this set of simulations is that they were performed at relatively low energies. At some energy, the kind of effects we are seeing should ultimately vanish as we approach the classical limit. The still open question is where. We have also done calculations on the $w/H = 0.6$ stadium for energies about an order of magnitude larger than those used for the preceding figures. The results are shown in Fig. 6.16. In Fig. 6.16(a), we plot G versus E/Δ for $1240 < E/\Delta < 1270$. Note the large values of conductance being attained here : $G \sim 16e^2/h$. Despite this, the resonant behavior is still apparent. The open dot wave functions coinciding with the resonances marked "b", "c", "d", and "e" are shown in the corresponding panels below. These represent four members (2 through 5) of the repeating "shell" found at much lower energies as shown in Fig. 6.15(c). In contrast, to the lower energy resonances, the scars are less well defined as the background that they are superimposed over is obviously making a stronger contribution.

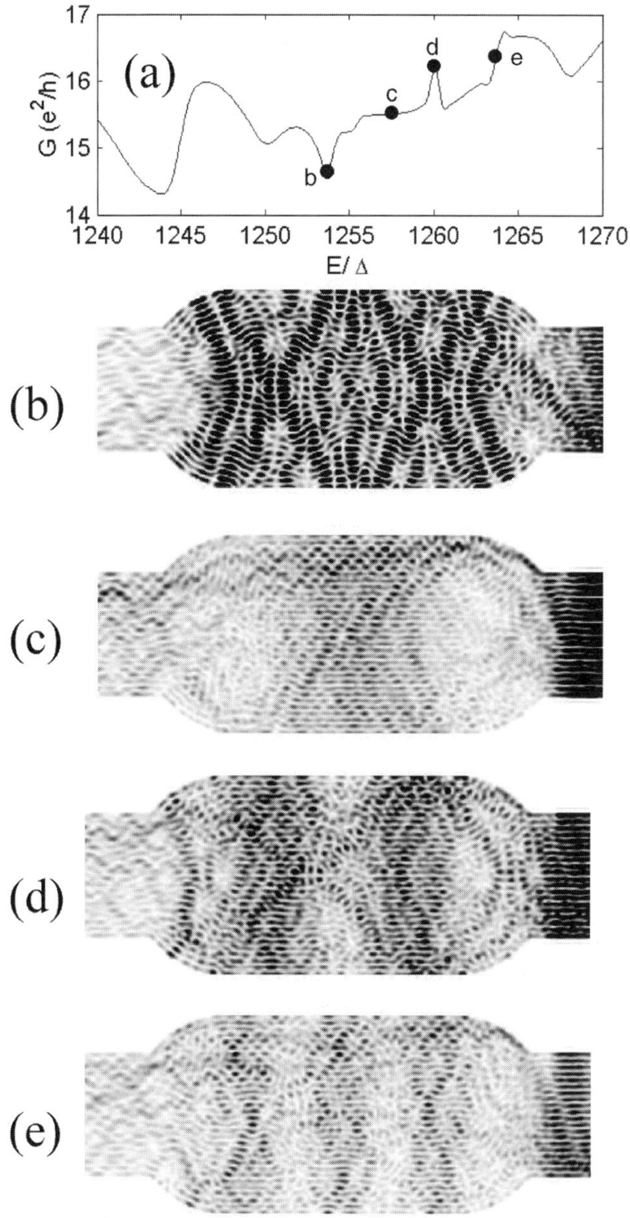

Fig. 6.16. (a) G versus E/Δ for the same open stadium but at much higher energies. Panels (b), (c), (d) and (e) illustrate the open stadium wave functions corresponding to the appropriately marked resonances in (b).

6. The Effect of Finite Temperature and Ensemble Averaging

6.1. Survival of the Diamond Orbit Feature at Finite Temperature

We have demonstrated that opening quantum dots to the external environment is not enough *in itself* to fully destroy the discrete energy level structure. As long as the dot is defined well enough to provide some measure of confinement, certain eigenstates should always survive. However, the calculations we have shown thus far have completely neglected thermal effects, and moreover we have only briefly touched on the effects of phase breaking. Needless to say, those additional broadening mechanisms are always present when experiments are performed. To simulate the effect of this averaging on the transport, we adopted the following approach [81,82]. One can account for thermal smearing and dephasing in combination by introducing an effective temperature, T^*, such that $T^* > T$, the actual temperature. The choice of T^*, is ideally determined by appealing to the results of experiment (see section 7 following). At a given Fermi energy (E), the energy averaged magneto-conductance may then be computed by solving the following integral numerically:

$$G_{av}(E) = \int G(E')\left[-\frac{df(E'-E)}{dE'}\right]dE' \tag{30}$$

While Eq. (30) shows that G_{av} is determined by convolving the resistance with the derivative of the Fermi–Dirac distribution function, it is sufficient in practice to integrate over an energy window which centered on the Fermi energy and is a few $k_B T^*$ wide.

In general, the kind of very sharp conductance resonances we have been focusing on are not typically resolved experimentally because of the effects of broadening. Does this mean that one can ignore their contribution to the physics of dots in real world situations? The answer to this question is an emphatic *no*. In Fig. 6.17(a) we show a shaded contour plot of $G_{av}(E,B)$ versus E and B for a $c = h = 0.3\,\mu\text{m}$ square dot with $b = 0.05\,\mu\text{m}$ QPC openings (three propagating modes in this energy range), for broadening corresponding to $T^* = 0.5\,\text{K}$. Even with this thermal averaging, the regular cross-hatched pattern noted earlier for this structure is still apparent. What the resonances apparently do is act as a kind of framework that pins the overlying fluctuations. Thus, while averaging may obscure the actual sharp resonances, the periodic framework is still visible. Figures 6.17(b) and (c) show a comparison of a diamond resonance wave function before and after thermal averaging. For temperatures at which the regular striation pattern is still apparent in $G_{av}(E,B)$, the diamond orbit is still apparent in the wave function.

6.2. Ensemble Averages of the Conductance of a 0.3 μm Square Dot: Is the Resistance Peak at B = 0 Really Weak Localization?

As mentioned in section 2, it has been stated that performing thermal or other kinds of ensemble averaging is required to make a real comparison with the semiclassical predictions for quantum dot transport properties. A primary example is the commonly observed peak in the magnetoresistance about zero field, which has been associated with weak localization. In large disordered systems, such resistance peaks result from

(a)

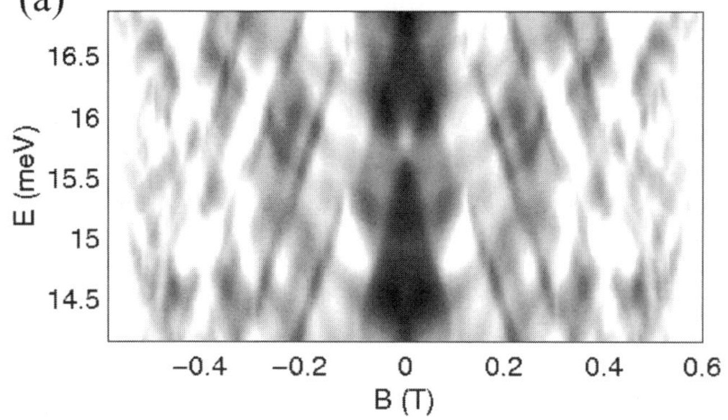

(b)

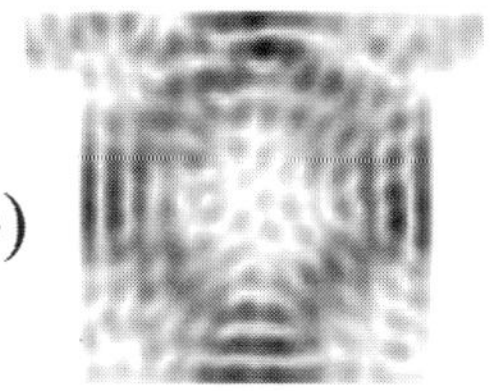

(c)

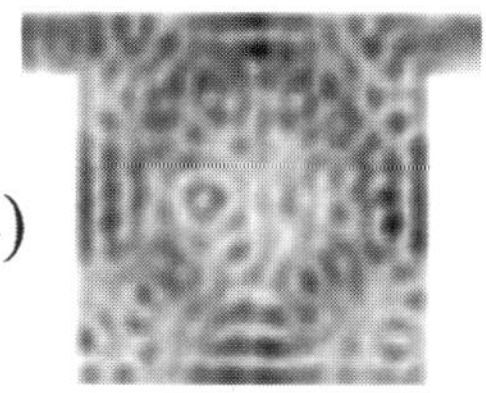

Fig. 6.17. (a) A shaded contour plot of the thermally averaged conductance $G_{av}(E, B)$ for a 0.3 μm square dot. The diamond resonance wave function without and with thermal averaging is shown in panels (b) and (c).

back-scattering of particle trajectories as they are scattered from the random potential [88–90]. These trajectories can close on themselves, and, together with a time-reversed orbit, cause destructive quantum interference, lowering the conductance. Finite fields break this time reversal symmetry, and result in a higher conductance. Thus a symmetric resistance peak appears at $B = 0$. It has been suggested that an analogous weak localization should also occur in ballistic quantum dots, where the scattering is now due to the confining boundaries. It is argued that the details of the line shape of the weak localization peak may be used to infer the dynamics of the electrons within the dot—chaotic dynamics should lead to a Lorentzian line shape, while regular dynamics should lead to a linear line shape, at least according to some semiclassical calculations [26,27]. Indeed, some measurements of differently shaped ballistic cavities have shown such a difference in the line shape of the resistance peak around zero magnetic field, when the data is averaged over several devices or over a range of gate voltage [16,17,91]. Unfortunately, *both* line shapes have also seen in a single dot, by varying either the gate voltage [54] or the temperature [17]. Moreover, in some experiments, circular cavities, which are expected to yield linear behavior, we found to produce Lorentzian line shapes [92,93]. These apparently conflicting results raise the questions: can a Lorentzian line shape be used as a reliable indicator of chaos? Should the above interpretation of the resistance peaks in ballistic dots even be regarded as weak localization? What we have found is that a peak in resistance at zero magnetic field can be a intrinsic property of the dot and one should not invoke weak localization as an explanation for the effect [70,71].

In this regard, two important elements of the semiclassical weak localization theory are: (i) it applies to a regime in which one could safely assume that the limit where $\Gamma/\Delta \gg 1$ is being approached, and, (ii) to obtain the actual line shapes, an integration over *all* wavenumbers, k, was performed. Our demonstration that narrow resonances can survive even in dots with a very large number of modes in the QPCs, means that actually achieving (i) is obviously much more problematic than previously thought. With (ii), any experimental comparison with this theory requires that an ensemble average be performed to ensure that a large region of k space is sampled, which is the reason why experiments have utilized, for example, multiple arrays of similarly configured devices, or gate voltage and/or thermal averages with a single device. In addition, single devices have also been constructed in which k could be independently varied [Keller 1996], and there have been experiments in which the quantum dots were subjected to thermal cycling in order to produce an averaging effect [54,64]. However, in many instances, the dots examined were relatively small and measurements were performed with only a few modes passing through the QPCs. Moreover, it is not clear from the semiclassical theory how much averaging is really sufficient, and *apparent* agreement with the theory has been achieved in many cases by performing very limited averaging, involving relatively small regions of k space [54,64,92–94]. Thus, it is questionable whether the criterion (ii) was even being satisfied in these experiments. Why then should a resistance peak be present in cases that do not satisfy the limits required by the weak localization theory? Why does one get a Lorentzian peak when a linear one is expected? The answers to these questions can be understood as a natural byproduct of the open dot resonance features.

The weak localization formula for the resistance peak in a chaotic cavity is given by the Lorentzian form [26,27]

$$R(B) = R_0 + \frac{\Delta_R}{1 + (2B/\alpha\phi_0)^2} \tag{31}$$

where ϕ_0 is the flux quantum, $\Delta_R = R(B = 0) - R_0$, and α^{-1} is a parameter that represents the area enclosed by a typical trajectory. It is assumed that the α^{-1} parameter can be substantially larger than the cavity size if the electron trajectories make many bounces before exiting. As mentioned above for a regular cavity, a linear line shape is expected in that case but there is no formula that can be used for fitting.

Figure 6.18(a) displays a shaded contour plot of $G(E, B)$ for a square dot with $c = h = 0.3\,\mu$m, and $b = 0.06\,\mu$m. Note that over the range of energy simulated, many features recur quite periodically. Diamond patterns are formed by sets of angled, almost parallel, lines of conductance resonances. At the central tips of these diamonds, subdivided triangular patterns occur (the clearest being at ~8 meV). As one might expect from this picture, a resistance peak at $B = 0$ is not a general feature of individual traces. Rather, one must perform averages (such as thermal averages) of conductance in order to obtain a peak in all cases. As mentioned above, there is always a natural averaging that goes on in experiment (one is never truly at $T = 0$ K), but this is not enough to ensure that a peak will always arise. In Fig. 6.18(b), a three dimensional plot of resistance, R, versus E and B, shows a result of such an averaging calculation. Each resistance trace shown here was obtained by averaging $G(E, B)$ over a window of

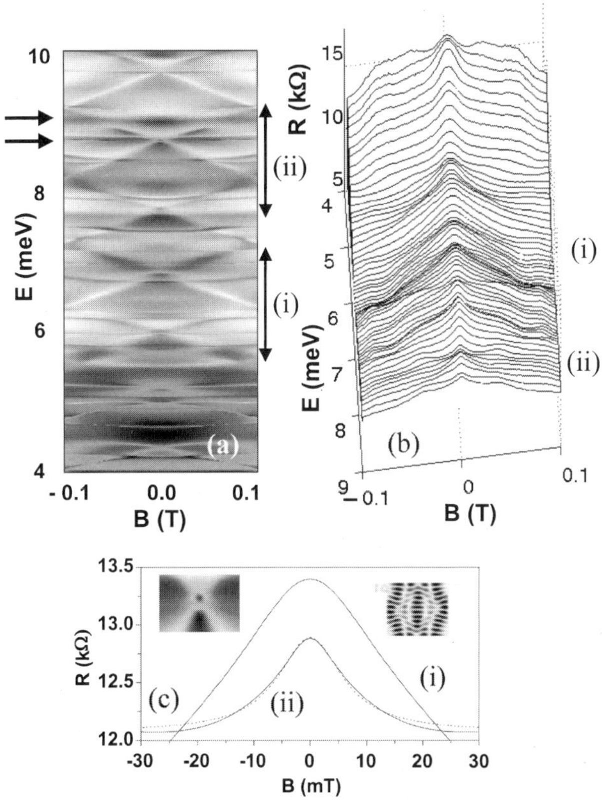

Fig. 6.18. (a) Another shaded contour plot of $G(E, B)$ for a 0.3 μm square dot. The lighter regions of shading correspond to higher values of the conductance G. (b) A series of traces of R versus B, obtained by averaging G, from panel (a), over a 1.5 meV window centered at energy E. (c) Two traces (R versus B) obtained by averaging over the ranges denoted (i) and (ii) by the vertical arrows in panel (a). The dotted line is a Lorentzian fit. The right inset is the wave function corresponding to the "eye-hole" resonance blown up in the left inset.

1.5 meV, which was sufficient to yield peaks over this entire range (we will discuss the issue of window size later). Note that the higher resistance traces shown in the upper part of the picture correspond to averages over the lower ranges of Fig. 6.18(a). While the presence of the peak is now universal, its line shape clearly evolves as a function of the energy. To emphasize this point, we focus on two of these traces in Fig. 6.18(c). The traces (i) and (ii) were obtained by averaging over the ranges indicated in panel (a). Despite having rounding at the top, trace (i) has a quite strong linear character over a wide range of field. On the other hand, trace (ii) quite closely follows the Lorentzian line shape form, as evidenced by the fitted curve (the dotted line). The (i) and (ii) markers beside Fig. 6.18(b) indicates where these traces fall in the group of average traces that are plotted.

For the fit of trace (ii), we have used $R_0 = 12.07\,\mathrm{k}\Omega$, $\Delta = 0.82\,\mathrm{k}\Omega$ and $\alpha\phi_0 = 6.82\,\mathrm{mT}$ in this formula. This gives $\alpha^{-1} \sim 7\,A$, where A is the dot area. Experimentally, line shape transitions have been observed experimentally in nominally square dots, with the transition occurring as V_g was varied to increase the resistance [54]. This change in line shape arises quite naturally in the present case because, by averaging over a different range of the energy spectrum, we have mapped different features into the resistance. With regards to trace (ii), a "hole" in the conductance that occurs at ~8.71 meV (lower arrow in Fig. 6.18(a)) is responsible for the Lorentzian peak. This "hole" is enlarged in the left inset of Fig. 6.18(c). It is an example of an antiresonance, one of the extremes noted of the Fano lineshape noted earlier (recall Fig. 6.9). The trough indicated by the upper arrow is what creates the plateau in this case. While region (i) contains some similar features, there are not as prominent, leading to the more uniform line shape after the average is completed.

Figure 6.19 shows more examples of how averaging over different regions yields different line shapes. In this case, we plot $G(E,B)$ for this dot as a three dimensional function. Figure 6.19(a) shows $G(E,B)$ over an energy range 5.9 meV $< E <$ 6.4 meV. In Fig. 6.19(b), the range is now 7.25 meV $< E <$ 7.75 meV and in (c) it is 7.3 meV $< E <$ 7.8 meV. Here, lighter shading corresponds to higher G. Figure 6.19(d) shows the ensemble averaged resistances for the ranges shown in Figs 6.19(a)–(c). Clearly, the line shape is dependent on the features which occur in the

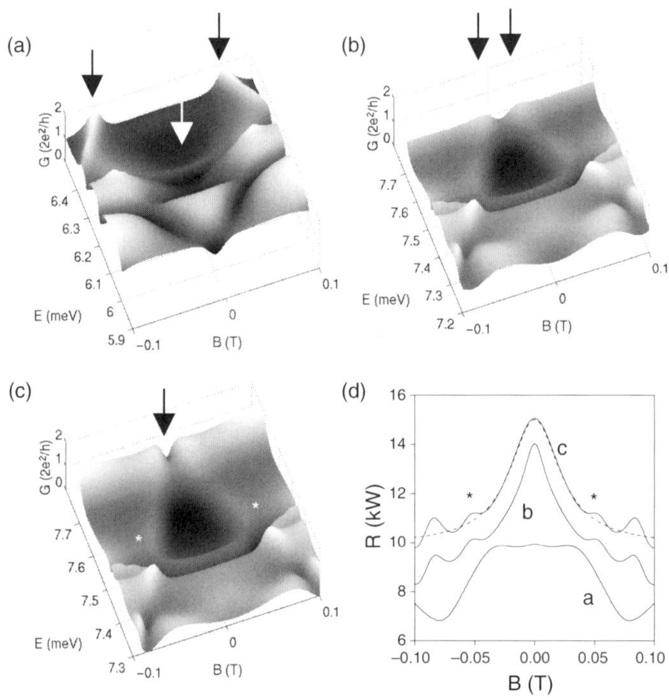

Fig. 6.19. (a) A surface plot of $G(E,B)$ for a 0.3 μm square dot with two modes in the QPCs. $G(E,B)$ is plotted in (b) and (c) as well, but for different energy ranges. (d) R versus B, where curves a, b and c are the resistances obtained by averaging over the conductance traces in (a), (b) and (c) respectively. Curve c has been shifted upward by 2 kΩ for clarity.

region over which the average has been done. Trace "a" shows a very tiny peak situated on a plateau. The plateau is set by the flat portion indicated by the white arrow as well as the horizontal "notches" which occur in the energy range just below it. The minima on either side of the plateau are primarily a reflection of the resonance lines indicated by the black arrows. Trace b displays a Lorentzian line shape quite similar to the result discussed in the previous figure, but it occurs here for a different reason. At the top end of the range, there are two conductance resonance lines that are close to merging (these are indicated by the arrows). In between is a rounded dip in the conductance at $B = 0$T. It is the shape of this dip which ultimately yields the Lorentzian shaped peak that occurs in trace b. This peak shape is quite similar to that obtained by averaging over an array of stadium shaped quantum dots (see Figs 6.1(a) and 6.2(a)). The dashed line is a fit to equation (3) with $R_0 = 9.9\,\mathrm{k\Omega}$, $\Delta = 5.1\,\mathrm{k\Omega}$, and $\alpha\phi_0 = 51$ mT. Now, α^{-1} has been reduced to $\sim A$, much smaller than the value obtained for the example in Fig. 6.18(c). If one holds to the interpretation of α^{-1} as an average area enclosed by the trajectories, one would be forced to conclude that there has been a dramatic change in the dynamics of the dot. However, as our dot geometry is constant, this is extremely unlikely. Interestingly, both circular cavities studied experimentally by Berry et al. [92] gave Lorentzian line shapes with $\alpha^{-1} \sim 3A$ (note that the circle, like the square, is supposed to be regular and should give linear behavior). Lee et al. [93] also found that the experimental data for their circular cavity could be fitted with a Lorentzian, but using $\alpha^{-1} \sim A$. The square dots studied experimentally by Bird et al. [54] also gave $\alpha^{-1} \sim A$. The issue of line shape width is something that we will return to later in this section.

The differences between trace "b" and trace "c" illustrate yet another line shape transition. Specifically, to borrow the terminology of Chang et al. [17], trace "c" has a more "triangular" appearance and closely resembles the line shape obtained by averaging over arrays of circular quantum dots (see Figs 6.1(b) and 6.2(b) of [17]). Comparing Fig. 6.19(b) to Fig. 6.19(c), the energy range in the latter has been allowed to extend a little higher, so that the merging resonance lines have come even closer together. Thus, the central conductance dip that appears in Fig. 6.19(c) (marked by an arrow) is considerably sharper than the one that appears in Fig. 6.19(b). The resistance peak shown in trace "c" is thus sharper and more linear looking in appearance as a result. As mentioned above, line shape transitions have been observed experimentally by varying the gate voltage. Varying gate voltage has a similar effect as shifting the energy spectrum, as we will show in the next section. Another important set of conductance features is indicated by the "*" marks on Fig. 6.19(c). At these points, linear striations intersect with parabolic ones, resulting in a pair of particular deep minima in the conductance. It is these features which produce the shoulders in the ensemble averaged resistance trace c, whose positions are indicated by the "*" marks in Fig. 6.19(d).

It is evident from trace "*a*" of Fig. 6.19(d) that a distinct resistance peak need not arise after averaging. What is the *minimum* requirement on the averaging to see a *distinct* peak? Fig. 6.19(c) shows essentially the top half of one of the repeating diamond patterns in the conductance alluded to earlier. While it is difficult to make complete generalizations, we have found that an ensemble average that includes at least half of a diamond is guaranteed to yield a true peak, rather than a plateau or a minimum. It is worth making a note of the combination of features that lead to a peak that are depicted in Fig. 6.19(c)—the narrow conductance minimum at the top of the energy range shown, the parabolic trench intersected by the linear striations and the relatively flat conductance in the triangular region defined by these resonance lines.

The three dimensional pictures shown in Fig. 6.19 reveal several different types of resonant phenomena, with resonant maxima, resonant minima, and combinations thereof all being present. These are all different manifestations of the Fano resonances discussed in the previous section. Recalling Fig. 6.15, where we demonstrated a direct correlation between such Fano resonances and peaks in the DOS, it thus becomes interesting to explore the relationship between the DOS, as obtained via an eigenstate decomposition, and the occurrence of a resistance peak. This is done in Fig. 6.20, for which a square dot with $c = h = 0.3\,\mu\text{m}$, and $b = 0.08\,\mu\text{m}$ was used. There were

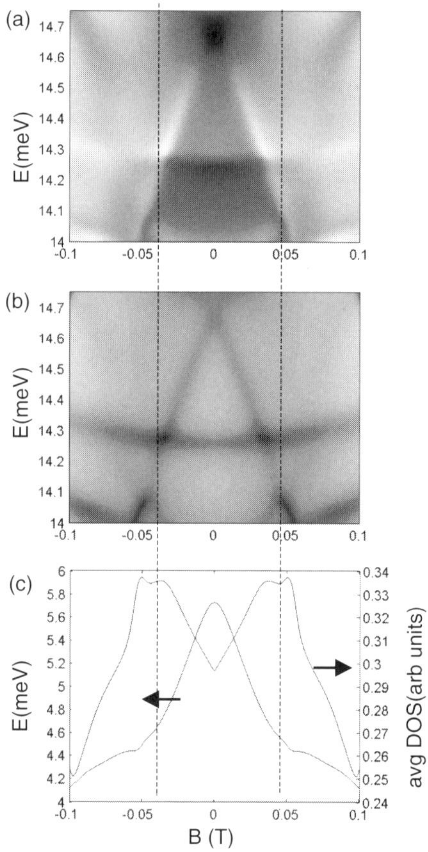

Fig. 6.20. (a) A shaded contour plot of $G(E, B)$ for a $0.3\,\mu\text{m}$ square dot with five modes in the QPCs. The lighter regions of shading correspond to higher values of the conductance G. (b) A shaded contour plot of the summed eigenstate decomposition, $\Sigma_n |C_n(E, B)|^2$ (essentially the DOS) for this dot. Note here darker shading indicates higher values. (c) The average resistance R and average DOS are plotted versus B, with averaging done over the 14.0–14.75 meV energy window shown in (a) and (b).

5 propagating modes in the energy range considered, 14 meV $< E <$ 14.75 meV. Fig. 6.20(a) displays a shaded contour plot of $G(E, B)$, with white corresponding to $G \sim 3.5e^2/h$ and black to $G \sim 1.5e^2/h$. In Fig. 6.20(b), using a T shaped cavity to obtain a basis set of eigenstates, we display a shaded contour plot of the DOS as represented by the decomposition $\Sigma_n |C_n(E, B)|^2$. Here, black represents the peak value (cut off at 0.75) and white the minimum (a value of zero). Comparing (a) and (b), it is evident that the familiar triangular patterns in $G(E, B)$ are quite clearly reproduced in the DOS. Figure 6.20(c) shows the resistance obtained from $G(E, B)$ averaging over the energy range displayed. We also plot the corresponding average DOS, $\langle \Sigma_n |C_n(E, B)|^2 \rangle$. From this picture, it is evident that the local peak in the average resistance coincides with a local minimum in the average DOS. Moreover, the onset of the "shoulders" of the resistance peak concides with local maxima in the average DOS (note the dashed lines). These in turn are produced by the presence of resonance line crossings in the energy range considered.

6.3. Confirmation of these Results in Larger Dots with Different Geometries

All the previous averaging results were for a 0.3 μm dot. However, very similar results can be obtained with larger dots. Figure 6.21 shows simulations of a $c = h = 0.5$ μm square dot. Here, the QPCs supports 4–5 propagating modes over the energy range shown in the $G(E, B)$ plot of Fig. 6.21(a) ($b = 0.065$ μm). As with pictures shown for the smaller dot, one sees groups of linear and parabolic striations in the conductance, with crossings occurring at $B = 0\,T$. Here we have done three averages over the ranges indicated by (i), (ii) and (iii). Trace (i) in Fig. 6.21(b) shows a narrow "linear" resistance peak in the center. As was the case in Fig. 6.18(c), this feature is a reflection of the near merging of resonance lines, in this case at the bottom of the energy averaging range shown in region (i), as indicated by the lowest white arrow. Trace (ii) shows a somewhat wider linear peak. In this case, the feature is largely a result of the deep slit-like minimum in the conductance indicated by the middle white arrow. Trace (iii) shows a Lorentzian peak on top of a small plateau. The dashed line shows the fit to equation (3) with $R_0 = 6.15\,k\Omega$, $\Delta = 0.85\,k\Omega$, and $\alpha\phi_0 = 25$ mT. The situation that yields this peak is similar to that shown in Fig. 6.18(c). Specifically, the peak corresponds to an "eye hole" in the conductance indicated by the uppermost white arrow. As in that previous example, this point corresponds to a resonant state. Significantly, it is exactly this combination of a narrow "Lorentzian" peak and a wide "linear" peak occurring in the same dot that has been observed *experimentally* in a dot of size comparable to the one used in these simulations [54].

One issue we have ignored thus far is that of straight through trajectories, i.e. those trajectories that travel between the two QPCs without any reflections. Recalling Fig. 6.3(b), QPCs generally produce a collimation effect causes the electrons to enter dot at quantized angles with respect to the y-axis, which automatically reduces the number of straight-through trajectories very significantly. Thus for the dot configurations we have examined thus far, this effect causes the majority of the electrons to be aimed towards the bottom right-hand corner of the dot, rather than at the exit QPC. With regards to the semiclassical weak location theory, the effect of the straight-through

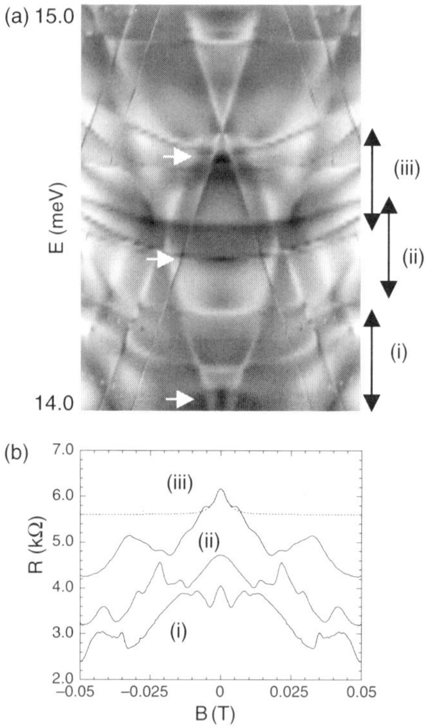

Fig. 6.21. (a) A shaded contour plot of $G(E, B)$ for a $0.5\,\mu m$ square dot with four modes in the QPCs. The lighter regions of shading correspond to higher values of the conductance G. (b) Three traces (R versus B) obtained by averaging over the ranges denoted (i), (ii) and (iii) by the vertical arrows in panel (a). For clarity, curves (ii) and (iii) has been shifted upward by $1\,k\Omega$ and $2\,k\Omega$, respectively. The dotted line is a Lorentzian fit.

trajectories is omitted because of mathematical complications they produce [26,27]. As a result, for comparison purposes, many experiments have utilized geometries in which the QPCs are misaligned or there is some barrier between the QPCs. However, several of the experiments which gave results that appeared to agree well with the theory involved geometries in which no such care was taken [54,93]. Ironically, misaligning the QPCs in this case would actually have an effect opposite to the desired one— the number of straight-through trajectories would actually increase.

Figure 6.22 illustrates how putting in a barrier between the QPCs to completely eliminate straight-through trajectories really does not change the qualitative results. Figure 6.22(a) shows $G(E, B)$ for a rectangular dot which has a semi-circular Sinai diffuser at the top which blocks the electrons from moving directly from the left QPC to the right one without bouncing. The geometry of this dot is shown in the right inset of Fig. 6.22(b). As in the previous example, $b = 0.065\,\mu m$ for the QPC, supporting 4–5 modes over the given energy range. The dot in this case is $h = 0.5\,\mu m$ tall and $c = 0.7\,\mu m$ long. The semi-circular diffuser, which is expected to generate chaotic scattering classically, has a radius of $0.15\,\mu m$. Despite this dot's comparatively large

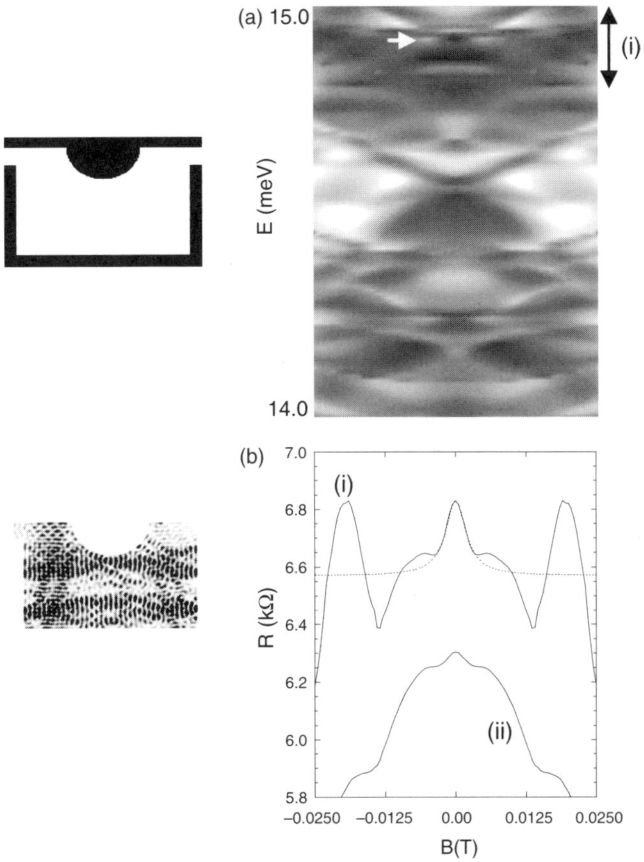

Fig. 6.22. (a) $G(E,B)$ is plotted for a 0.5 μm by 0.7 μm rectangular dot, with a semicircular Sinai diffuser of radius 0.15 μm included at the top of the dot to prevent straight-through trajectories from contributing. (b) Two traces (R versus B) obtained by averaging over the conductance. Trace (i) was obtained by averaging over the range indicated by the vertical arrow in panel (a). Trace (ii) was obtained by averaging over the entire range shown in panel (a). The dotted line is a Lorentzian fit. Left inset: the geometry of the dot considered. Right inset: the resonant state corresponding to the position of the white arrow in panel (a).

size and the ostensibly chaotic geometry, the striation pattern is qualitatively similar to that shown for the small, square dot in Fig. 6.18(a) (note the presence of both linear and parabolic striations). Indeed, as in that case and the above example, one can associate the presence of a Lorentzian line shape with a conductance resonance "eye-hole" (this is marked by the white arrow). The wave function associated with this resonance is shown in the right inset of Fig. 6.22(b). This resonant state reflects the sort of "bouncing ball" orbits noted previously. Ensemble averaging over the region marked (i) in Fig. 6.22(a) produces resistance trace (i) shown in Fig. 6.22(b), which we have fitted with a Lorentzian (dotted line) using $R_0 = 6.57\,\mathrm{k\Omega}$, $\Delta = 0.26\,\mathrm{k\Omega}$, and $\alpha\phi_0 = 43.5\,\mathrm{mT}$. Averaging over the entire range shown in Fig. 6.22(a) yields trace (ii) in Fig. 6.22(b). Thus, enlarging the averaging range has reduced the influence of this particular resonance, but a rounded peak occurs with roughly the same width. It should by noted that there are other B $= 0$ resonant minima that occur in this energy range, but their influence is weak compared to the "eye-hole" we have pointed out. This is likely because the Sinai diffuser tends to disrupt the "bouncing ball" orbits.

6.4. Results for Dots with Self-Consistent Confining Potentials: Ensemble Averaging Over Gate Voltage

Thus far, the dots simulated were confined by square well potentials. However, as mentioned in the introduction, it is well known that the actual potentials felt by electrons in quantum dots are rounded and softened due to self-consistent effects.

In order to ascertain a truly realistic picture for the actual shape of a quantum dot, we have also carried out a self-consistent simulations of these device structures [70,71,81,82,85]. The usual process of finding self-consistent solutions for confining potentials involves solving the 3D-Poisson equation, solving the 1D-Schrödinger equation and reconciling the potential and electron density in an outer iteration for self-consistency. Here, we use a standard 7-point finite difference approximation scheme and a variant of the Conjugate Gradient Squared method (CGS), the so-called Bi-Conjugate Gradient Stabilized method (Bi-CGSTAB) [95], for the numerical solution of the set of linear equations resulting from the discretization of the 3D-Poisson equation. This procedure leads to algebraic equations having a well defined structure, represented by the matrix equation $\mathbf{Ax} = \mathbf{b}$. Dirichlet boundary conditions are implied at the Schottky contacts, whereas Neumann boundary conditions are assumed at the artificial boundaries. Within the incomplete factorization scheme [96] used for the solution of the 3D Poisson equation, the original matrix $\mathbf{A}$ is decomposed into a product of lower and upper triangular matrices $\mathbf{L}$ and $\mathbf{U}$, respectively. This is achieved by modifying matrix $\mathbf{A}$ through the addition of a small matrix $\mathbf{N}$. Thus, one solves the modified system $\mathbf{LUx} = \mathbf{b} + \mathbf{Nx}$ by solving successively the matrix equations $\mathbf{LV} = \mathbf{b} + \mathbf{Nx}$ and $\mathbf{V} = \mathbf{Ux}$, where $\mathbf{V}$ is an auxiliary matrix. Fast and reliable convergence of this method is achieved by using incomplete factorization preconditioning. The convergence criterion for self-consistency is that the nodal potential energies of successive iterations differ by no more than 0.01 meV anywhere on the Poisson mesh. The energies of the bound states in the well and the corresponding electron density distribution along the growth direction are obtained from the solution of the one-electron 1D-Schrödinger equation within the density-functional formalism [97,98]. We approximate the exchange-correlation potential energy term by an interpolation formula developed by Hedin and Lundquist [99]. Because of the similar dielectric constants of the GaAs and the $Al_xGa_{1-x}As$ layer, the influence of the additional image term is insignificant and has been ignored in these simulations. The simulated heterostructure consists of 5 nm of undoped GaAs cap layer, 40 nm of $Al_xGa_{1-x}As(x = 0.26)$ Si-doped ($N_D = 1.5 \times 10^{18}\ \mathrm{cm}^{-3}$) barrier layer and 15 nm of undoped $Al_xGa_{1-x}As$ spacer layer grown on top of the GaAs buffer in which the 2DEG forms. Following Adachi [100], we take the bandgap difference as $\Delta E_G = E_G(Al_xGa_{1-x}As) - E_G(GaAs) = 1.55x + 0.37x^2$, and the conduction band offset as $\Delta E_c = 0.65 \times \Delta E_G$. For the electron effective mass in the $Al_xGa_{1-x}As$ layers we use $m^* = (0.067 + 0.083x)m_0$, and the relative permitivity is taken as

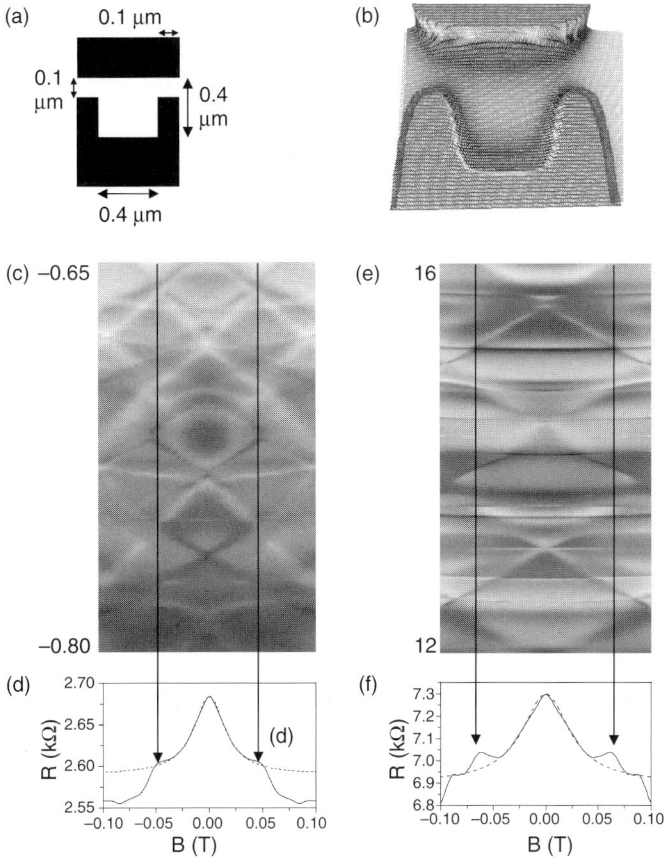

Fig. 6.23. (a) The top gate configuration for the dot used in the self-consistent simulations. (b) An example of a self-consistent potential that was obtained. (c) The conductance $G(E, V_g)$ for this dot. (d) R versus B, where R was obtained by averaging over the entire gate voltage range shown in (c). (e) $G(E, B)$ is plotted for the square dot used previously for Fig. 6.19. Note the higher energy range in this case. (f) R versus B, where R was obtained by averaging over the entire energy range shown in (e). The dotted lines in (d) and (e) are Lorentzian fits.

$\varepsilon_r = 13.18 - 3.12x$. The effective Schottky-barrier height is 0.7 eV. The binding energy of the deep donors is taken to be 50 meV. The impurities are assumed to be in thermal equilibrium with the semiconductor.

In Fig. 6.23, we show results for the dot with a self-consistent potential. For the gates on top of the heterostructure, we have simulated a dot that is *lithographically* defined as a 0.4 μm square with QPCs with openings that are 0.1 μm wide. This is shown schematically in Fig. 6.23(a). Of course, applying a bias reduces the dot size and the width of the QPCs. A picture of a typical potential obtained self-consistently is shown in Fig. 6.23(b). Note the bowl shaped region, whose effective *size* varies with V_g (the effective size of the order of 0.33×0.33 μm), although the basic shape does not change. The softening of the potential has important implications for the classical dynamics of electrons, making the behavior tend towards chaos [101]. However, even in systems whose classical dynamics is strongly chaotic, there can obviously be situations in open systems where the electrons may not dwell sufficiently long in the cavity region for the chaotic scattering to build up (the stadium results shown in the previous section immediately come to mind). With a Fermi energy fixed to 15 meV, the QPCs support 5–6 modes over the bias range considered, it is quite likely that this is the case here.

In Fig. 6.23(c), we plot $G(V_g, B)$, with lighter shading corresponding to higher G. As with the hardwall results, striations which reflect the underlying DOS are present, with the line crossings forming a "honeycomb" of diamonds. Thus, as suggested earlier, shifting gate voltage has a similar effect as altering the Fermi energy. The resistance, averaged over the entire gate voltage range shown, is plotted in Fig. 6.23(d). One obtains a resistance peak that appears linear. This to some extent is an artifact of the plotting. This trace can still be fitted quite well using Eq. (3). The dashed line is the fit using $R_0 = 2.59\,\text{k}\Omega$, $\Delta = 0.095\,\text{k}\Omega$, and $\alpha\phi_0 = 35\,\text{mT}$. Here, $\alpha^{-1} \sim A$. There are also two shoulders which are merely a reflection of the major non-zero crossings of striations in the upper panel (these are marked by arrows). Importantly, the average we have taken has included several units of the quasi-periodic diamond pattern. To further illustrate the similarities between gate voltage and energy variation, in Fig. 6.23(e), we plot $G(E, B)$ for $12 < E < 16\,\text{meV}$ again using the $0.3\,\mu\text{m}$ square dot used previously for Figs 18 and 19. However, in this energy range, there are 4 propagating modes compared to 2 in the ranges used earlier. Despite the increase in the number of propagating modes, one still sees the sort of diamond unit cells discussed at length in the previous section. In a similar manner to the soft potential dot, averaging over this large range produces a resistance peak that appears linear, but can actually be fitted relatively well with a Lorentzian form. The dashed line is the fit using $R_0 = 6.9\,\text{k}\Omega$, $\Delta = 0.31\,\text{k}\Omega$, and $\alpha\phi_0 = 55\,\text{mT}$. Here, as was the case for the soft dot, $\alpha^{-1} \sim A$. The shoulders as well coincide with major line crossings (note arrows). For both Figs 6.23(d) and 6.23(f) the resulting averaged resistance has saturated to a large extent because of the fact that several diamond units were included in the average. These results indicate that when the average is taken over a large range, the width of the central peak tends to be set by the major non-zero line crossings of the underlying spectrum. As should be evident from our previous results, averaging over smaller ranges tends to emphasize specific antiresonances, which generally yield much narrower line shapes.

6.5. Sensitivity of the Results to the Data Set Used for the Ensemble Averaging

Earlier, we discussed how two different experiments on circular cavities yielded significantly different results for α^{-1} (different by a factor of three). This of course could be because the dynamics of the dots in the two different experiments were indeed different (however, it should be mentioned that Berry et al. [92] got narrow peaks for two separate dots). It could also be due to an effect similar to what we have already discussed—changing the sampling range can dramatically effect the line width. Figure 6.24 indicates another possibility that we uncovered [70]. The dashed line is the resistance average previously shown in Fig. 6.23(d) for the soft potential dot. The solid line is the result of an "incomplete" average. In other words, to compute the average

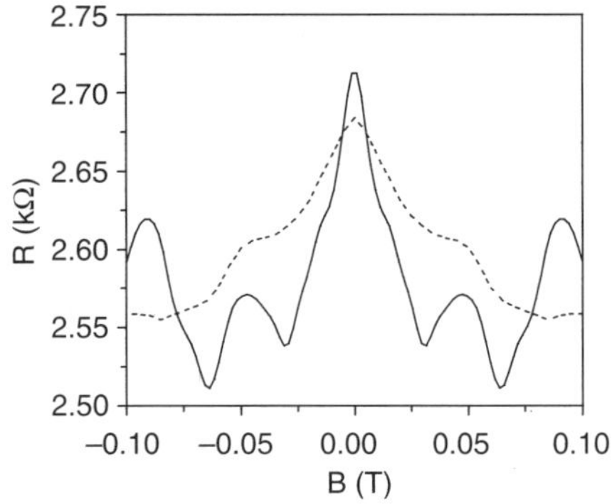

Fig. 6.24. (a) R versus B, where, for the dashed line, R was obtained by averaging over the entire gate voltage range shown in Fig. 6.23 (a) (this trace is the same as the solid line shown in Fig. 6.23(d)). The solid line represents the result obtained by doing an average over a limited number of traces.

in this case, we have only included 10 equally spaced traces ($\Delta V_g = 0.015$ V) out of the 151 total that were used to generate Fig. 6.23(c). As is evident, the peak is much narrower. While we have omitted it for the sake of clarity, a Lorentzian fit now yields $\alpha^{-1} \sim 3A$ in this case. In experiment, such incomplete or limited averaging can be unavoidable for practical reasons. Berry et al., who obtained the comparatively narrow peaks, averaged over only five traces to obtain their results.

7. Direct Comparisons of Theory with Experiment

7.1. Magneto-resistance Measurements on a Nominally Square Quantum Dot

The measurements that we shall discuss in this section were performed on dots defined on the surface of GaAs/AlGaAs heterojunction wafers, which were patterned into Hall bars to allow four-terminal resistance measurements to be made [81]. While we will concentrate on these structures, it is important to note that fully equivalent results have been also obtained by an in-plane gated structure in a InGaAs/InP heterojunction [102] and in *etched* GaAs/AlGaAs dots [103]. In both these cases, because of the differences in the way confinement is achieved, the resulting potentials were of the "hardwall" variety. Thus, in concurrence with the simulation studies discussed in the preceding sections, the experimental results we outline here are not unique to the soft walled potentials that are produced by electrostatic confinement.

For the dot illustrated in Fig. 6.24, the carrier density and mobility of the GaAs/AlGaAs wafer were found to be $4.5 \times 10^{11}\,\text{cm}^{-2}$ and $400{,}000\,\text{cm}^2/\text{Vs}$, respectively, at 4.2 K and after illumination with a red light-emitting diode. Magneto-resistance measurements were made at the base temperature of a dilution refrigerator, using standard lock-in detection and low constant current (≤ 1 nA) to avoid electron heating. In this particular study, a lithographically square dot of side 0.4 μm was used

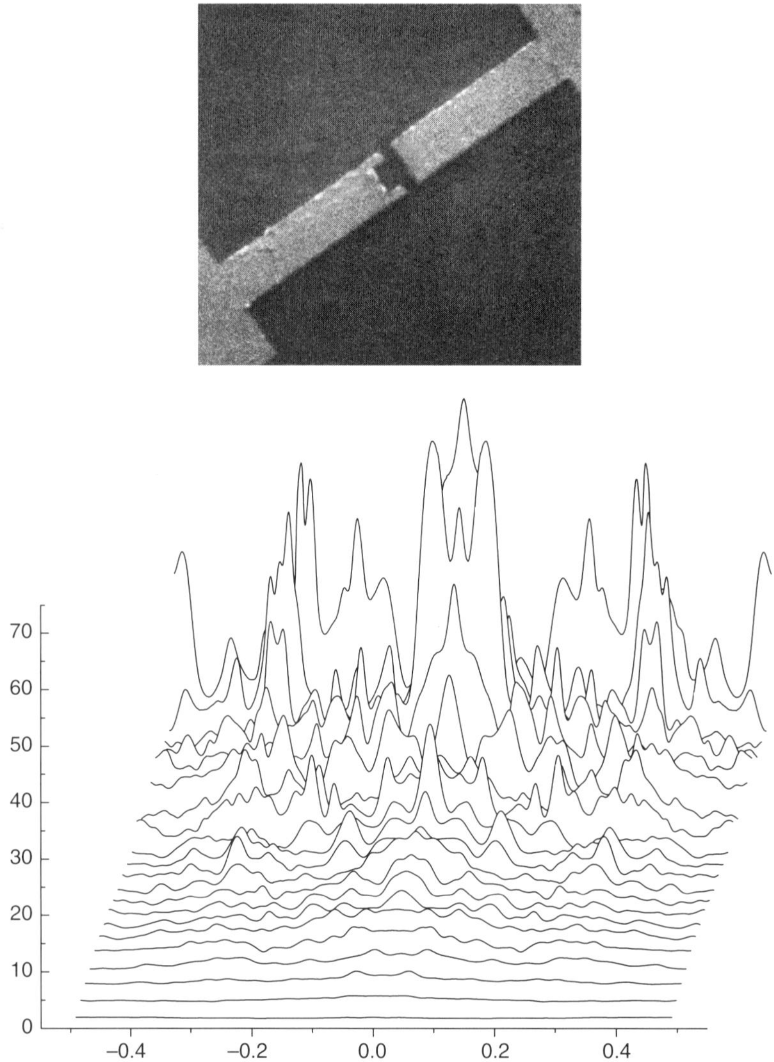

Fig. 6.25. Top panel: SEM micrograph of an actual dot studied experimentally. The central cavity has a lithographic dimension of 0.4 μm. Bottom panel: The measured variation of the low temperature (0.01 K) magneto-resistance of the 0.4 μm quantum dot with gate voltage. These magneto-resistance measurements were taken at twenty two unequally-spaced gate voltages ranging from −0.3610 V to −0.4335 V.

(Fig. 6.25 (top panel)). Due to the electrostatic fringing fields that develop around the gates, however, the actual size of the dot formed in the electron gas was somewhat smaller and was estimated to be of order 0.25–0.3 μm, based on the observation of Aharonov–Bohm oscillations in the edge state regime. Translating to an equivalent temperature scale, these sizes yield an average level spacing (Δ) in the dot ranging from 1–1.3 K.

In Fig. 6.25 (bottom panel), we show the magneto-resistance of a split-gate dot and its variation with gate voltage. Similar to the behavior reported in other experiments [104], a zero-field peak can be resolved in some of the traces, but the amplitude of this varies non-monotonically with gate voltage. Indeed, in certain cases it is difficult to resolve any evidence for a peak, whatsoever. Another feature of the data is the emergence of pronounced *side-peaks* at non-zero magnetic fields. As can be seen in Fig. 6.26, in which we plot selected traces from Fig. 6.25, these side-peaks may rise to resistance values comparable to the height of the zero-field peak itself. From studies of dots with lithographic dimensions ranging from 0.4 to 2 μm, both the non-monotonic evolution of the zero-field peak and the development of side-peaks at finite fields are found to be characteristic features of the low temperature magneto-resistance. In line with the discussion from previous sections, these distinct features are thought to result from magneto-probing of the discrete DOS of the open dots.

For the simulations used to compare with these particular experimental results, we used a confining potential obtained self-consistently, similar to those used to generate Fig. 6.23 in the previous section. Referring back to the discussion on thermal averaging and effective temperatures at the beginning of the last section, an *effective* temperature of 0.3 K was assumed in the simulations, which made the closest comparison with the experimental results. At these low temperatures, the phase breaking time (τ_ϕ) is of order 100 ps [29,55,64,103,104]. Assuming dephasing to give rise to an energy broadening of order $\hbar/k_B\tau_\phi$ ($\approx$0.1 K), and noting that remnant disorder in the dot will further mix its discrete spectrum, this choice of effective temperature is likely realistic. In this regard, simulation has indicated it that it is the order of magnitude, rather than the absolute value, of the effective temperature that is crucial to reproducing the behavior found in experiment [81].

As can be seen by comparing the experimental and theoretical curves in Fig. 6.26, the simulations were clearly successful in reproducing the main features of experiment. While we do not suggest that this agreement is perfect, we nonetheless believe it to be highly suggestive. In this regard, it is worth emphasizing that the experimental curves were obtained by varying *gate voltage*, which changes the size of the dot, and the width of its lead openings, in addition to modulating the electron density. In contrast, in these particular simulations, since we wished to explore broadening effects specifically, different *Fermi energies* have been used in a dot whose profile is kept *constant*. Consequently, we should not expect to obtain *perfect* agreement with the results of experiment. From Fig. 6.26, however, it is clear that the simulations are able to account very well for the overall magnitude and magnetic field scales of the features seen in the experimental curves.

We have mentioned already that conductance measurements of open quantum dots can be thought of as providing a selective spectroscopy of their DOS. Once again, the basic idea is that a magnetic field or gate voltage may be used to sweep discrete dot states past the Fermi surface, giving rise to an associated modulation of the conductance. The number of resonances that contribute to the conductance at any magnetic field is thus determined by the details of the density of states and the width of the window over which energy averaging is effective. At the low temperatures of interest here, the energy window is relatively narrow and the magneto-resistance should thus exhibit features associated with the movement of specific groups of dot resonances past the

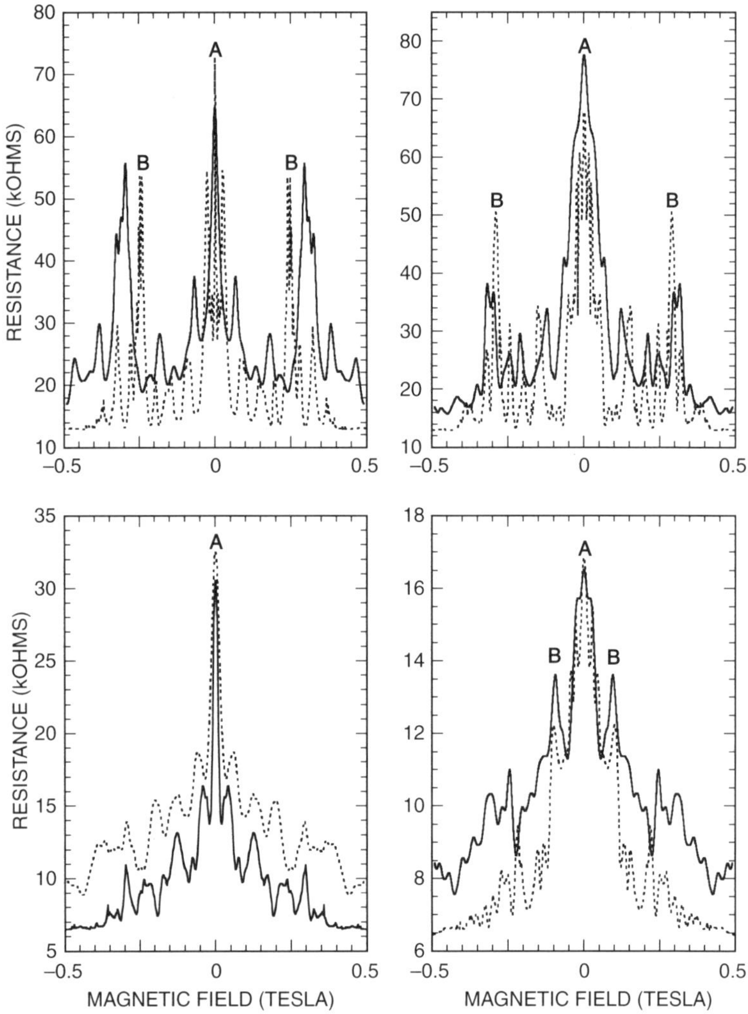

Fig. 6.26. In this figure, selected magneto-resistance curves from Fig. 6.25 are plotted (solid lines) and compared with the results of simulations (dashed lines). Upper left (experiment): $V_g = -0.4295$ V. Upper left (simulation): $V_g = -1.05$ V, $E_F = 15.75$ meV. Upper right (experiment): $V_g = -0.4260$ V. Upper right (simulation): $V_g = -1.05$ V, $E_F = 16.1$ meV. Lower left (experiment): $V_g = -0.4000$ V. Lower left (simulation): $V_g = -0.922$ V, $E_F = 15.1$ meV. Lower right (experiment): $V_g = -0.3885$ V. Lower right (simulation): $V_g = -0.922$ V, $E_F = 15.5$ meV. In these plots, the symbols A and B are used to denote the major features in the magneto-resistance (*see text for further discussion*).

Fermi surface. Referring back to Fig. 6.10 and Figs 6.18–6.23, it can be seen that the resonances tend to form boundaries between regions where the conductance is relatively slowly varying. In these flat regions, the conductance may be either high or low. When a low conductance region exists at zero magnetic field, a zero-field peak

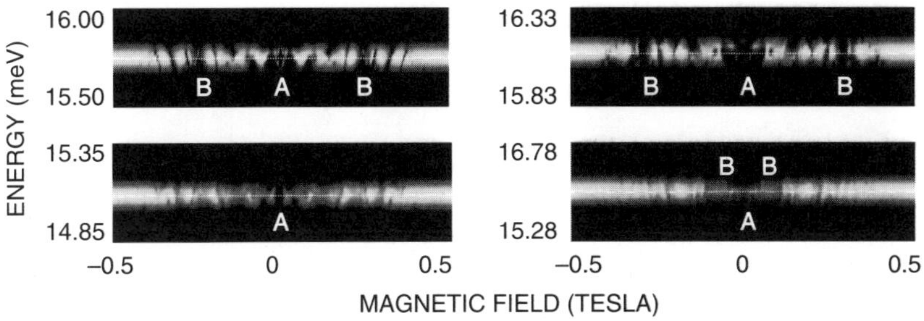

Fig. 6.27. Here, the conductance is convoluted with the derivative of the Fermi function and then plotted as a function of energy. The four figures correspond, in the same sequence, to the computed magneto-resistance plots shown in Fig. 6.26, while the symbols *A* and *B* denote the major magneto-resistance features seen in Fig. 6.26 (*see text for further discussion*).

will arise whose width is set by the resonance lines that form the boundary for that region. The large side-peaks noticed above occur for similar reasons. Conversely, one can obtain a zero field minimum when the Fermi energy lies in a region of relatively high conductance that is bounded by resonant minima. In the experiment here, the gate voltage is swept while the width of the energy window is (presumably) held constant. At each gate voltage, a unique portion of the density of states is therefore sampled, and it is this variation that is thought to give rise to the magneto-conductance modulations.

To further illustrate these notions, we show in Fig. 6.27 shaded conductance contours that were used to calculate the magneto-resistance curves in Fig. 6.26. These contours were obtained by convolving the magneto-conductance with the derivative of the Fermi function at a number of different energies. Brighter regions in the contours correspond to higher conductance and, at the low temperatures of interest here, only a narrow window contributes effectively to the energy average. An important feature of the contours is the series of bright and dark lines that pass through the Fermi level at different magnetic fields. These lines trace the evolution of the density of states as magnetic field and Fermi energy are varied. It is the presence of these lines that sets the position and widths of the major magneto-resistance peaks seen in experiment and theory. The width of the central peak found in the lower left panel of Fig. 6.27 (marked A), for example, is set by a pair of resonance lines that bound the local conductance minimum in the center of the lower left panel of Fig. 6.26. Examining the other panels in Fig. 6.26, it evident that similar features limit the width of the central peak in all other cases depicted in Fig. 6.27. Similarly, the side-peaks seen in the upper two plots of Fig. 6.27 result from the presence of a low conductance region that is bounded by resonances which cross the Fermi level near $\pm 0.3\,\mathrm{T}$ (Fig. 6.26, upper panels). Interestingly, the side-peaks are seen to be strongest in situations where the dot QPCs are configured to support just a few modes, a property that is apparent in both experiment and theory. Since the side-peaks might easily be mistaken as arising from classical magneto-focusing, it is perhaps worth emphasizing that such an effect can be confidently ruled out here. Firstly, the magnetic field values at which the peaks occur do not seem to correspond to any obvious commensurability condition in the dot. Secondly, the peaks rapidly wash out with increasing temperature (not shown here), while focusing-related features are known to persist to very much higher temperatures [105–108].

7.2. Periodic Conductance Fluctuations as a Function of Gate Voltage: Comparing Two Different Lead Orientations

We have also studied the effects that varying gate voltage has on the conductance fluctuations both experimentally and theoretically [82]. As part of this particular study, we were interested in seeing what effect, if any, changing the orientation of the QPCs would have. Thus in addition to continuing to examine a dot with QPCs aligned at the top boundary, we also considered a dot for which the QPCs are staggered, as shown in the micrograph at the top of Fig. 6.28.

The main panel of Fig. 6.28 shows the measured variation of conductance with gate voltage for the dot with staggered QPCs. Once again, of particular interest are the fine oscillations or fluctuations which ride on top of the monotonic background (Fig. 6.28, lower inset). The oscillations are observed over the entire range of gate voltage for which the dot is defined and persist to conductance values as high as $15\,e^2/h$. After subtracting the background from the total conductance variation, the series of regular oscillations shown in the upper inset of Fig. 6.28 is obtained. Note how the average amplitude of the oscillations does not vary significantly with gate voltage, *even when the leads are opened so that the dot conductance exceeds* $10\,e^2/h$. This characteristic was confirmed in studies of other dots, which showed similar oscillations to those discussed here. In the inset of Fig. 6.29, for example, we show similar behavior in a dot with aligned leads (we have already subtracted a background here, which varied from 1–8 e^2/h over the voltage range shown). A Fourier analysis of the oscillations in Figs 6.28 and 6.29 confirms their periodic nature and reveals the presence of a small number of dominant peaks at isolated frequencies (Figs 6.29 and 6.30). Crucially, these peaks occur at different *frequencies* in the two dot geometries, an issue we return to in detail below.

To account for the behavior seen in experiment as a function of gate voltage, the simulations of electron transport involved computing the self-consistent profile of the dots at more than 300 gate voltages that uniformly span the ranges considered in experiment [82]. In contrast to what was done for Fig. 6.26 and 6.27, calculations were performed for a fixed Fermi energy chosen by matching the carrier density in the reservoirs to the bulk value inferred from experiment.

Computed conductance oscillations are compared to experiment in the insets of Figs 6.28 and 6.29. The simulations account well for the frequency content found in experiment, a conclusion that is confirmed by comparing the Fourier spectra of the measured and computed oscillations (Figs 6.29 and 6.30). In particular, the *number* of Fourier peaks seen in experiment is matched precisely in the simulations, which also closely reproduce the frequencies at which the peaks occur. *These peaks in fact correspond to the frequencies at which specific sets of scar-like wavefunction features recur in the dots.* In the dot with aligned QPCs, for example, three different scar-like features are found to recur (Fig. 6.29, lower inset). The "bouncing-ball" features (labeled "1" and "2") recur at a frequency of $15\,\mathrm{V}^{-1}$, in good agreement with the position of the fundamental peak in Fig. 6.29. The third scar-like feature found in

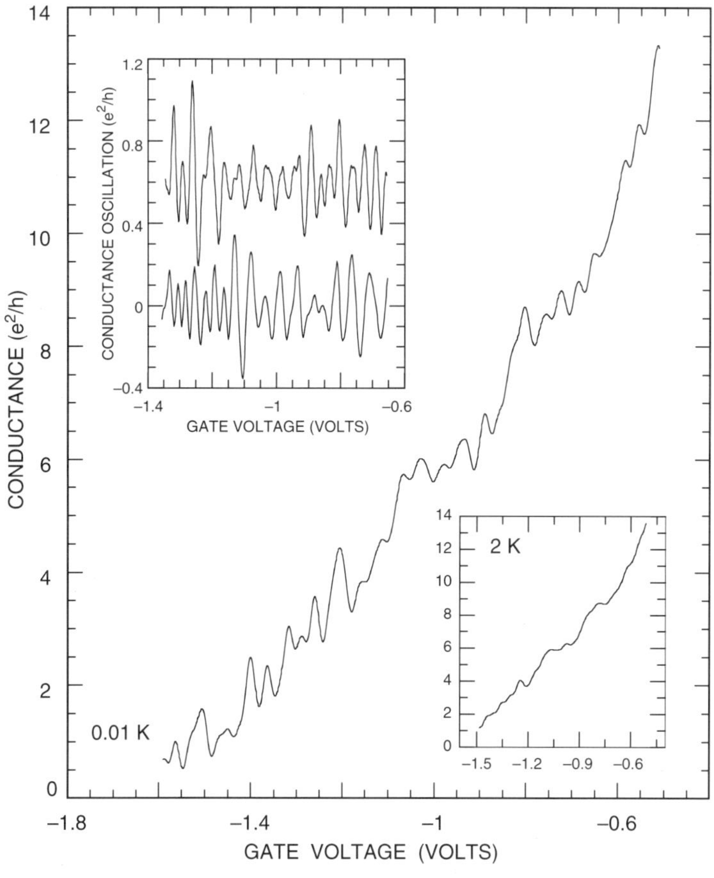

Fig. 6.28. Upper panel: SEM micrograph of another dot studied experimentally, one with a staggered QPC configuration. Main panel: conductance-gate voltage characteristic of the staggered dot. Lower inset: same plot, measured at 2 K. Upper inset: experimental and simulated conductance oscillations for the dot. The upper curve is experiment and is offset by $+0.6\,e^2/h$.

this dot ("3") shows a whispering-gallery form and recurs at a frequency of 40 V^{-1}, corresponding to the final major peak in Fig. 6.29. For the dot with staggered QPCs, the conductance oscillations in this are dominated by a single frequency component (Fig. 6.30). Similarly, the simulations show that *only one* scar recurs significantly in

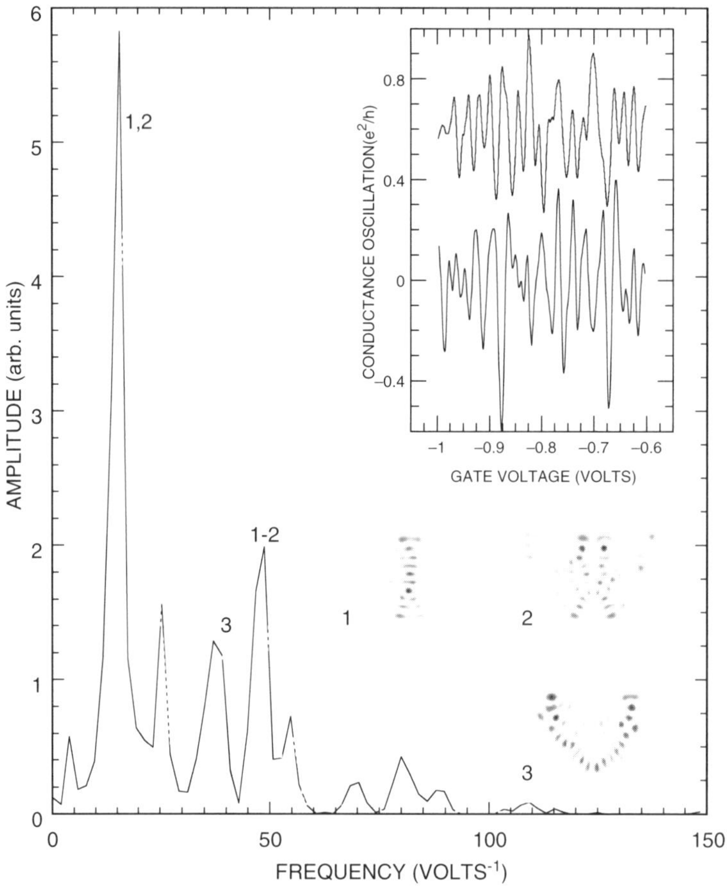

Fig. 6.29. Main panel: Fourier spectra of measured and computed oscillations in the aligned dot. Dashed line is theory. Upper inset: experimental and computed conductance oscillations for the dot. The upper curve is experiment and is offset by $+0.6\,e^2/h$. Lower inset: recurring scars of this dot. The numbers here show the association between specific scars and Fourier peaks and dark regions correspond to high probability.

this dot (Fig. 6.30, lower inset) and that the frequency of recurrence of this whispering-gallery feature matches that of the dominant peak observed in the Fourier spectra of the computed oscillations (Fig. 6.30).

The results above show that *quantum dots with different QPC alignments may be used to excite different resonant states, giving rise to measurable transport results.* These results are thus entirely consistent with the earlier theoretical results for square well confinement outlined in section 3 and Figs 6.6 and 6.7.

The self-consistent calculations show that increasing the negative gate bias reduces the effective size of the dots, but leaves their overall shape relatively unaffected. The conductance oscillations arise as this size variation forces successive dot states to sweep past the Fermi surface, thus modulating the coupling between these states and

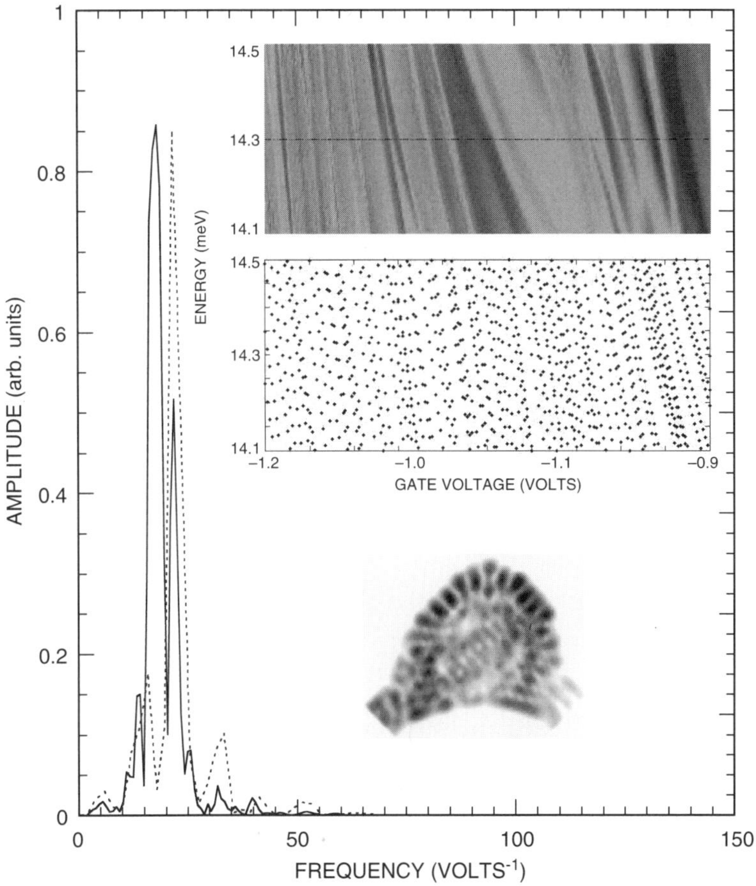

Fig. 6.30. Main panel: Fourier spectra of measured and computed oscillations in the staggered dot. Dashed line is theory. Lower inset: recurring scar found in this geometry. Upper inset: Variation of the staggered-lead dot conductance with gate voltage and energy (upper plot). Grayscale varies from 0–2 e^2/h and the dashed line shows the Fermi energy. Also shown (lower plot) is the evolution of the *closed* dot eigenstates with gate voltage. *See text for further details.*

the point contact QPCs. The connection between the eigenstates of the closed dot and the conductance of its open counterpart can be seen in the upper inset to Fig. 6.30, where we plot the energy spectrum of an *isolated* dot as a function of the gate voltage. These levels were computed using the self-consistent profiles for the dot with staggered leads, after closing the leads at their narrowest points. Note how these levels shift almost linearly over large ranges of gate voltage, which presumably accounts for the periodicity of the conductance oscillations. For comparison, the computed conductance variation for the *open* staggered dot is also shown in Fig. 6.30. While the linear striations that run through this grayscale plot follow the motion of certain closed-dot states, it is clear that *not all states of the isolated dot give rise to a marked modulation of conductance in the open dot*. Instead, this conductance provides a *filtered* probe of the density of states of the closed dot, again completely consistent with the viewpoint we have established in the previous sections.

Coulomb blockade may also give rise to conductance oscillations when the gate voltage is varied [11,109,110]. Previously, it was argued that this effect is obscured in small dots if their QPCs are configured to support one or more propagating modes [11]. However, Coulomb blockade has been found to persist in semi-open dots, in which one lead is configured as a tunnel barrier while the other transmits a *single* mode [111,112]. In yet another study, evidence for the persistence of Coulomb blockade to dot conductances as high as $4e^2/h$ has even been presented [110]. This latter observation seems to be unique to the very special dot geometry they considered, *which we emphasize is very different to the geometry studied here*. Instead, it is fairly well established from experiment that in dots with lead realizations similar to those studied here, the Coulomb blockade is dramatically suppressed when *both* leads are opened to support one mode [11,109]. Nonetheless, *the oscillations in the experiment here are found to persist to dot conductances as high as* $15e^2/h$ *with little noticeable variation in amplitude* (Fig. 6.28). The details of these oscillations are also well accounted for by what is essentially our single electron picture of transport. We therefore believe that the oscillations *do not* result from some remnant of the Coulomb blockade effect.

8. An Alternate Semiclassical Interpretation of Transport in Open Quantum Dots: Dynamical Tunneling

8.1. A Classical Phase Space Analysis of a Soft Walled Open Quantum Dot

In the preceding sections, we have provided theoretical and experimental evidence that long-lived resonant states corresponding to single eigenstates in an analogous closed system can play a highly significant role in determining the transport properties of very open quantum dots. Such resonant effects are beyond the range of applicability of the semiclassical theory of transport that has been typically applied to these systems. Besides the violation of the semiclassical requirements that the dots levels be completely broadened and that the orbits significant to transport be comparatively short, the scarring type effects we have observed in conjunction with the resonances indicate that it is *closed, periodic orbits, not connected to the QPCs* that are actually dominant. Importantly, in a purely classical open system, these periodic orbits would normally be *inaccessible* to any particle entering the system from the outside.

The results summarized thus far in this review, as well as those found elsewhere in other studies [101,113–121], indicate that hyperbolicity is an unusual dynamical feature in 2DEG quantum dots. Thus, for typical dots, the observed properties of the transport, such as the periodic fluctuations we have discussed at length, simply cannot be explained by RMT-like approaches. In regard to maintaining a semiclassical viewpoint of these results, it is necessary to take into account the quantum-mechanical tunneling of the electron through the KAM islands (dynamical tunneling), which corresponds to classically forbidden trajectories that are not taken into account in the usual semiclassical approach. We have in fact found that the periodic resonant features in the conductance of open dots can be correlated to situations in which a semiclassical

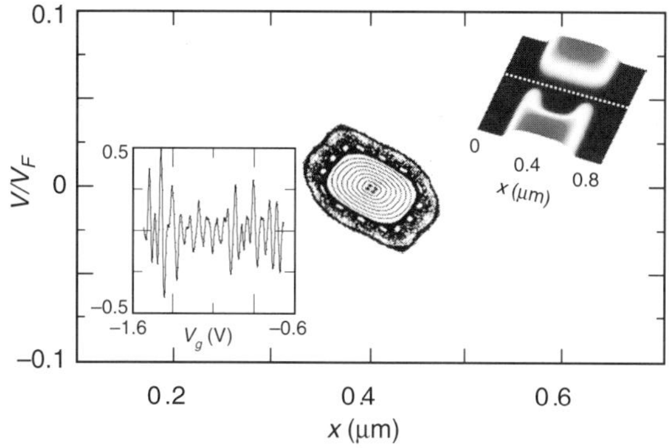

Fig. 6.31. Poincaré section for $V_g = -0.6\,\text{V}$, with Fermi energy $E = 14.3\,\text{meV}$. The section is taken at $y = 0.3\,\mu\text{m}$. Here, v is the x-component of the velocity, and is given in units of the Fermi velocity. The "chaotic sea" surrounding the island is the plot of one single trajectory that stays a long time near the island before escaping. The phase-space area of the island is about $\hbar$. The upper right inset shows the computed potential profile of the dot for a gate voltage of $-0.6\,\text{V}$. The lower-left inset shows the measured conductance fluctuations observed by varying the gate voltage.

quantization condition is satisfied for periodic orbits with low period, including stable orbits within the KAM islands [122]. To illustrate this, the example we shall use is the quantum dot with aligned QPCs that demonstrated periodic conductance fluctuations as a function of gate voltage, as shown in Fig. 6.29.

The self-consistently computed profile of this dot for a gate voltage of $-0.6\,\text{V}$ is shown in the upper right inset of Fig. 6.31. To study the classical dynamics as a function of gate voltage, we simulate the classical motion of electrons under the influence of the potentials generated previously for the quantum calculation, with the total energy of the electrons being set at the Fermi energy. We use a Poincaré section (this is taken along the dotted line drawn across the potential in the inset) to visualize the dynamics. For hyperbolic dynamics, all orbits are unstable, and all initial conditions escape the dot in a finite time (except for a set of null measure). The Poincaré section of almost every initial condition consists then of a finite number of points, and there are no stable orbits or KAM islands in phase space. However, in the case of this dot, we see that the dynamics is clearly non-hyperbolic, with a large KAM island dominating the phase space. This island is centered on a period-1 orbit that bounces back and forth vertically at the center of the dot, a bouncing-ball orbit. Clearly, RMT cannot be applied to this system. Given the results previously outlined in this review, we believe the features of Fig. 6.31 are typical of most quantum dots.

There are some important papers on the semiclassical theory of nonhyperbolic systems [101,113,115,117,119,120,123], but they also predict stochastic conductance fluctuations (with different statistical properties than in the hyperbolic case), and they fail to predict the regular fluctuations seen here. The reason for this is that the usual semiclassical theory only takes into account interference between the classically allowed trajectories, and hence ignores, completely, the possibility of electron tunneling into the KAM island (corresponding to diffraction in optics). This effect is negligible only if the electron de Broglie wavelength, λ_e , is much smaller than the size of the cavity. For dots with typical size of less than 1 μm, λ_e is of the order of one-tenth of the cavity length, and therefore tunneling cannot be neglected, as we will show. On the other hand, 1 μm is small enough that some important semiclassical concepts such as the Bohr-Sommerfeld quantization of periodic orbits can still be applied. We have come to believe that the observed periodic conductance fluctuations are, in fact, due to dynamical tunneling [124]. Specifically, there is a probability that an incoming electron tunnels inside the KAM island. If the electron's energy and the system's parameters are such that the semiclassical quantization condition (detailed below) is satisfied for a low-period stable periodic orbit within the island, there is a sharp resonance. As the system's parameters (in this example, the gate voltage) change, these resonances occur with a period given by the position of the peaks in the Fourier transform of the fluctuations. These arguments are made quantitative below. We note that dynamical tunneling is of great importance in many fields, in particular in cold-atom physics [125–127]. In fact, recent studies have demonstrated the existence of a mixed phase space in the atomic optical billiard [124], so that the results shown here may even be important for such systems.

8.2. Semiclassical Quantization: Quantifying the Relationship between the Resonant Behavior and Classical Periodic Orbits

Because KAM separatrices are classically impenetrable, the classical dynamics restricted to the island shown in Fig. 6.31 is that of a closed system. As noted in section 2, for closed systems, it is well-known from the expansion of the Gutzwiller trace formula that each stable orbit generates a series of delta functions in the density of states, corresponding to energies for which the resonance condition holds [33,36,128]:

$$S_{\text{eff}} = S + \frac{\omega}{2\pi}\left(m + \frac{1}{2}\right) + \frac{\lambda}{4\pi} = n \tag{32}$$

where $S = 1/\hbar \int \mathbf{p} \cdot d\mathbf{q}$ is the action along the periodic orbit in units of Planck's quantum, ω is the stability angle of the orbit, λ is the Maslov index and $m, n = 0, 1, 2, \ldots$ are integers. In a closed system, for each energy satisfying Eq. (32) for some m and n , there is a delta function in the density of states, corresponding to a discrete energy level. Our system is open, however. Just as an incoming electron can tunnel in, an electron within the island may tunnel out and escape. An electron within the island is thus in a metastable state, and may "decay" by escaping. This causes the peaks in the density of states to broaden, and their width is inversely proportional to the average time it takes for an electron to tunnel out of the island. If this time is not too short, the peaks will be sharp enough to be resolved, and their positions are given by Eq. (32).

Consider first the case $m = 0$. Note that is the longitudinal quantum number, and it counts the number of nodes in a resonant wave function along the orbit. Within the island of Fig. 6.31, there are infinitely many periodic orbits, but only the lowest-period ones are expected to be resolved (high-period orbits generate peaks that are too closely spaced to be resolved, even more so with the broadening of the levels caused by the quantum metastability of the system). Clearly, the most important orbit in the island is the period one bouncing-ball orbit corresponding to the fixed point in the center of the island. We have calculated S_{eff} as defined in Eq. (32) as a function of the gate voltage for this island. The result is shown in Fig. 6.32, and we see that the points fall reasonably well on a straight line, which corresponds to a periodic recurrence of the resonance, and a periodic oscillation in the conductance. Since a resonance happens each time S_{eff} goes through an integer, the (absolute value of) the slope of this straight line gives directly the semiclassical prediction for the frequency of the conductance oscillations. From the results shown previously, experiments yield a frequency of 15 V^{-1} while fully quantum mechanical simulation give a value of about 16 V^{-1}. Here, we find the semiclassical value for this frequency to be 16.4 V^{-1}, in remarkable agreement. This gives yet more evidence that this Fourier peak corresponds to recurring tunneling resonances, and cannot be explained by the usual semiclassical theory, where tunneling is ignored. Figure 6.32 shows a sequence of open dot wave functions, each corresponding to a different value of the gate voltage for which there was a resonance in the fully quantum mechanical simulation. Each wave function in this series has amplitude enhanced along a bouncing ball orbit. The recurrence frequency of the wave functions showing this pattern was determined to be 16 V^{-1}, in agreement with the preceding Fourier analysis as well as the semiclassical prediction. As the bouncing ball orbit is classically inaccessible from the outside, this gives more support that we are observing a manifestation of dynamical tunneling in this system.

We now consider the general case, when m is any positive integer. The second term in S_{eff} represents the quantization of the component of the motion transversal to the periodic orbit. This means that for each n there is actually a (theoretically infinite) set of resonances, labeled by n, similar to a vibrational band in a molecule. Assuming that S_{eff} changes linearly with the gate voltage V_{g} (which, as Fig. 6.32 shows, is a good approximation), we can estimate the separation between two resonances with consecutive transversal quantum numbers:

$$\Delta V_{\text{g}} \approx \frac{\omega}{2\pi \left| \mathrm{d}S_{\text{eff}}/\mathrm{d}V_{\text{g}} \right|} \tag{33}$$

where it is assumed that ω does not change much between the two resonances. Although ω depends on n, its values are found numerically to lie between 1 and 2 radians . Using the value of $\left|\mathrm{d}S_{\text{eff}}/\mathrm{d}V_{\text{g}}\right|$ derived from Fig. 6.32, this gives a ΔV_{g} between 0.01 and 0.02 V. Just as n counts the number of nodes along the orbit, m counts the number of nodes across it, so we expect from all this that for each n (corresponding to a fixed number of longitudinal nodes in the resonant wave function), there is a set of scar-like wave functions having $m = 0, 1, 2 \ldots$ transversal nodes, separated by a gate voltage interval of ΔV_{g}. In fact, such recurring pairs of wave functions were observed in the quantum mechanical simulation for the bouncing-ball orbit, separated by a ΔV_{g} of about 0.02 V, which agrees well with our prediction. One such pair is shown in the bottom inset of Fig. 6.32, and we see from this figure that they correspond to $m = 0$ and $m = 1$ (the left and center figures in the inset, respectively). They are also found for the other values of n, and they recur at the same period, as predicted. We stress that this

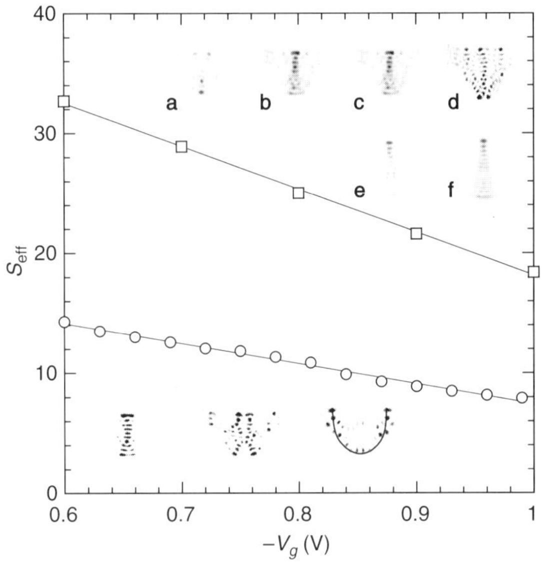

Fig. 6.32. Effective action as a function of the gate voltage, for the stable (circles) and unstable (squares) orbits. The series of insets at the top correspond to a series of resonant wave functions obtained by a full quantum mechanical simulation. The members of the this sequence labeled a-f correspond to the respective gate voltages −0.928 V (a), −0.862 V (b), −0.807 V (c), −0.755 V (d), −0.700 V (e), and −0.640 V (f). The bottom row of insets shows a pair of closely spaced wave functions corresponding to the stable bouncing-ball orbit (left and center images), as well as a wave function scarred by the unstable orbit that have been overlayed (right image).

phenomenon cannot be explained without tunneling, since it requires that the electrons have access to the stable orbit, which is classically (and also semiclassically) forbidden. The concentrated wave function corresponding to higher values of m ($m = 2, 3, 4\ldots$) are not resolved by the simulation, presumably because they have a short life-time; from Fig. 6.32, we see that the $m = 1$ wave function is quite "fat", and as m increases it becomes even more so, since the number of nodes increases. For $m > 1$, we can expect that the wave function will have a large overlap with that outside of the island, which corresponds to orbits that escape quickly, thus making these resonances very short-lived. As a consequence, they are very much broadened, and unresolved by the simulation.

Although we focused on the stable orbits, unstable orbits are also present in the system, and they contribute to the density of states. A resonant wave function corresponding to the main unstable orbit is displayed as the right image in the bottom inset of Fig. 6.32. A classical trajectory analysis suggests that the orbit giving rise to this "whispering-gallery" scar-like feature is guided by the soft walls around the perimeter of the lower section of the dot, bouncing from the upper wall at two points, located close to the two lead openings. The orbit in question has been drawn over the wave function in the figure inset. The resonant condition unstable periodic orbits is given by Eq. (32) without the ω term [128]. This means that unstable orbits do not give rise to the subband of resonances associated with m. The Fourier transform of the conductance oscillations, as shown in Fig. 6.29, yields a peak at $V_g \sim 37V^{-1}$, corresponding to an unstable periodic orbit of period 1. Figure 6.32 shows a plot of S_{eff} versus V_g for this orbit, and from the slope we get a recurrence frequency of 36.3 V^{-1}, again in very good agreement with the experimental result. The scar-like wave functions related to these resonances are seen in the quantum-mechanical simulation, but no subband is seen, confirming our predictions. The other main periodicities found in the conductance correspond to harmonics of the main stable and unstable resonances studied above. Note that according to Hufnagel et al. [120], isolated resonances are predicted to arise from the chaotic part of the phase-space outside the islands (see Fig. 6.31), but these can be detected in the limit $\hbar \rightarrow 0$, i.e. larger quantum dots [81].

9. Summary

We have shown that transport through open quantum dots may be mediated by single eigenstates, even when the leads allow several propagating modes. Moreover, there is no simple transition between "closed" and "open" regimes and one can have "insulating" behavior, i.e. tunneling through resonant states, overlaying a "metallic" background of completely broadened levels. The presence of these resonances tends to lead to conductance fluctuations that are periodic in magnetic field or gate voltage. Additional level broadening should arise at the finite temperatures where experiment is performed. However, at temperatures below a degree Kelvin, the magnitude of this additional broadening should be sufficiently small to allow the effects discussed here to be resolved, and the excellent agreement between theory and experiment we have summarized here strongly suggests that this is indeed the case. Even with a significant amount of thermal or other types of ensemble averaging, peaks in the magneto-resistance occur that bear the fingerprint of this underlying resonance structure. Averaging does not destroy this dependence, in large part because so many features recur periodically. These results have important implications. In particular, the RMT based semiclassical approach [19,21,22,26–30,42] commonly applied to open dots has a far more limited range of validity than previously thought, as it assumes a completely broadened level structure a priori. Since it ignores the resonant structure that can actually dominant the fluctuations, any predictions based on this approach must be called into question. It should be noted that similar conclusions have been reached by Blomquist and Zozoulenko [129]. Besides producing quantum simulation results similar to those shown here, they have done a further analysis of the traditional semiclassical approach and they have found that it has additional *major* difficulties, such as failing to satisfy the requirement for current conservation.

In regards to semiclassical predictions, the one example we have focused on is the so-called weak localization peak in the magneto-resistance. We have found that the mere presence of a particular line shape, such as linear or Lorentzian, by itself cannot be used to infer the dynamics within the dot (such as regular or chaotic). An entire range of lineshapes can be produced by a single dot regardless of its geometry. Indeed, whether we use a confining potential whose classical dynamics is expected to be regular (the square), chaotic (the stadium) or mixed (soft walls), we find the same general resonance behavior. This convergence of behavior is likely attributable to the QPCs always creating an intermediate situation that can neither be purely regular, since the QPCs inevitably perturb the system, or purely chaotic, since the QPCs let the electrons following chaotic trajectories to escape after a comparatively short time. To explain the resonance effects we have discussed here on a semiclassical level, one must take into account dynamical tunneling of electrons into and out of *periodic* orbits that are trapped inside the dot.

In closing, we should mention that much of the physics we have covered in this review is not unique to quantum dots. *Resonance trapping* is a term that has been applied to such phenomena when observed in other physical systems, and an extensive literature exists on the subject (the paper by Rotter et al. [130] is a good starting point). Amongst other examples, it has been noted in the context of nuclear physics [131] and microwave cavities [132]. In regard to the latter, a very detailed comparison between microwave cavities and quantum dots has been performed [133]. In that study, the microwave cavity shape chosen was based on one of the self-consistent quantum dot potentials used here in section 4 and section 5. The behavior was found to be extraordinarily similar, with periodic fluctuations occurring in the microwave transmission, and the corresponding appearance of resonant states with wave function amplitude maximized along classical periodic orbits that recurred with the same periodicity as the fluctuations. Importantly, in these microwave studies, the resonant wave functions were actually mapped out *experimentally*, thus providing even further validation of the ideas we have expressed in this review.

10. Acknowledgment

Work supported by the Office of Naval Research.

References

1. D.K. Ferry and S.M. Goodnick, *Transport in Nanostructures*. Cambridge University Press, Cambridge, (1997).
2. T.J. Thornton, M. Pepper, H. Ahmed, D. Andrews, and G.J. Davies, *Phys. Rev. Lett.* **56**, 1198 (1986).
3. B.J. van Wees, H. van Houten, C.W.J. Beenakker, J.G. Williamson, L.P. Kouwenhoven, D. van der Marel, and C.T. Foxon, *Phys. Rev. Lett.* **60**, 848 (1988).
4. D.A. Wharam, T.J. Thornton, R. Newbury, M. Pepper, H. Ahmed, J.E.F. Frost, D.G. Hasko, and D.C. Peacock, *J. Phys.* C **21** L209 (1988).
5. J.A. Nixon and J.H. Davies, *Phys. Rev.* B **41**, 7929 (1990).
6. M. Stopa, *Phys. Rev.* B **53**, 9595 (1996).
7. K. Richter, D. Ullmo, and R.A. Jalabert, *Phys. Rev.* B **54**, R5219 (1996).
8. A. Grabert and M.H. Devoret, (eds.) *Single Charge Tunneling*, vol. 294 of *NATO Advanced Study Institute, Series B: Physics*. Plenum, New York, (1991).
9. P.L. McEuen, E.B. Foxman, U. Meirav, M.A. Kastner, Y. Meir, and N.S. Wingreen, *Phys. Rev. Lett.* **66**, 1926 (1991).
10. R.C. Ashoori, H.L. Stormer, J.S. Weiner, L.N. Pfeiffer, S.J. Pearton, K.W. Baldwin, and K.W. West, *Phys. Rev. Lett.* **68**, 3088 (1992).

11. N.C. van der Vaart, A.T. Johnson, L.P. Kouwenhoven, D.J. Maas, W. de Jong, M.P. de Ruyter van Steveninck, A. van der Enden, and C.J.P.M. Harmans, *Physica* B **189**, 99 (1993).
12. F.R. Waugh, M.J. Berry, D.J. Mar, R.M. Westervelt, K.L. Campman, and A.C. Gossard, *Phys. Rev. Lett.* **75**, 705 (1995).
13. N.C. van der Vaart, S.F. Godijn, Y.V. Nazarov, C.J.P.M. Harmans, J.E. Mooij, L.W. Molenkamp, and C.T. Foxon, *Phys. Rev. Lett.* **74**, 4702 (1995).
14. A. Yacoby, M. Heiblum, D. Mahalu, and H. Shtrikman, *Phys. Rev. Lett.* **74**, 4047 (1995).
15. S. Tarucha, D.G. Austing, T. Honda, R.J. van der Hage, and L.P. Kouwenhoven, *Phys. Rev. Lett.* **77**, 3613 (1996).
16. C.M. Marcus, A.J. Rimberg, R.M. Westervelt, P.F. Hopkins, and A.C. Gossard, *Phys. Rev. Lett.* **69**, 506 (1992).
17. A.M. Chang, H.U. Baranger, L.N. Pfeiffer, and K.W. West, *Phys. Rev. Lett.* **73**, 2111 (1994).
18. J.P. Bird, K. Ishibashi, Y. Aoyagi, and T. Sugano, *Phys. Rev.* B **50**, 18678 (1994).
19. M.W. Keller, O. Millo, A. Mittal, and D.E. Prober, *Surf. Sci.* **305**, 501 (1994).
M.W. Keller, A. Mittal, J.W. Sleight, R.G. Wheeler, D.E. Prober, R.N. Sacks, and H. Shtrikmann, *Phys. Rev.* B **52** R14336 (1995).
20. M. Persson, J. Petterson, B. von Sydow, P.E. Lindelof, A. Kristensen, and K.F. Berggren, *Phys. Rev.* B **52**, 8921 (1995).
21. R.A. Jalabert, H.U. Baranger, and A.D. Stone, *Phys. Rev. Lett.* **65**, 2442 (1990).
22. H.U. Baranger, D.P. DiVincenzo, R.A. Jalabert, and A.D. Stone, *Phys. Rev.* B **44**, 10637 (1991).
23. C.M. Marcus, R.M. Westervelt, P.F. Hopkins, and A.C. Gossard, *Chaos* **3**, 643 (1993a).
24. C.M. Marcus, R.M. Westervelt, P.F. Hopkins, and A.C. Gossard, *Phys. Rev.* B **48**, 2460 (1993b).
25. C.M. Marcus, R.M. Westervelt, P.F. Hopkins, and A.C. Gossard, *Surf. Sci.* **305**, 480 (1994).
26. H.U. Baranger, R.A. Jalabert, and A.D. Stone, *Phys. Rev. Lett.* **70**, 3876 (1993a).
27. H.U. Baranger, R.A. Jalabert, and A.D. Stone, *Chaos* **3**, 666 (1993b).
28. H.U. Baranger and P.A. Mello, *Europhys. Lett.* **33**, 465 (1996).
29. R.M. Clarke, I.H. Chan, C.M. Marcus, C.I. Duruöz, J.S. Harris Jr., K. Campman, and A.C. Gossard, *Phys. Rev.* B **52**, 2656 (1995).
30. W.A. Lin and R.V. Stone, *Phys. Rev.* B **53**, 3638(1996).
G. Casti and B. Chirikov, *Physica D.* **86**, 230 (1995).
31. M. Brack and R.K. Bhaduri, *Semiclassical Physics*. Addison-Wesley, Reading, MA (1997).
32. H.-J. Stöckmann. *Quantum Chaos: An Introduction*. Cambridge University Press (1999).
33. M.C. Gutzwiller, *Chaos in Classical and Quantum Mechanics*. Springer-Verlag, New York (1990).
34. E. Ott, *Chaos in Dynamical Systems*. Cambridge University Press (1993).
35. L.A. Bunimovich, *Comm. Math. Phys.* **65**, 295 (1979).
36. M.C. Gutzwiller, *J. Math. Phys.* **12**, 343 (1971).
37. M.V. Berry and M. Tabor, *Proc. Roy. Soc. (London)* A **349**, 101 (1976).
38. M.V. Berry and M. Tabor, *J. Phys.* A **10**, 371 (1977).
39. K. Nakamura, *Quantum Chaos: A New Paradigm of Non-Linear Dynamics*. Cambridge University Press (1993).
40. P.A. Lee, A.D. Stone, and H. Fukuyama, *Phys. Rev.* B **23**, 6851 (1987).
41. D.J. Thouless, *Phys. Rev. Lett.* **39**, 1167 (1977).
42. L. Wirtz, J.-Z. Tang, and J. Burgdörfer, *Phys. Rev.* B **56**, 7589 (1997).
43. R. Landauer, *IBM J. Res. Develop.* **1**, 223 (1957).

44. R. Landauer, *Phil. Mag.* **21**, 863 (1970).
45. M. Büttiker, Y. Imry, R. Landauer, and S. Pinhas, *Phys. Rev.* B **31**, 6207 (1985).
46. M. Büttiker, *Phys. Rev.* B **33**, 3020 (1988).
47. H.U. Baranger and A.D. Stone, *Phys. Rev.* B **40**, 8169 (1989).
48. T. Usuki, M. Saito, M. Takatsu, R.A. Kiehl, and N. Yokoyama, *Phys. Rev.* B, **52**, 8244 (1995).
49. D.Y.K. Ko and J.C. Inkson, *Phys. Rev.* B **38**, 9945 (1988).
50. Y. Ochiai, A.W. Widjaja, N. Sasaki, R. Akis, D.K. Ferry, J.P. Bird, K. Ishibashi, Y. Aoyagi, and T. Sugano, *Phys. Rev.* B **56**, 1073 (1997).
51. R. Akis, D.K. Ferry, J.P. Bird, and D. Vasileska, *Phys. Rev.* B **60**, 2680 (1999a).
52. R. Akis, D.K. Ferry, and J.P. Bird, *Phys. Rev.* B **54**, 17705 (1996a).
53. R. Akis, J.P. Bird, and D.K. Ferry, *J. Phys. Condens. Matter* **8**, L667 (1996b).
54. J.P. Bird, D.M. Olatona, R. Newbury, R.P. Taylor, K. Ishibashi, M. Stopa, Y. Aoyagi, T. Sugano, and Y. Ochiai, *Phys. Rev.* B **52**, R14336 (1995a).
55. J.P. Bird, K. Ishibashi, D.K. Ferry, Y. Aoyagi, T. Sugano, and Y. Ochiai, *Phys. Rev.* B **52**, 8295 (1995b).
56. J.P. Bird, D.K. Ferry, R. Akis, K. Ishibashi, Y. Aoyagi, T. Sugano, and Y. Ochiai, *Europhys. Lett.* **35**, 529 (1996).
57. S.W. McDonald and A.N. Kaufman, *Phys. Rev. Lett.* **42**, 1189 (1979).
58. E.J. Heller, *Phys. Rev. Lett.* **53**, 1515 (1984).
59. E.J. Heller, P.W. O' Connor, and J. Gehlen, *Phys. Scr.* **40**, 354 (1989).
60. E.J. Heller and S. Tomsovic, *Phys. Today*, **46**, 38 (July 1993).
61. L. Kaplan and E.J. Heller, *Ann. Phys.* **264**, 171 (1998).
62. E.B. Bogolmolny, *Physica* D **31**, 169 (1988).
63. Z.-L. Ji and K.F. Berggren, *Phys. Rev.* B **45**, 6652 (1992).
64. J.P. Bird, R. Akis, D.K. Ferry, D.M. Pivin Jr., K.M. Connolly, R.P. Taylor, R. Newbury, D.M. Olatona, Y. Ochiai, Y. Okubo, K. Ishibashi, Y. Aoyagi, and T. Sugano, *Chaos Solitons Fractals* **8**, 1299 (1997).
65. G. Neofotistos, R. Lake, and S. Datta, *Phys. Rev.* B **43**, 2442 (1991).
66. I.V. Zozoulenko, R. Schuster, K.F. Berggren, and K. Ensslin, *Phys. Rev.* B **55**, R10209 (1997).
67. R. Akis, D.K. Ferry, and J.P. Bird, *Phys. Rev. Lett.* **79**, 123 (1997a).
68. R. Akis and D.K. Ferry, *Semicond. Sci. Tech.* **13**, A18 (1998a).
69. R. Akis, D.K. Ferry, and J.P. Bird, *Phys. Rev. Lett.* **81**, 1745 (1998b).
70. R. Akis, D.K. Ferry, J.P. Bird, and D. Vasileska, *Phys. Rev.* B **60**, 2680 (1999b).
71. R. Akis, D. Vasileska, J.P. Bird, and D.K. Ferry, *J. Phys. Condens. Matter* **11**, 4657 (1999c).
72. V. Fock, *Z. Phys.* **47** 446, (1928).
73. C.G. Darwin, *Proc. Camb. Phil. Soc.* **27**, 86 (1930).
74. R. Akis, D.K. Ferry, and J.P. Bird, *Jpn. J. Appl. Phys.* **36**, 3981 (1997b).
75. R. Akis, D.K. Ferry, and J.P Bird, *J. Comp. Electron.* **1**, 9 (2002c).
76. H. Linke, J.P. Bird, J. Cooper, P. Omling, Y. Aoyagi, and T. Sugano, *Phys. Rev.* B **56**, 14397 (1997a).
77. H. Linke, J.P. Bird, J. Cooper, P. Omling, Y. Aoyagi, and T. Sugano, *Phys. Stat. Sol.* **204**, 318 (1997b).
78. Z. Shao, W. Porod, and C.S. Lent, *Phys. Rev.* B **49**, 7453 (1994).
79. J. Göres, D. Goldhaber-Gordon, S. Heemeyer, M.A. Kastner, H. Shtrikman, D. Mahalu, and U. Meirav, *Phys. Rev.* B **62**, 2188 (2000).
80. U. Fano, *Phys. Rev.* **124**, 1866 (1961).
81. J.P. Bird, R. Akis, and D.K. Ferry, *Phys. Rev.* B **60**, 13676 (1999a).

82. J.P. Bird, R. Akis, D.K. Ferry, D. Vasileska, J. Cooper, Y. Aoyagi, and T. Sugano, *Phys. Rev. Lett.* **82**, 4691 (1999b).
83. R. Akis, J.P. Bird, and D.K. Ferry, *Microelectron. Eng.* **63**, 277 (2002b).
84. J.L. Vega, T. Uzer, and J. Ford, *Phys. Rev.* B **48**, 3414 (1993).
85. A. Kudrolla and S.P. Sridhar, *J. Phys.* **48**, 459 (1997).
    R. Akis, J.P. Bird, D.K. Ferry, and D. Vasileska, *Physica* E **7** 745 (2000).
86. R. Akis, J.P. Bird, and D.K. Ferry, *Appl. Phys. Lett.* **81**, 129 (2002).
87. H. Alt, C. Dembowski, H.-D. Gräf, H. Hofferbert, H. Rehfeld, A. Richter, and C. Schmit, *Phys. Rev.* E **60**, 2851 (1999).
88. B.L. Altshuler, A.G. Aronov, and B.V. Spivak, *Pis'ma Zh. Eksp. Teor. Fiz.* **30**, 248 (1979); *JETP Lett.* **30**, 248 (1979).
89. G. Bergmann, *Phys. Repts.* **107**, 3 (1984).
90. S. Chakravarty and A. Schmid, *Phys. Rep.* **140**, 193 (1986).
91. I.H. Chan, R.M. Clarke, C.M. Marcus, K. Campman, and A.C. Gossard, *Phys. Rev. Lett.* **74**, 3876 (1995).
92. M.J. Berry, J.A. Katine, R.M. Westervelt, and A.C. Gossard, *Phys. Rev.* B. **50**, 17721 (1994).
93. Y. Lee, G. Faini, and D. Mailly, *Phys. Rev.* B. **56**, 9805 (1997).
94. R.P. Taylor, R. Newbury, R.B. Dunford, P.T. Coleridge, A.S. Sachrajda, and J.A. Adams, *Phys. Rev.* B **51**, 9801 (1995).
95. H.A. Van der Vorst, *SIAM J. Sci. Stat. Comput.* **13**, 631 (1992).
96. G.V. Gadiyak and M.S. Obrecht, in *Simulation of Semiconductor Devices and Processes: Proceedings of the Second International Conference* (eds. K. Board and D.R.J. Owen), Pineridge Press, Swansea, UK (1986).
97. P. Hohenberg and W. Kohn, *Phys. Rev.* **136**, B864 (1964).
98. W. Kohn and L.J. Sham, *Phys. Rev.* **140**, A1133 (1965).
99. L. Hedin and B.I. Lundqvist, *J. Phys. C: Solid St. Phys.* **4**, 2064 (1971).
100. S. Adachi, *J. Appl. Phys.* **58**, R1 (1985).
101. R. Ketzmerick, *Phys. Rev.* B **54**, 10841(1996).
102. K.M. Connolly, D.P. Pivin, D.K. Ferry, and H.H. Wieder, *Superlatt. Microstruct.* **20**, 307 (1996).
103. D.P. Pivin, A. Andresen, J.P. Bird, and D.K. Ferry, *Phys. Rev. Lett.* **82**, 4687 (1999).
104. A.G. Huibers, M. Switkes, C.M. Marcus, K. Campman, and A.C. Gossard, *Phys. Rev. Lett.* **81**, 200 (1998).
105. R.P. Taylor, A.S. Sachrajda, J.A. Adams, P.T. Coleridge, and P. Zawadzki, *Phys. Rev.* B **47**, 4458 (1993).
106. P. Bøggild, A. Kristensen, H. Bruus, S.M. Reimann, and P.E. Lindelof, *Phys. Rev.* B **57**, 15408 (1998).
107. L. Christensson, H. Linke, P. Omling, P.E. Lindelof, I.V. Zozoulenko, and K.F. Berggren, *Phys. Rev.* B **57**, 12306 (1998).
108. P.D. Ye and S. Tarucha, *Phys. Rev.* B **59**, 9794 (1999).
109. C. Pasquier, U. Meirav, F.I.B. Williams, D.C. Glatti, Y. Jin, and B. Etienne, *Phys. Rev. Lett.* **70**, 69 (1993).
110. C.-T. Liang, M.Y. Simmons, C.G. Smith, G.H. Kim, D.A. Ritchie, and M. Pepper, *Phys. Rev. Lett.* **81**, 3507 (1998).
111. S.M. Cronenwett, S.M. Maurer, S.R. Patel, C.M. Marcus, C.I. Duruoz, and J.S. Harris, *Phys. Rev. Lett.* **81**, 5904 (1998).
112. I.L. Aleiner, and L.I. Glazman, *Phys. Rev.* B **57**, 9608 (1998).
113. T. Geisel, R. Ketzmerick, and O. Schedletzky, *Phys. Rev. Lett.* **69**, 1680 (1992).

114. A.P. Micolich, R.P. Taylor, R. Newbury, J.P. Bird, R. Wirtz, C.P. Dettmann, Y. Aoyagi, and T. Sugano, *J. Phys.: Condens. Matter* **10**, 1339 (1998).
115. A.S. Sachrajda, R. Ketzmerick, C. Gould, Y. Feng, P.J. Kelly, A. Delage, and Z. Wasilewski, *Phys. Rev. Lett.*, **80**, 1948 (1998).
116. R.P. Taylor, A.P. Micolich, R. Newbury, J.P. Bird, T.M. Fromhold, J. Cooper, Y. Aoyagi, and T. Sugano, *Phys. Rev.* B **58**, 11107 (1998).
117. B. Huckestein, R. Ketzmerick, and C.H. Lewenkopf, *Phys. Rev. Lett.* **84**, 5504 (2000).
118. Y. Takagaki, M. ElHassan, A. Shailos, C. Prasad, J.P. Bird, D.K. Ferry, K.H. Ploog, Lin L.-H., N. Aoki, and Y. Ochiai, *Phys. Rev.* B **62**, 10255 (2000a).
119. Y. Takagaki and K.H. Ploog, *Phys. Rev.* E **62**, 4804 (2000b).
120. L. Hufnagel, R. Ketzmerick, and M. Weiss, *Europhys. Lett.* **54**, 703 (2001).
121. A.P. Micolich, R.P. Taylor, A.G. Davies, J.P. Bird, R. Newbury, T.M. Fromhold, A. Ehlert, H. Linke, L.D. Macks, W.R. Tribe, E.H. Linfield, D.A. Ritchie, Y. Cooper, J. Aoyagi, and P.B. Wilkinson, *Phys. Rev. Lett.* **87**, 036802 (2001).
122. A.P.S. de Moura, Y.C. Lai, R. Akis, J.P. Bird, and D.K. Ferry, *Phys. Rev. Lett.* **88**, 236804 (2002).
123. P. Seba, *Phys. Rev.* E **47**, 3870 (1993).
124. A. Kaplan, N. Friedman, M. Andersen, and N. Davidson, *Phys. Rev. Lett.* **87**, 274101 (2001).
125. W.K. Hensinger, H. Häffner, A. Browaeys, N.R. Heckenberg, K. Helmerson, C. McKenzie, G.J. Milburn, W.D. Phillips, S.L. Rolston, H. Rubinsztein-Dunlop, and B. Upcroft, *Nature* **412**, 52 (2001).
126. D.A. Steck, W.H. Oskay, and M.G. Raizen, *Science* **293**, 274 (2001).
127. S. Tomsovic, *Phys. Scr.* **T90**, 162 (2001).
128. W.H. Miller, *J. Chem. Phys.* **63**, 996 (1975).
129. T. Blomquist and I.V. Zozoulenko, *Phys. Rev.* B **64**, 1935301 (2001).
130. I. Rotter, E. Persson, K. Pichugin, and P. Seba, *Phys. Rev.* E **62**, 450 (2000).
131. S.E. Muraviev, I. Rotter, S. Shlomo, and M.H. Urin, *Phys. Rev.* C **59**, 2040 (1999).
132. E. Persson, I. Rotter, H.-J. Stöckmann, and M. Barth, *Phys. Rev. Lett.* **85**, 2478 (2000).
133. Y.H. Kim, M. Barth, H.-J. Stöckmann, and J.P. Bird, *Phys. Rev.* B **65**, 165317 (2002).
134. T. Ando, *Phys. Rev.* B **44**, 8017 (1991).

Chapter 7

A Review of Fractal Conductance Fluctuations in Ballistic Semiconductor Devices

R. Taylor[1], R. Newbury[2], A. Micolich[2], M. Fromhold[3], H. Linke[1], G. Davies[4], T. Martin[1], and C. Marlow[1]

[1] *Department of Physics, University of Oregon, Eugene, OR 97403-1274, USA*
[2] *School of Physics, University of New South Wales, Sydney, 2052, Australia*
[3] *School of Physics and Astronomy, Nottingham University, Nottingham NG7 2RD, UK*
[4] *Cavendish Laboratory, Cambridge University, Cambridge, CB3 OHE, UK*

1. Introduction

The ability to scale device sizes to below a micro-meter has profound implications for electron conduction in semiconductor systems. For conventional circuits, the reduced component size offers the rewards of higher packing densities and faster operating speeds. However, in terms of the search for new classes of electronic devices, designed to replace the transistor as the basic component of electronic circuits, the prospect of current flow across sub-micron distances holds even greater potential. Traditional current flow concepts, in which electron propagation along the device's length is modeled as a classical diffusion process, can no longer be applied. Both the classical and quantum mechanical transmission characteristics may ultimately be harnessed to produce revolutionary modes of device functionality. Intimately coupled to these objectives of applied physics, sub-micron devices also provide a novel environment for the study of a rich variety of fundamental semiconductor physics.

Fueled by these considerable goals, many techniques for constructing sub-micron devices have emerged over the last twenty years. One of the more successful approaches adopts the cornerstone of the microelectronics industry—metallic surface patterns—and reduces their length scales through the use of electron-beam rather than optical lithography [1–3]. In particular, since its first demonstration in 1986 [4], surface gates have frequently been used to define corresponding patterns in the two-dimensional electron gas (2DEG) at the interface of AlGaAs/GaAs heterostructures [1]. Whereas this "top-down" construction technique (where component devices are formed within a macroscopic material) can routinely define devices with feature sizes of 50 nm, more recent "bottom-up" technologies (where atomic components are assembled to form a device), are capable of constructing even smaller device sizes. Nevertheless, the gated AlGaAs/GaAs system remains successful due to its capacity to define small devices in the high quality environment provided by the host semiconductor material. In particular, surface gate technology remains the technique of choice for studies of "ballistic" electron conduction, where the device size is smaller than the average distance ℓ between scattering sites in the material.

When first observed in 1988 [5,6], the appeal of the ballistic conduction regime was thought to lie in its remarkable simplicity: through the use of sophisticated semiconductor growth and lithography techniques, devices could be constructed in which the electrons were unhindered by material-induced scattering events [1,7]. The electrons followed straight trajectories, allowing experiments originally planned for vacuum to be performed in a solid-state environment! A famous example of this concept is the semiconductor "billiard". Analogous to a billiard table, electrons in the 2DEG move along a two-dimensional plane bounded by shaped walls. Because the bounded region is smaller than ℓ, electrons passing between the entrance and exit (formed by openings in the walls) follow ballistic trajectories shaped predominantly by the walls (see Fig. 7.1(a)). In the early 1990s, experimental [8,9] and theoretical [10,11] research focused on a comparison of two distinct billiard shapes—stadia and circles. The reasoning behind this choice was that circular cavities were expected to produce stable trajectories, whilst the combination of curved and straight walls of the stadium would generate chaotic electron trajectories exhibiting an exponential sensitivity to initial conditions [12]. Furthermore, these ground-breaking studies moved beyond classical chaos by investigating "quantum chaos"—the quantum mechanical behavior of the classically chaotic electrons [13].

This exploration of quantum chaos in electronic conduction serves as the starting point for the experiments presented in this chapter. In our experiments, the semiconductor billiards are cooled to milli-Kelvin temperatures to reduce the electron scattering events (e.g. with phonons and other electrons) that disrupt the coherence of the electrons' quantum mechanical wave properties. When the electrons reach a temperature of 30 mK, the characteristic phase-coherence length of the waves is in excess of 30 μm,

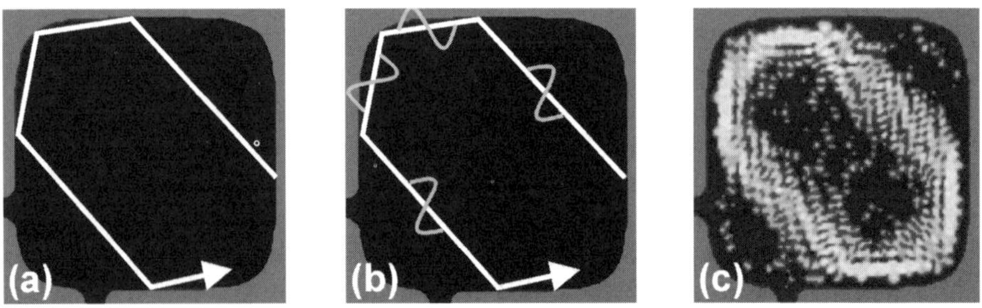

Fig. 7.1. (a) A schematic representation of a classical electron trajectory within a billiard. (b) In the semi-classical regime, electron waves travel along the trajectories. (c) A simulation showing the intensity variation of a quantum wave.

allowing electrons to maintain coherence whilst traversing billiards with typical sizes of 1 μm. Traditionally, this billiard size is regarded as optimal for studies of quantum chaos. Whilst being sufficiently small to ensure phase coherence and therefore that the quantum wave-character of the electrons plays a crucial role in the conduction process, it also ensures that the ratio S of the billiard size to the electron Fermi wavelength (~50 nm) is relatively large. Electron transport within the billiard is then semi-classical rather than fully quantum-mechanical. A crucial consequence of the semi-classical regime is that the electron waves can be pictured as moving along the classical trajectories (see Fig. 7.1(b,c)) [7,10,11]. As they do so, the waves accumulate phase and this determines the wave interference between pairs of trajectories that intersect to form closed loops [7]. This interference process can be investigated experimentally by tuning the phase through the application of a small magnetic field B perpendicular to the plane of the billiard [7]. According to the Aharonov–Bohm effect, each loop generates a periodic oscillation in the conductance with a magnetic field period ΔB inversely proportional to area enclosed by the loop [7,14]. Because a typical billiard supports a distribution of loop areas, the magneto-conductance $G(B)$ is composed of oscillations with many different periods. The resulting fluctuations can then be viewed as "magneto-fingerprints" of the classical trajectories because they are sensitive to the precise, microscopic distribution of loop areas in the billiard [8,9,15]. In this way, the underlying classical chaos in the trajectories is expected to manifest itself in the quantum behavior of the billiard—i.e. quantum chaos.

We find that patterns observed in the magneto-conductance fluctuations repeat at increasingly fine magnetic field scales, a phenomenon that has since been labeled as fractal conductance fluctuations (FCF) [16]. Furthermore, two distinct forms of fractal behavior are observed—exact self-affinity (where the patterns observed at increasingly fine field scales repeat exactly) and statistical self-affinity (where the patterns simply follow the same statistical relationship at different scales). Statistical self-affinity describes many of nature's patterns, ranging from clouds to coastlines [17,18]. In contrast, although exact self-affinity has been the subject of mathematical studies for over one hundred years, observations of this form of fractals in physical systems remain rare [17,18]. The billiards investigated therefore represent a unique physical environment in which both forms of fractal can be observed. Furthermore, fractal studies in natural environments are "passive" in the sense that there is no experimental interaction with the system being investigated. In contrast, the semiconductor billiard represents an artificial environment in which the conditions can be controlled and adjusted in a systematic fashion. We exploit this control to address fundamental questions about transitions in fractal behavior. In particular, we investigate how fractal behavior evolves between its two distinct forms (exact and statistical self-affinity) and also how fractals deteriorate into non-fractal behavior as the process generating them is gradually suppressed. These central issues cannot be addressed using other physical systems and this unique opportunity has generated considerable interest from the fractals and chaos research communities.

The primary motivation for this research, however, remains firmly anchored in semiconductor physics and the exploration of electron conduction properties of sub-micron electronic devices. Almost fifteen years on from the first observation of ballistic conduction in semiconductor devices, their basic properties are not as simple as initially hoped. Despite the continued use of the descriptive title of "billiard", these devices do not follow the well-understood rules of their name-sake! Magneto-conductance measurements play a central role in many studies of sub-micron semiconductor physics, yet there is no theoretical basis for the observed fractal behavior. In contradiction to initial expectations and theory, fractal conduction fluctuations are not restricted to the semi-classical conduction regime. Instead, they appear to be a generic feature of ballistic conduction in small devices, extending across the spectrum of classical, semi-classical and quantum conduction regimes.

Far from trivial, this preservation of fractal character is quite remarkable. Fractals are a holistic phenomenon: if the contribution of just one trajectory loop is changed, then *all* the other loops supported by the billiard have to adjust in order to preserve the fractal scaling properties of the conductance. In this sense, fractal conductance fluctuations are a sensitive probe of the underlying dynamic properties of *all* the ballistic trajectories determining the conduction process. In this chapter, we will present a review of the experimental quantification of this fractal phenomenon, along with recent and surprising experimental developments, in the hope that these novel observations will attract further theoretical interest in the relationship between the fractals and the chaotic dynamics that generate them, leading to an improved understanding of electron conduction processes at the sub-micron scale.

2. The Semiconductor Sinai Billiard: Can Chaos be Controlled with the "Flick of a Switch?"

We began our investigations in 1994 with the question, "can chaos be controlled with the flick of a switch?" Two years earlier, the experiment that pioneered the study of quantum chaos in semiconductor billiards had detected a significant difference in the magneto-conductance fluctuations generated by two billiards—one was a circular billiard designed to support stable trajectories, the other a stadium designed to support chaotic trajectories [8]. A natural step forward in these investigations would be to see if the same effects could be observed in a *single* device: could the billiard parameters be adjusted to introduce or remove chaotic dynamics from the quantum conduction process? At this time, the semiconductor community was eagerly exploring analogies between ballistic electrons and ray optics, where electrostatic surface gates were being used to "steer" ballistic trajectories through reflection, refraction or collimation [7]. We therefore proposed the idea of a "hybrid" billiard where certain sections of the billiard wall would be shaped to support stable trajectories whilst others would have curved surfaces designed to diverge the trajectories and generate chaotic sensitivity. The ballistic beam injected into the billiard could then be directed to either the chaotic or stable region of the billiard by tuning the voltage applied to a "steering" gate. In the experiment that evolved from this principle, chaos was introduced into the quantum conduction by directing a collimated beam of electrons at the curved surface located around the exit of the billiard [19]. Evolving this principle further still, rather than deflecting beams from straight to circular walls, why not introduce the chaos by inserting a circular object into the flow of ballistic trajectories?

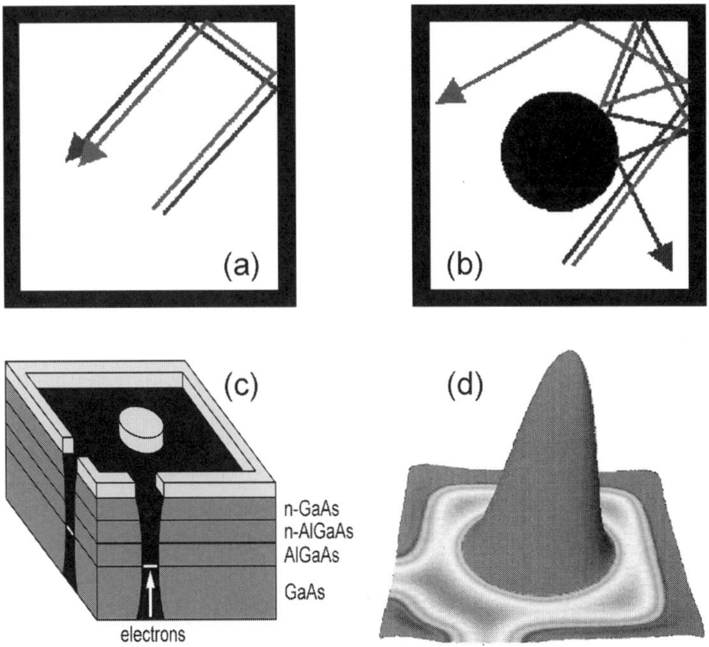

Fig. 7.2. (a) Two electron trajectories launched with similar initial conditions. Shaped by collisions with the square billiard, these stable trajectories do not diverge significantly. (b) In contrast to the square billiard, the same two trajectories rapidly diverge in the Sinai billiard due to scattering events with the circle. (c) A schematic representation of the surface gate technique used to define the Sinai billiard in the 2DEG located below the surface at the AlGaAs/GaAs interface. The medium gray regions indicate the depletion regions under the gates. (d) Self-consistent calculations of the soft-wall potential landscape expected for the Sinai billiard defined by surface gates. Potential energy (vertical axis) is plotted as a function of position within the billiard.

The very same idea had been explored mathematically twenty-five years earlier by a Russian chaologist called Sinai [20]. Sinai presented a theoretical investigation of the trajectories of classical particles and how collisions with straight and curved walls affected the dynamics of the particles. In particular, the geometries shown in Fig. 7.2(a,b) have since become a model system for the theoretical demonstration of chaotic trajectories. Whereas the "empty" square billiard shown in Fig. 7.2(a) supports stable trajectories, introduction of the circle at its center transforms the geometry into the so-called Sinai billiard shown in Fig. 7.2(b). This simple transition has a profound effect on the billiard's scattering dynamics. The convex surface of the circular scatterer acts as a "Sinai diffuser" producing diverging trajectories, whilst the straight outer walls bounce the trajectories back towards the diffuser to repeat the diverging process over and over. The result is to produce trajectories with an exponential sensitivity to initial conditions—the signature of chaotic behavior [12,20].

Possible systems in which this elegant method for generating chaos could be achieved include microwave cavities [21], atom optics [22,23] and semiconductor billiards. The most flexible construction technique for semiconductor billiards is shown in the schematic representation of Fig. 7.2(c). Within the semiconductor heterostructure shown, a 2DEG is located at the interface between the GaAs and AlGaAs layers. The billiard is defined in this 2DEG using the patterned metallic gates (light gray) deposited on the heterostructure surface. A negative gate bias defines depletion regions (medium gray), forming the billiard walls in the regions of the 2DEG directly below the gates [1,7]. In contrast to the physical walls of microwave cavities, these electrostatic walls can be switched on and off and their sizes varied by tuning the gate bias, allowing an evolution in the billiard geometry. In particular, the central circle can be switched off, allowing the transition from the Sinai billiard to the empty square shown in Fig. 7.2(a,b).

Figure 7.3 shows the device design used to construct the semiconductor Sinai billiard in our collaboration with the National Research Council in 1995 [24]. The major technological challenge concerned establishing electrical contact to the 0.3 μm diameter "inner" circular gate shown in Fig. 7.3(b). This was achieved using a bridging interconnect technology developed by one of us (RPT) in the early 1990s [25,26]. Shown schematically in Fig. 7.3(c), the fabrication details of this technique are described in detail elsewhere [1,25,26]. The proposed device operation was as shown in Fig. 7.3(d–g): by applying a negative bias V_O to the three "outer" gates to form a micron-sized square billiard, the circle would then be introduced and enlarged by applying an increasingly negative bias V_I to the inner gate. Because the host material's ℓ value of 25 μm was significantly larger than the billiard size, the trajectories of electrons traversing the billiard were expected to be ballistic and to be profoundly affected by this change in geometry. Furthermore, as required by Sinai's model, these electrons would scatter off the walls specularly and elastically [1,7]. Thus the evolution in billiard geometry shown in Fig. 7.3(d–g) was expected to be accompanied by Sinai's transition from stable to chaotic dynamics outlined in Fig. 7.2(a,b).

The proposed experiment would not, however, be the first study of a physical system in which chaotic particles could be tuned. Indeed, transitions had already been achieved in relatively basic systems. For example, chaos had been introduced into the flow of water droplets by adjusting the aperture of a faucet [27]. However, the proposed "chaotic transistor" [28] offered a number of significant advantages over the dripping faucet as follows: (i) The precision associated with tuning the electrostatic gates was superior to adjustments of the faucet aperture. (ii) The chaos would be introduced into a flow of electrons, raising the possibility of novel electronic applications. For commercial transistors, electrostatic surface gates serve as a switch to modulate the flow of electrons through the device. For the Sinai billiard, the proposed gating operation was considerably more refined—tuning the voltage applied to the gates would induce a controlled evolution in the scattering dynamics of the electrons. By introducing chaos and the associated sensitivity of the trajectories, small controlled changes to the device could lead to large changes in output, leading to efficient switching operations [28]. (iii) In contrast to the classical water droplets, it would be possible to harness the electrons' wave properties and study the device's quantum operation as chaos was "switched on" [29].

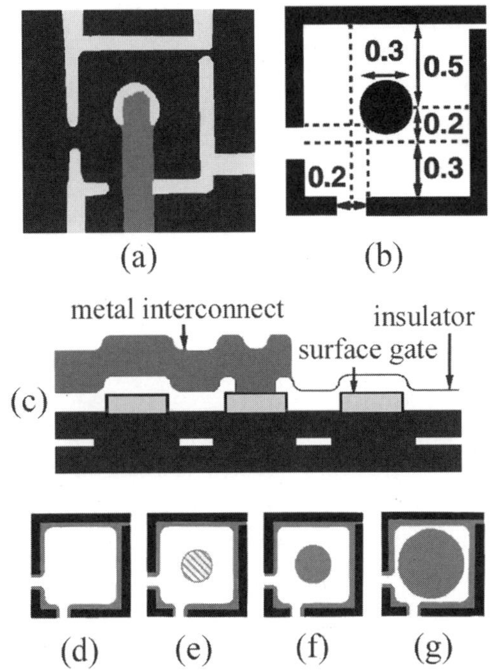

Fig. 7.3. (a) A scanning electron micrograph of the Sinai billiard device. (b) A schematic representation of the surface gate geometry. The lithographic dimensions are in microns. The dashed lines are guides indicating the relative position of the central gate. (c) A cross-section showing the bridging interconnect that traverses the billiard above a thin layer of insulator. The bridge contacts to the central circular gate through a hole etched in the insulator in the region directly above the central gate. (d)–(g) Schematic representations of four regimes of device geometry: (d) the "empty" square billiard, (e) the square with a partially depleted central region, (f) a Sinai billiard formed by the circle at the center of the square billiard, (g) a Sinai billiard featuring a larger circle.

3. The Experimental Observation of Exact Self-Affinity

Directed by these specific aims, initial experiments focused on the Sinai geometry. Characterization studies showed that the presence of the central gate could be minimized by setting the bias V_{I} to $+0.7$ V. A circular depletion region could be formed under the gate by reducing the bias to 0 V and for negative V_{I} the radius R of this depletion circle increased according to the expression $R = R_{\mathrm{g}} - (8 \times 10^{-8})V_{\mathrm{I}}$ (where R_{g} is the circle's lithographic radius) [28]. With the circle activated, we observed that the magneto-conductance fluctuations clustered around two distinct magnetic field scales: "fine" (f) fluctuations superimposed on "coarse" (c) fluctuations [24,29,30]. This is shown in Fig. 7.4(left), where V_{O} has been tuned to set the number of conducting modes, n, in the entrance and exit openings to be 7 and V_{I} had been tuned to give $R = 0.37\,\mu$m. This figure was generated by first measuring the c fluctuations (top) using a magnetic

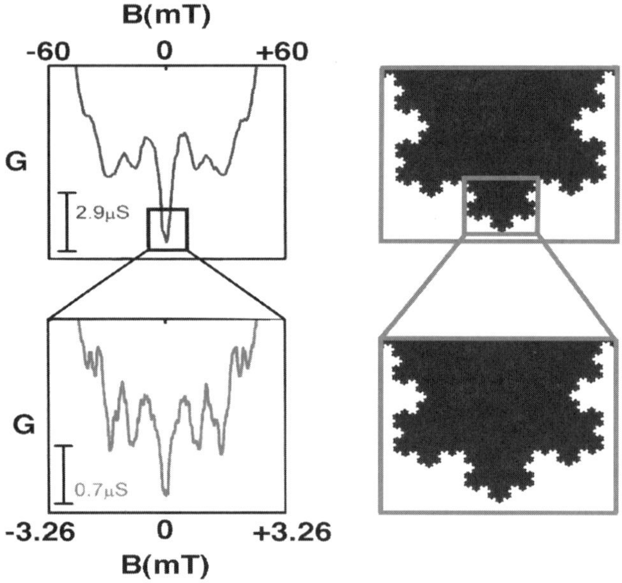

Fig. 7.4. Left: course (top) and fine (bottom) structure observed in the magneto-conductance $G(B)$ measured for $n = 7$. See text for details. Right: the repeating patterns of the Koch Snowflake.

field resolution of 1 mT and then concentrating on a narrower magnetic field range and adopting a finer resolution of 0.008 mT to measure the f fluctuations (bottom). For each level, $G(B)$ was measured for both magnetic field directions (signified in the figure by a change in the sign of B) in order to distinguish between the measurement signal and measurement noise—due to the Onsager relationships [7], the signal was symmetric about $B = 0$ T and thus the observed minor deviations from symmetry originated from noise.

We were able to suppress both the c and f fluctuations by increasing V_O to a value sufficiently high to pinch off the channel around the circle, indicating that both sets of fluctuations were generated by circulating trajectories [24,30]. Furthermore, we were also able to suppress both sets of fluctuations by raising the electron temperature from 50 mK (the temperature used in the measurements in Fig. 7.4 (left)) to 4 K to remove electron phase coherence [24,30], indicating that both sets of fluctuations were generated by wave interference associated with these circulating trajectories. However, their shared qualities extended even further—the patterns observed in the c and f fluctuations appeared to be remarkably similar. To dismiss the possibility that this similarity was simply coincidence, we varied billiard parameters (such as using V_O to adjust n) to induce a change in the c fluctuations and found that the f fluctuations evolved in a similar fashion [30]. This similarity was investigated across an array of fifty bias settings [31]. The observed repetition of patterns observed on different scales clearly bears a close resemblance to mathematical patterns such as the Koch Snowflake shown in Fig. 7.4 (right) [17,18]. To distinguish this exact repetition of the pattern from patterns that merely follow the same statistics at different scales (see later), we labeled the observed behavior as "exact" self-affinity (ESA).

We confirmed the presence of ESA using a mathematical comparison of the fine scale conductance fluctuations, $\delta G_f(B)$, and the coarse scale conductance fluctuations, $\delta G_c(B)$. To extract $\delta G_c(B)$ from the c scale conductance trace, $G_c(B)$, we used the definition $\delta G_c(B) = G_c(B) - G_c(B = 0)$, and also applied the equivalent definition for $\delta G_f(B)$. For the observed patterns to exhibit ESA, the $\delta G_f(B)$ trace should be a scaled version of the $\delta G_c(B)$ trace and therefore it should be possible to select conductance and field scaling factors, λ_G and λ_B, such that $\delta G_c(B)$ and $\lambda_G \delta G_f(\lambda_B B)$ are nominally identical traces. To quantify this ESA, we introduced a correlation function F [32]:

$$F = 1 - \frac{\sqrt{\langle \{\delta G_c(B) - \lambda_G \delta G_f(\lambda_B B)\}^2 \rangle}}{N} \tag{1}$$

The averaging <> was performed over 100 magnetic field points between ± 50 mT. N was introduced as a normalization constant calculated by averaging 1000 values of the expression $\sqrt{\langle \{X(B) - Y(B)\}^2 \rangle}$, where $X(B)$ and $Y(B)$ were functions that generated random number distributions over the 100 magnetic field points [32]. Because $\delta G_c(B)$ and $\delta G_f(B)$ were symmetric about $B = 0$ T, $X(B)$ and $Y(B)$ were therefore reflected about $B = 0$ T to ensure the same basic symmetry as the data. The amplitude ranges of $X(B)$ and $Y(B)$ were equated to that of $\delta G_c(B)$. The role of N was to set $F = 0$ when $\delta G_c(B)$ and $\lambda_G \delta G_f(\lambda_B B)$ were randomly related traces and $F = 1$ if the two traces were mathematically identical. In this way, F could be used to identify similarities in patterns observed at the two different field scales. Fig. 7.5(b) is a "scale factor map", showing F as a function of λ_G and λ_B for the case of $R = 0.37\,\mu$m. A clear maximum of $F = 0.94$ occurs at $\lambda_G = 6.1$ and $\lambda_B = 26.4$. Indeed, all scale

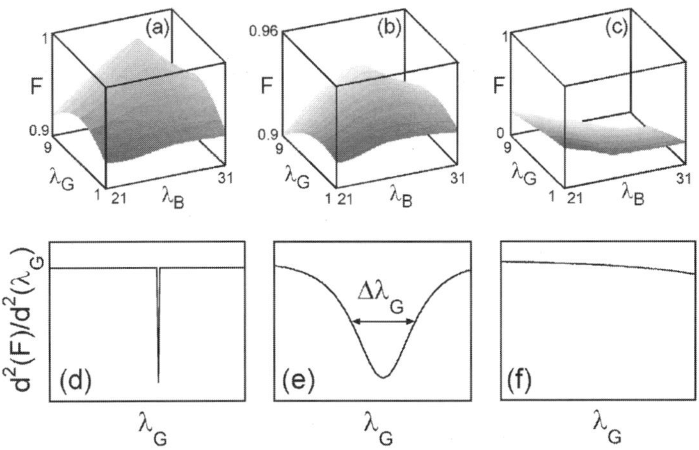

Fig. 7.5. (a), (b) and (c) are scale factor maps, F versus (λ_G, λ_B) for $V_O = -0.55$ V. (a) Circle activated, ideal case; (b) circle activated, experimental case; (c) circle de-activated, experimental case. (d,e,f) are second derivatives of F versus λ_G. $V_O = -0.52$ V.

factor maps obtained for the Sinai billiard (measured for different n values etc) reveal a single peak characterized by F as high as 0.97 [33]. These maps are characteristic of ESA, indicating a striking similarity between the c and f fluctuations. Note, however, that this experimentally-observed ESA is not mathematically perfect. Figure 7.5(a) is a scale factor map calculated for the ideal case of ESA, where the f fluctuations are a mathematically-generated replica of the c fluctuations. The peak rises to a maximum of $F = 1$ at a singular point in the (λ_G, λ_B) map. This singularity is demonstrated in Fig. 7.5(d), where the second derivative of F features a δ function at the peak's maximum. Compare this ideal ESA behavior to the experimentally-observed ESA shown in Fig. 7.5(b): although centered around the same (λ_G, λ_B) point, the experiment's F peak has a slightly lower value and, as indicated in Fig. 7.5(e), the peak is less sharp. Ideal ESA was not achieved in the experiment due to contributions from a narrow range of scaling factors rather from a unique pair, and the full widths at half maximum extracted from the second derivative plots quantified this range as $(\Delta\lambda_G, \Delta\lambda_B) = (2.7, 1)$ [33]. We will return to this observation in the next section.

Having established that the two sets of fluctuations (c and f) exhibited ESA, the next step was to determine if they were also fractal. To be fractal, the fluctuations should not be limited to two magnetic field scales but instead should continue to cluster at increasingly fine and increasingly course scales, building up a cascade of levels spanning many field scales. Furthermore, as with the Koch Snowflake, there should be a constant magnification factor separating neighboring levels. For our fluctuations, this would be set by the magnetic field scale factor λ_B. For the data of Fig. 7.4 (left), λ_B was found to have a value of 18.6. Therefore, for the magneto-conductance to be fractal, we should be able to "zoom" into the f level by a magnification factor of *precisely* 18.6 and observe an additional cluster of fluctuations. This was found to be the case and we labeled this as the ultra-fine (uf) level [29,32]. Similarly, we zoomed out from the c level by precisely 18.6 and found the ultra-course (uc) level [34,35]. A full discussion of the ESA analysis and visual inspections of the uf and uc fluctuations can be found elsewhere [35]. Significantly, we found that, in addition to the common field scaling factor λ_B, the four levels of fluctuations (uf, f, c, uc) were described by a common conductance scaling factor λ_G, as required for fractal patterns [12,17,18]. The final step for establishing fractal behavior involves an examination of the relationship between the two constant scaling factors, λ_B and λ_G. These factors relate to a crucial parameter for characterizing fractal patterns called the fractal dimension D. Whereas F quantifies the similarity between the patterns observed at different magnifications, D quantifies their scaling relationship [17,18]. For Euclidean patterns, dimension assumes the familiar integer values: for a smooth line, D has a value of 1, whilst for a completely filled area its value is 2. However, for a fractal pattern, the repeating structure at different magnifications causes the line to begin to occupy area. D then has a fractional value between 1 and 2 and, as the complexity of the repeating structure increases, its value moves closer to 2. To confirm that the four levels observed in $G(B)$ were fractal it was therefore necessary to show that their scaling relationship was described by a fractional value of D.

To calculate D from the scaling relationship of the four observed levels, we constructed scaling plots such as the one shown in Fig. 7.6(a). For simplicity, here we show the analysis performed on one selected feature within each cluster of fluctuations—the

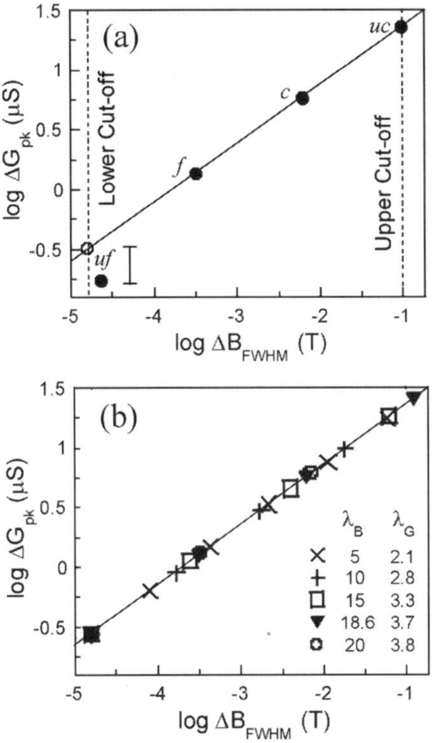

Fig. 7.6. (a) Scaling properties of the levels observed in the experimental data based on the central peak height ΔG_{pk} and full width at half maximum ΔB_{FWHM}. The four points (uf, f, c and uc) represent a single hierarchy described by $\lambda_{\text{B}} = 18.6$. (b) A scaling plot showing four additional λ_{B} values and their associated hierarchies.

peak in resistance observed around $B = 0$ T. We quantified the peak's scaling properties using ΔB_{FWHM} (the full width at half maximum) and ΔG_{pk} (the peak's height) and obtained the four filled circles shown in Fig. 7.6(a) [32,35]. With the exception of the uf data point (discussed below), the data condenses onto the power law line shown. Note that, as expected, the four levels lie at equal increments along the power law line, indicating that the conductance amplitudes and magnetic field periods of consecutive levels can be related by the common field and conductance scaling factors $\lambda_{\text{B}} = \Delta B_m / \Delta B_{m+1}$ and $\lambda_{\text{G}} = \Delta G_m / \Delta G_{m+1}$ respectively (here the indices $m = 1$–4 have been assigned in order of decreasing scale—i.e. the uc level corresponds to $m = 1$ and uf to $m = 4$). Furthermore, these scaling factors are related by the power law relationship $\lambda_{\text{G}} = (\lambda_{\text{B}})^{\beta}$. Such a scaling behavior is defined as fractal if the fractal dimension $D = 2 - \beta$ lies in the range $1 < D < 2$ [12]. Allowing for uncertainties in the uf point (see below), the data points in Fig. 7.6(a) all lie on a line whose gradient β gives $D = 1.55$, confirming that the four observed clusters of fluctuations are fractal.

The dashed vertical lines in Fig. 7.6(a) provide the experimental observation limits of this fractal behavior and we label these the upper and lower "cut-offs". The lower cut-off was determined by the magnetic field resolution limit for the experiment. Although the separation between measured magnetic field points was 0.008 mT, a minimum of three data points was required to observe a feature in $G(B)$. Thus the lower cut-off for the experiment corresponds to an interval ΔB of 0.016 mT and this magnetic field value is indicated by the left dashed line. Note that the uf fluctuations lie on this resolution limit. The uf trace consists of only 46 data points compared to the 788 data points within the equivalent f trace and the distortion caused by the limited resolution of uf has been used to account for the slight difference in ΔB_{FWHM} (0.004 mT) between the uf point's actual position (filled circle) and its anticipated position on the power law line (open circle) [35]. As indicated by the 2 Ω adjacent bar, the difference in ΔG_{pk} between the measured and anticipated position of the uf point lies within the noise limit of the experiment (0.05% of the signal, corresponding to ~3 Ω). As a consequence of these measurement limitations, it was not possible to observe further levels of clustered fluctuations at finer field scales beyond the uf level.

At the other end of the scaling plot, the calculation of the position of the upper cut-off was based on the magnetic field at which the cyclotron radius of the electron becomes smaller than the billiard size. Above this field limit a transition to electron transport via skipping orbits occurs and the Sinai diffuser is no longer expected to induce chaos in the trajectories in the manner shown in Fig. 7.2(b). Accordingly, in the experiment the magnetic field traces were only measured up to this B field, and this sets the maximum possible ΔB value. Note that the uc trace lies at the upper cut-off limit. As expected, further levels of clustered fluctuations at field scales larger than the uc level were not observed. The range of observation of fractal behavior defined by the upper and lower cut-offs shown in Fig. 7.6(a) is 3.7 orders of magnitude in ΔB [34]. This scaling range is well in excess of the majority of observations of fractals in other physical systems—a recent survey revealed a typical range of less than 1.5 orders [36]. This extended range of observation makes the Sinai billiard an ideal system in which to investigate the scaling relationships of fractal phenomena.

4. The Interpretation of Exact Self-Affinity

The primary aim of the "chaotic transistor" experiment was to induce chaos in the classical trajectories by activating the Sinai diffuser, and then to investigate how the classical chaos affected the conductance in the quantum regime. However, although the classical trajectories were expected to follow the transition outlined in Fig. 7.2(a,b), there were crucial differences between Sinai's original concept and the experiment. Firstly, whereas the theoretical Sinai billiard of Fig. 7.2 was "closed", the semiconductor Sinai billiard of Fig. 7.3 incorporated openings in the walls to inject the electrons and thus facilitate current flow. This is a significant difference because current injection processes typically play a major role in determining the conduction properties of ballistic devices [37]. Secondly, Sinai's model was purely classical and offered no prediction for how the chaos would translate in the quantum regime. Consequently, at the time of the experiment, there was considerable debate over the precise form in which the quantum chaos would emerge in the measurements. One possibility was that the Sinai billiard would reveal similar magneto-conductance characteristics to the chaotic stadium investigated in the original studies of quantum chaos [8]. In this original experiment, the amplitude of the fluctuations had been shown to have an exponential dependence on the period ΔB. Applying a semi-classical model based on the Aharonov—Bohm concept discussed earlier, this dependence was interpreted as being a consequence of an exponential distribution of classical trajectory loops within the billiard. This result was appealing because an exponential distribution of loop areas was thought to be a universal signature of chaotic billiards [38]. However, the Sinai billiard's fractal dependence of fluctuation amplitude on ΔB stood in sharp contrast to the stadium's "universal" signature of chaos. Why, then, was the Sinai billiard—the model system for demonstrating chaos—producing fractal properties in the conductance fluctuations?

A plausible explanation emerged soon after the experiments were completed. The 2DEG of the semiconductor Sinai billiard was located 163 nm below the surface gates and, as a consequence, our simulations showed that the electrostatic walls differed from Sinai's model in a crucial respect. Whereas Sinai's model billiard was described by so-called "hard" walls with vertical energy profiles, the depletion regions penetrating the 2DEG defined a billiard with electrostatic walls featuring approximately parabolic energy profiles [39]. Figure 7.2(d) shows the "soft" wall profile for the semiconductor Sinai billiard of Fig. 7.3(a,b). A comparison of Figs 7.2 and 7.3 reveals the striking difference between the straight shapes of the surface gate pattern and the smoother, rounded shapes of the billiard walls that form in the 2DEG. Such simulations signaled that modeling of semiconductor ballistic transport was destined to be more subtle than initially anticipated. Mapping of the ballistic trajectories would require more than knowledge of the surface gate pattern—details of the wall's precise energy profile at the 2DEG would also be necessary. Although this revelation emphasized an intrinsic limitation of semiconductor billiards, a positive side to this discovery quickly emerged—the trajectories predicted for "realistic" soft-walled billiards displayed remarkably intricate and exotic properties compared to those predicted for their "ideal" hard-walled counterparts! In particular, soft-wall billiards were predicted to generate "mixed" electron dynamics featuring *both* stable and chaotic trajectories [16,39,40]. Significantly, mixed systems are well-known for generating complex behavior such as fractals at the borders between stability and chaos.

Our first approach to establishing a link between the mixed trajectory dynamics and the measured ESA was to model the *precise* soft-wall profile of the semiconductor Sinai billiard—including perturbations from impurities, defects and surface states [39]. To detect any underlying order in the classical trajectories generated from this realistic model of the potential profile, we then mapped out the velocities and positions of the electrons [39]. A typical Poincaré plot is shown in Fig. 7.7 and, significantly, reveals structure occurring at many magnifications. This raises the possibility that the fractal character of the conductance fluctuations is essentially classical in origin and that the role of the quantum wave interference is to serve as a conduction process that is sensitive to this repetition of the classical structure. Due to the remarkably rich character of the structure, determination of the individual trajectories that generate the observed ESA has proven difficult. In response, we adopted a parallel approach to the understanding of the ESA that treated the conductance fluctuations as a "fingerprint" of the contributing trajectories. By fitting the fluctuations, we were able to identify a possible set of trajectories that generate the precise experimental features of the

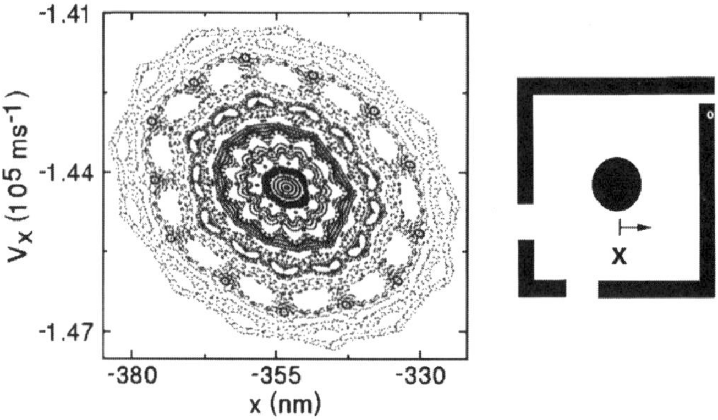

Fig. 7.7. A Poincaré section for the Sinai billiard, showing that structure in the position (x) versus velocity (v_x) phase space plot occurs at many magnifications. The x position is indicated in the schematic diagram of the billiard.

ESA [41]. These trajectories form loops similar to those predicted for the hard-wall Sinai billiard [42]. The effect of the soft-wall profile of the measured semiconductor billiard appears to be to modify the coefficients that determine the relative contributions of these loops to the conductance. We found that if we modified these coefficients so that they followed a Weierstrass scaling relationship (a function well-known in mathematics for generating ESA) [12,43] then the experimentally observed fluctuations could be reproduced with remarkable accuracy [41]. However, what remains unclear is precisely how the realistic soft-wall profiles produce this modification.

Whereas a detailed explanation for the origin of the observed ESA proved to be elusive, the experiment served to demonstrate the differences between ideal and realistic semiconductor billiards. Following the experiment, both the Sinai geometry and the "empty" square geometry were pictured as mixed trajectory systems rather than the purely stable and purely chaotic systems predicted for the two equivalent hard-wall geometries of Fig. 7.2(a,b). Whereas insertion of the central circle had originally been modeled in terms of a Sinai diffuser converting stable to chaotic trajectories, this gating operation was now pictured as a "plunger" gate, fine-tuning the relative contributions of the stable and chaotic trajectories within the mixed trajectory system. In particular, the plunger was viewed as a trajectory "selector." Within this picture, on activation of the plunger, large parts of the Poincaré plot would become exclusion regions and all but a selected set of trajectories would be removed from the billiard system. De-activation would re-introduce the forbidden trajectories, producing Poincaré plots of increased complexity. Consequently, the transition from Sinai billiard to empty square was predicted to be accompanied by a sharp increase in complexity in the observed conductance fluctuations.

This model of trajectory selection was developed further in terms of an associated selection of the scaling factors (λ_B, λ_G). Whereas a full mathematical framework is presented elsewhere [33,44], here we summarize the essential results using the scaling plots shown in Fig. 7.6. As discussed earlier, the ESA observed for the Sinai billiard was generated by a single hierarchy of points ($m = 1, 2, 3, 4, \ldots$) located along the line with gradient $\beta = 2 - D$. The four points each satisfy the power law relationship $\lambda_G = (\lambda_B)^\beta$ and are separated by equal spacings along the line determined by $\lambda_B = 18.6$. According to the model, as the circle's size is reduced, introduction of extra trajectories leads to an increase in the number of allowed values of λ_B. For each allowed λ_B value there exists a hierarchy of points ($m = 1, 2, 3, 4, \ldots$) lying on the same power-law line with a spacing set by the λ_B value. For simplicity, in Fig. 7.6(b) we have shown just four additional λ_B values and the associated hierarchies. As the circle is reduced further, and the number of allowed λ_B values increases, the power law line gradually becomes filled with points at all ΔB values. This is expected to have a striking impact on the appearance of the conductance fluctuations. Firstly, since each data point represents a cluster of fluctuations observed at a particular ΔB magnification (see Fig. 7.4), then fluctuations should be observed at *all* magnifications for the empty square. Secondly, fluctuations belonging to different hierarchies coincide at the same ΔB magnification causing them to merge and lose the exact repetition characteristic of ESA. This predicted behavior is confirmed in the magneto-conductance fluctuations shown in Fig. 7.8 (left) measured for the square billiard geometry. The observed fluctuations are strikingly different from those of the Sinai billiard shown in Fig. 7.4 (left). As predicted, the fluctuations no longer cluster around selected magnetic field scales but instead are evident at *all* ΔB scales. Secondly, regardless of the magnification factor chosen, the patterns observed at different scales are clearly different: ESA is no longer observed.

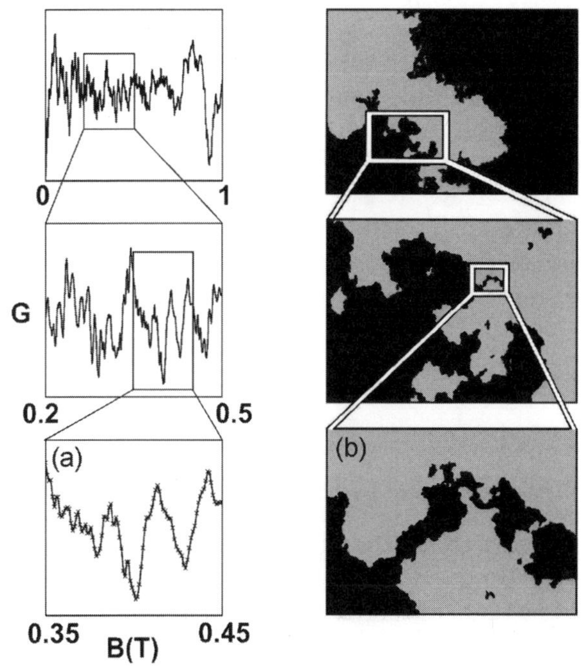

Fig. 7.8. (a) Statistical self-affinity for an empty billiard. Similar to the coastline shown in (b), $G(B)$ does not reveal an exact repetition of patterns at different magnifications.

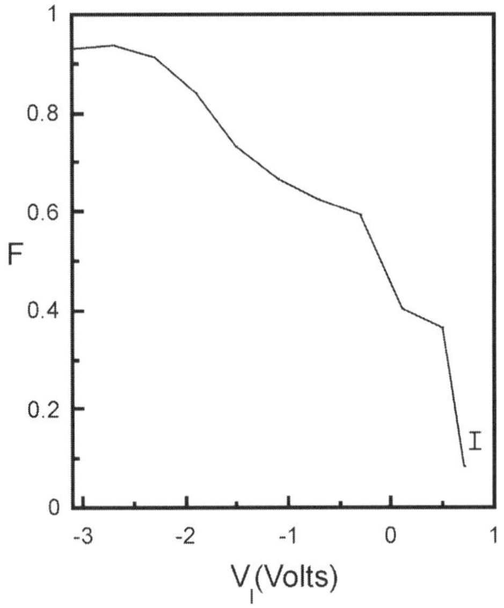

Fig. 7.9. The transition from the Sinai billiard to the empty square billiard, charted by F versus V_I. The trace is taken for $n = 7$.

In Fig. 7.9 we have used the correlation analysis to examine the suppression of ESA during the transition from the Sinai billiard to the "empty" square billiard.[1] As noted earlier, for the largest circle radius investigated in the experiment ($R = 0.37\,\mu m$), a narrow range of scaling factors was already present rather than the unique pair required for ideal ESA. It appears, therefore, that at $R = 0.37\,\mu m$ the circle was not quite large enough to act as a "perfect" scale factor selector. The full widths at half maximum extracted from the second derivative plots of Fig. 7.5(e) quantified this range in scale factors as $(\Delta\lambda_G, \Delta\lambda_B) = (2.7, 1)$. This range produced the small degradation in ESA from the ideal case of $F = 1$ to the observed value of $F = 0.94$ for $R = 0.37\,\mu m$. In Fig. 7.9, $R = 0.37\,\mu m$ corresponds to the bias $V_I = -3.1$ V. Moving from left to right, Fig. 7.9 shows that an increase in V_I towards 0 V was accompanied by a rapid fall in F, indicating a dramatic loss in ESA as the circle size was reduced and the range of allowed scale factors was increased. For $V_I = +0.7$ V (corresponding to the minimized presence of the circle) the correlation has fallen to less than $F = 0.1$. The scale factor map for the empty billiard is shown in Fig. 7.5(c,f) and this fails to yield the well-defined single maximum found in the equivalent plot of Fig. 7.5(b,e) for the Sinai billiard. Instead, the plot reveals a smooth contour with an average value of F well below 0.2 for all values of (λ_G, λ_B).[2] This scale factor map is a signature of a $G(B)$ trace that no longer contains an exact repetition of patterns at different field scales.

[1] The error bar in Fig. 7.9 represents the difference between the F value obtained by correlating two nominally-identical $\delta G_c(B)$ traces taken in opposite magnetic field directions and the expected value of $F = 1$. Therefore any deviations in F greater than this indicator can be regarded as significant.

[2] As for all plots, the observed background has a slope set by the condition $(\lambda_G, \lambda_B, F) = (1, 1, 1)$. For more details see [33].

5. The Observation of Statistical Self-Affinity

In the last section, we saw that the model of the central circle as a trajectory selector, developed during 1996–1997, successfully describes the suppression of ESA observed in the conductance fluctuations during the transition from the Sinai billiard to the empty square billiard. The model also offers further crucial predictions for the transition. Because all the hierarchies shown in the scaling plot of Fig. 7.6(b) lie on the same power-law line, the scaling properties of all the fluctuations are described by a common D value [32,33,44]. Therefore, according to the model, fractal behavior should be preserved during the transition. This can be demonstrated mathematically as follows [33]. Starting with the analysis of Fig. 7.6(a), we saw that the fractal scaling behavior of ESA is described by $\lambda_G = (\lambda_B)^\beta$, where $\lambda_B = \Delta B_m/\Delta B_{m+1}$ and $\lambda_G = \Delta G_m/\Delta G_{m+1}$ respectively. If we perform an equivalent analysis replacing ΔG_m (the conductance amplitude of a selected feature in the mth cluster) with $\langle\{\delta G(B) - \delta G(B + \Delta B_m)\}^2\rangle$ (the variance of $\delta G(B)$ with ΔB_m for the mth cluster), we obtain an analogous ESA power law expression:

$$\frac{\langle\{\delta G(B) - \delta G(B + \Delta B_m)\}^2\rangle}{\langle\{\delta G(B) - \delta G(B + \Delta B_{m+1})\}^2\rangle} = \left(\frac{\Delta B_m}{\Delta B_{m+1}}\right)^\gamma \quad \text{where } \lambda_B = \left(\frac{\Delta B_m}{\Delta B_{m+1}}\right) \tag{2}$$

and $D = 2 - \gamma/2$. Note that this equation, representing the scaling behavior shown in Fig. 7.6(a), has a unique λ_B value and holds only for ΔB values corresponding to the cluster field scales ΔB_m. In contrast, for the scaling behavior represented in Fig. 7.6(b), this condition is relaxed: the system has a continuous range of λ_B values and, consequently, the power law holds for all values of ΔB. The power law expression describing Fig. 7.6(b) therefore has the form:

$$\langle\{\delta G(B) - \delta G(B + \Delta B)\}^2\rangle \propto (\Delta B)^\gamma \tag{3}$$

where $D = 2 - \gamma/2$. Significantly, this form of equation is well-known within fractal studies. Commonly referred to as "fractional Brownian statistics", this scaling relationship has been used to describe a diverse range of natural fractal phenomena [17,45]. As shown in the coastline picture of Fig. 7.8 (right), fractals observed in nature don't exhibit ESA [17,18]. Rather than an exact repetition, the coastline patterns observed on different scales simply follow the same statistical relationship—"statistical self-affinity" (SSA).

The relatively simple mathematical steps outlined above therefore predict that the evolution from the Sinai to square geometry should be accompanied by a transition from ESA to SSA in the magneto-conductance fluctuations. If this prediction turned out to be correct then soft-walled semiconductor billiards would represent a truly remarkable environment for the study of fractal behavior—transitions between the two forms of fractals (ESA and SSA) had never been observed in a physical system. Furthermore, this possibility of observing SSA was particularly intriguing because, a year earlier, an independent theoretical investigation had also predicted SSA for soft-walled semiconductor billiards [16]. This elegant theory proposed that structure at the boundaries between chaotic and stable regions of the mixed trajectory system might be described in terms of an infinite hierarchy of Cantori [12,17,19] and that the Cantori would strongly influence the wave chaos properties of the soft-wall billiards [16]. The presence of Cantori was predicted to produce a power-law distribution of classical loop areas and, through a semi-classical wave interference process, this would be detected as fractal scaling of the conductance fluctuations [16]. Specifically, the relationship between the measured D and the power law exponent γ was predicted to be $D = 2 - (\gamma/2)$.

To confirm that the fractal character of the magneto-conductance fluctuations had indeed been preserved during the Sinai to square transition, we had to demonstrate that the scaling relationship of the fluctuations could still be described by a fractional value of D. For the ESA observed in the Sinai billiard, the D value was calculated by identifying specific features in the fluctuations and then plotting their sizes at different magnifications (see Fig. 7.6(a)). However, because there was no longer an exact repetition of features, this approach was not possible for SSA. Rather, it was necessary to calculate how the statistical qualities of the fluctuation patterns scaled. To do this, we employed four well-established methods for fractal detection and found them to be consistent [46]. One approach, based on Eq. (3), is to obtain a linear fit to the plot of $\log[\langle\{\delta G(B) - \delta G(B + \Delta B)\}^2\rangle]$ versus $\log(\Delta B)$. However, here we will concentrate on a more traditional method of fractal analysis known as the "box-counting" technique [17,18,46]. A comparison of the relative advantages of the two techniques can be found elsewhere [46]. The box-counting technique is shown in the schematic representation of Fig. 7.10(a). The magneto-conductance trace is covered with a computer-generated mesh of identical squares (or "boxes"). The statistical qualities of the fluctuation patterns can then be determined by analyzing which squares are occupied by the trace (shaded in Fig. 7.10(a)) and which are empty. These statistics can then be compared at different magnifications by reducing the square size in the mesh. In particular, D can be obtained by calculating the number of occupied squares in the mesh, $N(\Delta B)$, as a function of square size ΔB. For fractal behavior, $N(\Delta B)$ scales according to $N(\Delta B) \sim \Delta B^{-D}$ where $1 < D < 2$ [17,18]. Therefore, by constructing a scaling plot of $-\log N(\Delta B)$ against $\log \Delta B$, the fractal behavior is detected as a straight line and quantified by extracting D from the gradient. This scaling plot, shown in Fig. 7.11, is equivalent to the one shown in Fig. 7.6(a) that we used for the case of ESA.[3]

In 1997 we published results reporting that the square billiard geometry generates SSA [47]. As predicted, fractal behavior was preserved during the transition from the Sinai billiard to the empty square. The transition from $R = 0.37\,\mu\text{m}$ to (nominally) $R = 0\,\mu\text{m}$ was accompanied by a smooth reduction in D by approximately 10%. We note that, although the presence of the circle was minimized at $V_I = +0.7$ V, we cannot exclude the possibility that remnants of the circle remained. Therefore, to confirm that this second form of fractal behavior, SSA, does not require the presence of a circle, we constructed a billiard with an identical geometry to the one shown in Fig. 7.3 but without the central circular gate. Figure 7.11 shows a typical scaling plot obtained for

[3] Note that the exponent of the power law relationship used in Fig. 7.6(a) is $2 - D$ whilst the exponent used in Fig. 7.11 is $-D$. The reason for this difference is as follows. D is a "covering" dimension—it measures the amount of area covered by the trace at different magnifications (see [17,18]). To convert the box count N into a covered area, N should be multiplied by the box area $(\Delta B)^2$. This then converts the $-D$ exponent used in Fig. 7.11 into the $2 - D$ exponent used in Fig. 7.6(a).

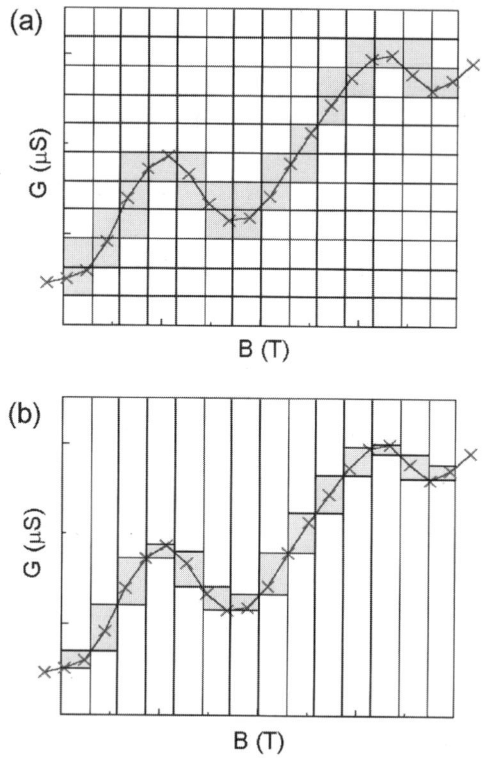

Fig. 7.10. Schematic representations of (a) the box counting technique and (b) the Variation method.

this billiard ($T = 50\,\text{mK}$ and $n = 6$) and this confirms that the completely empty billiard generates SSA. The dashed straight line is a guide to the eye, indicating that the data follow the fractal scaling relationship between the magnetic field scales ΔB_2 and ΔB_3. The values of these upper and lower cut-offs are determined using the derivative plots shown in Fig. 7.11 (a,b). The field scale marked ΔB_1 represents the magnetic field resolution limit. However, the data remain non-fractal up to ΔB_2 due to the dominance of measurement noise over signal for these small fluctuations in $G(B)$. Note, therefore, that for both ESA (Fig. 7.6(a)) and SSA (Fig. 7.11), the lower cut-off is set by noise restrictions. Just as with the case of ESA, the upper cut-off for observing SSA is set by the emergence of non-fractal (skipping orbit) conduction processes at higher fields. This change in conduction processes is marked as B_{cyc} in Fig. 7.12. This limits the length of the data trace to $B < 150\,\text{mT}$ and this is marked as ΔB_4 in Fig. 7.11. The upper cut-off in fact occurs at the field scale marked ΔB_3, which is slightly below ΔB_4. The field scale ΔB_3 forms the upper cut-off rather than ΔB_4 because it is necessary to fit at least 50 squares in the mesh in order to achieve sufficient box-counting statistics to resolve the fractal behavior [17,18,46].

Figure 7.11(c) focuses on the data lying within the fractal scaling regime and shows the individual data points. To produce this plot we employed two refinements

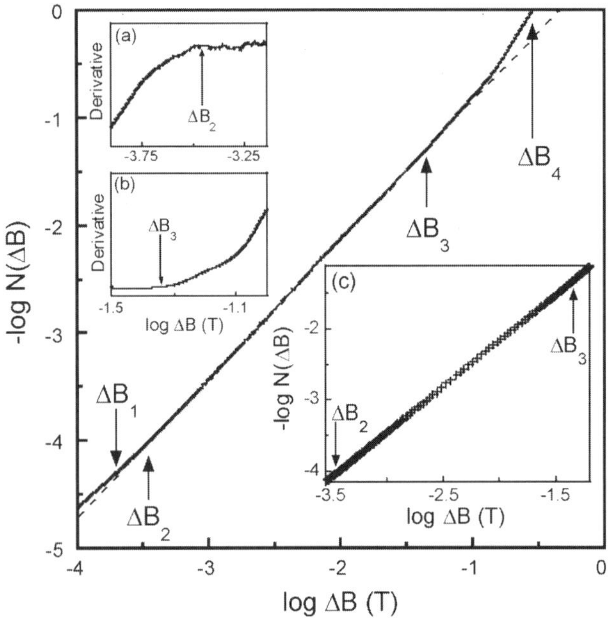

Fig. 7.11. The scaling plot for an empty billiard. The data (solid) follows a straight line (dashed line) between the magnetic field scales ΔB_2 and ΔB_3. Insets (a) and (b) are derivative plots of the data and (c) shows the individual data points (for clarity only every tenth point is shown).

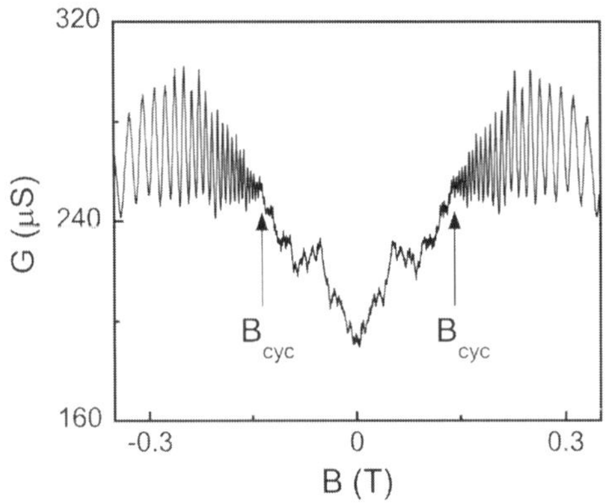

Fig. 7.12. A plot of $G(B)$ showing the change in conductance behavior above the magnetic field B_{cyc} at which the cyclotron radius fits within the billiard size. Above this field, electron transport is described by skipping orbits and the conductance is non-fractal. B_{cyc} therefore forms the ΔB_4 cut-off shown in Fig. 7.11.

to the box-counting method—the Variation method (for which the data point density increases at small ΔB) and the Horizontal Structured Elements method (for which the density increases at large ΔB) [48]. Here we briefly discuss the Variation method, which is compared to the basic "box-counting" method in the schematic of Fig. 7.10.

Strictly speaking, the technique of extracting D should calculate the minimum number of non-overlapping boxes required to cover the data trace. Although the boxes shown in Fig. 7.10(a) are not overlapping, the rigidity provided by the mesh leads to a small over-estimation of the number of boxes required to cover the data trace. We remove this error using the Variation method, where vertical columns of width ΔB are used rather a mesh of squares of size ΔB. The total shaded area shown in Fig. 7.10(b) is then calculated and divided by the unit box size $(\Delta B)^2$ to obtain the minimum number of boxes required to cover the trace. By adopting the Variation and Horizontal Structured Elements methods rather than the simple box-counting method we remove errors in the D value, and by combining the two methods on one scaling plot we achieve a high data point density across all ΔB values, ensuring a reliable detection of fractal scaling. The gradient of the dashed line shown in Fig. 7.11 corresponds to $D = 1.29$.

Whereas we use plots such as Fig. 7.11 to quantify "fractal scaling"—the statistical relationship between patterns observed on different magnification scales—we also use plots such as Fig. 7.13 to investigate "fractal displacement"—the statistical relationship between patterns observed at different magnetic field locations along the trace. In Fig. 7.13(d) we show an experimental $G(B)$ trace ($n = 3$) that exhibits SSA with $D = 1.28$. To quantify how the statistical qualities of patterns vary along the $G(B)$ trace, we introduce the concept of a "local" fractal dimension D_L. Whilst D is the "universal" fractal dimension, extracted from a scaling plot that assesses the full length of the

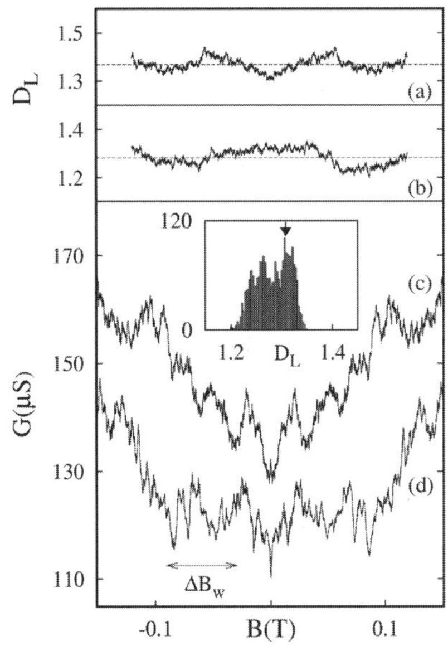

Fig. 7.13. Investigations of fractal displacement. (a) A plot of local fractal dimension D_L versus window position for the simulated $G(B)$ trace shown in (c). The D_L value "meanders" around the D value indicated by the dashed line. (b) A similar plot for the experimental $G(B)$ trace shown in (d). The histogram plots the distribution of D_L around the D value (indicated by the arrow) for the experimental case.

$G(B)$ trace, D_L is extracted from a scaling plot that assesses only a selected region of the trace. The window size of this region is indicated by the horizontal bar next to the experimental $G(B)$ trace ($\Delta B_W = 60\,\text{mT}$). Figure 7.13(b) shows the D_L value plotted as a function of the B field location of the window's center along the experimental $G(B)$ trace. The D_L value "meanders" around the D value of the trace (indicated by the dashed line). We stress that this variation in D_L is not a random variation due to measurement restrictions associated with fitting over a narrow range in the scaling plot (this random variation is significantly smaller and can be seen superimposed on the "meander" variation). The histogram shows the distribution of D_L values around the D value (indicated by the arrow). Due to the limited number of points, it is not possible to determine the exact distribution.[4] We note, however, that a $G(B)$ trace (Fig. 7.13(c)) generated by fractional Brownian statistics [12,17,45] with an identical D value to the experimental trace has a very similar meander (see Fig. 7.13(a)) to that revealed by the experimental data. This behavior of the local dimension D_L meandering around a constant D value is significant—the fact that D does not evolve with magnetic field indicates that, over the range considered ($-150 < B < 150\,\text{mT}$), the changing field is not altering the dynamics of the classical loops. This result is in contrast to other studies, where application of a magnetic field has been pictured in terms of altering the mixed character of the system and enhancing the fractal conductance [49].

6. The Classical to Quantum Transition: How do Fractals "Disappear?"

Since the initial Sinai experiment in 1995, a number of research groups have observed SSA in soft-walled semiconductor billiards [49–51]. Whereas most of these studies have varied the electron phase by sweeping magnetic field, others have used a back-gate to vary the phase through a changing Fermi energy E_F [49]. In our own experiments spanning the period from 1995 to 2000, we measured SSA for an unexpectedly wide range of billiard parameter values. Furthermore, SSA has even been observed in gold nano-wires [52]. This prevalence of SSA suggests a remarkably robust phenomenon, prompting us to ask a fundamental question—how do fractals deteriorate into non-fractal behavior as the generating process is gradually suppressed? Whereas this question could, in principle, be applied to any fractal system, the semiconductor billiard is of particular interest for two distinct reasons. Firstly, many studies of fractals are "passive", where rather than actively interacting with the process generating the fractals, the experiments are limited to a simple monitoring role [18]. In contrast, the parameters thought to influence the fractal behavior in the billiard system can be adjusted with precision, allowing a systematic study. Secondly, whereas most of nature's fractals are classical in origin [17], the billiard's fractals were thought to arise from a semi-classical process (see sections 1 and 5). We planned therefore to suppress

[4] The only way to increase the number of points plotted in the distribution is to decrease the window size. However, this increases the random variations in D_L due to the narrower fitting range used in the scaling plots.

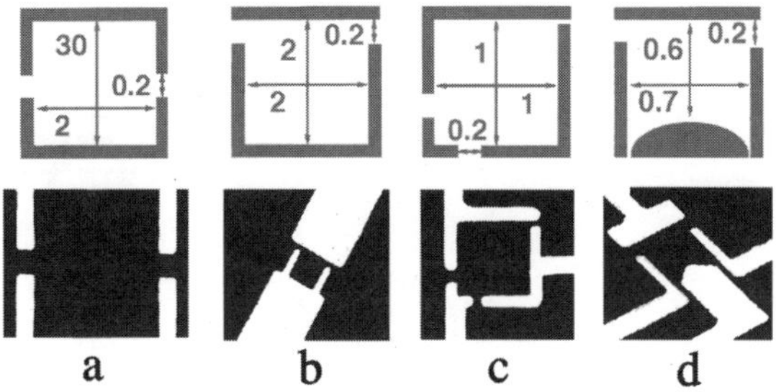

Fig. 7.14. Schematics of the gate patterns (not to scale) together with the equivalent scanning electron micrographs for four (*a–d*) of the seven billiards used to investigate the classical-quantum transition. The dimensions are in μm. Billiards *e*, *f* and *g* have the same geometry as *b* but are different sizes (see Table 7.1).

Table 7.1. A summary of the billiard parameters discussed in section 6 of the text

Device	A (μm^2)	T (K)	ℓ (μm)	n	τ_q(ps)	Q
a	60.0	0.03	25	4	16.6	0.003
b	4.0	0.03	4.4	2	109.6	0.30
c	1.0	0.03–4.2	12	2–6	2.8–182.9	0.012–1.99
d	0.22–0.09	0.03	5.5	2	75.0–87.9	4.34–9.05
e	1.0	0.03–4.2	4.4	2	24.9–113.2	0.02–1.23
f	0.36	0.03–2.5	4.4	2	20.1–98.7	0.08–2.96
g	0.16	0.03	4.4	2	83.4	5.67

this semi-classical process using controlled transitions towards fully classical and fully quantum–mechanical conduction.

To span the full experimental range of classical, semi-classical and quantum conduction in a comprehensive fashion, in 2000 we re-visited all of the data we had accumulated over the previous five years and also performed new experiments in the parameter ranges previously unexplored [53]. Figure 7.14 and Table 7.1 summarize the properties of the seven billiard investigated. Billiards with similar Fermi wavelengths (typically 40 nm) but different sizes (ranging from 0.4 μm to 30 μm billiard width) were studied at different temperatures (ranging over two orders of magnitude from 0.02 to 2 K). We started by considering a typical micron-sized billiard (device *e*) where the ratio S of the billiard size to the Fermi wavelength was 22, ensuring semi-classical transport [10,11,16]. The magneto-conductance measured at $T = 30$ mK (corresponding to an electron phase coherence length of 30 μm) is shown in the fourth (from top) trace of Fig. 7.15 and reveals the complex fluctuation patterns expected for FCF behavior. The remaining traces in Fig. 7.15 demonstrate the effect of varying T and the area enclosed by the billiard A. An increase in T reduces the electron phase coherence length and is therefore expected to induce more classical conduction. In contrast, a decrease in A reduces S and should therefore increase the importance of purely quantum–mechanical transport processes. It is clear from Fig. 7.15 that the complexity of the measured fluctuations decreases by increasing T (moving to the upper traces) or

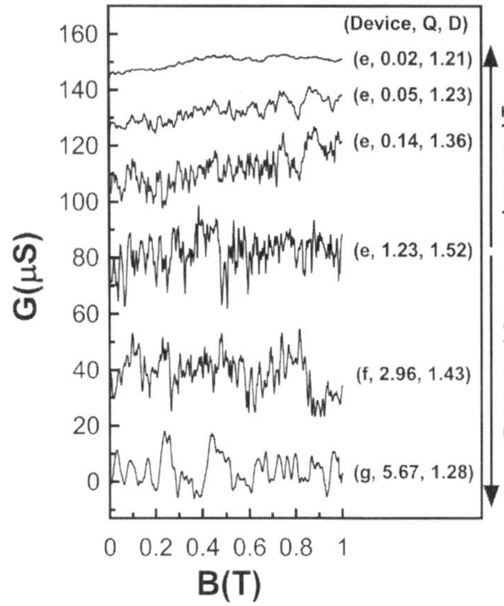

Fig. 7.15. Magneto-conductance fluctuations measured for billiards *e*, *f* and *g* (see trace labels) with $n = 2$ and $\ell = 4.4\,\mu m$. From top to bottom the (T, A) values for the traces are $(4.2\,K, 1.0\,\mu m^2)$, $(1.4\,K, 1.0\,\mu m^2)$, $(0.48\,K, 1.0\,\mu m^2)$, $(0.03\,K, 1.0\,\mu m^2)$, $(0.03\,K, 0.36\,\mu m^2)$ and $(0.03\,K, 0.16\,\mu m^2)$. The right-hand brackets indicate (Device, Q, D) (see text for the definition of Q). The traces are offset vertically for clarity.

reducing A (moving to the lower traces obtained from the smaller billiards f and g). This suggests that deviations from semi-classical conduction within the billiard induce a profound change in the scaling properties of the fluctuations.

At first, this observation seemed to be in agreement with theoretical expectations: as a semi-classical phenomenon, the FCF should smooth out in the classical and quantum regimes [16,50]. To quantify the observed changes in scaling behavior we employed the Variation analysis to calculate D during the transition from classical to quantum–mechanical conduction. As T and A were adjusted to suppress the semi-classical process, the most anticipated scenario for the evolution of FCF towards a non-fractal trace was the one illustrated in Fig. 7.16(a)—the value of D was expected to remain constant whilst the range of ΔB over which fractal behavior was observed decreased to zero [16,50]. In other words, the transition would be marked by a gradual deterioration in fractal observation, shifting the lower cut-off towards the higher cut-off, until the fractal scaling disappeared altogether. For example, consider raising T to induce the transition to the classical regime. In the traditional semi-classical picture, the ΔB spectrum of conductance fluctuations is directly related to the distribution of areas enclosed by classical trajectories (for a given trajectory, the Aharonov–Bohm effect predicts ΔB to be inversely proportional to the enclosed area) [14]. Increasing T renders the longer trajectories incoherent, preventing the loops with the largest enclosed areas from contributing to the FCF process. Thus the fluctuations with small ΔB should be suppressed first, leaving the large ΔB fluctuations relatively unaffected, as indicated in Fig. 7.16(a). Similarly, in the semi-classical to quantum transition, the deterioration

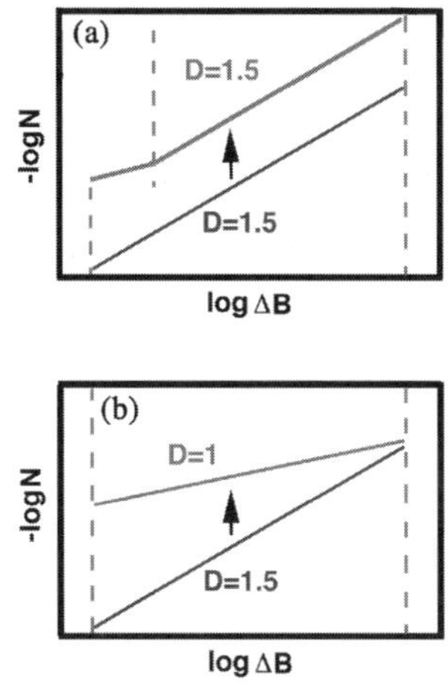

Fig. 7.16. Schematic representations of scaling plots showing (a) the scenario of fractal deterioration anticipated from semi-classical theory, and (b) the measured evolution in fractal behavior.

in FCF behavior is predicted to set in when the semi-classical theory becomes invalid. This occurs when the electron states inside the billiard start to become resolved (as determined by the Heisenberg time, $\tau_H = m^* A/h$, where m^* is the electron effective mass).[5] As the billiard area is made smaller and consequently τ_H also becomes smaller, loops with traversal times longer than τ_H will no longer contribute to the semi-classical process [16,50,53,54]. According to the Aharonov–Bohm effect, these longer loops correspond to small ΔB values. Thus, as indicated in Fig. 7.16(a), during the transition to fully quantum behavior, the fluctuations with small ΔB should be suppressed first, leaving the large ΔB fluctuations relatively unaffected.

In sharp contrast to these expectations, the fractal analysis revealed a more remarkable behavior, illustrated in Fig. 7.16(b), where the value of D decreased *gradually* towards the non-fractal value of 1 in both the classical and quantum limits! [55] Due to its increased sensitivity to small changes, we demonstrate this behavior using the derivative technique introduced in Fig. 7.11(a). The derivative plot of Fig. 7.17 shows the effect of changing T. In this figure, the D values (measured for the data of Fig. 7.15) are labeled next to the respective traces. The key observation is that the fractals turned out to be remarkably robust to deviations from the semi-classical regime. The fluctuations did *not* lose their fractal character (as defined by their fractional D value) as long as fluctuations could be observed. Rather than the expected deterioration (Fig. 7.16(a)), a smooth evolution in the fractal scaling properties was observed (Fig. 7.16(b)), involving an evolution of the entire ΔB spectrum [53,55].

[5] The billiard's average energy level spacing ΔE_S is calculated by dividing the energy of the highest occupied level (the Fermi energy) by the total number of occupied energy levels and this equals $2\pi\hbar^2/m^* A$ (see [54]). By definition, the Heisenberg time is then given by $\tau_H = \hbar/\Delta E_S$ (see [16,50,54]).

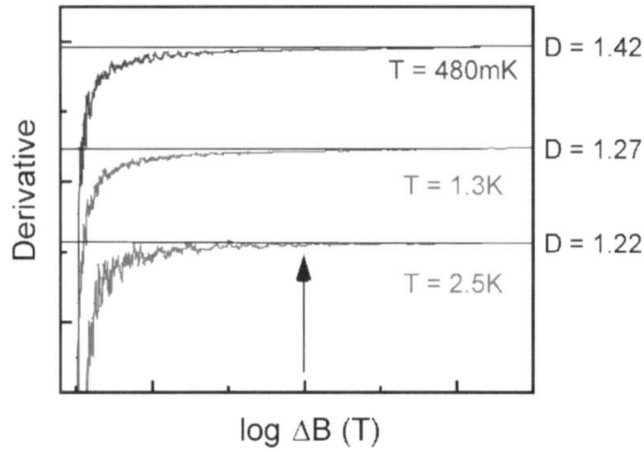

Fig. 7.17. Derivative of $-\log N$ versus $\log \Delta B$ for three temperatures (480 mK, 1.3 K and 2.5 K) showing that the lower cut-off (indicated by the arrow) remains constant but D changes as a function of temperature.

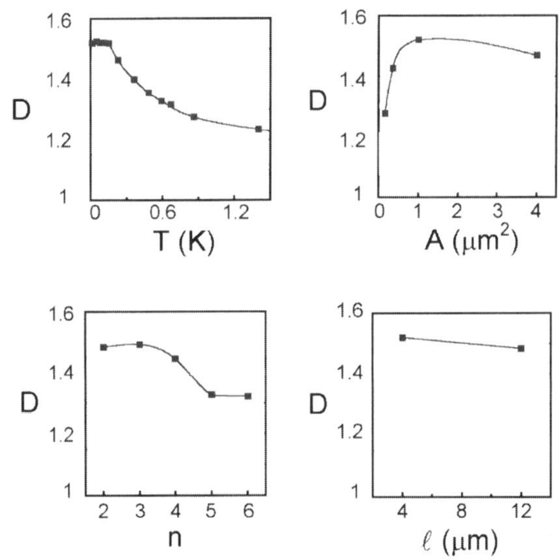

Fig. 7.18. D plotted against T, A, n and ℓ. ($A = 1\,\mu\text{m}^2$, $n = 2$, $\ell = 4.4\,\mu\text{m}$) for the T plot, ($T = 30\,\text{mK}$, $n = 2$, $\ell = 4.4\,\mu\text{m}$) for the A plot, ($T = 30\,\text{mK}$, $A = 1\,\mu\text{m}^2$, $\ell = 4.4\,\mu\text{m}$) for the n plot and ($T = 30\,\text{mK}$, $A = 1\,\mu\text{m}^2$, $n = 2$) for the ℓ plot.

To further confirm this unusual behavior, we next considered *all* four adjustable experimental parameters (T, A, n and ℓ) for the seven billiards presented in Fig. 7.14 and Table 7.1. As demonstrated in Fig. 7.18, the plots of D versus T, A, n and ℓ were all found to have different forms. However, we found that D changed in a *universal* way when plotted against a more general parameter Q that quantifies the transition towards resolvable, quantized energy levels within the billiard. We define Q as the ratio of the average energy level spacing ΔE_S to the average energy level broadening

ΔE_B. The condition $Q = 0$ therefore corresponds to $\Delta E_S = 0$ or $\Delta E_B = \infty$. Q can then be increased either by increasing ΔE_S or decreasing ΔE_B, improving the resolution of the discrete energy levels. We calculate ΔE_S to equal $2\pi\hbar^2/m^*A$ (see footnote 5 on p. 301). ΔE_B is given by the quadrature summation of the billiard's two characteristic energy broadening widths—thermal broadening kT and the intrinsic energy level broadening $\hbar/\tau_q$ originating from phase-breaking scattering events that limit the lifetime τ_q of the quantum states.[6] Thus:

$$Q = \frac{\Delta E_S}{\Delta E_B} = \frac{(2\pi\hbar^2)/(m^*A)}{\sqrt{(\hbar/\tau_q)^2 + (kT)^2}} \tag{4}$$

To extract τ_q from the data, we used a well-established technique, commonly applied in semiconductor billiards, that analyses the correlation field of the data as a function of magnetic field [57]. In Fig. 7.19(a–c), we demonstrate how τ_q was

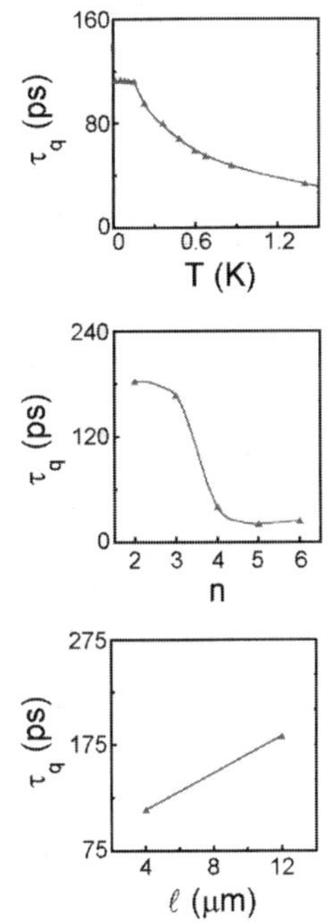

Fig. 7.19. τ_q plotted against T, n and ℓ. ($A = 1\,\mu m^2, n = 2, \ell = 4.4\,\mu m$) for the T plot, ($T = 30\,mK, A = 1\,\mu m^2, \ell = 4.4\,\mu m$) for the n plot and ($T = 30\,mK, A = 1\,\mu m^2, n = 2$) for the ℓ plot.

[6] Guided by the Boltzmann picture of conductance, we determine ΔE_B from a convolution of the two energy broadening distributions—the scattering-broadened density of states and the derivative of the Fermi–Dirac function (see [56]). We find that, for the parameter ranges considered, ΔE_B can be most closely approximated by the quadrature summation of the two characteristic widths kT and $\hbar/\tau_q$.

varied using the experimental parameters. Reducing T induced the rise in τ_q shown in Fig. 7.19(a). The saturation at low T is a common observation in semiconductor billiards [54,55,57,58] and in our experiments coincided with the condition $kT < \Delta E_S$. In Fig. 7.19(b), the gate bias V_g has been tuned in order to change n. We found that τ_q decreased for larger values of n due to an increased coupling to the external 2DEG environment [53,57,58]. We also varied ℓ by constructing billiards using different quality materials and found that τ_q scaled with ℓ (Fig. 7.19(c)). In addition to varying ΔE_B through changes in τ_q and T, we also varied ΔE_S through changes in A. To investigate a large range of A values, we constructed billiards with different physical areas. Tuning V_g was also used to change A. In particular, the side-gate featured in billiard d (see Fig. 7.14) could be used to reduce A to less than 20% of the physical area enclosed by the gates. Thus, in summary, we varied Q using all four available billiard parameters: T, A, n and ℓ. Table 1 shows the ranges of Q achieved.

In Fig. 7.20(a), all of the D measurements from the seven different billiards are plotted against Q. In contrast to the distinct forms of the D variations shown in Fig. 7.18, all of the D data condense onto a single "universal" curve when plotted as a function of Q [53]. Therefore it is clear that Q, which gives a measure of how well the energy levels of the billiard are resolved, is a particularly useful parameter for charting the variations in D. Moving from left to right across Fig. 7.20(a), D rises and then falls. This trend can be seen in the raw FCF data by moving down through Fig. 7.15. In the extreme limit of $Q = 0$, D assumes the non-fractal value of 1. As Q is increased, D rises smoothly until it reaches a peak value of 1.52 at $Q = 1$. Further increases in Q are accompanied by a gradual decrease in D. Extrapolating the downward trend in D, we find that to obtain $D = 1$ would require $Q = 11$. By this stage, the billiard width matches that of the openings and the device resembles a long channel.

It is important to emphasize that Q is an empirical parameter, identified from experiments rather than based on theory. Q is distinctly different to S, the parameter traditionally associated with theories charting evolutions in the semi-classical nature of a system. As discussed in the Introduction, S is the ratio of the billiard size to the electrons' Fermi wavelength. A large S value corresponds to fitting a large number of

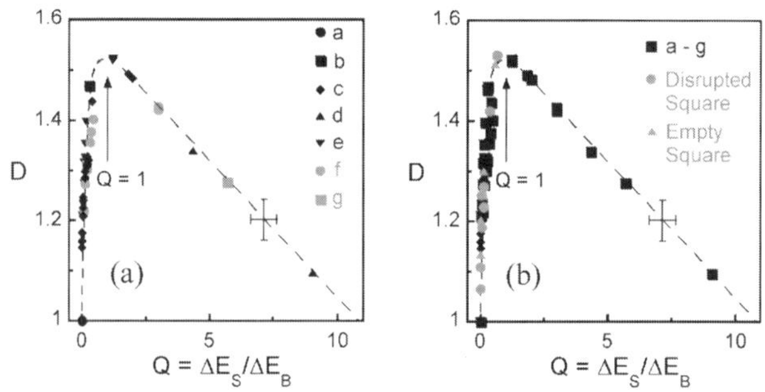

Fig. 7.20. "Q curves" generated by plotting D against Q. (a) The Q curve for the original seven single-layer billiards, labeled a to g. The error bars indicate the largest possible uncertainties in D and Q. (b) The D values of the empty square and disrupted square are plotted against Q along with the original seven billiards (a to g). The dashes are guides to the eye.

wavelengths along a typical trajectory—a necessary condition for most semi-classical models.[7] Whereas S is related to the ratio of Fermi energy to the energy level spacing, Q is the ratio of energy level spacing to energy level broadening. Thus, billiard parameters which alter the energy broadening (e.g. τ_q and T) can be used to alter Q whilst leaving S unaffected. Similarly, parameters that alter the Fermi energy can be used to alter S whilst leaving Q unaffected.[8] This difference in their dependences on billiard parameters can be used to establish that that Q rather than S is indeed the relevant parameter for charting the evolutions in D shown in Fig. 7.20(a). For example, billiards with the same Q values but substantially different S values have the same measured D values. However, data points with the same S values but different Q values have different measured D values. Having established that Q is the relevant parameter for charting the evolutions in D, in the next section we will discuss some unusual and surprising consequences of this result.

7. The Role Played by the Billiard Walls

A central characteristic of the ballistic scattering regime is that the device walls play the dominant role over material-related scattering in determining the device conduction processes. The expectation is that geometry-related factors such as billiard shape and positioning of the entrance and exit should have a profound influence on quantum interference effects such as FCF. The results represented in the "Q curve" of Fig. 7.20(a) stand in contradiction to this picture. Different billiard shapes and orientations of the openings were purposely used for the billiards examined (see Fig. 7.14). Yet Fig. 7.20(a) demonstrates that these factors do not influence D. For example, the bottom gate of device d (see Fig. 7.14) not only changes A but also the billiard's shape from that of a square to a thin rectangle, yet the induced evolution in D is well-charted by Q—a parameter which is insensitive to billiard shape. According to the Q curve, two billiards with radically different geometries will generate identical D values provided Q is set to be the same. To emphasize this result we examined the two billiards shown in Fig. 7.21. The two billiards were defined within one micrometer of each other on the same AlGaAs/GaAs heterostructure to ensure that they were described by identical material-related parameters. The only difference, therefore, between the two devices is the geometry—in particular, the finger-shaped extension incorporated into the gate design of the right-hand billiard. By raising the temperature to 4 K to remove quantum interference effects, the classical magneto-conductance of each billiard was measured. As expected, the finger extension radically altered the distribution of the classical trajectories, leading to very different traces. Furthermore, when the devices were cooled to 20 mk and the FCF emerged on top of the classical background, there were clear

[7] It is relatively easy to show that S is inversely proportional to the effective Planck's constant of the system. A necessary condition of semi-classical pictures is that the effective $\hbar$ must be significantly smaller than $\hbar$ (see [16]).

[8] Electron density is the parameter most commonly used to change the Fermi energy. Adjustments of electron density will therefore lead to direct changes in S. Note, however, that changes in electron density also indirectly affect Q—electron density alters the screening properties of the 2DEG, leading to changes in τ_q and thus ΔE_B.

Fig. 7.21. A scanning electron micrograph of two surface gate patterns deposited on the same heterostructure, used to demonstrate that radically different billiard geometries generate the same Q curve. The square pattern is 1 μm across.

differences in the individual features of the two sets of fluctuations. However, the statistics of the two FCF traces, as charted with D, were found to be identical. This is demonstrated in Fig. 7.20(b), where the data for both devices condense onto the *same* "Q curve".

If the shape of the walls doesn't influence the fractal properties, perhaps the potential profile of the walls isn't important either? This defines a critical question in our investigations because the soft character of the profile is known to generate the mixed stable-chaotic trajectories believed to produce the FCF. More specifically, D is predicted to have a critical sensitivity to the *precise* profile [16,50]. To investigate this FCF dependence on profile, we designed a unique "double-layer" billiard system in which a common set of surface-gates define billiards in two 2DEGs located at different depths beneath the heterostructure surface. Using the procedure outlined below, the two billiards have nominally identical geometries but different soft-wall profiles, making them an ideal system for studying the relationship between quantum interference phenomena and wall profile.

The soft-wall profile of billiards is known to be sensitive to both surface-gate bias V_g and 2DEG depth z [59,60]. Previously, FCF were investigated as a function of V_g due to the simplicity of adjusting V_g compared to z [50]. However, adjustments to V_g also modify A and n, both of which are known from the previous section to induce independent changes in the FCF. This obscures the relationship between the profile and the measured FCF. The "double-layer" billiard was designed to overcome this problem. Figure 7.22(a) shows the device, which used a common set of surface gates to define billiards in "shallow" ($z = 90$ nm) and "deep" ($z = 140$ nm) 2DEGs. Note that the previous billiards examined in section 6 (see Fig. 7.14 and Table 7.1) were defined in shallow 2DEGs. Fabrication details (including the "isolation" gate technique that allowed the upper and lower 2DEGs to be measured separately) are discussed elsewhere [61]. Prior to activating the billiard gates, we matched the electron densities of the two 2DEGs by adjusting the bias applied to a back-gate [61]. The matched electron densities were 2.85×10^{15} m^{-2} (corresponding to a matched E_F of 10.2 meV and a Fermi wavelength of 50 nm) and the electron mobilities in the shallow and deep 2DEGs were 130 and 110 m^2/Vs respectively (corresponding to ℓ values of 12 μm and 10 μm respectively).

Because of the 50 nm separation between the 2DEGs, electron tunneling and interaction effects were expected to be negligible [62,63]. In Fig. 7.23(a), the shallow

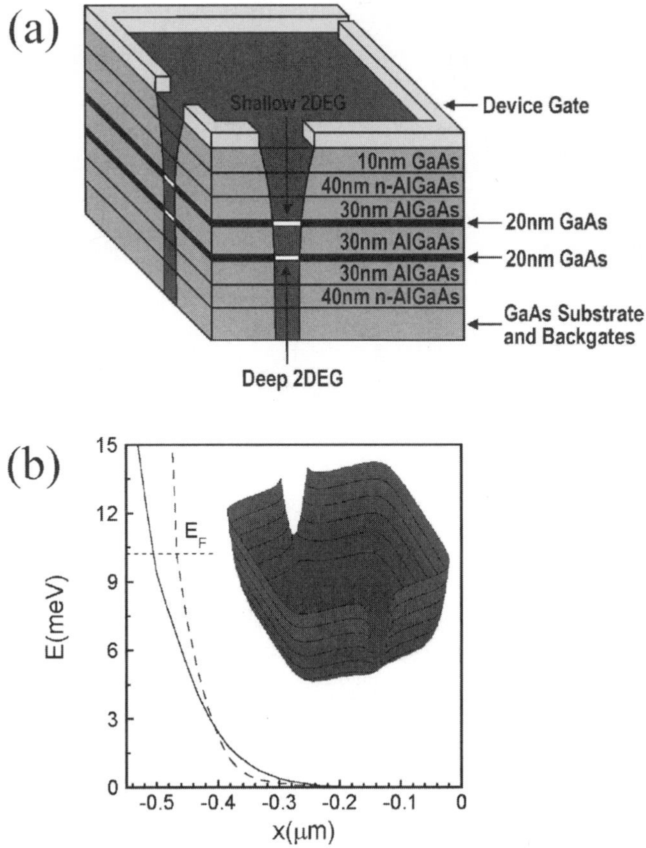

Fig. 7.22. (a) A schematic representation of the double-layer AlGaAs/GaAs heterostructure system in which a common set of gates are used to define billiards in the "shallow" and "deep" 2DEGs. (b) (Inset) The confining potential of a soft-wall billiard. The potential energy (plotted vertically) has an approximately parabolic profile (plotted against spatial location) at the billiard edges and a flat bottom in the central region of the billiard. (b) (Main) Self-consistent profile models for the shallow (solid line) and deep (dashed line) billiards with $n = 6$. Energy E is plotted against spatial location x, where $x = 0$ corresponds to the center of the billiard. A 16% difference is predicted for the areas of the two billiards.

billiard is shown to exhibit *identical D* behavior to the shallow billiards in single-layer systems, confirming that the FCF are not affected by coupling effects between the shallow and deep billiards. We now describe the procedure for measuring the FCF shown in Fig. 7.24(b). First, the deep 2DEG was electrically disconnected from the measurement circuit (using the isolation gates) [61] in order to measure just the shallow billiard. The biases of the three billiard surface gates (see Fig. 7.22(a)) were then individually tuned to form a billiard in the shallow 2DEG with $n = 2$ and the FCF for this shallow billiard were then measured (upper trace of Fig. 7.24(b)). The shallow 2DEG was then disconnected and the three gate biases were re-tuned to form a billiard in

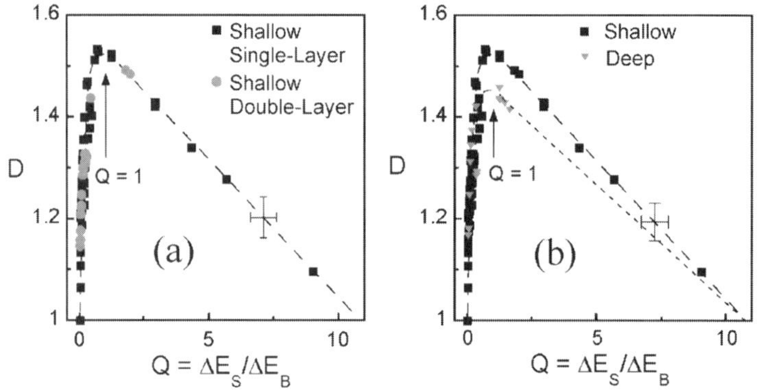

Fig. 7.23. "Q curves" generated by plotting D against Q. (a) The D values for the shallow double-layer billiard are plotted against Q, along side the D values for the nine single-layer billiards of Fig. 7.20(b). (b) The D values of the deep billiard are compared to the D values shown in (a).

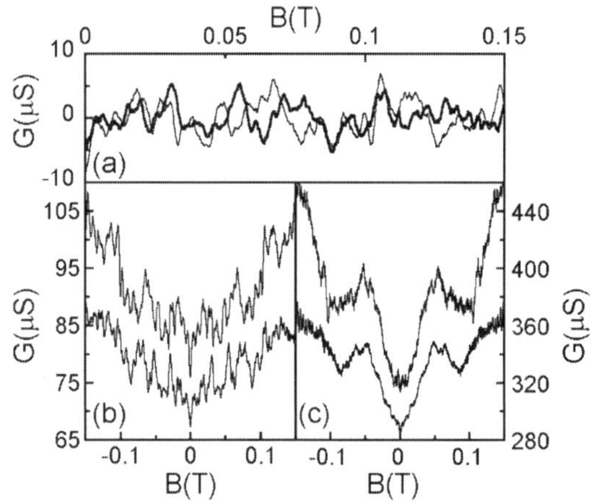

Fig. 7.24. (a) An overlay of FCF traces (with classical backgrounds removed and for $B > 0$ only) from the deep and shallow billiards for $n = 2$. $G(B)$ traces for the shallow (upper traces) and deep (lower traces) billiards at (b) $n = 2$ and (c) $n = 8$. The shallow billiard traces are offset by 20 μS and 5 μS respectively for clarity. For all traces $T = 20$ mK.

the deep 2DEG with $n = 2$ and the FCF for this deep billiard were measured (lower trace of Fig. 7.24(b)). Using this technique, the two billiards (shallow and deep) were independently measured for identical n values and had A values (measured using the high field Aharonov–Bohm effect)[7] matched to within 14%, a difference that previous studies (section 6) indicated will produce less than a 1% change in D.

In terms of geometry (n, A and gate shape), the two billiards were, therefore, essentially matched. Crucially, however, the shallow billiard profile was softer than that of the deep billiard due to the smaller V_g required to define it [64]. The profiles, calculated using self-consistent numerical simulations, are shown in Fig. 7.22(b). The profile gradient P at E_F differs by a factor of three between the two billiards. Note also that the difference in profile is not limited to the edge of the billiard. Figure 7.22(b) shows that the two profiles differ by at least 0.5 meV (corresponding to 5% of E_F) across more than a quarter of the width of the billiard. Given the predicted critical sensitivity to profile [16,50], this is expected to significantly re-distribute the classical trajectories. Our double-layer billiard strategy of using a common set of surface gates offers significant advantages over other possible approaches to this experiment. One alternative approach, adopted in [50], simply uses the changes in V_g applied to a single-layer billiard to induce variations in P. However, this does not allow independent control of n and P: e.g. our simulations show that a change in P of only a factor of two is accompanied by a change in n from 2 to 6. Another approach would be to use two single-layer billiards of different lithographic sizes and to apply different V_g to the two billiards to induce the same A and n but differing P. However, this approach introduces problems associated with lithographic variations in the two sets of gate patterns defining the two billiards and, since the mixed electron dynamics are extremely sensitive to the precise billiard geometry [16,50], this is not desirable.

Magneto-conductance $G(B)$ traces measured at $T = 20$ mK are shown in Fig. 7.24(b,c) and reveal FCF superimposed on a classical background. For each of the two n settings, the similarity between the classical structure of the deep and shallow billiards confirms that both have the same nominal geometry (i.e. size and shape). To better compare the actual FCF, in Fig. 7.24(a) we present an overlay of the traces in Fig. 7.24(b) with their classical backgrounds removed. It is evident that the differing wall profile induces a significant change in the precise details of the FCF, and intuitively, one might expect that the fractal scaling has changed, as predicted by theory [16,50]. We investigated the effect of the change in profile on two FCF parameters—the ΔB range over which fractal scaling was observed and D. The locations of the upper and lower limits of fractal scaling matched for the two billiards, and this is expected since neither limit is related to the soft-wall profile (see section 5). In Fig. 7.23(b), the Q curve obtained for the deep billiard is compared to that of the shallow billiards. In the regime $Q < 1$, the D values for the deep and shallow billiards agree within experimental uncertainties. In other words, although the difference in billiard profile is sufficient to induce changes in the individual features of the FCF, the statistical characteristics of the FCF have not changed. At $Q = 1$, the D values obtained for the deep billiard (with the harder profile) are lower than those measured for the shallow billiards. This raises the possibility that the deep billiard is following a slightly different Q curve and that this difference might persist in the fully quantum regime of $Q > 1$ (the dashed line indicates a possible scenario for the second Q curve). Although technically challenging, we plan to investigate smaller sized billiards to compare deep and shallow billiards in the $Q > 1$ regime.

8. Conclusions

In this chapter, we have reviewed our experimental investigations of the fractal behavior observed in the low temperature magneto-conductance of semiconductor billiards. We found that the fractal properties could be detected over scaling ranges of up to 3.7 orders of magnitude in ΔB—substantially larger than the observational ranges of typical physical systems. As a consequence, we were able to determine the parameters that quantify the fractal properties (such as the fractal dimension D and similarity correlation F) with great accuracy. We showed that the billiards could be used to induce two distinct forms of fractals—exact and statistical self-affinity. We exploited the high degree of control provided by the electrostatic gate technique to study the experimental conditions for which each form of fractals is observed. Because this existence of two forms of fractals in one physical system is rare, these experiments provided a unique environment in which the fundamental evolution in fractal form could be studied.

Having noted the scientific appeal of semiconductor billiards as a physical system in which to study fractal phenomena, the subtlety of the relationship between the fractal conductance and the dynamics of the electrons should also be stressed—after over five years of investigation, there is no theoretical framework that can fully explain the observed properties of either the exact or statistical self-affinity. The generation of exact self-affinity using the central "plunger" gate of the Sinai billiard (see Fig. 7.3) is linked to the formation of a scattering region at the billiard's center. However, this central scatterer does not act as a Sinai diffuser, as originally intended. According to Sinai's theory, the Sinai diffuser would have had the same effect on the billiard's dynamics if the curved surface had been incorporated into the billiard side-wall rather than at the center. Yet device *d* (see Fig. 7.14), which features a curved side-gate, generates statistical rather than exact self-affinity. This confirms that the role of the circular gate in Fig. 7.3 is more subtle than simply introducing a curved surface into the billiard's geometry—its insertion *at the billiard's center* is crucial for the generation of exact self-affinity. Due to the complexity of the mixed chaotic-stable trajectory system revealed in the Poincaré plots, the plunger gate's action on individual trajectories is not fully understood. However, from the discussions of section 4, it is clear that the central scatterer acts as a "trajectory selector".[9] Removal of the scatterer introduces extra sets of trajectories into the billiard, which together with the original trajectories, lead to the observed statistical self-affinity.

This statistical self-affinity appears to be a generic property of "empty" billiards. The evolution in statistical self-affinity charted by the Q curve extends across a remarkably comprehensive range of billiard conditions. In particular, in following the Q curve, the fractal behavior persists well beyond the specific conditions required by existing semi-classical theory, which links the fractal fluctuations to the areas enclosed by the classical trajectories [16,50]. The region of the Q curve where the semi-classical theory *is* valid centers around $Q = 1$ (in this region both τ_q and S are sufficiently large for the semi-classical theory to hold), where the curve peaks at $D = 1.52$. Significantly, this peak value of 1.52 matches the maximum value predicted by the semi-classical theory. Therefore, across this narrow range of the Q curve, the experimental observations match the semi-classical predictions for D. However, the manner in which D evolves on moving away from the $Q = 1$ region reveals dramatic discrepancies between the experiment and the predictions of the semi-classical theory.

[9] This picture is in agreement with recent theoretical investigations that reproduced the exact self-affinity (see [65]). In this theory the selected trajectories generating the exact self-affinity are a set of self-similar unstable periodic orbits.

According to the semi-classical theory, the mixed chaotic-stable system of classical trajectories features a power-law distribution of loop areas and D is directly related to the exponent γ of this distribution according to the expression $D = 2 - (\gamma/2)$ (see section 5). Crucially, D should therefore depend only on parameters that affect γ through re-arrangements of the loop distribution. According to this theory, D should not depend on temperature T or parameters that determine τ_q. However, this central prediction is *not* in agreement with the experimental observations, where D depends on both T and τ_q through Q (see Eq. 4). Again in conflict with our experiments, the dependence on A is predicted by the semi-classical theory to be erratic since it will lead to trajectories with different classical power-law exponents. However, the experimental observations charted by Eq. 4 reveal a smooth, well-defined evolution of D as a function of A. In addition to contradicting the original semi-classical predictions, Eq. 4 also stands in contrast to more recent semi-classical modeling predicting that, instead of depending on Q, D should depend on the ratio of the width of the openings to that of the billiard [66]. In summary, although the semi-classical picture leads to an elegant explanation for how fractals might be generated by a semiconductor billiard, it fails to explain essential characteristics of the observed FCF, suggesting that a very different process is responsible.

It is worth emphasizing that our experimental findings are not just in conflict with existing theory but, more generally, with what one might have expected intuitively. In the past, a simple semi-classical picture of electron transport has been found to be an extremely powerful tool for the intuitive understanding of the ΔB spectrum of quantum conductance fluctuations in ballistic cavities. The results presented in Fig. 7.20 thus raise questions, not only regarding the physical origin of FCF, but also more fundamentally about our understanding of ballistic quantum transport. As discussed in relation to Fig. 7.20 (see section 6), the observed gradual change in D is in sharp contrast to expectations based on the Aharonov–Bohm effect—it indicates that the whole ΔB spectrum is evolving with temperature, otherwise the fractal scaling relationship could not be preserved. The Q curve of Fig. 7.20 predicts further counter-intuitive behavior. Starting at low temperatures (high Q in Fig. 7.20), the Q curve predicts that a rise in temperature will be accompanied by a rise in D, indicating that the contributions of the large ΔB (small area) fluctuations are suppressed more rapidly than the small ΔB (large areas) fluctuations. This is again in contradiction to expectations based on the Aharanov–Bohm effect. These unusual trends have no simple explanation! What is clear, however, is that the resolution of electron states, as described by the empirical parameter Q, is central to the evolution process. Furthermore, in identifying Q, we emphasize that D is determined only by conditions *within* the billiard rather than the electron injection properties. Billiard opening parameters, such as n, only affect D indirectly through their effect on τ_q and therefore Q.

More recently, a number of theoretical papers have started to explore beyond the original semi-classical theory of mixed dynamics [67–71]. However, in general, modifications to the original theory predict the appearance of individual resonances in the conductance fluctuations rather than the evolution in fractal behavior charted by the Q curve. In particular, why the fractal behavior persists into the fully quantum regime of conduction remains a mystery. Whereas a theory for fractal fluctuations in the fully quantum regime has been published [70] this was developed for the case of strong

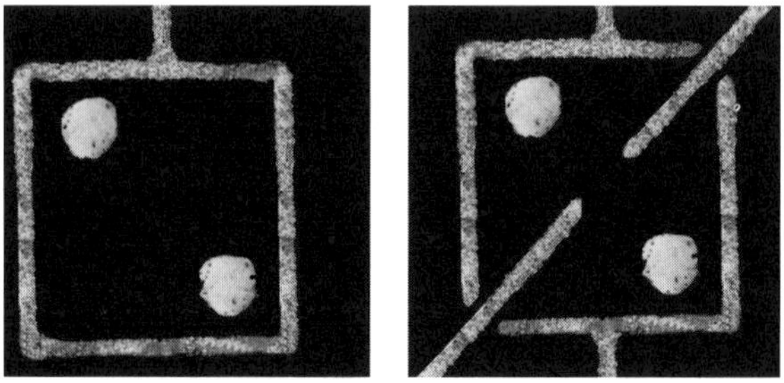

Fig. 7.25. Schematic representations of future billiard designs in which Ohmic spikes (circular metallic patterns) will be used to inject electrons directly into the confined regions of billiards defined by surface gates (grainy patterns).

localization and therefore is not applicable to our billiards. At the present stage of our understanding of FCF, where theoretical and experimental results are inconsistent, it is critical to identify parameters that are accessible to both theory and experiment. Current and future experiments feature two complementary lines of investigation, each of which addresses one such parameter.

Specifically, we are studying the two fundamental processes of ballistic transport: the mechanism by which the electrons are injected into the ballistic device and the manner in which they interact with the device walls. In the first study, electrons will be injected directly into the middle of the billiard rather than through entrances in the walls [72]. This technique uses the phase-coherent sub-micron ohmic contacts shown in Fig. 7.25. Electrons injected from the contacts will allow the study of quantum phenomena such as FCF of an "isolated" billiard (i.e. a billiard which is no longer linked to the outside 2DEG environment via openings). This is a very different approach to "closed" billiard studies than those carried out by closing the wall openings, where electron conduction enters the tunneling regime. The central aim is to determine if the generic fractal character of quantum conduction extends to this new form of semiconductor billiard. The second study builds on our investigations of wall profile discussed in section 7. In addition to using the double-layer billiard to define billiards of varying softness, we are investigating billiards defined in InGaAs/InP heterostructures. Simulations for these devices indicate that the walls have hard profiles and these experiments will therefore determine whether softness is necessary for the generation of the FCF. We note that preliminary investigations of InGaAs/InP billiards indicate that magnetoconductance fluctuations are fractal and that their D values condense onto the same Q curve as AlGaAs/GaAs billiards, further emphasizing the generic character of this remarkable phenomenon.

Acknowledgments

The authors thank A. Sachrajda, P. Coleridge, Y. Feng, M. Pepper, C. Dettmann, J. Cooper, D. Ferry, Y. Aoyagi, T. Sugano, L. Macks, B. Tribe, E. Linfield, D. Ritchie, A. Ehlert, I. Shorubalko, P. Omling, C. Tench and P. Wilkinson for their valued assistance.

References

1. R.P. Taylor, The role of surface gate technology for ALGaAs/GaAs nanostructures, *J. Nanotechnol.* **5**, 183 (1994).
2. *Microstructural Engineering* (ed. R.P. Kramer), Delft University Press, Amsterdam, pp. 381 (1981); *The Physics and Fabrication of Microstructures and Microdevices* (ed. M.J. Kelly and C. Weisbuch), Springer-Verlag, Berlin, pp. 36 (1986).
3. H.G. Craighead, A. Scherer, and M.L. Roukes, Nanofabrication, in *The Physics and Technology of Submicron Structures* (ed. Heinrich, Bauer, and Kuchar), Springer, Heidelberg (1988).
4. T.J. Thornton, M. Pepper, H. Ahmed, D. Andrews, and G.J. Davies, One-dimensional conduction in the 2D electron gas of a GaAs-AlGaAs heterojunction, *Phys. Rev. Lett.* **56**, 1198–1201 (1986).
5. B.J. van Wees, H. van Houten, C.W.J. Beenakker, J.G. Williamson, L.P. Kouwenhoven, D. van der Marel, and C.T. Foxon, Quantized conductance of point contacts in a two-dimensional electron gas, *Phys. Rev. Lett.* **60**, 848–851 (1988).
6. D.A. Wharam, T.J. Thornton, R. Newbury, M. Pepper, H. Ahmed, J.E.F. Frost, D.G. Hasko, D.C. Peacock, D.A. Ritchie, and G.A.C. Jones, One-dimensional transport and the quantisation of the ballistic resistance, *J. Phys.* C **21**, L209–212 (1988).
7. C.W.J. Beenakker and Van H. Houten, Quantum transport in semiconductor nanostructures, *Solid State Phys.* **44**, 1 (1991).
8. C.M. Marcus, A.J. Rimberg, R.M. Westervelt, P.F. Hopkins, and A.C. Gossard, Conductance fluctuations and chaotic scattering in ballistic microstructures, *Phys. Rev. Lett.* **69**, 506–509 (1992).
9. A.M. Chang, H.U. Baranger, L.N. Pfeiffer, and K.W. West, Weak localization in chaotic versus nonchaotic cavities: A striking difference in the line shape, *Phys. Rev. Lett.* **73**, 2111–2114 (1994).
10. R.A. Jalabert, H.U. Baranger, and A.D. Stone, Conductance fluctuations in the ballistic regime: A probe of quantum chaos?, *Phys. Rev. Lett.* **63**, 2442–2445 (1990).
11. H.U. Baranger, R.A. Jalabert, and A.D. Stone, Weak localisation and instability in ballistic cavities, *Phys. Rev. Lett.* **70**, 3876–3879 (1993).
12. E. Ott, *Chaos in Dynamical Systems*. Cambridge University Press, Cambridge (1993).
13. For a review see K. Nakamura, *Quantum Chaos—A New Paradigm of Non-linear Dynamics*. Cambridge Universty Press, Cambridge (1993).
14. Y. Aharonov and D. Bohm, Significance of electromagnetic potentials in the quantum theory, *Phys. Rev.* **115**, 485 (1959).
15. R.P. Taylor, P.C. Main, L. Eaves, S. Thoms, S.P. Beaumont, and C.D.W. Wilkinson, Universal conductance fluctuations in the magnetoresistance of submicron-size GaAs wires, *Surf. Sci.* **196**, 52–55 (1988).
16. R. Ketzmerick, Fractal conductance fluctuations in generic chaotic cavities, *Phys. Rev.* B **54**, 10841–10844 (1996).
17. B.B. Mandelbrot, *The Fractal Geometry of Nature*. Freeman and Company, New York (1977).
18. J.-F. Gouyet, *Physics and Fractal Structures*. Springer-Verlag, New York (1996).
19. R.P. Taylor, R. Newbury, R.B. Dunford, P.T. Coleridge, A.S. Sachrajda, and J.A. Adams, A tunable ballistic electron cavity exhibiting geometry induced weak localization, *Superlatt. Microstruct.* **16**, 317 (1994); R.P. Taylor, R. Newbury, R.B. Dunford, P.T. Coleridge, A.S. Sachrajda, and J.A. Adams, Classical and weak localization processes in a tunable ballistic electron cavity, *Phys. Rev.* B **51**, 9801 (1995); R.P. Taylor, R. Newbury, A.S. Sachrajda, P.T. Coleridge, P. Zawadzki, R.B. Dunford, Y. Feng, J.P. Bird, C.R. Leavens,

J.M. Cadogan, J.A. Adams, P.J. Kelly, M. Davies, and S. Brown, The use of wide ballistic cavities to investigate local weak localization processes induced by geometric scattering, *Semicond. Sci. Technol.* **11**, 1 (1996).
20. Y.G. Sinai, Dynamical systems with elastic reflections, *Russian Mathematical Surveys* **25**, 137 (1970).
21. E. Doron, U. Smilansky, and A. Frenkel, Experimental demonstration of chaotic scattering of microwaves, *Phys. Rev. Lett.* **65**, 3072–3075 (1990).
22. V. Milner, J.L. Hanssen, W.C. Campbell, and M.G. Raizen, Optical billiards for atoms, *Phys. Rev. Lett.* **86**, 1514–1517 (2001).
23. A. Kaplan, N. Friedman, M. Anderson, N. Davidson, Observation of islands of stability in soft wall atom-optics billiards, *Phys. Rev. Lett.* **87**(274101), 1–4 (2001).
24. R.P. Taylor, R. Newbury, A.S. Sachrajda, Y. Feng, N. Zhu, H. Guo, P.T. Coleridge, A. Delage, P.J. Kelly, Z. Wasilewski, and P. Zawadski, Geometry-induced quantum interference: A continuous evolution from square to Sinai billiard, *Superlatt. Microstruct.* **20**, 297–305 (1996); and R.P. Taylor, R. Newbury, A.S. Sachrajda, Y. Feng, P.T. Coleridge, Z. Wasilewski, N. Zhu, and H. Guo, The transition from a square to a Sinai billiard, in *World Scientific 2* (ed. M. Scheffler and R. Zimmermann), 1549–1552 (1996).
25. R.P. Taylor, J.A. Adams, M. Davies, P.A. Marshall, and R. Barber, Fabrication of nanostructures with multi-level architecture, *J. Vac. Sci. Technol.* B **11**, 628 (1993).
26. R.P. Taylor, Y. Feng, A.S. Sachrajda, J.A. Adams, and M. Davies, Fabrication and characterization of multi-level lateral nanostructures, *Surf. Sci.* **305**, 648–651 (1994).
27. R. Shaw, *The dripping faucet as a model chaotic system*. Aerial Press, Santa Cruz (1984).
28. R.P. Taylor, A.P. Micolich, R. Newbury, C. Dettmann, and T.M. Fromhold, Fractal Transistors, *Semicond. Sci. Technol.* **12**, 1459–1464 (1997).
29. T.M. Fromhold, Fractal resistance in a transistor, *Nature* **386**, 124 (1997).
30. R.P. Taylor, R. Newbury, A.S. Sachrajda, Y. Feng, P.T. Coleridge, C. Dettmann, N. Zhu, H. Guo, A. Delage, P.J. Kelly, and Z. Wasilewski, Self-similar magnetoresistance in a semiconductor Sinai billiard, *Phys. Rev. Lett.* **78**, 1952–1955 (1997).
31. R.P. Taylor, R. Newbury, A.S. Sachrajda, Y. Feng, and P.T. Coleridge, Tunable semiconductor Sinai billiards, *IEEE J.* 171–173 (1997) (ISBN 0-7803-3374-8/97).
32. R.P. Taylor, A.P. Micolich, R. Newbury, and T.M. Fromhold, Correlation analysis of self-similarity in semiconductor billiards, *Phys. Rev.* B *Rapid Commun.* **56**, R12733–R12756 (1997).
33. R.P. Taylor, A.P. Micolich, R. Newbury, J.P. Bird, T.M. Fromhold, J. Cooper, Y. Aoyagi, and T. Sugano, Exact and statistical self-similarity in semiconductor billiards: A unified picture, *Phys. Rev.* B **58**, 11107–11110 (1998).
34. R.P. Taylor, A.P. Micolich, R. Newbury, T.M. Fromhold, and C.R. Tench, Observation of fractal conductance fluctuations over three orders of magnitude, *Aus. J. Phys.* **52**, 887–893 (1999).
35. R.P. Taylor, A.P. Micolich, R. Newbury, T.M. Fromhold, A. Ehlert, A.G. Davies, C.R. Tench, J.P. Bird, H. Linke, W.R. Tribe, E.H. Linfield, and D.A. Ritchie, Semiconductor billiards: A controlled environment to study fractals, in *Nobel Symposium on Quantum Chaos* (eds. K.F. Berggren and S. Aberg), World Scientific, Singapore, ISBN 981 02 4711-7, and The Royal Swedish Academy of Sciences, 41–49 (2001).
36. D. Avnir, O. Biham, and O. Malcai, Is the geometry of nature fractal?, *Science* **279**, 39 (1998).
37. See, for example, J.P. Bird, Recent experimental studies of electron transport in open quantum dots, *J. Phys.: Condens. Matter* **11**, R413 (1999).
38. R.V. Jensen, The signature of chaos, *Chaos* **1**, 101 (1991).

39. T.M. Fromhold, C.R. Tench, R.P. Taylor, A.P. Micolich, and R. Newbury, Self-similar conductance fluctuations in a Sinai billiard with mixed chaotic phase space: Theory and experiment, *Physica* B **249–251**, 334–338 (1998).
40. Y. Takagaki and K.H. Ploog, Numerical simulation of fractal conductance fluctuations in softwall cavities, *Physical Review* B **61**, 4457–4461 (2000); Relation between stable orbits and quantum transmission resonance in ballistic cavities, *Phys. Rev.* E **62**, 4804–4808 (2000).
41. A.P. Micolich, R.P. Taylor, R. Newbury, T.M. Fromhold, and C.R. Tench, The Weierstrass function as a model of self-similarity in a semiconductor Sinai billiard, *Europhys. Lett.* **49**, 417–423 (2000).
42. S. Kawabata and K. Nakamura, Altshuler Aronov Spivak effect in ballistic chaotic Agharonov-Bohm billiards, *J. Phys. Soc. Japan* **65**, 3708 (1996).
43. G.H. Hardy, *Transcripts Am. Math. Soc.* **17**, 301 (1916).
44. A.P. Micolich, R.P. Taylor, J.P. Bird, R. Newbury, T.M. Fromhold, J. Cooper, Y. Aoyagi, and T. Sugano, Scale factor mapping of statistical and exact self-similarity in billiards, *Semicond. Sci. Technol.* **13**, A41–43 (1998).
45. R.P. Taylor, Levy flights, in *Encyclopedia of Non-linear Science* (ed. A. Scott), Fitzroy-Dearborn, London (2003).
46. A.P. Micolich, R.P. Taylor, R. Newbury, J.P. Bird, R. Wirtz, C.P. Dettmann, Y. Aoyagi, and T. Sugano, Geometry induced fractal behavior in a semiconductor billiard, *J. Phys.: Condens. Matter* **10**, 1339–1347 (1998); R.P. Taylor, A.P. Micolich, T.M. Fromhold, and R. Newbury, Comment on fractal conductance fluctuations in a soft wall stadium and a Sinai billiard, *Phys. Rev. Lett.* **83**, 1074 (1999).
47. A.P. Micolich, R.P. Taylor, J.P. Bird, and R. Newbury, Electron behavior in AlGaAs/GaAs square quantum dots, *Proc. N.Z. Aus. Inst. Phys. Annu. Condens. Matter Phys. Meeting* 1037–1214 (1997), ISSN TP20.
48. B. Dubuc, J.F. Quiniou, C. Roques-Carnes, C. Tricot, and S.W. Zucker, Evaluating the fractal dimension of profiles, *Phys. Rev.* A **39**, 1500 (1989).
49. Y. Takagaki, M. ElHassan, A. Shailos, C. Prasad, J.P. Bird, D.K. Ferry, K.H. Ploog, L.H. Lin, N. Aoki, and Y. Ochiai, Magnetic field controlled electron dynamics in quantum cavities, *Phys. Rev.* B **62**, 10255–10260 (2000).
50. A.S. Sachrajda, R. Ketzmerick, C. Gould, Y. Feng, P.J. Kelly, A. Delage, and Z. Wasilewski, Fractal conductance fluctuations in a softwall stadium and a Sinai billiard, *Phys. Rev. Lett.* **80**, 1948–1951 (1998).
51. Y. Ochiai, L.H. Lin, K. Yamamoto, K. Ishibashi, Y. Aoyagi, T. Sugano, J.P. Bird, D. Vasileska, R. Akis, and D.K. Ferry, Low temperature magnetotransport in ballistic quantum dots and wires, *Semicond. Sci. Technol.* **13**, A15 (1998).
52. H. Hegger, B. Huckestein, K. Hecker, M. Janssen, A. Freimuth, G. Reckziegel, and R. Tuzinski, Fractal conductance fluctuations in gold nanowires, *Phys. Rev. Lett.* **77**, 3885–3888 (1996).
53. A.P. Micolich, R.P. Taylor, A.G. Davies, J.P. Bird, R. Newbury, T.M. Fromhold, A. Ehlert, H. Linke, L.D. Macks, W.R. Tribe, E.H. Linfield, D.A. Ritchie, J. Cooper, Y. Aoyagi, and P.B. Wilkinson, Evolution of fractal patterns during a classical-quantum transition, *Phys. Rev. Lett.* **87**, (036802), 1–4 (2001).
54. R.M. Clarke, I.H. Chan, C.M. Marcus, C.I. Duruoz, J.S. Harris, K. Campman, and A.C. Gossard, Temperature dependence of phase breaking in ballistic quantum dots, *Phys. Rev.* B **52**, 2656 (1995).
55. A.P. Micolich, R.P. Taylor, R. Newbury, J.P. Bird, Y. Aoyagi, and T. Sugano, Temperature dependence of the fractal dimension of magnetoconductance fluctuations in a mesoscopic semiconductor billiard, *Superlatt. Microstruct.* **25**, 157–161 (1999).

56. J.M. Ziman, *Electrons and Phonons; The Theory of Transport Phenomena in Solids*. Clarendon Press, Oxford (1960).
57. J.P. Bird, K. Ishibashi, D.K. Ferry, R. Newbury, D.M. Olatona, Y. Ochiai, Y. Aoyagi, and T. Sugano, Phase breaking in ballistic quantum dots: A correlation field analysis, *Surf. Sci.* **361/362**, 730–734 (1996).
58. D.P. Pivin, A. Andresen, J.P. Bird, and D.K. Ferry, Saturation of phase breaking in an open ballistic quantum dot, *Phys. Rev. Lett.* **82**, 4687–4700 (1999).
59. Y. Sun, G. Kirczenow, A.S. Sachrajda, and Y. Feng, An electrostatic model of split-gate quantum wires, *J. Appl. Phys.* **77**, 6361 (1995).
60. O.E. Raichev and F.T. Vasko, Edge-induced modification of tunnel mixing and resistance resonance in double-quantum-well strips, *Phys. Rev.* B **57**, 12350 (1998).
61. A.P. Micolich, R.P. Taylor, A.G. Davies, T.M. Fromhold, H. Linke, R. Newbury, A. Elhert, L.D. Macks, W.R. Tribe, E.H. Linfield, and D.A. Ritchie, Dependence of fractal conductance fluctuations on soft-wall profile in a double-layer semiconductor billiard, *Appl. Phys. Lett.* **80**, 4381–4383 (2002).
62. N.P.R. Hill, J.T. Nicholls, E.H. Linfield, M. Pepper, D.A. Ritchie, G.A.C. Jones, Y.K.H. Ben, and K. Flensberg, Correlation effects on the coupled plasmon modes of a double quantum well, *Phys. Rev. Lett.* **78**, 2204–2207 (1997).
63. T.J. Gramila, J.P. Eisenstein, A.H. Macdonald, L.N. Pfeiffer, and K.W. West, Mutual friction between parallel two-dimensional electron systems, *Phys. Rev. Lett.* **66**, 1216–1219 (1991).
64. K.J. Thomas, M.Y. Simmons, J.T. Nicholls, D.R. Mace, M. Pepper, and D.A. Ritchie, Ballistic transport in one-dimensional constrictions formed in deep two-dimensional electron gases, *Appl. Phys. Lett.* **67**, 109–111 (1995).
65. A. Budiyono and K. Nakamura, Self-similar magnetoconductance fluctuations induced by self-similar periodic orbits (submitted).
66. E. Louis and J.A. Verges, Self-similar magnetoconductance fluctuations in quantum dots, *Phys. Rev.* B **61**, 13014–13020 (2000).
67. B. Huckerstein, R. Ketzmerick, and C.H. Lewenkopf, Quantum transport through ballistic cavities: Soft vs hard quantum chaos, *Phys. Rev. Lett.* **84**, 5504–5507 (2000).
68. R. Ketzmerick, L. Hufnagel, F. Steinbach, and M. Weiss, New class of eigenstates in the generic hamiltonian systems, *Phys. Rev. Lett.* **85**, 1214–1217 (2000).
69. G. Casati, I. Guarneri, and G. Maspero, Fractal survival probability fluctuations, *Phys. Rev. Lett.* **84**, 63–66 (2000).
70. G. Benenti, G. Casati, I. Guareri, and M. Terraneo, Quantum fractal fluctuations, *Phys. Rev. Lett.* **87**, 014101-1-4 (2001).
71. A.P.S. de Moura, Y.C. Lai, R. Akis, J.P. Bird, and D.K. Ferry, Tunnelling and nonhyperbolicity in quantum dots, *Phys. Rev. Lett.* **88**, (236804), 1–4 (2002).
72. R.P. Taylor, R. Newbury, A.S. Schrajda, Y. Feng, P.T. Coleridge, M. Davies, and J.P. McCaffrey, An investigation of current injection properties of ohmic spikes in nanostructures, *Superlatt. Microstruct.* **24**, 337–345 (1998); Can ohmic spikes define quantum systems?, *Japan. J. Appl. Phys.* **36**, 3964–3967 (1997).

Chapter 8

Electron Ratchets—Nonlinear Transport in Semiconductor Dot and Antidot Structures

H. Linke[1] and A. M. Song[2]

[1] *Department of Physics, University of Oregon, Eugene, OR 97403-1274, USA*
[2] *Department of Electrical Engineering and Electronics, UMIST, Manchester M60 1QD, UK*

1. Introduction

All materials at finite temperature store a substantial amount of energy in the form of kinetic energy of electrons, atoms or molecules. It is an old and tempting idea to convert this undirected, thermal motion (heat) into directed, useful motion (work) by rectifying the random motion of particles [1,2]. However, we know that the attempt to do so is doomed to fail: in thermal equilibrium, work can not be extracted from heat—this is the essence of the Second Law of thermodynamics. The situation is different when the system is kept away from thermal equilibrium: then, the Second Law does not apply, and thermal motion can be rectified. A trivial way of doing this is to apply a bias voltage across a quantum point contact, i.e., a narrow constriction that limits electron flow [3,4]. The potential difference opens a gap in the occupation of states on either side of the point contact, disturbing thermal equilibrium. As described by the Landauer equation [5,6], it is essentially the voltage-induced lack in balance of thermal motion of the electrons, that leads to a net current towards the lower electro-chemical potential.

In the case of a voltage across a point contact, the current is driven by a steady-state gradient. A more subtle approach to rectifying thermal motion, that does not require a time averaged force or gradient, is used by so-called Brownian motors, often also called "ratchets" [7–10]. The essence of a Brownian motor is best explained by the so-called pulsating ratchet (also called "flashing ratchet") illustrated in Fig. 8.1 [10–12]. In this system, Brownian particles are subject to a time-dependent, asymmetric, periodic potential (the ratchet). In the simplest case, the potential can be in one of two states, either flat ("off") or saw-tooth shaped ("on"). When the potential is off at first, and is then suddenly switched on, energy is added to the system and particles will during a transient period of time move towards the nearest local potential minimum. Due to the

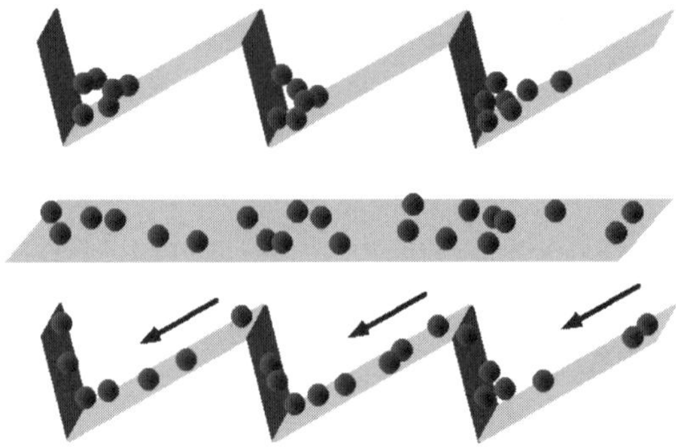

Fig. 8.1. A flashing ratchet (pulsating ratchet) consists of Brownian particles in a potential that periodically switches between two states: flat ("off") or saw-tooth shaped ("on"). During the on-phase, Brownian particles concentrate around the potential minima. During the off-phase, the particles diffuse isotropically. Periodic or stochastic switching between both states induces a time-averaged particle current to the left.

asymmetry of the saw-tooth potential, the nearest minimum is on average to the left. Periodic or stochastic switching of the potential (on a time scale comparable to the diffusion time between neighboring potential minima) will therefore result in a net particle flow towards the left, although no net force acts on the particles (for an instructive Java applet, see [13]). When the switching frequency is chosen to be commensurate with the characteristic diffusion time, the current can flow against a small external force.

The pulsating ratchet illustrates all the characteristic properties of a Brownian motor [14]:

1. A spatially asymmetric potential (the ratchet) breaks symmetry and defines a preferred direction of motion;
2. There is an absence of detailed balance (i.e. the system is not in thermal equilibrium; see [15]). However, no macroscopic, external forces are used—the spatial and temporal averages of all forces and gradients are zero;
3. Thermal noise plays a prominent role; for instance, in the case of the pulsating ratchet, no pumping effect would occur in the absence of thermal diffusion. In other types of ratchets, thermal noise may lead to effects such as a reversal of the current direction (see, e.g. Fig. 8.11 in section 2.2.1) [10].

It is interesting to note that one of the original driving forces for the study of Brownian motors comes from biology. Inside living cells, protein assemblies called "molecular motors" perform mechanical tasks such as exerting force (for instance during muscle contraction), pumping ions across membranes, and transporting materials in cells [16,17]. In understanding molecular motors one has to take into account their substantial Brownian motion due to frequent collisions with surrounding water molecules. The concept of the pulsating ratchet integrates random motion into its operational principle (Fig. 8.1), and was originally developed as a minimalist's model for the workings of some molecular motors [8,9,11,12,15,18]. However, the principle of ratchets and Brownian motors is by no means limited to biology: in the last few years, artificial Brownian motors have been realized in a variety of different systems, including colloidal particles in electrostatic potentials [12,19], optically confined particles [20], mercury droplets [21], cold atoms [22], electrons in semiconductors [23–28], and molecules in lipid bilayers [29] and in aqueous solution [30] on patterned surfaces. For a representative overview of experimental work on Brownian motors, see the special issue [31].

In the present chapter we will review ratchets based on electrons in semiconductors [23–28]. Semiconductors have several distinct experimental advantages for ratchet studies. Firstly, because modern semiconductor heterostructure materials have a typical impurity density so low that a device on the micron scale is normally free of any impurities, electrons experience the shape and symmetry defined by the device geometry. State-of-the-art nanotechnology therefore offers a variety of ways to expose electrons to a tailored, controllable asymmetric potential or geometry. A second advantage is that electron currents can be measured with very high accuracy. Most importantly, however, semiconductor heterostructures and the use of low-temperature techniques allow to expand the study of ratchets from the classical regime to the semiclassical and quantum regimes. Quantum effects such as electron-wave interference, quantum tunneling, and energy quantization lead to novel behavior and to functionality not observable in most other ratchet systems.

Most of the experiments discussed in the present chapter are easiest understood in the context of so-called "rocking" ratchets. A rocking ratchet consists of particles in an asymmetric potential that is periodically and symmetrically tilted (Fig. 8.2). The effective height of each potential barrier changes upon tilting of the potential, and is lower for tilt towards the gentler slopes of the barriers (to the right in Fig. 8.2). In the simplest case, when only thermal excitation over the barriers is taken into account (while quantum effects, such as tunneling through the barriers, are neglected), a net current to the right will be observed after averaging over both tilt directions, because thermal excitation of particles over the barrier is more probable during tilt to the right when the effective barrier height is lowest [32,33].

Expressed in the language generally used to describe electron transport in semiconductors, a rocking ratchet is essentially a non-linear rectifier. Different types of

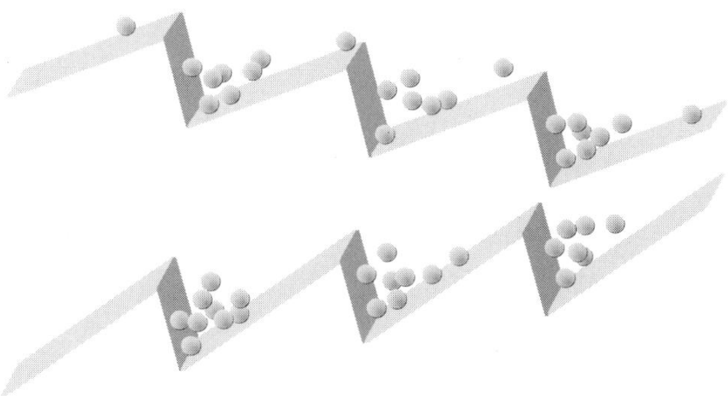

Fig. 8.2. A rocking ratchet is an asymmetric potential that is tilted symmetrically and periodically. Due to the asymmetry of the barriers, the currents induced during the two half cycles are different, and, averaged over a full cycle, a net (time-averaged) current flows. In the situation shown here, the barrier height is lower for tilt to the right, and classical, thermal particles would move on average to the right. When quantum effects—such as tunneling through the barriers—are taken into account, the net current direction can reverse direction (see section 2.2).

rocking ratchets are distinguished by the way non-linear effects are induced. In the low-voltage, linear-response regime of electron transport, the device conductance, $G = I/V$, is independent of the voltage V and electric conduction is symmetric by definition. Rectification—i.e. a net current $I_{\text{net}} = (\frac{1}{2})[I(V) + I(-V)]$—is represented by the odd orders of non-linearity (G_1, G_3, etc.) in the expansion $I(V) = G_0V + G_1V^2 + G_2V^3 + \ldots$ Two conditions must be fulfilled to observe rectification. Firstly, the system must lack a symmetry axis prependicular to the direction of current flow. Secondly, the voltage must have an effect on the device's transport properties, inducing non-linear $I(V)$ behavior. Many rectification effects observed in semiconductor structures are fully described by the notion of non-linear rectification. However, as will become clear in this chapter, in other cases the thermal motion of electrons leads to subtle effects, such as a reversal of the current direction induced by temperature. It is instructive to discuss these effects in the context of ratchets.

The chapter is structured as follows. In section 2 we describe a number of low-temperature experiments that used quantum effects to generate non-linear behavior and rectification. In particular, so-called *quantum dot ratchets*—described in section 2.1—are phase-coherent electron billiards that have no symmetry axis perpendicular to the direction of current flow. We will show that quantum conductance fluctuations sensitively respond to an applied electric field, and a small ac voltage produces a rectified current. A *tunneling ratchet*, consisting of a series of asymmetric tunneling barriers is reviewed in section 2.2. In this device, the barrier's transmission coefficients depend on the electron energy, and the rectified current reverses direction when the temperature is lowered [27]. Tunneling ratchets can also be operated as *heat pumps* or *heat engines* [34,35]. This behavior is described in section 2.3 where we also discuss the use of quantum dots as energy filters, allowing the operation of quantum Brownian heat engines arbitrarily close to Carnot efficiency in principle. In section 3 we leave the experimental limitation to very low temperatures behind and review experiments on non-linear effects in asymmetric, triangular *antidot structures*. Experiments on both individual multi-terminal devices, called ballistic rectifier, and on macroscopic, *artificial nanomaterials* will be discussed (section 3.2), including a semiclassical model accounting for the observations (section 3.3). The pronounced non-linear effects generated by the ballistic rectifiers and the nanomaterials are comparable to that of a conventional diode, but the working principle is completely different. Without using any doping junction or barrier structure, the rectifying (ratchet) effect is a result of the intentionally broken symmetry in the device geometry and the voltage-induced electron collimation effect [25,28]. The smallest devices have been shown to work at room temperature and at frequencies up to at least 50 GHz. Promising practical *applications* such as rectification, microwave detection, and higher harmonic generation will be discussed in section 3.4. Finally, in section 4, we give a brief outlook on expected future developments.

2. Non-Linear Rectification in the Quantum Regime

2.1. Quantum Dot Ratchets

2.1.1. Non-linear Rectification in the Phase-Coherent Regime A quantum ratchet is a ratchet in which quantum effects lead to new behavior or functionality as compared to

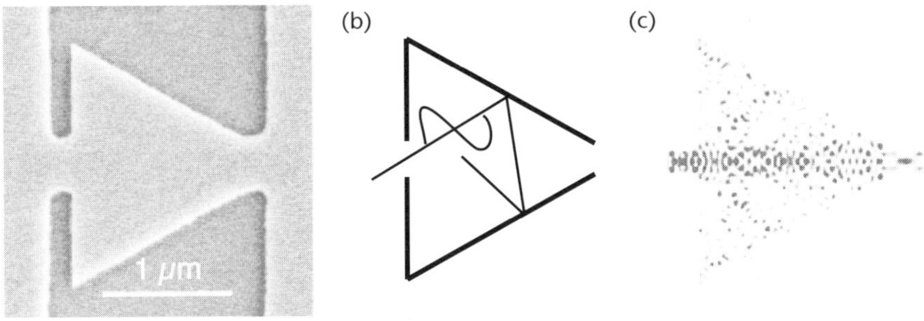

Fig. 8.3. (a) Electron microscope image of a triangular quantum dot (top-view). The darker areas are shallow wet-etched trenches that interrupt a two-dimensional electron gas located about 100 nm below the surface. (b) At low temperatures electrons move along ballistic trajectories, similar to a game of billiards. When electron transport is phase coherent, electrons can be pictured as semiclassical wave-packets that are equipped with a phase and that move along classical trajectories. (c) In a fully quantum-mechanical description the Schrödinger equation is solved and electron quantum states are calculated (calculation by Igor Zozoulenko in [40]).

the ratchet's classical counter part [10,36]. The construction of a quantum ratchet thus requires, (i) a controlled (a-)symmetry of the potential sensed by particles; (ii) particles subject to quantum effects such as wave interference or tunneling; (iii) absence of detailed balance; and (iv) sensitivity to noise, such as thermal noise. In the following, we will address these requirements while introducing the concept of a quantum dot ratchet—i.e. an asymmetrically shaped electron cavity that exhibits ratchet behavior in the phase-coherent regime of transport.

(i) One of the simplest ways to expose electrons to an asymmetric potential is to use geometrical confinement. In Fig. 8.3 we show a top-view scanning electron micrograph of a device designed to achieve this. The darker areas are wet-etched trenches that serve to interrupt a two-dimensional electron gas (2DEG) located at the GaAs/AlGaAs interface of a heterostructure, typically 30–100 nm below the surface. The etched trenches electrically isolate the inner part of the triangle—the electron cavity—from the surrounding 2DEG areas, except for contact openings at the tip and in the center of the base of the triangle. The dimension of the cavity is typically of the order of a micrometer, which at low temperatures ($T < 10$ K) is much smaller than the characteristic length scales for electron scattering. In this so-called ballistic limit of transport, the dynamics of electrons is well described by a single-particle picture, in which electrons move along straight trajectories between boundary collisions [5,37] (Fig. 8.3(b)). Because of the similarity with a game of billiards, ballistic 2D cavities are often referred to as *electron billiards*. By designing the shape and symmetry of the cavity we can control the symmetry of the potential experienced by the electrons.

(ii) The phase coherence length, l_ϕ—the length scale over which the wave-character of electrons (the phase memory) is preserved [5,37]—is limited by inelastic scattering events, such as electron–electron or electron–phonon interaction (it is important to note that elastic scattering with the billiard walls does not destroy phase coherence). l_ϕ decreases quickly with increasing temperature and is typically less than a micrometer at about 5 K. However, at sub-Kelvin temperatures electron transport across an electron billiard generally is phase coherent. By tuning the temperature around 1 K one can therefore switch from an experimental situation well described by a classical, single-particle picture of the electron dynamics, to a semi-classical picture where electrons have a phase but move on classical trajectories (Fig. 8.3(b)), or even to a fully quantum mechanical, wave description of transport (Fig. 8.3(c)). The experimental signature of transport in the phase coherent regime is the fluctuations observed when the electron phase is tuned using an experimental variable such as a small external magnetic field. Conductance fluctuations are caused by the constructive or destructive interference of electron waves traveling on pairs of semiclassical trajectories. Electron billiards in which quantum conductance fluctuations are observed are often referred to as *quantum dots*.

(iii), (iv) Finally, due of their small size, mesoscopic electron devices can very easily be driven into non-linear response because transport via electron quantum states is very sensitive to external perturbations [38], and quantum transport effects are extremely sensitive to thermal noise. Our experimental system thus fulfills all four requirements for the study of quantum ratchets mentioned above.

Electron transport through a quantum dot is usually described by a Landauer equation [5,6] which views the two point contacts as the ends of a one-dimensional (1D) wave guide connected to source and drain electron reservoirs. The electron cavity is represented as a scattering potential $U(V)$ with energy-dependent transmission probability $t(\varepsilon, U(V))$. The current flowing between reservoirs is then given by the product of the conductance of the 1D wave-modes ($2e/h$ per fully occupied wave mode) and $t(\varepsilon, U(V))$. It can be written as [5,6]

$$I(V) = \frac{2e}{h} \int t(\varepsilon, U(V)) M(\varepsilon, V)[f_S(\varepsilon) - f_D(\varepsilon)] d\varepsilon \tag{1}$$

where $M(\varepsilon, V)$ is a step function describing the integer number of 1D wave-modes available for transport at each energy, ε. The equilibrium Fermi distribution functions, f_S and f_D, describe the occupation of energy levels in the source- and drain-electron reservoirs, respectively, with electrochemical potentials, μ_S and μ_D, where $\mu_S = [\mu_D + eV]$, and $e = -1.6 \times 10^{-19}$ C is the charge on an electron. The bracket $[f_S - f_D]$ is the Fermi window and represents the temperature- and voltage-dependent energy range over which net flow of electron takes place [5].

When the device conductance $G = I/V$ of an electron billiard is measured as a function of a magnetic field, one can distinguish classical and quantum behavior. In Fig. 8.4 we show measurements of the magneto-conductance of a triangular dot in linear response, i.e., with no dc bias voltage applied. The bold and thin lines were recorded at $T = 0.3$ K and $T = 5$ K, respectively. The data recorded at the higher temperature can be explained in great detail in terms of commensurability effects due to a classical, billiard-ball like motion of the electrons inside the cavity [39]. The quickly varying conductance fluctuations emerging at the lower temperature, in contrast, are related to electron-wave interference [40] (in an alternative description, the fluctuations reflect the shell structure of electron states inside the dot [41–46], see Fig. 8.5).

By subtracting the data recorded at the higher temperature from the low-temperature data, the quantum conductance fluctuations can be isolated to a good approximation (Fig. 8.4, right-hand axis). The rms value of the quantum fluctuations is about $0.15\, e^2/h$,

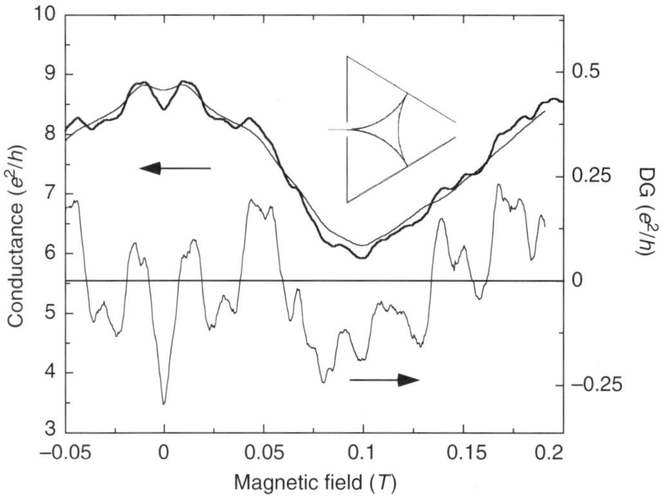

Fig. 8.4. Conductance of a triangular quantum dot as shown in Fig. 8.3(a) measured as a function of a perpendicular magnetic field. The thin and bold lines are data taken at $T = 5$ K and $T = 0.3$ K, respectively (left axis). The conductance minimum at about $B \approx 0.09$ T is caused by a classical commensurability effect, namely by the trajectory shown in the inset (for details, see [39]). By subtraction of the high temperature data from the low temperature data (shown with respect to the left-hand vertical axis) one can, to a good approximation, isolate the quantum conductance fluctuations, $\Delta G = [G(0.3\text{K}) - G(5\text{ K})]$ (for details, see [40]).

which is a typical value for experimentally observed magneto-conductance fluctuations at temperatures around 300 mK. The magnetic field scale of magneto-conductance fluctuations is given by the flux required to significantly change the electron states inside the cavity, $(h/e)/A$, where A is the effective area of the device cavity (about 1.2 μm^2), which yields a value of a few millitesla. Classical effects, by comparison, change much more slowly with magnetic field. The typical field scale is determined by the field that markedly bends classical, ballistic electron orbits (illustrated in Fig. 8.4), i.e. when the cyclotron radius $r = m * v_F/eB(v_F$ is the electron velocity at the Fermi energy) [37] becomes comparable to the lateral size of the device. This happens typically at about 50–100 mT for micron scale devices. The significant differences in the dependence on magnetic field and on temperature between classical and quantum behavior can be employed to distinguish classical and quantum effects in the non-linear response regime.

To see how one can use quantum dots to rectify an ac voltage, one needs to identify mechanisms by which electron interference induces a non-linear current–voltage dependence. Consider first the limit of small voltages (linear response). Figure 8.5(a) shows schematically the bottom of the conduction band (the lowest allowed electron energy) along the symmetry axis of a triangular quantum dot like the one in Fig. 8.2. The variation of the band represents the confinement energy in the point contacts and inside the dot. The bias voltage is assumed to be negligible, such that there is no potential drop between the electro-chemical potentials (visible as the maximum occupied energies, and identical to the Fermi energy for $V = 0$) in the source and drain reservoirs.

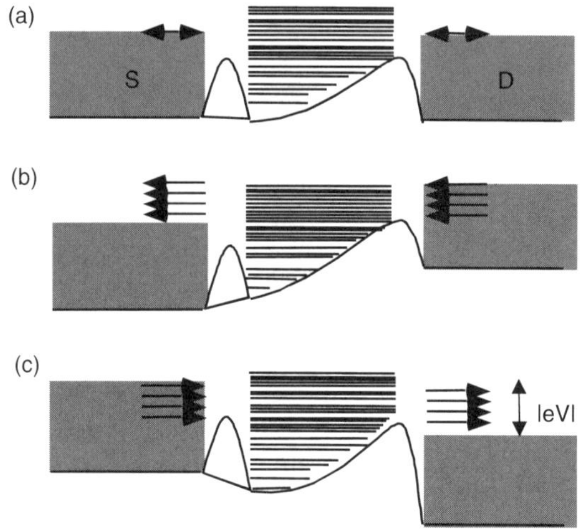

Fig. 8.5. Cartoon illustration of electron transport through a triangular quantum dot. The shape of the conductance band edge inside the dot illustrates the effect of the spatial confinement energy inside the triangular dot (not to scale). The horizontal lines inside the dot indicate the modulation of the density of states due to overlapping, broadened single electron states. Shading indicates the energy distribution functions in the electron reservoirs forming the source (S) and drain (D). (a) For very small voltages (linear response) the transmission probability (the conductance) is independent of the absolute value and the sign of the voltage. (b) and (c) Illustrate the non-linear response regime where the potential and the electron states depend on the applied voltage. When the potential is not inversion symmetric, the electrical current flows for positive and negative voltage via different sets of electronic states, and rectification (i.e. a ratchet effect) is observed.

At very small voltages ($|eV| \ll \mu_F, kT$), transport through the dot is via the electron states within a few kT of the Fermi energy [5]. In this regime, Eq. (1) can be written as $I = GV$, where $G = [2e^2/h]M(\mu_F)t(\mu_F)$ does not depend on voltage and μ_F is the equilibrium Fermi energy. The electron states inside the dot contributing to the current are those at the Fermi energy, and are the same for both current directions, such that conductance is symmetric.

The potential drop between the reservoirs induced by a finite bias voltage (Fig. 8.5(b,c)) distorts the 2D dot potential U. Consequently, the exact configuration of electron states inside the dot changes as the voltage is tuned [27,47]. Because of the asymmetry, the distortion depends, in general, on the sign of the voltage ($U(V) \neq U(-V)$), such that—depending on the current direction—different sets of electron states carry the current. The transmission probablity of the dot, $t(\varepsilon, U(V))$, therefore depends on the sign of the voltage, causing non-linear and non-symmetric behavior of the conductance (see Eq. (1)). The second difference from the linear reponse regime is that at finite bias voltage not only states within a few kT contribute to transport through the dot. Instead, the width of the Fermi window, $[f_S(\varepsilon) - f_D(\varepsilon)]$, which describes the energy range over which electrons contribute to the conductance, increases to a width approximately given by $(|eV| + 4kT)$. The exact position of this window relative to the conductance band bottom inside the dot depends on how the voltage drop is distributed over the device [48]. When the two point contacts are different, rectification is induced, because a different range of electron states will contribute to the current for the two signs of the voltage [27,47].

Experimental data for the differential conductance, $\partial I(V)/\partial V$, are shown in Fig. 8.6. In the experiment, a dc bias voltage V was added to a small ac voltage

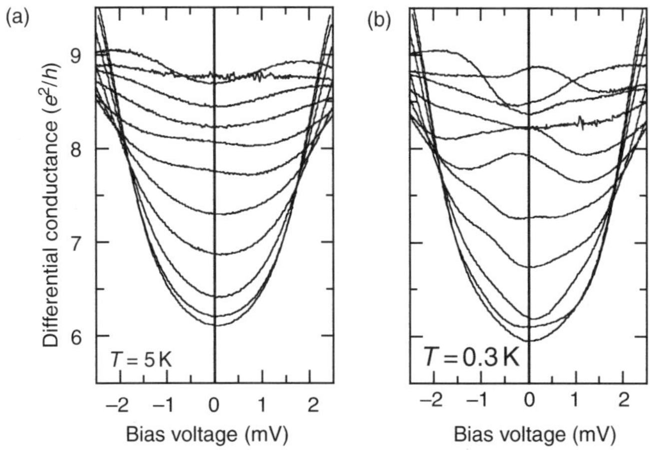

Fig. 8.6. (a) Differential conductance versus dc bias voltage V measured at a temperature of 5 K at, from the top-down, magnetic fields between zero and 100 mT with a relative spacing of 10 mT. (b) When the temperature is lowered to 0.3 K additional structure emerges in the non-linear conductance because of quantum interference effects. By subtracting the data taken at the lower temperature from the data taken at 5 K, the quantum effects can be isolated (see Fig. 8.7). Figure from [47].

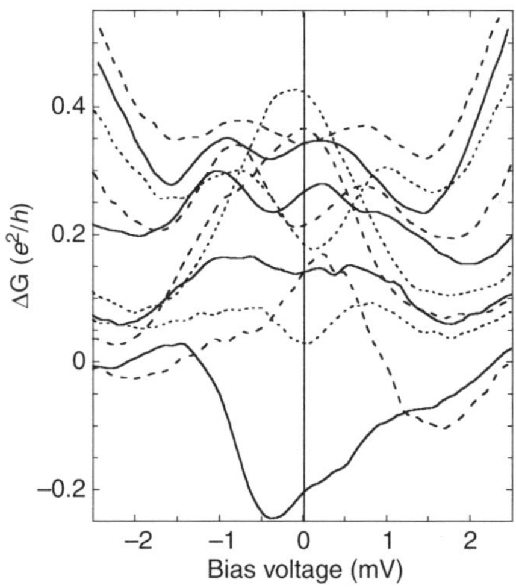

Fig. 8.7. The difference $\Delta G(V) = [G(V, T = 0.3\text{ K}) - G(V, T = 5\text{ K})]$ obtained by subtracting the data in Fig. 8.6(a) ($T = 5$ K) from corresponding data in Fig. 8.6(b) ($T = 0.3$ K). The curves have been offset by $+0.05e^2/h$ from one another for clarity. Figure from [47].

(frequency 10–100 Hz), and the differential resistance was measured as a function of V using standard phase-locking techniques. The data shown in Fig. 8.6 were recorded at a series of different magnetic fields, ranging from 0 to 100 mT, at a relative separation of 10 mT. At $T = 5$ K (Fig. 8.6(a)), when quantum effects are expected to be suppressed, the non-linear conductance is in all cases found to increase monotonically with increasing bias voltage. The conductance value at zero bias voltage changes with magnetic field in agreement with Fig. 8.4 and the shape of the $G(V)$ curves depends qualitatively only little on magnetic field. At the lower temperature ($T = 0.3$ K, Fig. 8.6(b)), however, when quantum effects are expected to appear, the non-linear behavior becomes richer and varies qualitatively as a function of magnetic field. The non-linear quantum effects can be isolated by subtracting the high temperature data from those recorded at low temperature (Fig. 8.7). Crucially, the non-linear quantum effects (that is, the conductance fluctuations measured in the non-linear regime) are in general not symmetric with respect to zero bias voltage. The orientation of the asymmetry depends on the magnetic field strength, because different electron states contribute at different magnetic fields. Importantly, the magnetic field scale of these changes (a few mT) is the same as that of magneto-conductance fluctuations. This scale is thus consistent with magnetic-field induced modifications to quantum interference but not with classical effects.

2.1.2. Role of the Dot Geometry The results presented in the preceding section establish the existence of a rectifying effect in the conductance of an asymmetric quantum dot. However, rectification in mesoscopic devices had been observed previously, when it had been attributed not to intentionally asymmetric geometries, but to the random configuration of material impurities [49–53]. An important question is therefore whether the origin of rectification in our devices indeed is the geometry of the dot. One way of answering this question is to check whether the conductance fluctuations are *symmetric* with respect to a geometric symmetry axis of the device. This would then be evidence that it is the device geometry that determines the symmetry of ballistic conductance, and not unintentional deviations from this geometry.

The data shown in Fig. 8.8 were obtained from a triangular billiard that did have a symmetry axis with respect to the current direction, but that had no symmetry axis parallel to the direction of the current (see insets). Therefore, any conductance fluctuations should be symmetric with respect to zero bias voltage—i.e. $G(V, B = 0) = G(-V, B = 0)$. On the other hand, the fluctuations are *not* expected to be symmetric with respect to B, that is $G(V, B) \neq G(V, -B)$, because different electron trajectories participate in the current, as is illustrated in Fig. 8.8. Indeed, as shown in the figure, these expected symmetry relations can be observed. To see this, compare characteristic features of fluctuations observed for $(V > 0, B > 0)$ with features observed for $(V < 0, B < 0)$ (or, equally, compare $(V > 0, B < 0)$ with $(V < 0, B > 0)$). This is particularly clear for $B > \pm 0.2$ T where the magnetic field is strong enough to alter the classical trajectories. Observations such as this one confirm that the nominal device geometry does indeed determine the symmetry of conductance fluctuations in the non-linear regime [54].

In measurements as shown in Figs 8.6–8.8, a complication can arise due to the fact that the experimental set-up used to study the device may introduce an asymmetry. In

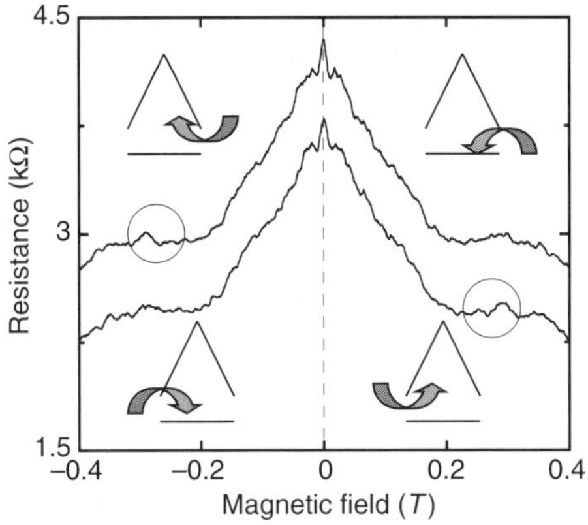

Fig. 8.8. Magneto-resistance data measured at bias voltage of $V = +2$ mV (upper curve) and $V = -2$ mV (lower curve) for a triangular quantum dot with a symmetry axis perpendicular to the current direction. The dot is defined by wet etching in an InP/InGaAs quantum well. The cartoons illustrate the expected symmetry relations and indicate the direction of electron flow and the effect of a magnetic field on the semi-classical electron trajectory. Note that because the data were obtained under finite bias conditions, each individual trace is not symmetric in magnetic field (in the linear response regime, $R(B)$ is symmetric in B). However, when the bias voltage is reversed, the reversed curve is found (one characteristic feature that illustrates this is highlighted by circles), demonstrating that the nominal symmetry of the quantum dot determines the electron dynamics. Data courtesy of Colleen Marlow and Ivan Shorubalko.

particular, when gates are used to define the structure or to adjust the Fermi energy, the reference point of the gate voltage is often identical with the drain contact used for the current. When a finite bias voltage is used, this can lead to asymmetric non-linear behavior that is not related to the asymmetry of the device, but to that of the measurement circuit. This circuit-induced asymmetry is essentially equivalent to bias-dependent depletion in a field-effect transistor [55,56]. In a symmetric device, where the conductance should be symmetric with respect to bias voltage, such an effect introduces asymmetric non-linear behavior and is thus easily detected. The device used in Fig. 8.8 had a very small dependence of conductance on top-gate voltage, and no circuit-induced asymmetry was evident. The situation is more complicated when an asymmetric device is used, and non-symmetric conductance with respect to zero bias is expected. It is then possible to distinguish between nonsymmetries related to the device symmetry and those related to the asymmetry of the circuit, by exploiting symmetry relations upon reversal of the grounding configuration. That technique is described by Löfgren et al. [54].

The demonstrated relationship between device asymmetry and the *existence* of rectification does not mean, however, that one can precisely predict the *direction* of rectification of a given quantum dot device. The reason is that electron-wave interference is extremely sensitive to variations of the Fermi energy, and to the exact distribution of charges forming the electrostatic potential. Both these parameters depend on the cooling process, and can not be predicted in detail. While this means that the sign of rectification for a specific set of parameters is not controllable in the device design, it is very easy in an experiment to adjust the direction of rectification, once it is established. One can control the current direction using any experimental parameter that changes the electron phase and/or the density of states inside the quantum dot. Such parameters are: the device shape (which can be varied using an external gate), the Fermi energy (using a top-gate [26,47]), or a small magnetic field (Figs 8.7, 8.8). Also the amplitude of the applied ac voltage can be used to reverse the sign of rectification [26].

2.2. Tunneling Ratchets

Quantum effects in ratchets were first studied theoretically by taking into account tunneling through the energy barriers in a rocking ratchet (Fig. 8.9) [36]. In the classical regime of transport, the current in an adiabatically rocked ratchet is in the direction of the gentler slopes of the potential, because the barrier heights are lower upon linear tilt in this direction. In the quantum regime however, when tunneling contributes to the current, not only the *height*, but also the *shape* of the barrier is important. This leads to qualitatively new behavior, namely a net current direction that depends on temperature. This behavior can be understood intuitively as follows (Fig. 8.9): when the saw-tooth barriers are tilted to the left, towards the steeper slopes of the potential, the barriers deform to be thinner at a given energy below the barrier top than for tilt to the right. Consequently, the (time-averaged) net tunneling current will be to the left, in a direction opposite to that of the net classical current due to excitation over the barriers. Which of the two contributions—tunneling through, or excitation over the barriers—is larger, depends on the energy distribution of the particles. In the high temperature limit, the net current is in general to the right due to thermal excitation. In the low temperature limit, the tunneling current to the left can be larger than the thermally excited current, and the total current can reverse direction as the temperature is decreased [27,36,57].

2.2.1. Experiment To realize a tunneling ratchet one can use asymmetric tunneling barriers formed by a series of quantum point contacts. The etched trenches visible in

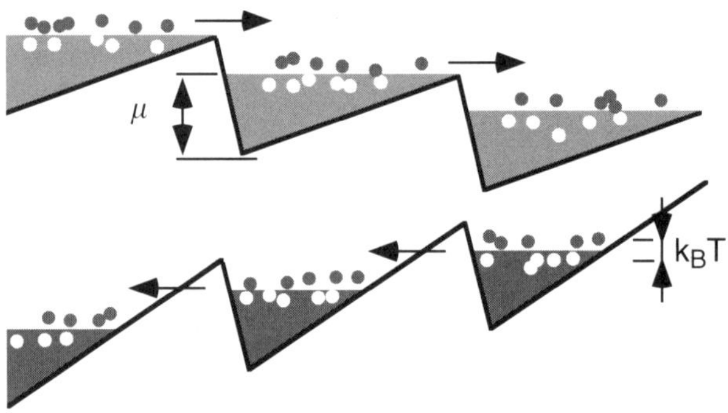

Fig. 8.9. A tunneling ratchet is a rocking ratchet in which tunneling through the barriers is taken into account. The indicated distribution of energies assumes fermions (electrons) at low temperature (electro-chemical potential μ). Tilting the ratchet potential deforms the barriers. Thermal excitation over the barriers occurs predominantly for tilt to the right where the barrier height is smaller. Tunneling is more important for tilt to the left where the barriers become thinner.

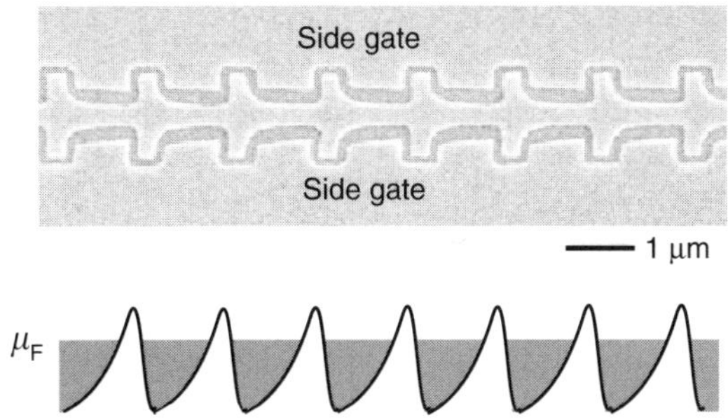

Fig. 8.10. An electron tunneling ratchet. The darker areas in the scanning electron micrograph (top) are etched trenches patterned by electron beam lithography and shallow wet etching, forming an electron-wave guide in a 2DEG located under the surface. The lower figure indicates the ratchet potential experienced by one-dimensional electrons moving along the wave guide.

the scanning electron micrograph shown in Fig. 8.10 electrostatically deplete a 2DEG situated about 100 nm underneath the device surface and form a narrow channel with asymmetric, funnel-like constrictions.

An electron can access a constriction only when the electron energy exceeds the confinement energy $\hbar^2 k_y^2/2m^*$, where $k_y = 2\pi/\lambda_y$ is the component of the wave vector perpendicular (direction y) to the confining walls, and m^* is the effective electron mass in the crystal [5,37]. The wavelength of electrons at the Fermi energy, λ_F, is about 40 nm (typical for a 2DEG in a GaAs/AlGaAs heterostructure). In other words, an electron moving along the channel experiences each of the periodic constrictions as an asymmetric energy barrier. The lithographic width of this channel has been chosen to be about 100 nm at the narrowest points and the electronic channel width can be varied further by using the electron sheet regions parallel to the channel as side-gates. Figure 8.10 shows the energy variation of the conductance band bottom (the lowest 1D wave mode) along the channel when the narrowest parts of the channel are only about $\lambda_F/2$ wide, such that the confinement energy at these points is of the order of the Fermi energy. Transport in the channel is then possible only by thermal excitation over the energy barriers, or by tunneling through the barriers.

Using contacts situated to the left and right of the channel as source and drain contacts, a voltage V can be applied along the channel, effectively "tilting" the potential and changing the potential shape. The resulting current I is determined by the barriers' reflection and tunneling coefficients and is sensitive to the precise shape of these barriers, such that $I(V) \neq -I(-V)$. In order to "rock" the ratchet we here used a square-wave source–drain voltage of amplitude V_0. The frequency of order 100 Hz was chosen to be much slower than all electronic time-scales, such as energy relaxation times (adiabatic rocking). The electronic system was therefore in a steady state at all times, and the net current induced by the rocking is given by $I_{\text{net}} = 0.5[I(V_0) + I(-V_0)]$.

Temperature dependence of the current direction in a tunneling ratchet is detectable when electrons at energies just above and below the top of the barriers contribute to the current. To achieve this condition in the experiment, the barrier height was set to approximately match the Fermi energy $\mu_F = 11.8$ meV, as estimated from the device conductance (see Eq. (1)). Further, the rocking voltage, V_0, is chosen such that, by varying T, the width of the Fermi window, about $(|eV_0| + 4kT)$, can be varied over the energy range around the barrier top where quantum corrections to the transmission probability are important. Calculations show that this energy range extends about 2 meV above and below the barrier maximum (inset to Fig. 8.13). In this way, one finds sets of values for the rocking voltage and the side-gate voltage, where the net current direction depends on temperature. This reversal can then be observed directly by sweeping the temperature, while all other parameters, including the shape and height of the potential barriers, are kept constant. The data shown in Fig. 8.11(a) were obtained using a rocking voltage $V_0 = 1$ mV corresponding to a voltage drop of 0.1 mV over each barrier (less than 1% of μ_F) in the device shown in Fig. 8.10, which consisted of 10 ratchet barriers in total. The net current was positive at 4 K but reversed its direction as the temperature was reduced to 0.4 K. To detect any contribution to I_{net} not related to the asymmetry of the device, each experimental sweep was carried out a second time with the source and drain leads physically interchanged using a switch. If care is taken to ensure that no significantly non-linear elements are in the path of the measurement current, and that the sum of Ohmic resistances (such as leads, ballast resistors and contacts pads) on either side of the device are balanced, then rectifying effects that are not due to the asymmetry of the device, such as circuit-induced symmetry (also called self-gating [55,56]), can be removed by adding the two data sets [54].

2.2.2. Model To quantitatively model the net current as a function of temperature it is necessary to estimate the shape of a single energy barrier at zero bias voltage from the lithographic shape of the channel and the lateral electron confinement energy (Fig. 8.12, for details see [27]). Furthermore, we need to estimate the shape of the barrier at finite positive or negative voltage, that is, we need to know the spatial distribution of the potential drop between reservoirs. A rigorous theory would require self-consistent treatment of the device potential based on the three-dimensional Schrödinger equation under finite bias conditions. Here we assume a spatial distribution of the voltage drop that is proportional to the local derivative of the barrier (Fig. 8.12). This model is based on the notion that a more rapid spatial potential variation leads to stronger wave

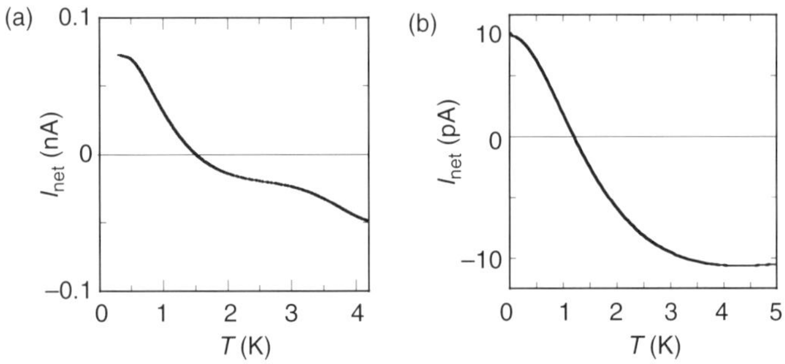

Fig. 8.11. (a) Measured net electric current, induced by "rocking" the ratchet potential of Fig. 8.10 by a square-wave voltage of amplitude 1 mV, versus temperature. (Data from [27].) (Note that positive electrical current corresponds to a particle current from right to left in Fig. 8.10.) (b) Calculated net current for the barrier shown in Fig. 8.12 and for a rocking voltage of amplitude 0.5 mV and Fermi energy 11.7 meV (see section 2.2.2 and Eq. (2)).

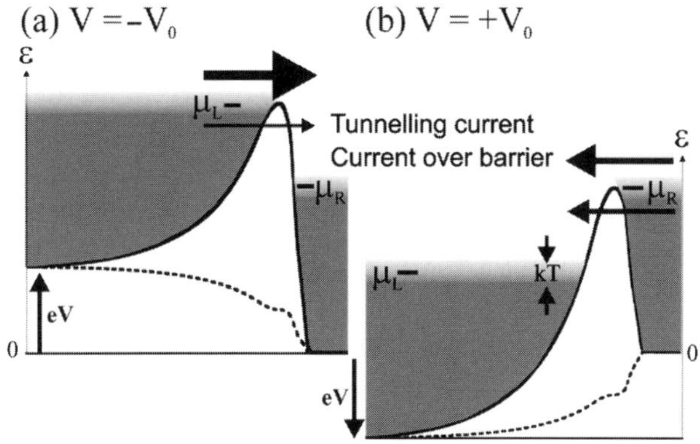

Fig. 8.12. Model of a single ratchet barrier of the device in Fig. 8.10. The shape of the barrier is based on the lithographic shape of the electron wave-guide and has a maximum barrier height of 12 meV at zero bias. To obtain the barrier at finite negative (a) or positive (b) voltage, an assumption for the spatial distribution of the voltage drop needs to be made (dashed line). The arrows indicate the flow directions of electrons (electron flow from right to left corresponds to positive electrical current). Figure by Tammy Humphrey [34].

reflection, and therefore a local voltage drop [48]. This particular choice has the side effect that the barrier height changes with voltage in a symmetric manner (the height of the ratchet barrier is the same for positive and negative voltage), resulting in the suppression of a classical contribution to the net current. Crucially, however, when the barrier is tilted to the right ($V < 0$), it deforms to be thicker at a given energy under the barrier top than if tilted to the left ($V > 0$). Consequently, the energy composition of the current is different for $V > 0$ and $V < 0$: a thicker barrier reduces the probability for tunneling through the barrier, but, at the same time, makes it easier for electrons with high energy to cross over the barrier, because the smoother shape reduces wave reflection. The electron flow above the barrier is thus larger for $V < 0$ than for $V > 0$, while the flow of electrons that tunnel through the barrier is larger for $V > 0$ than for $V < 0$.

To calculate the resulting current we assume that transport across the energy barrier is ballistic and elastic (that is, the electrons do not change their energy while traversing the barrier), and we consider electron transport in the lowest 1D wave mode only ($M = 1$ in Eq. (1)). Using the estimated barrier shapes at $V > 0$ and $V < 0$, the transmission $t(\varepsilon)$ is then calculated for each of the two voltages by solving the 1D Schrödinger equation (see inset to Fig. 8.13), and the dc current for each voltage can be calculated using Eq. (2). The (time-averaged) net current induced by a square-wave voltage switching between $\pm V_0$ can be written as [27]:

$$I_{\text{net}}(V_0) = \frac{e}{h} \int_0^\infty \Delta t(\varepsilon, V_0) \Delta f(\varepsilon, V_0) \mathrm{d}\varepsilon \tag{2}$$

where $\Delta f(\varepsilon, V_0) \equiv [f_L(\varepsilon, V_0) - f_R(\varepsilon, V_0)]$ is the "Fermi window" and is shown in Fig. 8.13 for $V_0 = 0.5$ mV, $\mu_F = 11.7$ meV and three different temperatures. The term

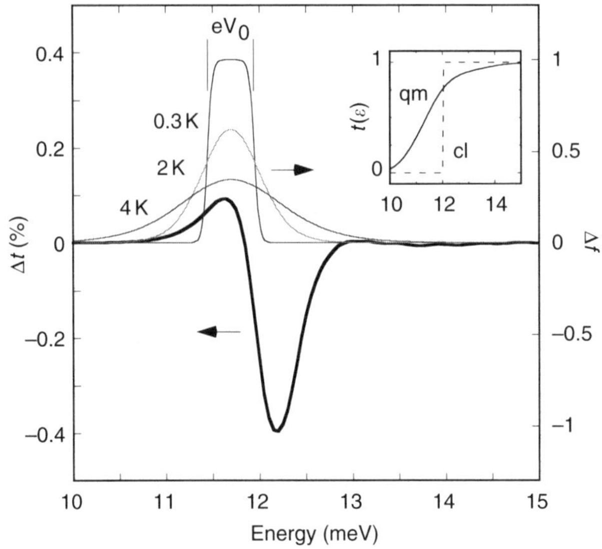

Fig. 8.13. Model of net current in the ratchet of Fig. 8.10. The bold line is $\Delta t(\varepsilon)$, the difference between the transmission functions for positive and negative bias voltage for the potential barrier in Fig. 8.12 (see Eq. (2)). Thin lines are the Fermi windows $\Delta f(\varepsilon)$, that is, the range of electron energies contributing to the current for the temperatures indicated. The net current is given by the integrated product $\Delta t(\varepsilon)\Delta f(\varepsilon)$ (see Eq. (2)). At $T = 0.3$ K the electron flow within Δf samples an area where Δt is negative. At 4 K, Δf samples an energy range where Δt is predominantly positive, and the current reverses direction. The inset shows the classical and quantum mechanical transmission functions $t(\varepsilon)$ for a barrier height of 12 meV. Quantum mechanical corrections extend about 2 meV above and below the barrier maximum.

$\Delta t(\varepsilon, V_0) \equiv [t(\varepsilon, V_0) - t(\varepsilon, -V_0)]$ is the difference between the transmission probabilities for positive and negative voltages. For $V_0 = 0.5$ mV, we find that $\Delta t(\varepsilon)$ is of the order of 10^{-3}, for energies within a few meV of the barrier height of $E = 12.0$ meV (Fig. 8.12). Electrons with energies smaller than the barrier height are more likely to tunnel from right to left ($V = V_0$) when the barrier becomes thinner, than from left to right ($V = -V_0$) when the barrier becomes thicker. This results in Δt being positive in this energy range. For energies above the barrier height the situation is reversed: electron wave reflection is stronger for steep, thin barriers (for $V = V_0$) than for the smoother, thicker barrier ($V = -V_0$), and Δt is negative in that energy range. As Δf is adjusted (through changing T, V_0 or the Fermi energy) to sample the Δt curve where it is negative rather than positive, the net current reverses direction (Fig. 8.11(b)).

The model just described concentrated on the quantum mechanical contributions to the net current. It is important to note, however, that a classical contribution is also possible if the barrier height (and not only the barrier shape) depends on the sign of the applied voltage. This is generally the case if a linear potential tilt is assumed, as in Fig. 8.9. It can be expected that a full self-consistent analysis of the barrier shape in our experimental device will yield some dependence of the barrier height on voltage direction, which would lead to a finite classical ratchet current in addition to the quantum-mechanical current contributions considered here.

2.3. *Quantum-Dot Based Brownian Heat Engines*

2.3.1. Tunneling Ratchets Act as Heat Pumps As described in the previous section, in a tunneling ratchet electrons at higher energies and electrons at lower energies flow in opposite direction on average. A direct consequence of this behavior is that tunneling ratchets can be interpreted as heat pumps [34]. This is easiest to see by considering the situation where the net particle flow due to tunneling, flowing at low energies, is exactly counter-balanced by the thermally excited net current flowing at higher energies (Fig. 8.12). In this situation, where the net particle flow is zero ($I_{\text{net}} = 0$ in Fig. 8.11), the ratchet clearly pumps heat. In the more general case, when the net particle flow is finite, the heat flow due to the ratchet's heat pumping action is superimposed on the net energy current that trivially accompanies any finite particle current. For clarity, the following analysis will assume a situation where $I_{\text{net}} = 0$.

We consider again the single ratchet barrier of Fig. 8.12, connecting two 2D electron reservoirs, R and L. The heat added when one electron is transferred to a reservoir with chemical potential μ is given by $\Delta Q = (\Delta U - \mu)$, where ΔU is the change in internal energy and is given by the electron's energy ε. The change in heat in R and L, respectively, upon transfer of one electron from the left to the right is then given by $\Delta Q_{\text{R/L}} = \pm(\varepsilon - \mu_{\text{R/L}})$. We replace the electron charge, *e*, by a factor of $\Delta Q_{\text{R/L}}$ inside the integral of Eq. (2), and can write the steady state heat current at a given bias voltage V :

$$q_{\text{R/L}} = \pm\frac{2}{h}\int_0^\infty (\varepsilon - \mu_{\text{R/L}})t(\varepsilon, V)\Delta f(\varepsilon, V)\text{d}\varepsilon \tag{3}$$

Note that $q_{\text{R}}(V)$ (the heat current entering R) exceeds $-q_{\text{L}}(V)$ (the heat current leaving L) by the externally supplied Joule heating power, *IV*. The *net* (time-averaged) heat current into the right and left reservoirs is $q_{\text{R/L}}^{\text{net}}(V_0) = 0.5[q_{\text{R/L}}(V_0) + q_{\text{R/L}}(-V_0)]$ [34]. It is illustrative to write this equation in the form

$$\begin{aligned} q_{\text{R/L}}^{\text{net}} &= \tfrac{1}{2}[q_{\text{R}}^{\text{net}} + q_{\text{L}}^{\text{net}}] \pm \tfrac{1}{2}[q_{\text{R}}^{\text{net}} - q_{\text{L}}^{\text{net}}] \\ &= \tfrac{1}{2}\Omega \pm \tfrac{1}{2}\Delta E \end{aligned} \tag{4}$$

which can be interpreted as follows: $\Omega = q_{\text{L}}^{\text{net}} + q_{\text{R}}^{\text{net}}$ is the electrical power input (Joule heating), averaged over one cycle of rocking. Ω is essentially the "waste heat" generated by operating the ratchet as a heat pump, and is proportional to the average electron current flow [34]. $\Delta E = q_{\text{R}}^{\text{net}} - q_{\text{L}}^{\text{net}}$ is the heat pumped from the left to the right sides of the device due to the energy sorting properties of the ratchet. In a sense the ratchet acts as a badly constructed refrigerator because it pumps heat from one reservoir to another, but deposits waste heat in both reservoirs [34]. The performance of the ratchet as a heat pump can be quantified by a coefficient of performance $\chi = |\Delta E|/\Omega$. When $\chi < 1$, both reservoirs are heated ($\Omega > |\Delta E|$), but one more than the other. Cooling of one reservoir takes place when $\chi > 1$. However, a numerical analysis of the experimental

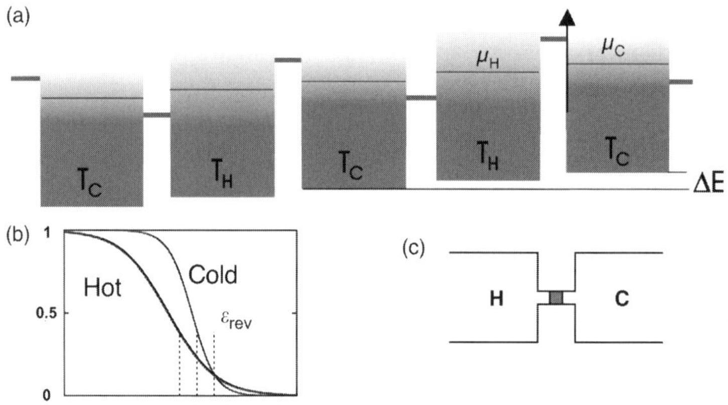

Fig. 8.14. (a) Principle of a Brownian heat engine consisting of a sequence of electron reservoirs alternating between temperatures T_H and $T_C (T_H > T_C)$. The shading indicates occupation of states as a function of energy according to the Fermi–Dirac distribution (chemical potentials are indicated by thin lines). A linear potential gradient of ΔE per period is applied. Bold lines indicate "energy filters" electrically linking the reservoirs at particular energies. (b) Fermi–Dirac distribution functions of two electron reservoirs with different temperatures and chemical potentials. The crossing point of the two distribution functions defines the energy ε_{rev}. (c) The energy filter could be realized in principle by a quantum dot with strongly separated energy states embedded into 1D channels connecting two reservoirs [35,58].

device shown in Fig. 8.10 predicts values for χ on the order of only 1% [34]. This is because the barriers transmit electrons in a wide range of energies (all of which contribute to heating), while the rectified current accounts for only a few percent of the heat flow.

2.3.2. Reversible Brownian Heat Engines Based on Quantum Dots Quantum dots could be used to substantially improve the performance of "quantum heat pumps", or to construct highly efficient "quantum Brownian heat engines" [58,59]. A "Brownian heat engine" is a ratchet that uses local temperature gradients in an asymmetric potential to move particles against an external force (see [2,60–68] and references therein). Fig. 8.14(a) illustrates the principle: an electron heat engine could consist of a periodic sequence of alternating hot and cold, isolated regions of a two-dimensional electron gas. A macroscopic potential gradient of ΔE per period represents an external force against which the engine can do work. The reservoirs are thermally isolated except for the heat carried by electron current in and out of each reservoir. The reservoirs are electrically interconnected via "energy filters"—for instance, resonant tunneling structures based on tunable quantum dots embedded into connecting 1D channels (Fig. 8.14(c)). These filters allow electron transfer between reservoirs only in a specific energy range that can in principle be adjusted by changing the quantum dot properties. We further assume that electrons enter one reservoir at the same energy at which they left the other and, once inside a reservoir, become part of the thermal distribution.

Figure 8.14(b) illustrates the effect of an energy filter on the electron flow between two reservoirs, for instance the two reservoirs denoted H (electrochemical potential

μ_H, temperature T_H) and $C(\mu_C, T_C)$ in Fig. 8.14(a). In the general case, where the reservoirs have different temperatures and different chemical potentials, their Fermi-Dirac occupation functions cross at one particular energy ε_{rev}. As indicated in the figure we assume $T_H > T_C$ and $\mu_H < \mu_C$. Below ε_{rev}, the probability for an electron state to be occupied is higher in the colder reservoir than in the hot reservoir. Spontaneous, net electron flow at this energy will therefore be from the cold reservoir to the hot reservoir, regardless of the direction of the potential gradient. Correspondingly, flow will always be from hot to cold at energies above the crossing energy. A filter positioned above ε_{rev} in Fig. 8.14(b) will therefore result in current flowing against the potential gradient, using thermal energy from the hot reservoir (Seebeck effect). A filter positioned below ε_{rev} results in cooling of the cold reservoir using electric energy (Peltier effect). By appropriately choosing the position of each filter at each interface of the setup shown in Fig. 8.14(a), the mode of operation can be chosen: the ratchet-like arrangement can operate either as a Brownian heat engine (using the temperature variations to produce work against the potential gradient) or as a Brownian refrigerator (using the potential gradient to remove heat from the colder reservoirs).

When an ideal filter (i.e. a filter that allows electron flow in an infinitesimally small energy range only) is positioned exactly at ε_{rev}, electron transfer between the two reservoirs is reversible [35]. To see this, consider the change of heat in C associated with transfer of an electron at energy ε from C to H. It is given by $\Delta Q_C = -(\varepsilon - \mu_C)$, where the negative sign in front of the bracket indicates removal of heat. The heat added to H is $\Delta Q_H = (\varepsilon - \mu_H)$ and exceeds $-\Delta Q_C$ by ΔE, because the electron picks up kinetic energy in the electric field between reservoirs. The overall increase in entropy due to the transfer of the electron is

$$\Delta S = \frac{\Delta Q_C}{T_C} + \frac{\Delta Q_H}{T_H} = \frac{-(\varepsilon - \mu_C)}{T_C} + \frac{(\varepsilon - \mu_H)}{T_H} \tag{5}$$

Significantly, Eq. (1) yields $\Delta S = 0$ for $\varepsilon = \varepsilon_{rev}$, where

$$\varepsilon_{rev} \equiv \frac{\mu_C T_H - \mu_H T_C}{T_H - T_C} \tag{6}$$

is the energy where the two Fermi distributions have the same value: $f_C(\varepsilon_{rev}) = f_H(\varepsilon_{rev})$. One may say that at this particular energy the two reservoirs behave as if they were in thermal equilibrium with each other. If two reservoirs were connected via an ideal energy filter that was transparent for electrons at ε_{rev} and at no other energy, no time-averaged particle or heat current would occur spontaneously. The warm bath would not cool, and the voltage would drive no current! On the other hand, Humphrey et al. [35] have shown that by positioning *all* energy filters in Fig. 8.14(a) arbitrarily close to ε_{rev} at each of the interfaces, the operation of the Brownian engine can be brought arbitrarily close to Carnot efficiency. This result is significant because the proposed system represents the first Brownian heat engine that can theoretically achieve the thermodynamically ideal efficiency—in most other Brownian heat engines considered so far, efficiency is very low due to irreversible heat flow carried by particles making contact with different heat baths [63–67]. The key property of the quantum-dot based heat engine is the energy filtering process, which makes use of the electron's wave properties. For a discussion of the technical feasibility of such a refrigerator/engine, see the work by Edwards et al. [59].

3. Nonlinear Transport in Antidot Structures

In section 2 we described a number of nonlinear electron transport phenomena in the quantum regime of transport. In modeling the observed effects, it was suitable to focus on the effect that the voltage had on the form of the scattering potential. The resulting changes in electron transport properties were then described in terms of the voltage-induced changes in the transmission coefficient. This is a somewhat different approach from the one used in the following section, where semi-classical and classical electron transport phenomena in multi-terminal, ballistic, asymmetric antidot structures are described. In modeling these high-temperature effects, the device's scattering potential—usually represented by asymmetric antidots—is taken as voltage-independent. Nonlinear effects are accounted for in terms of voltage-dependent velocity distributions of the electrons participating in the transport. Before turning to specific experiments, we will briefly discuss this regime of nonlinear ballistic electron transport.

3.1. Nonlinear Ballistic Transport

In a traditional, macroscopic semiconductor device—which is much larger than the electron elastic scattering length (mean free path, l_e)—the electron transport is diffusive. Electrons undergo a large number of scattering events with randomly distributed scatterers such as impurities and phonons, and electric current consists of electrons diffusing in an electric field. At a finite electric field (or applied voltage), an electron is accelerated between two scattering events, and the velocity component of the electron in the opposite direction of the electric field increases constantly until the next scattering event. Since the current carried by the electron equals its velocity multiplied by the electron charge e, the current carried by any electron is time dependent even in a constant electric field. This appears to be in contradiction to Ohm's law, which states a linear relation between current and voltage (or electric field). However, Ohm's law holds only on the *macroscopic* level where one can average over a large number of scattering events. The large number of scattering events counteract the acceleration effect of the electric field, and the average velocity of all electrons is then proportional to the strength of the electric field. In a *mesoscopic* semiconductor device that is smaller than the electron mean free path, however, electrons going from one electric contact (often also called electric lead, terminal, or carrier reservoir) to another no longer undergo any scattering event, except with the well-defined device boundary. The electron transport is hence called ballistic [5,37,69]. For a narrow channel with a constant width, this means that even at a finite applied voltage, the electric field inside the channel has to be zero to fulfil the requirement of the conservation of current: the current flowing into one end of the channel must be the same as the current flowing out of the channel at the other end. Therefore, inside the channel, current flows in the absence of a finite electric field, just like in a superconductor. This remains to be true, regardless of the length and width of the channel, as long as the channel is shorter than the electron mean free path. Nevertheless, to drive a current through the entire device, a finite voltage has to be applied and the voltage will only drop at the entrance and exit of the channel, where resistance arises because most of the electrons are back-scattered and only a fraction of the electrons can enter the channel [70]. Clearly, *local quantities* such as resistivity and conductivity can no longer be defined in a ballistic semiconductor device, and only the total resistance or conductance can be used. As described by the Landauer–Büttiker formula, [6,71,72], the conductance of a mesoscopic semiconductor device is determined by *nonlocal quantities*: the transmission probabilities of the ballistic electrons going from one electric contact to another.

While there have been a great number of studies on the ballistic electron transport in the linear regime, only a small number of experimental [24–27,53,55,56,73–82] and theoretical [83–89] investigations have been performed in the nonlinear regime. Nonlinear transport is particularly important in mesoscopic structures. First of all, a very small applied voltage suffices to drive the electron transport into the nonlinear regime. This is because of the small device dimension and the fact that in general the applied voltage drops quite locally in the regions where the dimension of the electron path is changed, such as the opening and exit of a narrow channel, as mentioned above. Secondly, if the device dimension is comparable to the electron wavelength, quantum effects will take place, which are in general very sensitive to the applied voltage. For example, in a narrow channel or a device with a narrow constriction, the number of the transverse quantum confinement modes that electrons occupy, N, will determine the maximum conductance, $G_{\max} = 2Ne^2/h$ (assuming spin degeneracy) [3,4]. An applied voltage will change both the phase of the electrons and the electron distribution among the transverse quantum confinement modes, leading to a change in the conductance, as shown in section 2. Furthermore, in a multiple-terminal ballistic device, the angular distribution of the electrons may largely determine the electron transmission probabilities between the terminals, and consequently also the transport properties of the device. In contrast to the electron transport in a macroscopic semiconductor device, where the large number of scattering events completely randomize the direction of the electron velocity, a voltage applied to a ballistic device will change the angle of every electron equally. As will be shown in this section, this has a direct influence on the transmission probabilities and therefore also on the current distribution in the device, leading to a new type of nonlinear effect which does not exist in a macroscopic device.

Apart from the above nonlinear effects, it was very recently realized that the geometrical symmetry of a ballistic device can also have a very pronounced influence on the nonlinear device properties [24–27,78]. In many aspects, a nonlinear ballistic device is similar to a ratchet, which in general is a system that combines spatial asymmetry and non-equilibrium to generate directed transport (see section 1). In this context it is interesting to note that electron ratchets exist naturally. In particular, it was discovered in 1970s that some natural crystals can generate direct electron current under uniform illumination [90–92]. Referred to as the photogalvanic effect, the phenomenon has been identified to be a macroscopic manifestation of the absence of centro-symmetry in the elementary crystal structure, despite the macroscopic homogeneity of these crystals. In this case, light serves as a source of external fluctuation by imposing an unbiased, time-periodic electric field on the crystals. The generation of direct electron current can be viewed as an extremely fast rectification, or ratchet effect, at the frequency of the light. One difference to a one-dimensional ratchet as described in section 2 is that the generated electron current may flow in a direction different from, or even perpendicular to, that of the driving electric field of the light. This is due to the broken symmetry of the microscopic potential in two or three dimensions.

Recently, biomolecular separation devices consisting of arrays of symmetry-breaking microscopic obstacles, similar to those found in photogalvanic crystals, were designed and demonstrated [29, 93–95]. These experiments show that asymmetric geometries may provide the same nonlinear functionality as asymmetric potentials in creating a guided flow of particles.

In this section we will also describe how the high spatial-resolution of modern semiconductor fabrication technologies can be used to realize individual electron ratchets as well as large areas of artificial "photogalvanic crystals". A series of experiments on nonlinear effects in asymmetric, triangular antidot structures are reviewed. The observed nonlinear effect is so strong that the devices are capable in rectification applications and hence called ballistic rectifiers. The devices, made by simply cutting pieces from the artificial materials, generally have four terminals. When either a regular ac voltage or a random signal is applied via two of the electrical contact leads, a direct voltage or current is generated in the perpendicular direction and can be measured via the other two terminals. Not only operating at cryogenic temperatures, the smallest devices and artificial materials have also been shown to work at room temperature and at frequencies up to at least 50 GHz. Promising possibilities for practical applications, such as rectification, microwave detection, and higher harmonic generation, will be discussed.

3.2. Low-Temperature Experiments

3.2.1. Ballistic Rectifiers Figure 8.15(a) is an atomic force microscope image of the central part of one of the first ballistic rectifiers. The dark areas in Fig. 8.15(a) are the regions that are etched away and are, therefore, nonconductive for the electrons. As a result, the triangular antidot is defined in the cross junction formed by the two narrow channels (labeled "source" S and "drain" D) and the two wide channels (labeled "upper" U and "lower" L) [25]. The triangular antidot acts as an artificial scatterer which, as shown in the following, induces a pronounced nonlinear effect. Standard electron-beam lithography and wet chemical etching are used in the fabrication. A modulation-doped GaAs–AlGaAs heterostructure, containing a two-dimensional electron gas (2DEG) located 37 nm below the wafer surface, was used to fabricate the devices. The electron density is about 5×10^{11} cm^{-2} and the mobility about 5×10^5 cm^2/Vs at the temperature $T = 4.2$ K. The corresponding electron mean free path $l_e = 5.8$ μm is longer than the central part of the device, meaning that the electron transport is ballistic and the specular electron scatterings from the etched boundaries dominate the transport properties at 4.2 K. It is therefore possible to alter the device properties in the ballistic electron transport regime by simply changing the device shape (as will be shown below). This is in contrast to the case of the diffusive electron transport in a device much longer than l_e, where the electrons undergo a large number of scattering events with, e.g. randomly

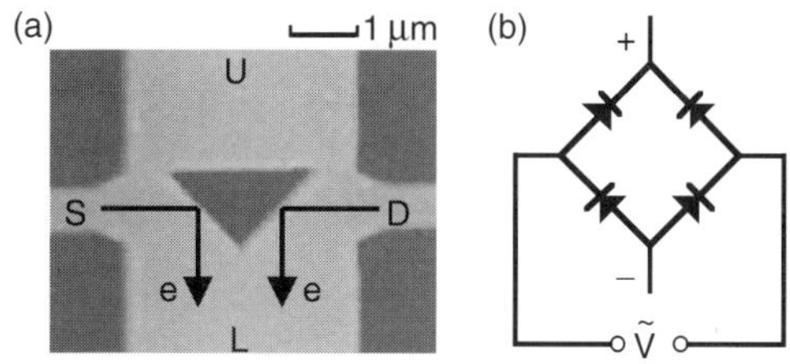

Fig. 8.15. (a) Atomic force microscope image of the central part of a ballistic rectifier. The dark areas were etched away, forming a triangular antidot in the middle and four terminals, S, D, L, and U. The current is applied via S and D, and the output voltage is measured between L and U. The device operates similarly to a bridge rectifier (b), but is based on a completely new working principle. From [25].

distributed impurities, and the electron transport is not sensitive to a change in the shape of the device.

In the experiments, a current is applied to leads S and D, and the output voltage is detected via L and U. The measurements are performed at $T = 4.2$ K and 77 K. The typical electron trajectories, illustrated by the arrows in Fig. 8.15(a), appear to suggest an accumulation of electrons in the lower lead and hence a negative voltage between L and U. Obviously, this cannot happen when the applied voltage or current is zero, since the same number of electrons will travel along the opposite directions of the arrows. For a nonzero applied current, $I_{SD} \neq 0$, however, as shown by a detailed model [79,96], the electron transmission along the arrows in Fig. 8.15(a) is changed by the applied electric field while the transmission in the reversed direction is virtually unaffected (details in section 3.3). This gives rise to a finite negative voltage between L and U, V_{LU}.

From the mirror symmetry along the central L–U axis, the output voltage V_{LU} should remain the same if the sign of the applied current I_{SD} is changed:

$$V_{LU}(I_{SD}) = +V_{LU}(-I_{SD}) \tag{7}$$

Correspondingly, $R_{SD,LU}(I_{SD}) = -R_{SD,LU}(-I_{SD})$, where $R_{SD,LU} = V_{LU}/I_{SD}$ is the four-terminal resistance. Such a picture is almost perfectly supported by the experimental V_{LU} versus I_{SD} curves at $T = 4.2$ K (solid line) and 77 K (dotted line) shown in Fig. 8.16. The slight deviation from Eq. (7) is attributed to the unintentional breaking of the desired symmetry along the L–U axis by the imperfection of the device fabrication. The imperfection is expected to contribute mainly a linear term to the V_{LU} versus I_{SD} curves, in a way similar to a change of one of the four resistors in an otherwise balanced Wheatstone resistor bridge.

Even though the device works best at liquid helium temperatures, a weaker rectification can also be observed at 77 K [dotted line in Fig. 8.16(a)]. The less pronounced nonlinear effect is attributed to the shorter elastic mean free path l_e (about 1 to 2 μm) at 77 K. Since l_e is comparable to the distance between the openings of the S or D channels and the antidot, the electron transport is only partially ballistic. Approximately half number of the electrons ejected out of S or D will experience at least one scattering event from, e.g., impurities or phonons, before they can reach the triangular antidot.

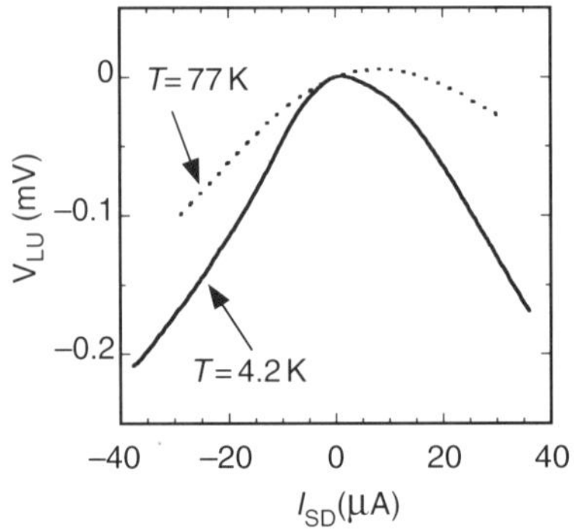

Fig. 8.16. The output voltage between the lower and upper leads, V_{LU}, as a function of the input current through the source and drain leads, I_{SD}, measured at $T = 4.2$ K (solid line) and 77 K (dotted line). Ideally, the curves should be perfectly symmetric with respect to the vertical axis at $I_{SD} = 0$. However, the imperfection in the fabrication causes the slight asymmetry in the curves. From [25].

This results in a reduced output between L and U. At room temperature, the device does not show any noticeable rectifying effect because the mean free path becomes much shorter (about a factor of 10) than the device size and the electron transport is diffusive. To realize room temperature operation, much smaller devices have been made using an InGaAs–InP quantum well wafer, which will be shown in section 3.4.

The nonlinearity in the I–V characteristics is remarkably different from that of a two-terminal nonlinear device, such as a diode, of which the nonlinearity always comes on top of a large linear term. For the ballistic rectifier with four-terminal geometry, there can be, however, no linear term presented in the relation between the output voltage and the applied current, as shown by Eq. (7). The striking nonlinearity is a result of the broken symmetry along the S–D axis and the mirror symmetry along the U–L axis. As will be shown in sections 3.3 and 3.4, the absence of a linear term has quite a few advantages in practical applications.

The above device demonstrates that new device functionalities can be realized by simply tailoring the shape of a ballistic device, because the only scattering events that the electrons undergo are with the designed device boundary. Based on a similar working principle to the ballistic rectifier, other types of devices can be designed. The inset of Fig. 8.17 shows schematically a device containing channels defined by gates rather than by chemical etching. Instead of introducing a symmetry-breaking triangular scatterer as in Fig. 8.15(a), the mirror symmetry along the S–D axis is broken by the horn-shaped U and L channels. The measured V_{LU} versus I_{SD} curves at different gate voltages V_{GU}, applied between the gates and the upper lead, are plotted in Fig. 8.17. As expected, virtually the same result is obtained when the gate voltage is applied using the lower lead as the ground [97]. Because of the imperfection of the fabrication, the curves are not perfectly symmetric along the $I_{SD} = 0$ axis. Nevertheless, it is obvious that by increasing the negative gate voltage from -1.2 V to -1.5 V, which effectively reduces the electronic width of the S and D channels, the nonlinear component of

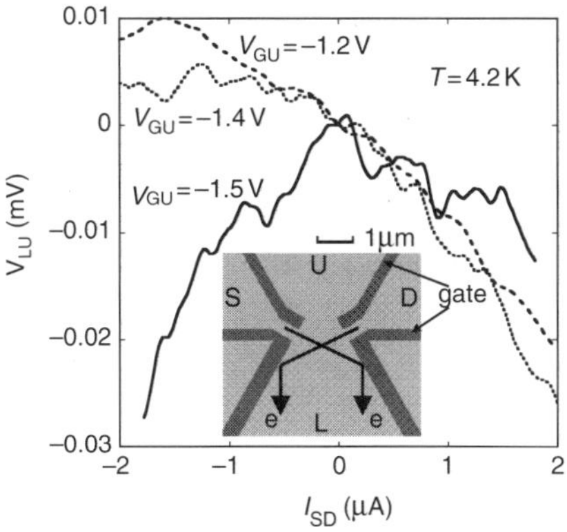

Fig. 8.17. V_{LU} versus I_{SD} curves at different gate voltages V_{GU} at $T = 4.2$ K, measured in a ballistic rectifier fabricated with gates rather than by chemical etching, as schematically shown by the inset. The dark areas in the inset are gates, to which the same gate voltage is applied. The arrows in the inset illustrate typical electron trajectories. From [97].

I–V curves increases. This phenomenon can be explained by the model (Eq. (16)) in section 3.3.

While the above experiments are performed using dc signals, it is expected, from the rectifying effect shown in Fig. 8.16 or 8.17, that if an ac voltage is applied to leads S and D, a negative average (dc) voltage should be observed between L and U as well. This will be demonstrated below using InP–InGaAs based devices. Actually, a direct voltage output has been observed even when an external noise signal is applied [98], showing a ratchet effect induced by the broken geometric symmetry as well as the guidance of ballistic electrons by the triangular antidot.

The devices shown in Figs 8.15 and 8.17 operate similarly to a bridge rectifier (see Fig. 8.15(b)), and therefore are called ballistic rectifiers. However, only a single device is used here rather than the four individual diodes that form a bridge rectifier. The mechanism is also entirely different since no doping junction or barrier structure along the current direction is used in the ballistic rectifier. Instead, the pronounced nonlinear effect is related to the geometric asymmetry and to the properties of ballistic electron transport. Furthermore, as will be shown in section 3.3, the ballistic rectifier has no intrinsic voltage or current threshold, and can be used for detection of very weak signals without the need of an external bias. High frequency experiments in section 3.4 also demonstrate that the ballistic rectifiers are capable of operating at frequencies up to at least 50 GHz.

3.2.2. Artificial Functional Materials Based on a working principle similar to that of the ballistic rectifier, a new type of artificial functional material has been constructed, which may be regarded as a type of artificial "photogalvanic" crystal. Figure 8.18 shows the atomic force microscope images of two examples of the artificial material that have been fabricated. Similar to the ballistic rectifiers, the artificial materials are fabricated using standard electron-beam lithography and wet chemical etching. In Fig. 8.18(a), the artificial material is constructed by simply periodically repeating a ballistic rectifier. The individual rectifiers are connected in series and parallel, forming an "artificial lattice". In principle, such an artificial material can be made over a whole wafer. One

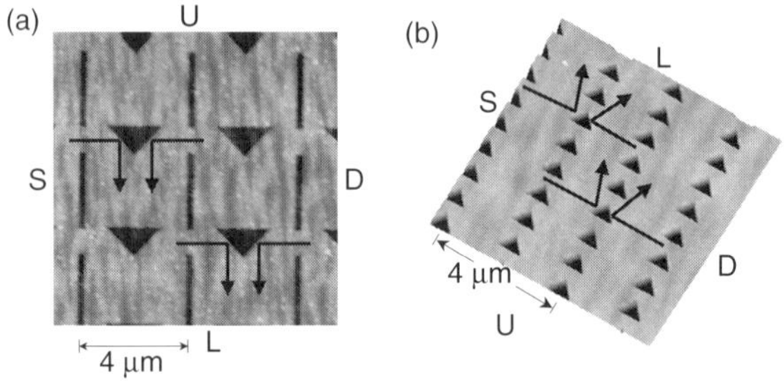

Fig. 8.18. Atomic force microscope images of artificial functional materials of two different designs, fabricated using a GaAs–AlGaAs heterostructure. Note that terminal L is on the upper-right side in (b). The arrows indicate the typical electron trajectories in the artificial materials.

apparent advantage is that the material is suitable for applications involving large signals, as a much higher voltage or current can be applied to the artifical material than to an individual rectifier. Compared with the fabrication of a diode which involves multiple-step mask alignment, the production of the nanomaterial (as well as the ballistic rectifier discussed above) involves only one step of lithography. Therefore, the recently invented nanoimprint technology [95] can be used to mass produce these materials. It allows to avoid the bottleneck of extremely long exposure time in standard electron-beam lithography, a challenging problem in producing a large number of nanodevices efficiently and economically. With the nanoimprint technology, once a master stamp (mold) is fabricated with, e.g., electron-beam lithography, it can be used repeatedly to replicate patterns over whole wafers of up to six inch in a matter of minutes [99]. Furthermore, for radio-frequency (GHz or even THz) applications, such as rf detection (the capability of the material to function at up to at least 50 GHz will be discussed in section 3.4), it is often desirable to have a detector with a large active region, in order to eliminate the need for building antennae. Moreover, as the nanomaterial is *intrinsically* nonlinear, individual devices can be made by simply cutting pieces from the material, according to the requirements for different applications.

A different artificial material, consisting of only triangular antidots, is shown in Fig. 8.18(b). In this material, not only the intentionally broken symmetry by the triangular antidots is important, but also the introduced offset between neighboring columns: the triangles are placed in such a way that neighboring triangles support each other by simultaneously scattering electrons and forming channels through which electrons are ejected, as shown by the arrows in Fig. 8.18(b).

An $80 \times 80\ \mu m^2$ piece of the artificial material shown in Fig. 8.18(b) has been measured at $T = 4.2$ K. To study how the nonlinear effect changes with the electron concentration or the electron Fermi wavelength, a layer of metal gate is evaporated on top of the artificial material after etching the triangular antidots. Negative gate voltages, V_{GU}, are applied between the gate and U to lower the carrier concentration of the 2DEG and to increase the Fermi wavelength of the electrons. The lowered carrier concentration is directly indicated by the increase of the two-terminal source–drain resistance R_{SD}, as shown in Fig. 8.19(c). From the figure, the pinch-off gate voltage is about -2.25 V.

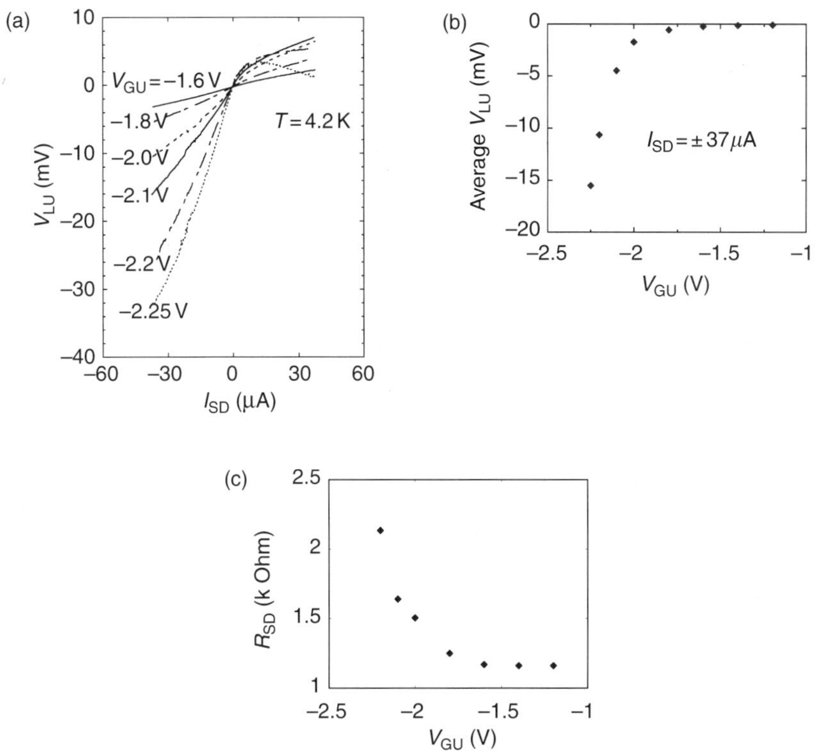

Fig. 8.19. (a) V_{LU} versus I_{SD} curves at different voltages between the gate G and terminal U, V_{GU}, measured in an 80×80 μm^2 piece of the artificial material shown in Fig. 8.18(b) at $T = 4.2$ K. (b) The average output voltage V_{LU} at $I_{SD} = +37$ μA and $I_{SD} = -37$ μA as a function of V_{GU}. (c) The two-terminal source–drain resistance R_{SD} as a function of V_{GU}.

Figure 8.19(a) shows the V_{LU} versus I_{SD} curves measured at different gate voltages V_{GU} at $T = 4.2$ K. Similar to the experimental curves of the ballistic rectifiers in Figs. 8.16 and 8.17, the curves are not perfectly symmetric with respect to the vertical axis at $I_{SD} = 0$, which is a consequence of the imperfection of the device fabrication. The overall downward bending of the curves clearly indicates the expected rectifying effect. To better illustrate the influence of the gate voltage on the rectifying effect, the average output V_{LU} at $I_{SD} = +37$ μA and $I_{SD} = -37$ μA is plotted as a function of the applied gate voltage V_{GU} in Fig. 8.19(b). The curve shows that the rectifying effect is enhanced dramatically by increasing the applied negative gate voltage. In particular, the average output increases most rapidly when V_{GU} is approaching the pinch-off gate voltage of about −2.25 V. This phenomenon can be very well explained by the model in section 3.3.

Apart from the potentials in practical high-frequency applications, which will be discussed in section 3.4, using such artificial photogalvanic crystals (with designed shapes of symmetry-breaking scatterers as well as the distance between them) may provide a useful way to further study the photogalvanic effect.

3.3. Semiclassical Model

The Büttiker–Landauer formalism [6,71,72], which treats transport in mesoscopic conductors as a transmission problem for carriers at the Fermi level, is widely used to describe the linear transport behavior of these conductors [100–106]. In the standard formula for a multi-terminal conductor at $T = 0$ (which is a different version of Eq. (2) for a multiple-terminal device),

$$I_\alpha = \frac{2e}{h} \sum_{\beta \neq \alpha} T_{\beta \leftarrow \alpha}(\mu_\alpha - \mu_\beta) \tag{8}$$

the transmission coefficient $T_{\beta \leftarrow \alpha}$ for electrons from terminal (contact) α to terminal β is not considered a function of the applied voltage, meaning that the formula holds only in the linear transport regime. Here, I_α is the current through lead α and μ_α is the chemical potential of the electrons in contact (carrier reservoir) α.

To model the nonlinear (rectifying) effect observed in the ballistic rectifiers and the artificial nonlinear materials, the Büttiker–Landauer formula is extended to the nonlinear regime. With this, a model, which provides an analytical description of the ballistic rectifying effect in the framework of classical ballistic electron transport, is obtained [79].

In general, the current through a lead α of a mesoscopic conductor, which is connected via perfect leads to a number of carrier reservoirs, can be written as

$$I_\alpha = \frac{2e}{h} \sum_{\beta \neq \alpha} \int \left[f(E - \mu_\alpha) - f(E - \mu_\beta) \right] \times T_{\beta \leftarrow \alpha}(E, B) \mathrm{d}E \tag{9}$$

Here, $f(E - \mu_\alpha) = [\exp(E - \mu_\alpha)/kT + 1]^{-1}$ is the Fermi–Dirac distribution function, and in general the transmission coefficient for carriers from lead α to lead β, $T_{\beta \leftarrow \alpha}(E, B)$, is a function of the electron energy E and magnetic field B. At $kT = 0$ and $B = 0$, Eq. (9) becomes

$$I_\alpha = \frac{2e}{h} \sum_{\beta \neq \alpha} \overline{T}_{[\beta,\alpha]}(\mu_\alpha - \mu_\beta) \tag{10}$$

Here, $\overline{T}_{[\beta,\alpha]} = \int_{\mu_\beta}^{\mu_\alpha} T_{\beta \leftarrow \alpha}(E) \mathrm{d}E / (\mu_\alpha - \mu_\beta)$ if $\mu_\alpha > \mu_\beta$, or $\int_{\mu_\alpha}^{\mu_\beta} T_{\alpha \leftarrow \beta}(E) \mathrm{d}E / (\mu_\beta - \mu_\alpha)$ otherwise. This shows that only the transmissions from the reservoirs with higher chemical potentials to the reservoirs with lower chemical potentials need to be considered.

Based on Eq. (10), the ballistic rectifying effect can be modeled in the framework of semiclassical ballistic electron transport. Similar to the model for ballistic electron transport in a cross junction by Beenakker and van Houten [107], the angular distribution of the ballistic electrons ejected from S and D is considered to largely determine the electron transmission probabilities into L or U leads. This is a particularly good approximation for this specific device geometry, because the transmission probabilities for electrons from S (or D) to L and U are virtually completely determined by the ejection angles of the electrons.

With no voltage applied, the angular distribution of the electrons ejected from a narrow constriction is given by

$$P(\theta) = \tfrac{1}{2}\cos\theta \qquad (11)$$

Here, θ is the ejection angle with respect to the channel direction. This is a property of ballistic electrons caused by adiabatic electron transport and is related to the device geometry [104,105,108,109]. Therefore, even at $V_{\mathrm{SD}} = 0$, the electrons ejected out of S are to some extent collimated as shown in Fig. 8.20(c).

In principle, to obtain the angular distribution of the ballistic electrons and the transmission coefficients at a finite I_{SD}, the potential profile in the device should be calculated self-consistently. Such a calculation is however very complex. Instead, it is noticed that the applied voltage mainly drops at the entrances and exists of the S and D channels [70]. Consider that at $V_{\mathrm{SD}} = 0$, the velocity components of a right-moving electron in the source channel are v_x and v_y, respectively, in the direction of the channel and in the perpendicular direction, as illustrated in Fig. 8.20(a). The angle of ejection, $\theta = \arctan(v_y/v_x)$, will change if a finite V_{SD} is applied. For example, if V_{SD} is negative, the current in the source channel $I_{\mathrm{S}} < 0$, meaning that the electrons in the source channel have a finite mean velocity or drift velocity, $\Delta v > 0$. As a good approximation, assume that the velocity component in the x direction of all the electrons in channel S is increased by Δv, while the velocity component in the perpendicular direction v_y is essentially not affected. The ejection angle of the electron changes from $\theta_0 = \arctan(v_y/v_x)$ at $I_{\mathrm{SD}} = 0$ to a smaller angle $\theta = \arctan[v_y/(v_x + \Delta v)]$, as shown in Fig. 8.20(b). Overall, this leads to a collimation effect induced by the applied

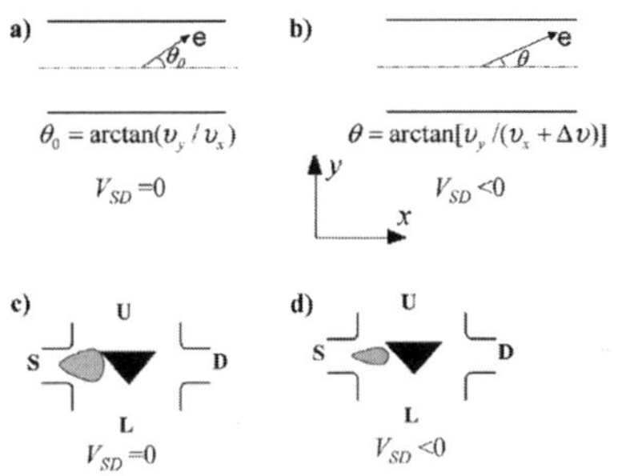

Fig. 8.20. (a) The sketch shows that at $I_{\mathrm{SD}} = 0$, an electron is moving in the source channel with the angle $\theta_0 = \arctan(v_y/v_x)$, where v_x and v_y are the velocity components along the direction of the channel and in the perpendicular direction. (b) When a negative voltage V_{SD} is applied, $I_{\mathrm{SD}} < 0$, corresponding to a finite drift velocity Δv of the electrons in the source channel. This results in a smaller ejection angle of the electron, i.e. $\theta = \arctan[v_y/(v_x + \Delta v)]$. Overall, this leads to a collimation effect induced by the applied voltage, and the angular distribution of the electron ejected out of S at $V_{\mathrm{SD}} = 0$, (c) becomes narrower at $V_{\mathrm{SD}} < 0$ as schematically shown in (d). Consequently, the electrons will more likely be scattered by the triangular antidot to the lower lead and less likely go through the gap between the antidot and the edge of the upper channel. This results in an accumulation of electrons in the lower lead and consequently a negative voltage V_{LU}.

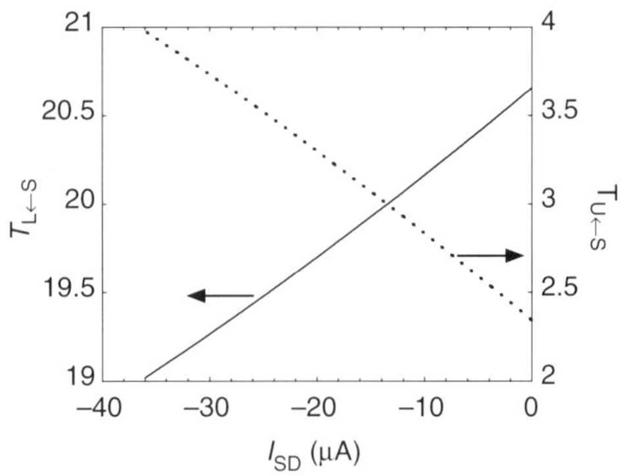

Fig. 8.21. The calculated transmission coefficients $T_{L\leftarrow S}$ and $T_{U\leftarrow S}$ as a function of the *negative* source–drain current at $T = 0$. From [79].

voltage, which effectively narrows the angular distribution of the electron ejected out of S at $V_{SD} < 0$, as schematically shown in Fig. 8.20(d).

With the narrower angular distribution of the electrons ejected out of S, the transmission probability for the ejected electrons to transmit from S to L will increase, since the electrons will have a greater chance to be scattered into the lower lead by the triangular antidot. In the mean time, the transmission probability from S to U will decrease as the percentage of the electrons that have a large enough ejection angle to transmit through the gap between the antidot and the edge of the upper channel will reduce. The above analysis shows that the transmission probabilities of electrons can be directly linked to the electron drift velocity and hence I_{SD}, as is calculated and shown in Fig. 8.21. This allows Eq. (10) to be written as

$$I_\alpha \approx \frac{2e}{h} \sum_{\beta \neq \alpha} \overline{T}_{[\beta,\alpha]}(I_{[\beta,\alpha]})(\mu_\alpha - \mu_\beta) \tag{12}$$

where $\overline{T}_{[\beta,\alpha]}(I_{[\beta,\alpha]}) = \overline{T}_{\beta\leftarrow\alpha}(I_\alpha)$ if $\mu_\alpha > \mu_\beta$ and $\overline{T}_{\alpha\leftarrow\beta}(I_\beta)$ otherwise [79].

With the values of the transmission coefficients determined, the V_{LU} versus I_{SD} characteristic is obtained by calculating the four-terminal resistance $R_{SD,LU} = V_{LU}/I_{SD}$. For the case of $V_{SD} < 0$,

$$R_{SD,LU} = \frac{h}{2e^2 D}[T_{L\leftarrow S}(I_S)T_{D\leftarrow U}(I_U) - T_{D\leftarrow L}(I_L)T_{U\leftarrow S}(I_S)] \tag{13}$$

where $I_S = I_{SD}$ and $I_L = I_U = 0$. Therefore, because the net current in L and U is zero, $T_{D\leftarrow L}$ and $T_{D\leftarrow U}$ are not sensitive to the applied current I_{SD} and their values at zero bias are employed. In Eq. (13), D is a function of the individual transmission coefficients, and is found to be insensitive to the applied voltage or current [110]. For the ballistic rectifier, taking into account of $T_{S\leftarrow D} = T_{D\leftarrow S} = 0$, it can be shown

$$D = N_{SD}[2T_{L\leftarrow S}T_{U\leftarrow S} + T_{L\leftarrow U}N_{SD}] \approx \frac{N_{SD}^2 N_{LU}}{3} \tag{14}$$

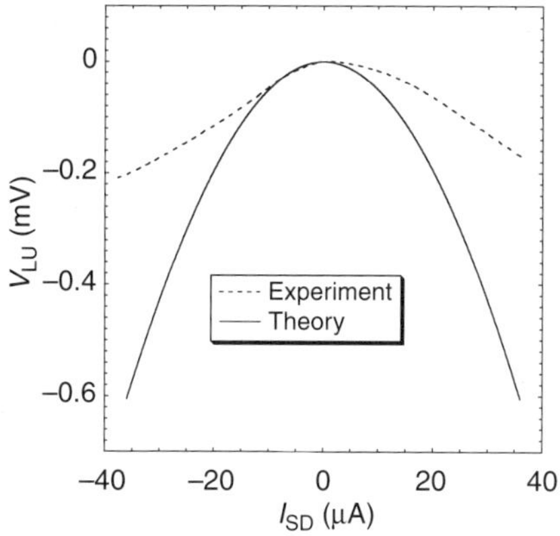

Fig. 8.22. Theoretical (solid line) and experimental (dashed line, the same as that in Fig. 8.16) V_{LU} versus I_{SD} curves for the device shown in Fig. 8.15(a). Although no fitting parameters have been used, the agreement between the experiment and the model is remarkably good.

For $V_{SD} < 0$, the V_{LU} versus I_{SD} characteristic is obtained from

$$V_{LU} = R_{SD,LU} I_{SD} = \frac{3h}{4e^2} \frac{\sin\theta_e - \sin\theta_0}{N_{LU}} I_{SD} \quad (15)$$

where $\theta_e = \theta_0 + \arcsin[(\Delta v/v_F)\sin\theta_0]$, which is also a function of I_{SD} because of the linear dependence of Δv on I_{SD}. $\theta_0 \approx \pi/4$ is the minimum ejection angle for an electron from S or D to be scattered by the triangular antidot to the lower lead, v_F is the Fermi velocity. N_{SD} and N_{LU} are the numbers of propagating modes in channels S and D and channels L and U, respectively [79]. The V_{LU} versus I_{SD} curve, calculated analytically, is shown by the solid line in Fig. 8.22. It is in very good agreement with the experimental result (dashed line) especially in the low current region. No adjustable parameters have been used at all in Eq. (15). At high currents, electron heating, not included in the above model, might reduce the rectifying effect. Moreover, the length of the L and U channels (5 μm) is comparable to the electron mean free path. This means that some electrons will be scattered by impurities or phonons and change directions before they can travel through the channels and enter the electrical contact regions (carrier reservoirs). This certainly causes a reduced output.

It is interesting to note that in the limit as $|I_{SD}| \to 0$, V_{LU} can be shown to have a quadratic response to I_{SD} [79]:

$$V_{LU} \approx -\frac{3\pi\hbar^2}{4e^3 E_F N_{SD} N_{LU}} I_{SD}^2 \quad (16)$$

Although the calculation is done under the condition of $T = 0$, it has been shown that Eq. (16) holds as well at a finite temperature for sufficiently small currents [111], meaning that there is no threshold or turn-on voltage for the ballistic rectifier. With respect to applications it is interesting to note that there are a number of additional practical advantages associated with the quadratic response, as will be discussed in more detail in section 3.4. Equation (16) indicates that $|V_{LU}|$ increases with reducing E_F, N_{SD}, and N_{LU}. This explains the gate voltage dependence of the output voltage V_{LU} shown in Fig. 8.17. With increasing the negative gate voltage, the S and D channels become narrower, leading to smaller N_{SD} and therefore the enhancement of the nonlinearity. Based on the above picture, the nonlinear ballistic electron transport at finite magnetic fields has also been well described [112].

Although so far the model is constructed semi-classically, the quantum effect, namely the transverse quantum confinement in all the four channels (S, D, L, and U), is of fundamental importance. From Eq. (16), the number of transverse quantum confinement modes in the channels are represented by N_{SD} and N_{LU}. If the electrons are solely regarded as classical billiard balls, both N_{SD} and N_{LU} will be infinite, leading to a zero output V_{LU}. Therefore, in the picture of purely classical ballistic electron scattering, the ballistic rectifier will not work. If the quantum effect is enhanced by, e.g., making a smaller device as will be shown later, N_{SD} and N_{LU} will reduce. From Eq. (16), the output of the ballistic rectifier will be substantially increased, showing that the device will work better when the wave nature of the electrons plays a more important role.

The rectifying effect observed in the artificial material shown in Fig. 8.19 can also be qualitatively explained in the same physical picture. Similarly to the voltage-induced collimation effect shown in Fig. 8.20, the angular distribution of the electrons ejected out of the narrow gaps between neighboring triangular antidots at $V_{SD} = 0$ (as illustrated in Fig. 8.23(a)) becomes narrower at $V_{SD} < 0$ (as schematically shown in Fig. 8.23(b)). The angular distribution is tilted upwards a little because of the asymmetry of the geometry [109,113]. The voltage-induced collimation effect leads to a higher probability for the electrons to be deflected downwards by the next triangular antidot on the side as schematically shown in Fig. 8.23(b) and a lower probability for the electrons to transmit upwards. Consequently, an accumulation of electrons at the lower contact is induced, which is manifested by a negative output voltage V_{LU}.

Although, strictly speaking, Eq. (16) does not hold for the artificial material, it can be used to qualitatively analyze the experimental results of the artificial material since the working principle is the same. For the material shown in Fig. 8.18(b), with decreasing the voltage on the top gate, the electron density and therefore E_F reduces. From Eq. (16), the output voltage $|V_{LU}|$ will increase. Moreover, the lowered E_F also means reduced electron occupations of the transverse confinement modes in the gaps

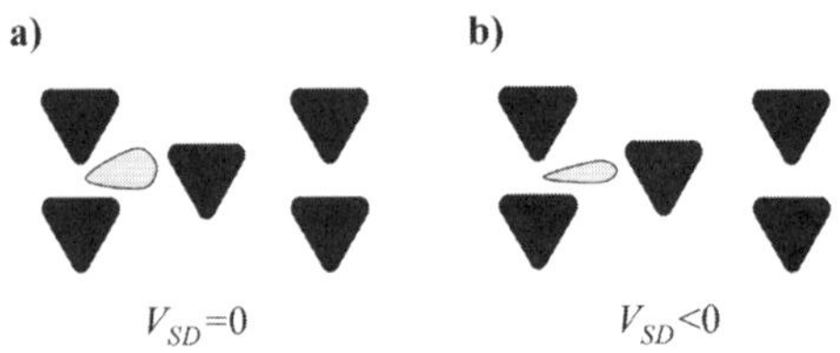

Fig. 8.23. Similarly to the voltage-induced collimation effect shown in Fig. 8.20, the angular distribution of the electrons ejected out of the narrow gaps between neighboring triangular antidots at $V_{SD} = 0$ (a) becomes narrower at $V_{SD} < 0$ as schematically shown in (b). The angular distribution is tilted upwards a little because of the asymmetry of the geometry.

between neighboring triangular antidots. This can be regarded as reducing N_{SD} in the Eq. (16), which also contributes to the dramatic increase of $|V_{LU}|$ in Fig. 8.19 as V_{GU} decreases (more negative).

3.4. Room-Temperature Devices and Experiments

3.4.1. Room-Temperature Ballistic Rectifiers As the miniaturization of conventional semiconductor devices is approaching the physical limits, great effort has been made to develop new types of nanodevices, which, from the application point of view, should operate at room temperature and GHz frequencies. The ballistic rectifier in Fig. 8.15(a) has already been demonstrated to work up to $T = 77$ K [25]. To function at room temperature, a device size smaller than l_e (about 100 nm) at $T = 300$ K is required. This is, however, very difficult if using a GaAs–AlGaAs heterostructure, due to the large depletion length (about a hundred nm) close to etched device boundary. The much shorter depletion length (below 30 nm) of the InGaAs–InP material system makes it possible to fabricate much smaller devices that work at room temperature [114].

Figure 8.24 shows the scanning electron microscope images of two room-temperature ballistic rectifiers with different sizes. The devices are fabricated from a modulation-doped $In_{0.75}Ga_{0.25}As$–InP quantum-well structure [115]. At room temperature, the carrier density of the two dimensional electron gas is 4.7×10^{15} m^{-2} and the mobility 1.2 m^2/Vs, corresponding to $l_e = 140$ nm. Since l_e is only slightly shorter than the distance from the S or D channel to the triangular antidot, a large number of electrons can still travel through this distance without being scattered. As shown by Hirayama et al. [116], if the electron transport is partially ballistic, a ballistic effect may still be observed, although the effect is weaker than at low temperatures. In our experiment, instead of a dc current, an ac voltage (1 kHz) is applied to the device while the dc (or average) voltage between the lower and upper terminals is measured. Indeed, rectification efficiency of about 14% at room temperature, roughly half of that at 100 K, is observed in the smallest ballistic rectifier, as shown in Fig. 8.25. This is actually one of the very few types of novel nanodevices that have so far been shown to work at room temperature.

To investigate the speed of the ballistic rectifiers, high-frequency experiments up to 50 GHz have been performed. Because of the in-plane nature of the ballistic rectifier,—the electrical contacts are laterally separated rather than placed on the surface and the back side (substrate),—the parasitic capacitance between contacts is substantially lower

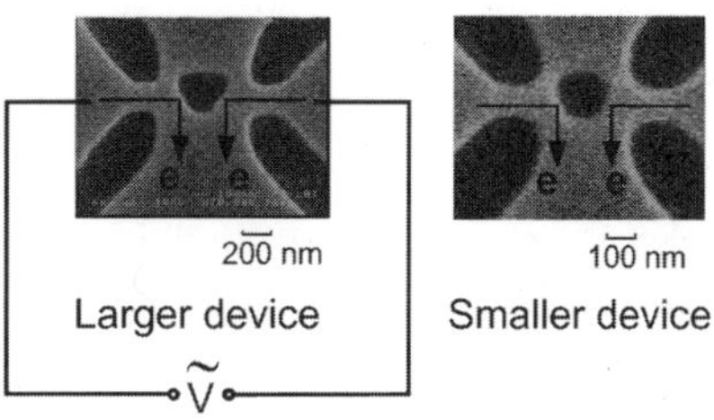

Fig. 8.24. Scanning electron microscope images of two room-temperature ballistic rectifiers with different sizes, fabricated from an InGaAs–InP heterostructure. From [114].

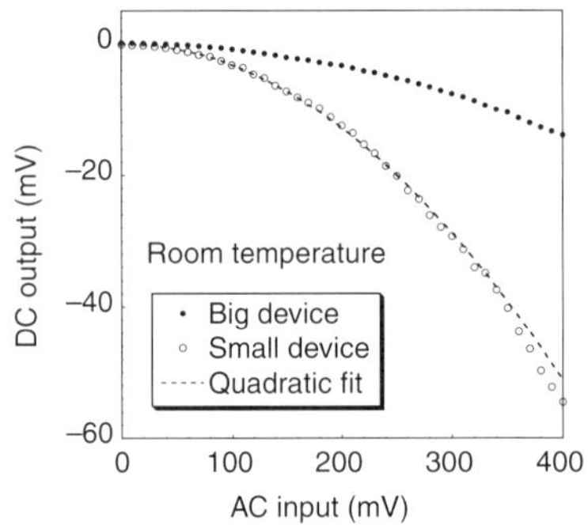

Fig. 8.25. Room-temperature operation of the ballistic rectifiers shown in Fig. 8.24. The frequency of the input ac signal is 1 KHz. The dashed line represent a quadratic fit to the experimental result of the smaller rectifier. From [114].

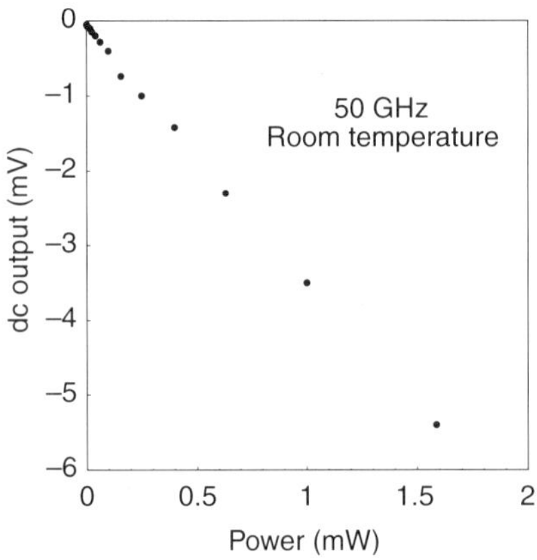

Fig. 8.26. The output dc voltage of the smaller ballistic rectifier shown in Fig. 8.24(a) versus the power of the applied 50 GHz signal at room temperature.

than in a conventional vertical device of the same size. Furthermore, the new working mechanism does not rely on any minority carrier diffusion or barrier structure, two factors that often limit the speed of conventional semiconductor diodes. The ballistic rectifier is therefore expected to function at very high frequencies. Figure 8.26 shows the dc output of the device versus the power of the 50 GHz signal at room temperature. The 50 GHz signal is transmitted to the device via a coplanar probe. The power (horizontal axis) in Fig. 8.26 is the output power from the signal source. Due to the impedance mismatch between the signal source (50 Ω) and the device (a few kΩ) as well as some power loss due to the cables at 50 GHz, effectively only a small fraction of the power from the signal source is applied to the device. It is estimated that the real sensitivity of the device at 50 GHz is a few hundred mV output per mW input, which is virtually as high as that of a commercial microwave detection diode [114]. The ballistic rectifiers have, however, not yet been specifically optimized. Much improvement can be expected in the future, by, for example, further reducing the device size. For a device about half of the one shown in Fig. 8.25, one can expect not only a much higher efficiency, but also a stable performance that is insensitive to temperature changes up to room temperature.

In Fig. 8.25, a fit of $y = ax^2$ (where y represents dc output voltage, x ac input voltage, and a a constant) is performed. The remarkably good fit supports the theoretical prediction of the quadratic response to input voltage, Eq. (16). This allows to generate second harmonic signals without producing third or higher harmonics. The polarization of the generated second harmonic signal (along the vertical direction) is perpendicular to that of the input signal (horizontal direction). This makes the separation from the input signal much easier. Because of the quadratic characteristic, one also expects that the dc output is a linear function of the applied microwave power, which is demonstrated in Fig. 8.26. Although we have not been able to test the devices at frequencies higher than 50 GHz so far, it is expected, from the working principle, the cutoff frequency to be much higher, possibly up to hundreds of GHz or even in the THz regime.

3.4.2. Room-Temperature Nanomaterials It is also possible to achieve room-temperature operations of the artificial material shown in section 3.2.2, by fabricating nanomaterials with "artificial lattice constants" comparable to, or shorter than, the electron mean free path. Figure 8.27(a) is an atomic force microscope image of one of the nanomaterials that operates at room temperature, fabricated from the same InGaAs–InP heterostructure. The artificial lattice constants of the nanomaterial are $a_x = 800$ nm and $a_y = 300$ nm in the lateral and vertical directions, respectively. Both the base and the height of the triangular antidots are about 150 nm. The experimental data at room temperature is shown in Fig. 8.27(b), in which the dc output between the lower and upper contacts is measured as a function of the applied low frequency (1 kHz) ac signal. The experiment is performed using a 30 μm $\times$ 30 μm piece of the nanomaterial. Note that the feature size of nanomaterial is still far from ideal, i.e. both a_x and a_y would ideally be shorter than l_e. The room-temperature operation shown in Fig. 8.27(b) may be greatly improved by fabricating nanomaterials with a smaller feature size.

Similar to a ballistic rectifier, the in-plane nature of the nanomaterial ensures that the parasitic capacitance between contacts is much lower than in a conventional vertical device of the same size. The nanomaterial is therefore capable of functioning at a very high speed, as is evident in high-frequency experiments. The curve in Fig. 8.28 shows the output dc voltage between the upper and lower contacts as a function of the rf power at 50 GHz, which is transmitted to the nanomaterial via a coplanar probe. The experiment is performed at room temperature. Similar results are obtained in other materials with different structural parameters.

Because of the experimental limitations, the materials have not been tested at frequencies higher than 50 GHz. However, from the working principle it is expected the cutoff frequency to be much higher, possibly up to hundreds of GHz or even in the THz regime. The large area, in combination with the in-plane layout of the material, enables straightforward and efficient coupling to high-frequency microwaves. Moreover, microwaves traveling along the normal direction of the nanomaterial surface yield the most efficient coupling, while an effective coupling has to be achieved in specific ways (including using antennae) in a conventional vertical device. From the material point of view, this is actually, to the best of our knowledge, the first artificial electronic nanomaterial that operates at room temperature.

(a)

(b)

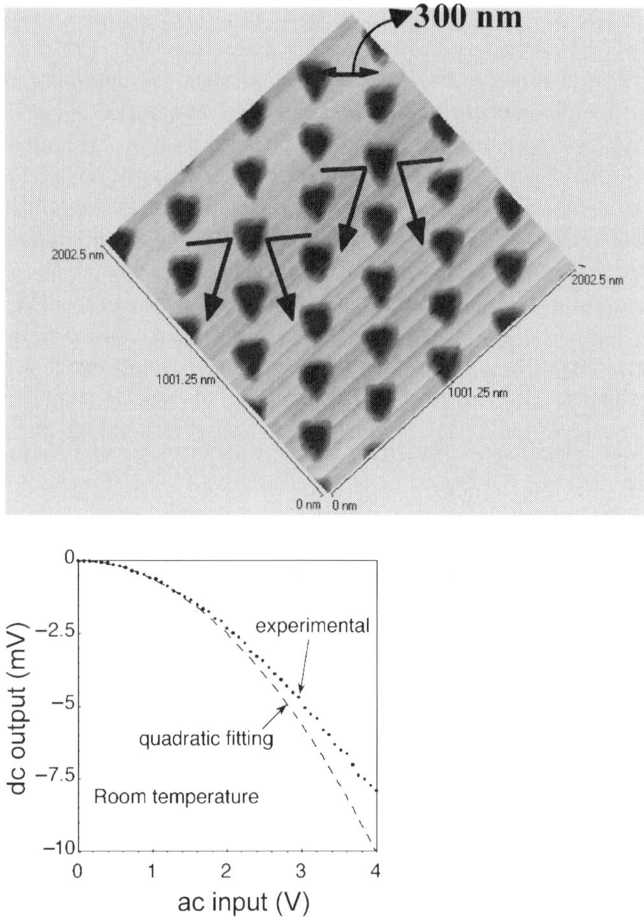

Fig. 8.27. (a) An atomic force microscope image of one of the nanomaterials that operates at room temperature. The arrows represent typical electron trajectories. The lattice constants are $a_x = 800$ nm and $a_y = 300$ nm in the lateral and vertical directions, respectively. Both the base and the height of the triangular antidots are 150 nm. (b) The experimental data of a 30 μm × 30 μm piece of the nanomaterial shown in (a) at room temperature, in which the dc output between the lower and upper contacts is measured as a function of the applied low frequency (1 kHz) ac signal. The dashed line represents a quadratic fit.

To investigate the ratchet effect, an external noise signal has been applied to the artificial material. As expected, a guided net flow of electrons along the upward direction is induced, which generates a negative direct voltage between L and U [98]. In many aspects, the ratchet effect in the artificial nanomaterial resembles the photovoltaic effect in some natural crystals [90–92]. The studies on the artificial material may have important implications for the further understandings of the photovoltaic effect.

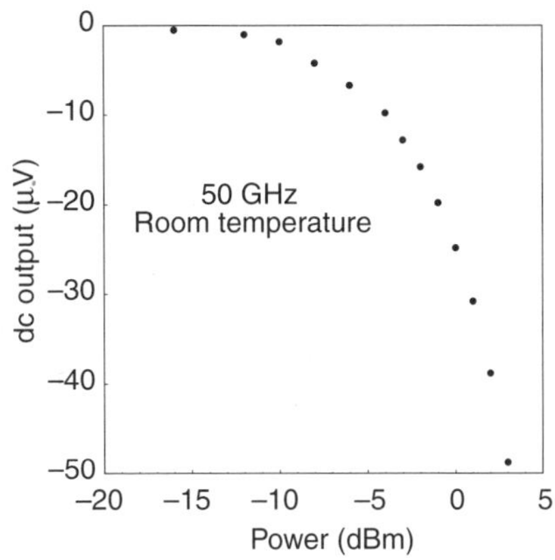

Fig. 8.28. The output dc voltage of the nanomaterial versus the power of the applied 50 GHz signal at room temperature. The rf signal is applied to the material via a coplanar probe.

4. Outlook

The body of work on nonlinear electron transport in dot and antidot structures reviewed in the past three sections covers the entire range from fundamental quantum ratchet effects to room-temperature, high-frequency electronic devices. Here we give a brief outlook on expected future developments in these different, yet connected directions of research.

4.1. Quantum Ratchets

The experimental exploration of quantum ratchets is still in its very early stages. The experiments reviewed in section 2 belong in the group of adiabatically rocked ratchets. In a way this is the regime that is most easily accessible to intuitive understanding. The condition of adiabatic (slow) changes of the potential tilt means that the system is in a steady state at each instant, and transient behavior can be neglected. In our experimental regime, electrons are "moderately damped"—they have an inelastic scattering length comparable to the size of one ratchet cell. In this regime one can neglect inelastic processes while an electron passes a barrier, but one can also assume that it will soon afterwards become part of the thermal distribution. This simplifies the analysis of a rocking ratchet enormously, because it is sufficient to analyse a single barrier. The periodicity of the ratchet device shown in Fig. 8.9 adds no new physics.

The theory community has already explored beyond this regime, considering non-adiabatic systems [117–119], fluctuating potentials [120] and semi-classical systems [121] (for an overview, see [10]). Future experimental work should address the so-called non-adiabatic regime, in which the potential is rocked at a frequency comparable to characteristic times of the particles, such as their escape time through ratchet barriers, or the rate of energy dissipation. For moderately damped classical particles in ratchets rocked on time scales comparable to the "time of flight" of a particle along one ratchet period, chaotic behavior of the current direction is predicted [122,123].

Chaos in quantum ratchets may give rise to novel signatures of quantum chaos in non-equilibrium systems [124]. An experimental realization of fast potential changes would allow the construction of so called pulsating ratchets, where the potential itself is modified, and no external forces are applied at any time. Semiconductor devices will remain a promising environment for experiments of new quantum ratchet systems, including semi-classical ratchets [121], ratchets in which quantum chaos may lead to new effects [124], and periodic ratchets small enough to allow the formation of mini-bands [125]. Recently, the group around Hans Mooij in Delft realized a perodic, rocking ratchet for vortices in superconductors with strong coupling effects in the deep quantum regime [131]. Other promising experimental systems for the study of quantum ratchet effects include optical ratchets for ultracold atoms [22] and ratchets for fluxons in superconductors [119, 126].

4.2. Room-Temperature Nanodevices

From the application point of view, there is substantial industrial interest in new types of nanodevices that operate at room temperature and at GHz or even THz frequencies. Since the invention of the integrated circuit in 1959, the remarkable progress of the semiconductor industry has been described by Moore's law: every 12 to 18 months, the integration density of integrated circuits doubles. While such a trend has been sustained for over three decades, it has been realized for many years that the miniaturization of conventional semiconductor devices will soon reach its limit, and the future generations of electronic devices will have to be built with nanoscience and technology and be based on completely new concepts. Despite great efforts in explorations of novel nanodevices that can be reproducibly mass fabricated as well as can function at room temperature, there is very little success. One important reason is that in order to work at room temperature, most electronic nanodevices proposed so far require a dimension of about a few nanometers, which is at the limit of even the best state-of-the-art facilities. For example, although the first single-electron transistor (SET) was realized many years ago [127], it is still very challenging to fabricate SETs that operate at room temperature [128,129]. Apart from the operation temperature, another key issue is the working speed, which is desired to be higher than 10 GHz for the next generation of integrated circuits. Great effort has been made to design nanodevices that operate in a high speed [130].

The above explorations of novel ballistic devices have provided working devices at room temperature as well as GHz frequencies. Since the required device dimensions are determined by the electron mean free path of the host material, which is typically about 150 nm in III–V heterostructures, such and other room-temperature ballistic devices can be reproducibly fabricated with the currently available techniques in the industry. Although ballistic electron transport has been studied for more than two decades, new concepts of room-temperature ballistic devices are still being discovered, such as the recently demonstrated nonlinear three-terminal junctions [80–82,89]. More and more research activities have shifted from the linear transport regime to the nonlinear regime, which is the research area where more exciting new device concepts are emerging. Currently, it is still very challenging to realize ballistic electron transport in silicon materials at room temperature, because of the very short mean free path (about 10 nm).

Such a precision of device fabrication is nevertheless expected to be achieved by the industry in the near future. Further exploration of other new types of ballistic devices, based on III–V as well as silicon materials, seems to be one of the most feasible routes to provide new building blocks for the next generation of electronics.

In addition to the semi-classical, ballistic effects discussed in detail in section 3, quantum effects are expected to be observed at elevated temperatures. For the ballistic rectifiers shown in Fig. 8.24, it was recently observed that the output voltage V_{LU} changed sign when the temperature was lowered to about 70 K. At even lower temperatures, it was found that the devices generated a pronounced oscillatory output as a function of the applied voltage. In the artificial nanomaterial shown in Fig. 8.27, very similar phenomena were discovered, but the output reversal persisted up to about 200 K. Further observations, such as the dependence on device size and the influence of illuminations on the devices at low temperatures, strongly suggest that the phenomena are a manifestation of quantum effects. A quantum model based on the transverse quantum confinement in the source–drain channel of the ballistic rectifiers or in the narrow gaps between neighboring triangular antidots in the nanomaterial has explained the experimental results [98]. Devices that display distinct quantum effects are expected to be ultra-fast and sensitive because of the wave nature and the phase coherence of electrons. With the rapidly improving nanotechnology, it is promising that by further reducing device dimensions, new types of room-temperature quantum devices might soon be realized and lead to exciting applications.

Acknowledgments

This research was supported by the Australian Research Council, the Deutsche Forschungsgemeinschaft (SFB 348), the Alexander von Humboldt Foundation, the Swedish Research Councils for Engineering and Natural Sciences, the Swedish Foundation of Strategic Research, and the European Commission through LTR research projects Q-SWITCH and NEAR. The bulk of the experimental work described here was carried out in laboratories of Lund University, Ludwig-Maximilians University München, and the University of New South Wales in Sydney. Writing of this chapter was supported by the University of Oregon (H.L.) and UMIST (A.M.S.). The authors thank Max Bichler, Tammy Humphrey, Jörg P. Kotthaus, Armin Kriele, Poul Lindelof, Axel Lorke, Anneli Löfgren, Stephan Manus, Colleen Marlow, Ivan Maximov, Richard Newbury, Pär Omling, Lars Samuelson, Werner Seifert, Weidong Sheng, Ivan Shorubalko, Martin Streibl, Anders Svensson, Richard Taylor, Werner Wegscheider, Hongqi Xu, and Herbert Zirath for their highly appreciated contributions to the work reviewed here.

References

1. M.V. Smoluchowski, Experimentell nachweisbare, der üblichen Thermodynamik widersprechende Molekularphänomene, *Phys. Z.* **13**, 1069 (1912).
2. R.P. Feynman, R.B. Leighton, and M. Sands, *The Feynman Lectures on Physics*, Vol. 1, Ch. 46. Addison Wesley, Reading, MA (1963).

3. D.A. Wharam, T.J. Thornton, R. Newbury, M. Pepper, H. Ahmed, J.E.F. Frost, D.G. Hasko, D.C. Peacock, D.A. Ritchie, and G.A.C. Jones, One-dimensional transport and the quantization of the ballistic resistance, *J. Phys. C. Solid State* **21**(8), L209–L214 (1988).
4. B.J. van Wees, H. van Houten, C.W.J. Beenakker, J.G. Williamson, L.P. Kouwenhoven, D. van der Marel, and C.T. Foxon, Quantized conductance of point contacts in a two-dimensional electron-gas, *Phys. Rev. Lett.* **60**(9), 848–850 (1988).
5. S. Datta, *Electronic Transport in Mesoscopic Systems*. Cambridge University Press, Cambridge (1995).
6. R. Landauer, Electrical resistance of disordered one-dimensional lattices, *Philos. Mag.* **21**, 863–867 (1970).
7. P. Hänggi and R. Bartussek, in *Nonlinear Physics of Complex Systems—Current Status and Future Trends* (eds. J. Parisi, S.C. Müller, and W. Zimmermann), Springer, Berlin (1996).
8. R.D. Astumian, Thermodynamics and kinetics of a brownian motor, *Science* **276**, 917 (1997).
9. F. Jülicher, A. Ajdari, and J. Prost, Modeling molecular motors, *Rev. Mod. Phys.* **69**, 1269 (1997).
10. P. Reimann, Brownian motors: Noisy transport far from equilibrium, *Phys. Rep.* **361**, 57 (2002).
11. J. Prost, J.-F. Chauwin, L. Peliti, and A. Ajdari, Asymmetric pumping of particles, *Phys. Rev. Lett.* **72**, 2652 (1994).
12. J. Rousselet, L. Salome, A. Ajdari, and J. Prost, Directional motion of Brownian particles induced by a periodic asymmetric potential, *Nature* **370**, 446 (1994).
13. R. Ketzmerick, M. Weiss, and F.-J. Elmer, Brownian motor, http://www.chaos.gwdg.de/java_gallery/brownian_motor/bm.html, http://monet.physik.unibas.ch/elmer/bm.
14. P. Reimann and P. Hänggi, Introduction to the physics of Brownian motors, *Appl. Phys.* A **75**, 169 (2002).
15. R.D. Astumian and P. Hänggi, *Phys. Today* **55**(11) 33–39 (November, 2002).
16. B. Alberts, D. Bray, J. Lewis, M. Raff, K. Roberts, and J.D. Watson, *Molecular Biology of the Cell*. Garland Publishing, New York (1994).
17. D. Bray, *Cell Movements—from Molecules to Motility*. Garland, New York (2001).
18. R.D. Astumian, Making molecules into motors, *Sci. Am.* 57, (July 2001).
19. J.S. Bader, R.W. Hammond, S.A. Henck, M.W. Deem, G.A. McDermott, J.M. Bustillo, J.W. Simpson, M.G.T., and J.M. Rothberg, DNA transport by a micromachined Brownian ratchet device, *Proc. Nat. Acad. Sci* (USA) **96**, 13165 (1999).
20. L.P. Faucheux, L.S. Bourdieu, P.D. Kaplan, and A.J. Libchaber, Optical thermal ratchet, *Phys. Rev. Lett.* **74**, 1504 (1995).
21. L. Gorre, E. Ioannidis, and P. Silberzan, Rectified motion of a mercury drop in an asymmetric structure, *Europhys. Lett.* **33**, 267 (1996).
22. C. Mennerat-Robilliard, D. Lucas, S. Guibal, J. Tabosa, C. Jurcak, J.-Y. Courtois, and G. Grynberg, Ratchet for cold rubidium atoms: The asymmetric optical lattice, *Phys. Rev. Lett.* **82**, 851 (1999).
23. H. Linke, P. Omling, H. Xu, and P.E. Lindelof, in *Proceedings of the 23rd International Conference on the Physics of Semiconductors* (eds. M. Scheffler and R. Zimmermann) World Scientific, Singapore 1593 (1996).
24. A. Lorke, S. Wimmer, B. Jäger, J.P. Kotthaus, W. Wegscheider, and M. Bichler, Far-infrared and transport properties of antidot arrays with broken symmetry, *Physica* B **249–251**, 312–316 (1998).

25. A.M. Song, A. Lorke, A. Kriele, J.P. Kotthaus, W. Wegscheider, and M. Bichler, Nonlinear electron transport in an asymmetric microjunction: A ballistic rectifier, *Phys. Rev. Lett.* **80**(17), 3831–3834 (1998).
26. H. Linke, W. Sheng, A. Löfgren, H.Q. Xu, P. Omling, and P.E. Lindelof, A quantum dot ratchet: Experiment and theory, *Europhys. Lett.* **44**(3), 341–347 (1998).
27. H. Linke, T.E. Humphrey, A. Löfgren, A.O. Sushkov, R. Newbury, R.P. Taylor, and P. Omling, Experimental tunneling ratchets, *Science* **286**(5448), 2314–2317 (1999).
28. A.M. Song, P. Omling, L. Samuelson, W. Seifert, I. Shorubalko, and H. Zirath, Room-temperature and 50 GHz operation of a functional nanomaterial, *Appl. Phys. Lett.* **79**(9), 1357–1359 (2001).
29. A.V. Oudenaarden and S.G. Boxer, Brownian ratchets: Molecular separations in lipid bilayers supported on patterned arrays, *Science* **285**, 1046 (1999).
30. C.F. Chou, O. Bakajin, S.W.P. Turner, T.A.J. Duke, S.S. Chan, E.C. Cox, H.G. Craighead, and R.H. Austin, Sorting by diffusion: An asymmetric obstacle course for continuous molecular separation, *P. Natl. Acad. Sci. USA* **96**(24), 13762–13765 (1999).
31. H. Linke (ed.), Ratchets and Brownian Motors: Basics, Experiments and Applications, Special issue, *Appl. Phys.* A **75**(2), 167–352 S (2002).
32. M.O. Magnasco, Forced thermal ratchets, *Phys. Rev. Lett.* **71**, 1477 (1993).
33. R. Bartussek, P. Hänggi, and J.G. Kissner, Periodically rocked thermal ratchets, *Europhys. Lett.* **28**, 459 (1994).
34. T.E. Humphrey, H. Linke, and R. Newbury, Pumping heat with quantum ratchets, *Physica* E **11**, 281 (2001).
35. T.E. Humphrey, R. Newbury, R.P. Taylor, and H. Linke, Reversible quantum Brownian heat engines for electrons, *Phys. Rev. Lett.* **89**, 116801 (2002).
36. P. Reimann, M. Grifoni, and P. Hänggi, Quantum ratchets, *Phys. Rev. Lett.* **79**, 10 (1997).
37. C.W.J. Beenakker and H. van Houten, in *Solid State Physics* Vol. 44 (eds. H. Ehrenreich and D. Turnbull), Academic Press, Boston (1991).
38. R. Landauer, in *Nonlinearity in Condensed Matter* (eds. A.R. Bishop, D.K. Campbell, P. Kumar, and S.E. Trullinger) Springer Verlag, Berlin (1987).
39. H. Linke, L. Christensson, P. Omling, and P.E. Lindelof, Stability of classical electron orbits in triangular electron billiards, *Phys. Rev.* B **56**, 1440 (1997).
40. L. Christensson, H. Linke, P. Omling, P.E. Lindelof, K.-F. Berggren, and I.V. Zozoulenko, Classical and quantum dynamics of electrons in open, equilateral triangular billiards, *Phys. Rev.* B **57**, 12306 (1998).
41. J.P. Bird, Spectral characteristics of conductance fluctuations in ballistic quantum dots: The influence of finite magnetic field and temperature, *Phys. Rev.* B **52**, 8295 (1995).
42. M. Persson, J. Pettersson, B.V. Sydow, P.E. Lindelof, A. Kristensen, and K.F. Berggren, Conductance oscillations related to the eigenenergy spectrum of a quantum dot in weak magnetic fields, *Phys. Rev.* B **52**, 8921 (1995).
43. S.M. Reimann, M. Brack, A.G. Magner, J. Blaschke, and M.V.N. Murthy, Circular quantum billiard with a singular magnetic flux line, *Phys. Rev.* A **53**, 39 (1996).
44. S.M. Reimann, M. Persson, P.E. Lindelof, and M. Brack, Shell structure of a circular quantum dot in a weak magnetic field, *Z. Phys.* B **101**, 377 (1996).
45. M. Brack and R.K. Badhuri, *Semiclassical Physics*. Addison-Wesley, Reading, Massachusetts (1997).
46. M. Brack, J. Blaschke, S.C. Creagh, A.G. Magner, P. Meier, and S.M. Reimann, On the role of classical orbits in mesoscopic electronic systems, *Z. Phys.* D **40**, 276 (1996).

47. H. Linke, W. Sheng, A. Svensson, A. Löfgren, L. Christensson, H. Xu, and P. Omling, Asymmetric nonlinear conductance of quantum dots with broken inversion symmetry, *Phys. Rev.* B **61**, 15914 (2000).
48. H. Xu, Theory of nonlinear ballistic transport in quasi-one-dimensional constrictions, *Phys. Rev.* B **47**, 15630 (1993).
49. R.A. Webb, S. Washburn, and C.P. Umbach, Experimental study of nonlinear conductance in small metallic samples, *Phys. Rev.* B **37**, 8455 (1988).
50. S.B. Kaplan, Structure in the rectifying behaviour of mesoscopic Si MOSFETs, *Surf. Sci.* **196**, 93 (1988).
51. P.A.M. Holweg, J.A. Kokkedee, J. Caro, A.H. Verbruggen, S. Radelaar, A.G.M. Jansen, and P. Wyder, Conductance fluctuations in a ballistic metallic point contact, *Phys. Rev. Lett.* **67**, 2549 (1991).
52. D.C. Ralph, K.S. Ralls, and R.A. Buhrmann, Ensemble studies of nonlinear conductance fluctuations in phase coherent samples, *Phys. Rev. Lett.* **70**, 986 (1993).
53. R. Taboryski, A.K. Geim, M. Persson, and P.E. Lindelof, Nonlinear conductance at small driving voltages in quantum point contacts, *Phys. Rev.* B **49**(11), 7813–7816 (1994).
54. A. Löfgren, H. Linke, C.A. Marlow, T.E. Humphrey, I. Shorubalko, P. Omling, R. Newbury, R.P. Taylor, P.E. Lindelof, L. Samuelson, I. Maximov, and W. Seifert, Symmetry of magneto-conductance fluctuations of quantum dots in the nonlinear transport regime, unpublished (2002).
55. N.K. Patel, J.T. Nicholls, L. Martin-Moreno, M. Pepper, J.E.F. Frost, D.A. Ritchie, and G.A.C. Jones, Evolution of half plateaus as a function of electric field in a ballistic quasi-one-dimensional constriction, *Phys. Rev.* B **44**, 13549 (1991).
56. A. Kristensen, H. Bruus, A.E. Hansen, J.B. Jensen, P.E. Lindelof, C. Marckmann, J. Nygård, C.B. Sorensen, F. Beuscher, A. Forchel, and M. Michel, Bias and temperature dependence of the 0.7 conductance anomalityin quantum point contacts, *Phys. Rev.* B **62**, 10950 (2000).
57. M. Brooks, Quantum clockworks, *New Sci.* 28, (January, 22, 2000).
58. H.L. Edwards, Q. Niu, and A.L. de Lozanne, Cryogenic refrigerator, *Appl. Phys. Lett.* **63**, 1815 (1993).
59. H.L. Edwards, Q. Niu, G.A. Georgakis, and A.L. de Lozanne, Cryogenic cooling using tunneling structures with sharp energy features, *Phys. Rev.* B **52**, 5714 (1995).
60. N.G.V. Kampen, Relative stability in nonuniform temperature, *IBM Res. Develop.* **32**, 107 (1987).
61. M. Büttiker, Transport as a consequence of state-dependent diffusion, *Z. Phys.* B **68**, 161 (1987).
62. R. Landauer, Motion out of noisy states, *J. Stat. Phys.* **53**, 233 (1988).
63. J.M.R. Parrondo and P. Espanol, Criticism of Feynman's analysis of the ratchet as a heat engine, *Am. J. Phys.* **64**, 1125 (1996).
64. K. Sekimoto, Kinetic characterization of heat bath and the energetics of thermal ratchet models, *J. Phys. Soc. Jap.* **66**, 1234 (1997).
65. I.M. Sokolov, On the energetics of a nonlinear system rectifying thermal fluctuations, *Europhys. Lett.* **44**, 278 (1998).
66. I. Derenyi and R.D. Astumian, Efficiency of Brownian heat engines, *Phys. Rev.* E **59**, R6219 (1999).
67. T. Hondou and K. Sekimoto, Unattainability of Carnot efficiency in the Brownian heat engine, *Phys. Rev.* E **62**, 6021 (2000).
68. J.M.R. Parrondo and B.J. de Cisneros, Energetics of Brownian motors: A review, *Appl. Phys.* A **75**, 193 (2002).

69. J. Imry, in *Directions in Condensed Matter Physics* (eds. G. Grinstein and G. Mazenko), 101–163 World Scientific, Singapore (1986).
70. S. Ulreich and W. Zwerger, Where is the potential drop in a quantum point contact? *Superlatt. Microstruct.* **23**(3–4), 719–730 (1998).
71. M. Büttiker, Four-terminal phase-coherent conductance, *Phys. Rev. Lett.* **57**(14), 1761–1764 (1986).
72. M. Büttiker, Symmetry of electrical-conduction, *IBM J. Res. Dev.* **32**(3), 317–334 (1988).
73. L.P. Kouwenhoven, B.J. van Wees, C.J.P.M. Harmans, J.G. Williamson, H. van Houten, C.W.J. Beenakker, C.T. Foxon, and J.J. Harris, Nonlinear conductance of quantum point contacts, *Phys. Rev.* B **39**(11), 8040–8043 (1989).
74. A. Palevski, C.P. Umbach, and M. Heiblum, High-gain lateral hot-electron device, *Appl. Phys. Lett.* **55**(14), 1421–1423 (1989).
75. R.I. Hornsey, A.M. Marsh, J.R.A. Cleaver, and H. Ahmed, High-current ballistic transport through variable-width constrictions in a high-mobility 2-dimensional electron-gas, *Phys. Rev.* B **51**(11), 7010–7016 (1995).
76. D.R.S. Cumming and J.H. Davies, Electron–electron scattering in a high purity mesoscopic conductor, *Appl. Phys. Lett.* **69**(22), 3363–3365 (1996).
77. A. Messica, A. Soibel, U. Meirav, A. Stern, H. Shtrikman, V. Umansky, and D. Mahalu, Suppression of conductance in surface superlattices by temperature and electric field, *Phys. Rev. Lett.* **78**(4), 705–708 (1997).
78. H. Ueno, K. Moriyasu, Y. Wada, S. Osako, H. Kubo, N. Mori, and C. Hamaguchi, Conductance through laterally coupled quantum dots, *Jpn. J. Appl. Phys.* **38**(1B), 332–335 (1999).
79. A.M. Song, Formalism of nonlinear transport in mesoscopic conductors, *Phys. Rev.* B **59**(15), 9806–9809 (1999).
80. K. Hieke and M. Ulfward, Nonlinear operation of the Y-branch switch: Ballistic switching mode at room temperature, *Phys. Rev.* B **62**(24), 16727–16730 (2000).
81. I. Shorubalko, H.Q. Xu, I. Maximov, P. Omling, L. Samuelson, and W. Seifert, Nonlinear operation of GaInAs/InP-based three-terminal ballistic junctions, *Appl. Phys. Lett.* **79**(9), 1384–1386 (2001).
82. L. Worschech, H.Q. Xu, A. Forchel, and L. Samuelson, Bias-voltage-induced asymmetry in nanoelectronic Y-branches, *Appl. Phys. Lett.* **79**(20), 3287–3289 (2001).
83. P.F. Bagwell and T.P. Orlando, Landauer conductance formula and its generalization to finite voltages, *Phys. Rev.* B **40**(3), 1456–1464 (1989).
84. L.I. Glazman and A.V. Khaetskii, Nonlinear quantum conductance of a lateral microconstraint in a heterostructure, *Europhys. Lett.* **9**(3), 263–267 (1989).
85. O. Heinonen and M.D. Johnson, Mesoscopic transport beyond linear-response, *Phys. Rev. Lett.* **71**(9), 1447–1450 (1993).
86. M.D. Johnson and O. Heinonen, Nonlinear steady-state mesoscopic transport—formalism, *Phys. Rev.* B **51**(20), 14421–14436 (1995).
87. M. Büttiker, Capacitance, admittance, and rectification properties of small conductors, *J. Phys. Condens. Mat.* **5**(50), 9361–9378 (1993).
88. S. Komiyama and H. Hirai, Two representations of the current density in charge-transport problems, *Phys. Rev.* B **54**(3), 2067–2090 (1996).
89. H.Q. Xu, Electrical properties of three-terminal ballistic junctions, *Appl. Phys. Lett.* **78**(14), 2064–2066 (2001).
90. A.M. Glas, D. von der Linde, and T.J. Negran, High-voltage bulk photovoltaic effect and the photorefractive progress in LiNbO3, *Appl. Phys. Lett.* **25**, 233–235 (1974).

91. V.I. Belinicher and B.I. Sturman, The photogalvanic effect in media lacking a center of symmetry, *Sov. Phys. Usp.* **23**(3), 199–223 (1980).
92. B.I. Sturman and V.M. Fridkin, *The Photovoltaic and Photorefractive Effects in Noncentrosymmetric Materials*. Gordon and Breach, Philadelphia (1992).
93. D. Ertaş, Lateral separation of macromolecules and polyelectrolytes in microlithographic arrays, *Phys. Rev. Lett.* **80**(7), 1548–1551 (1998).
94. T.A.J. Duke and R.H. Austin, Microfabricated sieve for the continuous sorting of macromolecules, *Phys. Rev. Lett.* **80**(7), 1552–1555 (1998).
95. S.Y. Chou, P.R. Krauss, and P.J. Renstrom, Imprint lithography with 25-nanometer resolution, *Science* **272**(5258), 85–87 (1996).
96. A.M. Song, A. Lorke, A. Kriele, J.P. Kotthaus, W. Wegscheider, and M. Bichler, Experimental and theoretical studies on nonlinear electron transport in ballistic rectifiers, in *Proceedings of the 24th International Conference on the Physics of Semiconductors* (ed. D. Gershoni), World Scientific, Singapore, 1348 (1998).
97. A.M. Song, Electron ratchet effect in semiconductor devices and artificial materials with broken centrosymmetry, *Appl. Phys.* B **67**(19), to be published May, 2002.
98. A. Löfgren, I. Shorubalko, P. Omling, and A.M. Song, Quantum behavior in nanoscale ballistic rectifiers and artificial materials, *Phys. Rev. B*, Vol. 67(19) 195–309 (2003).
99. B. Heidari, I. Maximov, E.L. Sarwe, and L. Montelius, Large scale nanolithography using nanoimprint lithography, *J. Vac. Sci. Technol.* B **17**(6), 2961–2964 (1999).
100. M.L. Roukes, A. Scherer, S.J. Allen, H.G. Craighead, R.M. Ruthen, E.D. Beebe, and J.P. Harbison, Quenching of the Hall-effect in a one-dimensional wire, *Phys. Rev. Lett.* **59**(26), 3011–3014 (1987).
101. G. Timp, H.U. Baranger, P. Devegvar, J.E. Cunningham, R.E. Howard, R. Behringer, and P.M. Mankiewich, Propagation around a bend in a multichannel electron wave-guide, *Phys. Rev. Lett.* **60**(20), 2081–2084 (1988).
102. H. van Houten, C.W.J. Beenakker, P.H.M. van Loosdrecht, T.J. Thornton, H. Ahmed, M. Pepper, C.T. Foxon, and J.J. Harris, Four-terminal magnetoresistance of a two-dimensional electron-gas constriction in the ballistic regime, *Phys. Rev.* B **37**(14), 8534–8536 (1988).
103. C.J.B. Ford, S. Washburn, M. Büttiker, C.M. Knoedler, and J.M. Hong, Influence of geometry on the Hall-effect in ballistic wires, *Phys. Rev. Lett.* **62**(23), 2724–2727 (1989).
104. C.W.J. Beenakker and H. van Houten, Magnetotransport and nonadditivity of point-contact resistances in series, *Phys. Rev.* B **39**(14), 10445–10448 (1989).
105. L.W. Molenkamp, A.A.M. Staring, C.W.J. Beenakker, R. Eppenga, C.E. Timmering, J.G. Williamson, C.J.P.M. Harmans, and C.T. Foxon, *Phys. Rev.* B **41**, 1274 (1990).
106. J. Spector, H.L. Stormer, K.W. Baldwin, L.N. Pfeiffer, and K.W. West, Electron focusing in two-dimensional systems by means of an electrostatic lens, *Appl. Phys. Lett.* **56**(13), 1290–1292 (1990).
107. C.W.J. Beenakker and H. van Houten, Billiard model of a ballistic multiprobe conductor, *Phys. Rev. Lett.* **63**(17), 1857–1860 (1989).
108. K.L. Shepard, M.L. Roukes, and B.P. Vandergaag, Direct measurement of the transmission matrix of a mesoscopic conductor, *Phys. Rev. Lett.* **68**(17), 2660–2663 (1992).
109. R.I. Hornsey, Angular distributions of electrons transmitted ballistically through tapered constrictions, *J. Appl. Phys.* **79**(12), 9172–9180 (1996).
110. H.U. Baranger and A.D. Stone, Quenching of the hall resistance in ballistic microstructures—A collimation effect, *Phys. Rev. Lett.* **63**(4), 414–417 (1989).
111. A.M. Song, S. Manus, M. Streibl, A. Lorke, J.P. Kotthaus, W. Wegscheider, and M. Bichler, A nonlinear transport device with no intrinsic threshold, *Superlatt. Microstruct.* **25**(1–2), 269–272 (1999).

112. A.M. Song, A. Lorke, J.P. Kotthaus, W. Wegscheider, and M. Bichler, Ballistic magneto-transport in a semiconductor microjunction with broken symmetry, *Superlatt. Microstruct.* **25**(1–2), 149–152 (1999).
113. R. Akis, D.K. Ferry, and J.P. Bird, Magnetotransport fluctuations in regular semiconductor ballistic quantum dots, *Phys. Rev.* B **54**(24), 17705–17715 (1996).
114. A.M. Song, P. Omling, L. Samuelson, W. Seifert, I. Shorubalko, and H. Zirath, Operation of InGaAs/InP-based ballistic rectifiers at room temperature and frequencies up to 50 GHz, *Japan J. Appl. Phys.* **40**(9AB), L909–L911 (2001).
115. P. Ramvall, N. Carlsson, P. Omling, L. Samuelson, W. Seifert, M. Stolze, and Q. Wang, $Ga_{0.25}In_{0.75}As/InP$ quantum wells with extremely high and anisotropic two-dimensional electron gas mobilities, *Appl. Phys. Lett.* **68**(8), 1111–1113 (1996).
116. Y. Hirayama and S. Tarucha, High-Temperature Ballistic Transport Observed in AlGaAs/InGaAs/GaAs small four-terminal structures, *Appl. Phys. Lett.* **63**(17), 2366–2368 (1993).
117. I. Goychuk, M. Grifoni, and P. Hänggi, Nonadiabatic quantum Brownian rectifiers, *Phys. Rev. Lett.* **81**, 649 (1998).
118. I. Goychuk and P. Hänggi, Quantum rectifiers from harmonic mixing, *Europhys. Lett.* **43**, 503 (1998).
119. M. Grifoni, M.S. Ferreira, J. Peguiron, and J. Majer, Multi-band quantum ratchets, *Phys. Rev. Lett.* **89**, 146801 (2002).
120. S. Yukawa, M. Kikuchi, G. Tatara, and H. Matsukawa, Quantum ratchets, *J. Phys. Soc. Japan* **66**, 2953 (1997).
121. H. Schanz, M.-F. Otto, R. Ketzmerick, and T. Dittrich, Classical and quantum hamiltonian ratchets, *Phys. Rev. Lett.* **87**, 070601 (2001).
122. P. Jung, J.G. Kissner, and P. Hänggi, Regular and chaotic transport in asymmetric periodic potentials: Inertia ratchets, *Phys. Rev. Lett.* **76**, 3436 (1996).
123. J.L. Mateos, Chaotic transport and current reversal in deterministic ratchets, *Phys. Rev. Lett.* **84**, 258 (2000).
124. H. Linke, T.E. Humphrey, R.P. Taylor, and R. Newbury, Chaos in quantum ratchets?, *Phys. Scr.* **T90**, 54 (2001).
125. K.N. Alekseev, E.H. Cannon, J.C. Mc Kinney, F.V. Kusmartsev, and D.K. Campbell, Spontaneous dc current generation in a resistively shunted semiconductor superlattice driven by a terahertz field, *Phys. Rev. Lett.* **80**, 2669 (1998).
126. E. Trias, J.J. Mazo, F. Falo, and T.P. Orlando, Depinning of kinks in a Josephson-junction ratchet array, *Phys. Rev.* E **61**(3), 2257–2266 (2000).
127. T.A. Fulton and G.J. Dolan, Observation of single-electron charging effects in small tunnel-junctions, *Phys. Rev. Lett.* **59**(1), 109–112 (1987).
128. H. Ishikuro, T. Fujii, T. Saraya, G. Hashiguchi, T. Hiramoto, and T. Ikoma, Coulomb blockade oscillations at room temperature in a Si quantum wire metal-oxide-semiconductor field-effect transistor fabricated by anisotropic etching on a silicon-on-insulator substrate, *Appl. Phys. Lett.* **68**(25), 3585–3587 (1996).
129. L. Zhuang, L.J. Guo, and S.Y. Chou, Silicon single-electron quantum-dot transistor switch operating at room temperature, *Appl. Phys. Lett.* **72**(10), 1205–1207 (1998).
130. R.J. Schoelkopf, P. Wahlgren, A.A. Kozhevnikov, P. Delsing, and D.E. Prober, The radio-frequency single-electron transistor (RF-SET): A fast and ultrasensitive electrometer, *Science* **280**(5367), 1238–1242 (1998).
131. J.B. Majer, J. Peguiron, M. Grifoni, M. Tusveld, and J.E. Mooij, Quantum ratchet effects for vortices, *Phys. Rev. Lett.* **90**, 056802 (2003).

Chapter 9

Single-Photon Detection with Quantum Dots in the Far-Infrared/ Submillimeter-Wave Range

O. Astafiev[1,2] and S. Komiyama[1]

[1] *Department of Basic Science, University of Tokyo, Komaba 3-8-1, Meguro-ku, Tokyo, 153-8902, Japan*

[2] *Crest, Japan Science and Technology Corporation (JST), Kawaguchi-shi, Saitama, 332-0012, Japan*

1. Introduction

Decreasing device size makes spatial and charge quantization play an explicit role. It was theoretically predicted long time ago that electron transport via tunneling through a small conductive island is strongly affected by charge quantization [1]. It took more than 10 years, however, before technology has been developed to make possible fabrication of such small structures in which the charge quantization is experimentally observable [2]. Devices with a small conductive island weakly coupled to two reservoirs are called single-electron transistors (SETs) [3,4]. Detailed theoretical analysis of the transport through SETs was developed in 1980s by Likharev and others [5,6]. At present, the technology for fabricating SETs, usually employing the technique of electron beam lithography (EBL), is well established.

A unique property of the SET is its extremely high sensitivity to the electrostatic environment as well as to the electromagnetic environment. Due to the sensitivity, SETs have been successfully applied to construct an electrometer detecting small and local charges: it has proved to be a powerful tool for a variety of purposes, such as charge transport in superconducting circuits [7,8], spectroscopic studies of artificial atoms [9,10], transport studies in low dimensional systems [4], and local imaging of electrostatic potentials [11,12,13]. In addition, SET is promising for further practical applications such as memory cells [14], logic elements of integrated circuits [15,16], basic elements for quantum computers (qubits) [17].

Among all, single-photon detectors in the far-infrared (FIR) through submillimeter wave (SMMW) range are supposed to be one of the most promising applications of SETs. In general, the FIR/SMMW region of electromagnetic spectrum is one of the richest areas for spectroscopic research, containing a variety of important spectra of matters, such as the quantum energy levels of semiconductor nanostructures, the energy gap of superconductors, and the rotational modes of molecules. Unfortunately, however, this spectral region is one of the most difficult areas for obtaining data. One reason is the lack of photon counters: Single photons in near-infrared range are readily tractable by semiconductor devices because band-gap electron–hole excitation can be exploited [18,19]. In the FIR/SMMW range, however, photons are difficult to handle on individual basis because typical photon energies are as small as meVs (millielectron volts), one part per one thousand of energies in the visible or near-infrared spectral region.

Prior to single-photon detection, a few different experimental works have been devoted to develop SET-based electromagnetic wave detectors. In one of the earliest works, an SET positioned on top of silicon substrate detected inter-subband photo-excitation in the substrate[20]. Another configuration of a superconducting SET demonstrated highly-sensitive detection of microwaves using quasi-particle photo-excitation at the superconducting energy gap[21]. In these attempts, however, the change of SET conductance induced by single events of photon absorption was too small to be detected on individual basis, either because of small coupling of SET's to induced charges or because of short lifetimes of the excited charges.

In this chapter, we describe physical mechanism and device configuration of single-photon detection in the FIR/SMMW range, which use SETs consisting of semiconductor quantum dots (QDs): Two different types of QD detectors will be described, where one is based on the cyclotron resonance (CR) at high magnetic fields [22–26] and the other utilizes the plasma resonance of a QD in the absence of magnetic field [27,28].

This article is organized as follows. Section 2 briefly describes characteristics of the SET. In Section 3, a basic principle of single-photon detection is discussed. In section 4 we describe a specific mechanism of single-photon detection in high magnetic fields. In section 5, a photon detection of SMMWs with parallel double QDs in the absence of magnetic fields is described. Technical aspects of both type detectors will be discussed in section 6. In section 7, we will summarize our review and briefly touch other possible directions for the single-photon detector improving and other perspective detector schemes.

2. Fundamental Characteristics of the SET

Transport characteristics of the SET are discussed in several review articles [3,4]. Below, we briefly summarize a few basic aspects of the SET operation that are necessary for describing single-photon detection. Letting C be the electrostatic capacitance between a conductive island and outside metallic surroundings, potential energy change caused by removing (adding) one electron from (to) the island is given by the charging energy, $E_{\mathrm{ch}} = e^2/C$. The island size determines the minimal value of C, which is $4\varepsilon\varepsilon_0 a$ for a conductive disk of diameter a, where ε is the dielectric constant of the surrounding media. At a low temperature such that $kT \ll E_{\mathrm{ch}}$, current can flow through the island only if its electrochemical potential lines up with that of external reservoirs

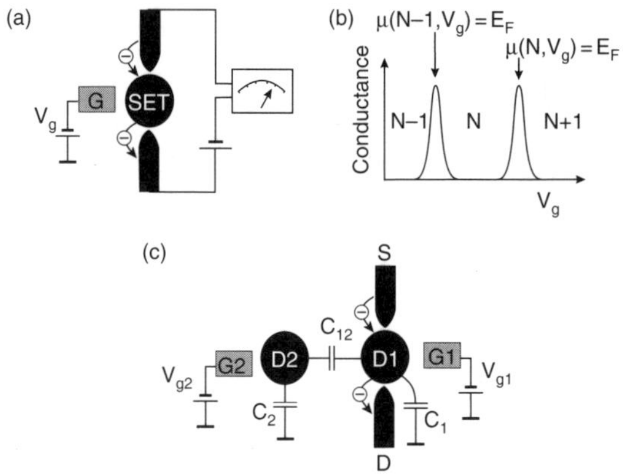

Fig. 9.1. (a) Schematic representation of an SET. (b) Conductance oscillations of an SET against the gate voltage V_g. (c) Schematic representation of a SET composed of parallel double QDs. Tunnel current is transmitted through D1, which is electrostatically coupled to D2. Gates G1 and G2 control the electrochemical potentials of D1 and D2, respectively.

(conductance resonance); otherwise, the number of electrons in the island is fixed and no current flows (Coulomb blockade regime). It leads to a typical SET operation at low temperatures, in which a series of equally spaced conductance-resonance peaks appear in the scan of a control gate voltage. The conductive island of $a = 1\ \mu\text{m}$ in a dielectric medium with $\varepsilon = 10$ yields $E_{\text{ch}} \approx 0.5\,\text{meV}$, which practically requires $T < 0.5\,\text{K}$ for this SET operation to be observable [4].

Figure 9.1(a) schematically shows an SET, in which the electrochemical potential of the island, μ, is controlled by a metallic control gate G, electrostatically coupled to the island through capacitance C_g. The electrochemical potential of the island with N electrons ($N \gg 1$) is written as

$$\mu(N, V_g) = E_N + E_{\text{ch}} \times N - \alpha e V_g \tag{1}$$

where E_N is the discrete energy levels due to confinement and $\alpha \approx C_g/C$. We will consider an island, for which $\Delta E_N = E_{N+1} - E_N \ll E_c$. It follows that μ is dominated by the electrostatic energy. As displayed in Fig. 9.1(b), the conductance through the SET exhibits equally spaced resonance peaks against V_g, with each peak corresponding to $\mu(N) = E_F$ with the Fermi energy E_F of the reservoir. In between two successive peaks, the number of electrons in the island is fixed (Coulomb blockade regime).

As described later, SETs serving as photon detectors are of a modified structure from the simplest one of Fig. 9.1(a): They consist of multi island structures. Suppose that M conducting islands ($i = 1, 2, \ldots, M$) containing N_i electrons, respectively, are electrostatically coupled mutually through capacitance C_{ij} and to the environment through C_i. Only the first island ($i = 1$) is tunnel-coupled to electron reservoirs. The electrochemical potential of each island i is linearly affected by $N_i (i = 1, 2, \ldots, M)$ if the control gate voltage for all the islands is fixed; viz.,

$$\mu_i = \sum E_{ij} N_j + \text{const} \tag{2}$$

where $E_{ij}(i, j = 1, 2, \ldots, M)$ are determined by the capacitance network defined by C_{ij} and C_i [23]. The SET conductance is determined by μ_1, where $E_{11} = E_{ch1}$ is the charging energy of the $i = 1$ island. The detectors described in sections 4 and 5 correspond, respectively, to SETs with $M = 3$ and $M = 2$.

The simplest example of multi-island SETs is a parallel double QDs system ($M = 2$) shown in Fig. 9.1(c) [29]. The first QD (D1), coupled to reservoirs through tunneling, is electrostatically coupled to the second QD (D2). When N_1 and N_2 change by ΔN_1 and ΔN_2, respectively, μ_1 of D1 shifts by

$$\Delta\mu_1(\Delta N_1, \Delta N_2) = E_{ch1} \times \Delta N_1 + E_{12} \times \Delta N_2 \tag{3}$$

where E_{ch1} and E_{12} are related to C_i and C_{12} through

$$E_{ch1} = e^2 \frac{C_2 + C_{12}}{C_1 C_2 + C_{12}(C_1 + C_2)} \quad \text{and} \tag{4a}$$

$$E_{12} = \frac{C_{12}}{C_2 + C_{12}} E_{ch1} \tag{4b}$$

Let us consider two particular cases that are relevant to the later discussion. First, D1 and D2 are strongly coupled ($C_{12}/C_2 \gg 1$) and an electron tunnels between them (D1 $\rightarrow$ D2, or $\Delta N_1 = -1$, $\Delta N_2 = 1$). Noting Eq. (3) and Eqs (4a), (4b), one finds that the electrochemical potential of D1 thereby changes by

$$\Delta\mu_1 = -\beta E_{ch1} \tag{5}$$

where $\beta = C_2/C_{12}$. Second, D1 and D2 are weekly coupled ($C_{12}/C_2 \ll 1$) and one electron is removed from D2 to a reservoir ($\Delta N_1 = 0$, $\Delta N_2 = -1$). The change, $\Delta\mu_1$, induced in D1 is also given by Eq. (5) with $\beta \approx C_{12}/C_2$.

As the control gate voltage, V_{g1}, for D1 is scanned, a conductance peak occurs at each time when $\mu_1(N_1, N_2, V_g) = E_F$, yielding a series of conductance peaks with a spacing given by $\Delta V_g = -E_{ch1}/(\alpha e)$ (Eq. 1). An electrochemical potential change, $\Delta\mu_1$, such as that one given by Eq. (5) can be detected as the conductance peak shift, $\Delta V_g = \Delta\mu_1/(\alpha e)$. In both cases discussed above, ΔV_g is given by

$$\Delta V_g = -\beta E_{ch1}/(\alpha e) \tag{6}$$

Thus, D1 operating in the SET regime works as an electrometer that senses a change in the electron number of neighboring conducting islands. Single-photon detection can be achieved if (i) a single event of photon absorption is converted to a change in the electron configuration among conducting islands and (ii) the time constant of measurements is short enough to resolve the effect.

3. Designing a Single-Photon Detector

Figure 9.2 shows a basic concept of a photon detector. It consists of three components: (i) an absorber of electromagnetic waves, (ii) a metallic antenna that couples incident

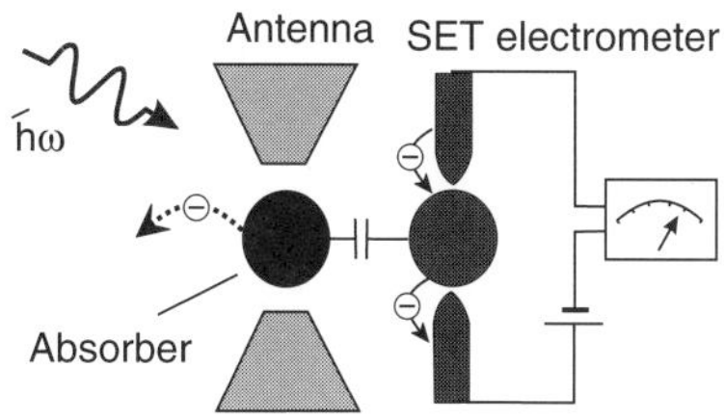

Fig. 9.2. Concept of a photon detector using an SET electrometer.

radiation to the absorber, and (iii) an electrometer (SET) detecting charge states of the absorber. The absorber of electromagnetic waves converts a photon to a charge. The absorber is a small conductive island, in which an electron is excited to escape from the island, for instance, to the neighboring electron reservoir. The number of electrons in the absorber (conducting island) thereby decreases by one, and this photo-induced ionization of the absorber by unit charge $e = 1.6 \times 10^{-19}$ C is sensed by the SET electrometer. Since the wavelength of FIR/SMMW radiation is in the range from 0.1 to 0.6 mm, being far larger than a sub-micron meter size of the absorber, the coupling efficiency of the absorber to the radiation is poor without antenna. A planar dipole antenna of about wavelength size improves the coupling, where impedance matching between the antenna and the absorber significantly affects the performance of the detector.

By considering different excitation mechanisms, one may design a wide variety of detectors based on the scheme of Fig. 9.2. We describe below two types of detectors: a FIR-photon-detector based on the CR and a SMMW-detector based on the plasma excitations (Kohn mode of plasma excitation). Both-type detectors are fabricated in GaAs/$Al_xGa_{1-x}As$ heterostructure crystals with two dimensional electron gas (2DEG) located below the crystal surface. In the heterostructure, electrons move freely along the interface between GaAs and $Al_xGa_{1-x}As$, while they are trapped at the interface, or the motion in the direction perpendicular to the interface is quantized. The conductive island (called quantum dot) can be formed in the 2DEG by negatively biased split metal gates deposited on top of it.

4. Detection in Magnetic Fields

In the FIR region, CR of a 2DEG provides a strong absorption mechanism. The CR yields strong FIR response in a large quantum Hall-device at integer fillings [30]. Especially, CR at edge states leads to long-lived electron–hole excitation [31] due to large spatial separation between edge states [32]. By using a QD in a quantum Hall regime, the CR makes it possible to detect single events of photon absorption.

4.1. QDs in High Magnetic Fields

In high magnetic fields, the energy spectrum of electrons in a QD is quantized into Landau Levels (LLs) as shown in Fig. 9.3(a). Each LL is split into spin-up and spin-down states. We consider such a magnetic field strength that the lowest orbital Landau

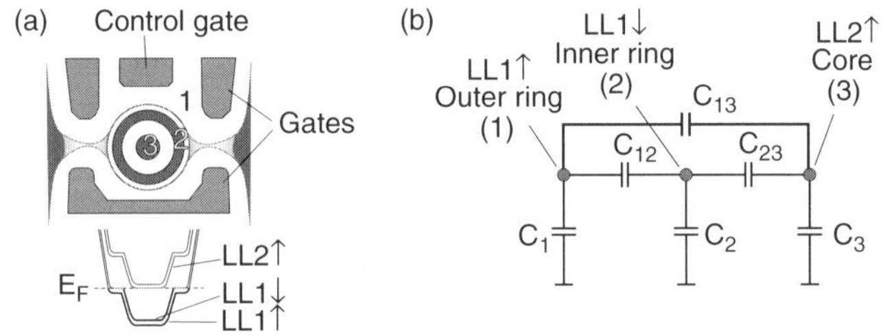

Fig. 9.3. (a) Schematic top-view of a QD in a high magnetic field ($2 < \nu < 3$). Due to Landau level (LL) quantization, separate conductive regions, two conductive rings ($i = 1$: LL1↑ and $i = 2$: LL1↓) and one inner core ($i = 3$: LL2↑) are formed. (b) The respective metallic regions are capacitively coupled to one another (C_{ij}) and to the surrounding (C_i).

level, LL1 (with two opposite spin polarizations) is filled, while the second Landau level, LL2, with the spin-up state is partially occupied with a small number of electrons. The filling factor of LLs, ν, is in the range between 2 and 3. Due to the confinement potential of the QD, the LLs increase their energies as the boundary of the QD is approached. At the energy position where the LL energies cross the Fermi level, LLs flatten due to the screening effect forming compressible regions. In the compressible regions, the electron states are partially occupied with electrons. The spin-up and the spin-down lowest LLs (LL1↑ and LL1↓) yield, respectively, the outer and inner ring-shaped metallic regions as schematically shown in Fig. 9.3(a). The partially occupied second LL (LL2↑) forms a compressible inner metallic core region [33].

These three metallic regions are spatially separated from one another by incompressible insulating strips with integer fillings. The tunneling probability among these three metallic regions, therefore, is negligibly small. Hence, the electrical conduction through the QD occurs via tunneling between the outermost metallic ring (LL1↑) and the electron reservoir. The conductance of the dot is accordingly determined by the electrochemical potential of LL1↑, μ_1, which is affected through capacitive coupling by the number of electrons on the other levels (LL1↓ and LL2↑). The system here thus corresponds to an SET with a triple islands structure ($M = 3$) discussed in section 2, where the metallic regions of LL1↑, LL1↓, and LL2↑ correspond to the conductive islands of $i = 1, 2$, and 3, respectively. An equivalent electrical circuit of the internal structure of the QD is shown in Fig. 9.3(b), where the inequality relations

$$C_{12} \gg (C_1, C_2) \gg C_3 \quad \text{and} \quad C_{12} \gg C_{23} \gg C_{13} \gg C_3 \tag{7}$$

generally hold for the present concentric geometry of relevant metallic regions.

The transport through QDs with a filling factor $2 < \nu < 3$ has been studied and analyzed in details in [23], where a matrix with elements E_{ij} appearing in Eq. (2) for the case of $M = 3$ is derived and compared with experiments. Without going into detail, we will briefly describe important consequences of this analysis below.

As the magnetic field increases, the degeneracy of each LL increases. When the total number of electrons on the dot is kept unchanged ($N_1 + N_2 + N_3 =$ constant), electrons redistribute themselves within the dot. Namely, they transfer from higher LLs towards lower LLs (LL2↑ → LL1↓ → LL1↑). This electron transfer occurs as a sequence of one-by-one electron tunneling, LL2↑ → LL1↓ or LL1↓ → LL1↑. The transition (LL2↑ → LL1↓) implies that $\Delta N_3 = -1$ and $\Delta N_2 = +1$ in Eq. (2), and causes μ_1 and μ_2 to increase by the same value,

$$\mu_1(N_1, N_2 + 1, N_3 - 1) - \mu_1(N_1, N_2, N_3) = (\mathrm{C}_3/\mathrm{C}_{23})E_{\mathrm{ch}} \tag{8a}$$

$$\mu_2(N_1, N_2 + 1, N_3 - 1) - \mu_2(N_1, N_2, N_3) = (\mathrm{C}_3/\mathrm{C}_{23})E_{\mathrm{ch}} \tag{8b}$$

where $E_{\mathrm{ch}} \approx e^2/(C_1 + C_2)$. The transition (LL1↓ → LL1↑) increases μ_1 by

$$\mu_1(N_1 + 1, N_2 - 1, N_3) - \mu_1(N_1, N_2, N_3) = (C_2/C_{12})E_{\mathrm{ch}} \tag{9}$$

Equations (8b) and (9) are similar to Eq. (5). The sign in Eq. (5) is negative because the process considered in this section is opposite to that discussed in section 2. Equation (9) is directly derived from Eq. (5) and Eqs (8a), (8b) follow from Eq. (5) under the condition of inequality relations from Eq. (7). Either change of μ_1 given by Eq. (8b) or Eq. (9) gives rise to a shift, $\Delta V_{\mathrm{g}} = \Delta\mu_1/(\alpha_1 e)$, in the conductance peak position, where $\alpha_1 = C_{\mathrm{g}1}/C_1$ with $C_{\mathrm{g}1}$ being the capacitance between the control gate and the outer ring. The peak shift induced by the one-electron transfer, LL2↑ → LL1↓ ($\Delta N_2 = +1$, $\Delta N_3 = -1$), is accordingly given by

$$\Delta V_{\mathrm{g}}(23) = (C_3/C_{23})/(\alpha_1 e) \tag{10a}$$

and that induced by the transfer, LL1↓ → LL1↑ ($\Delta N_1 = +1$, $\Delta N_2 = -1$) is given by

$$\Delta V_{\mathrm{g}}(12) = (C_2/C_{12})/(\alpha_1 e) \tag{10b}$$

4.2. Mechanism of Detection

Let us now consider effects of the CR excitation inside the dot. Cyclotron resonance excitation is not accompanied by spin flip. Therefore, the transition process is either LL1↑ → LL2↑ or LL1↓ → LL2↓, as shown by two vertical arrows in Fig. 9.4. In either process, the excited electron rapidly gives up its excess energy to the lattice, probably, within several nanoseconds, and falls to the core region of LL2↑ ($i = 3$). The excited hole similarly releases its energy and rapidly climbs up either to the outer ring LL1↑ ($i = 1$) or to the inner ring LL1↓ ($i = 2$). In the case when the hole is created in the outer ring LL1↑ ($i = 1$), it will eventually tunnel to the inner ring LL1↓ ($i = 2$) because the electrostatic energy prefers the configuration of ($\Delta N_3 = 1$, $\Delta N_2 = -1$) to the other ($\Delta N_3 = 1$, $\Delta N_1 = -1$). (This may be understood from Eqs (8a), (8b) showing equal electrochemical-potential change of the two rings in the case of ($\Delta N_3 = -1$, $\Delta N_1 = 1$).)

CR transition thus eventually leads to the intra-dot electron transfer LL1↓ → LL2↑ (i: 2 → 3) after a relatively fast relaxation process. This single-photon induced

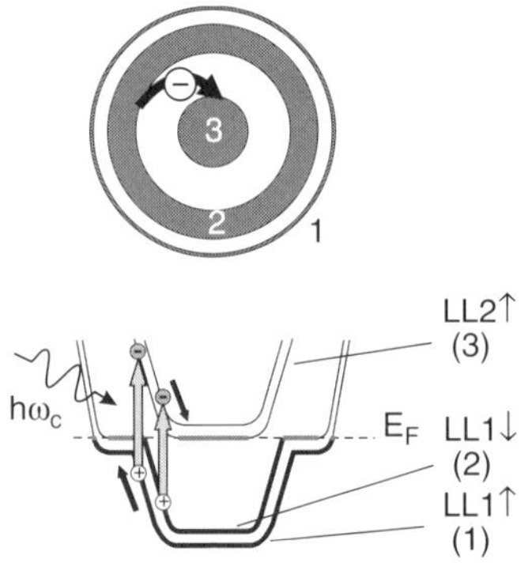

Fig. 9.4. Inter-LL excitation, and resulting internal polarization of the QD. Either excitation process of LL1↑ → LL2↑ or LL1↓ → LL2↓ ends up with an excess electron in LL2↓ and a hole in LL1↓ after relaxation. (See text.)

intra-dot polarization eventually causes the conductance resonance peak to shift by

$$\Delta V_g^{(CR)} = -(C_3/C_{23})E_{ch}/(\alpha_1 e) \tag{11}$$

according to Eq. (10a).

The intra-dot polarization in the above is featured by an excited electron and a hole that are spatially well separated with opposite spin polarizations. These features lead to an extremely long electron–hole recombination lifetime as will be discussed later. This long-lifetime character helps single FIR-photon detection to be feasible.

Transport studies on unequally populated electrons in edge and bulk states in large Hall bars have shown that electron tunneling between the highest occupied LL and the lower LLs is strongly suppressed[32]. The suppression is markedly pronounced as B increases to approach $\nu = 2$. This leads us to speculate that when an electron–hole pair is excited through CR in a QD its lifetime in the relaxed excited state can be particularly long in a relatively narrow B-range close to $\nu = 2$.

4.3. Experimental Set-Up

The QD samples studied are fabricated by means of electron-beam lithography on a GaAs/Al$_x$Ga$_{1-x}$As heterostructure crystal. The sheet carrier density and the mobility of the two dimensional electron gas (2DEG) are $\mu = 85\,\mathrm{m^2/Vs}$ and $n_s = 2.6\times10^{15}\,\mathrm{m^{-2}}$, respectively, at 4.2 K. A scanning electron micrograph of the sample is shown in Fig. 9.5. Metal gates deposited on top of the crystal surface laterally define a QD, which is weakly coupled to external 2DEG regions through tunnel barriers. The electrochemical potential of the QD is controlled by biasing the lower central gate. The metal gates along with the leads extend over about 100 μm length to form a planar dipole antenna that couples incident FIR radiation to the QD. An ac source–drain voltage, $V_{SD} = 25\,\mu A$, is applied at a frequency of 10 Hz ∼ 1 kHz, and conductance through the QD is studied through a standard lock-in technique. The lithographic size of the dot is $0.7\times0.7\,\mu\mathrm{m}^2$, while the effective conductive island size is about 0.5 μm in diameter, containing

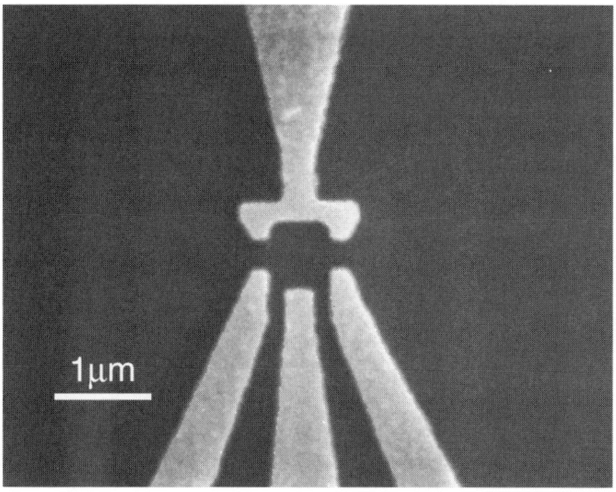

Fig. 9.5. Scanning electron microscope image of a QD fabricated on a GaAs/$Al_xGa_{1-x}As$ heterostructure crystal.

about 350 electrons. The charging energy of the QD, E_{ch}, is determined by the I–V characteristics to be about 4×10^{-4} eV, which corresponds to the capacitance of 0.4 fF. The QD is placed in a mixing chamber of a $^3He/^4He$ dilution refrigerator, the base temperature of which is 70 mK. Magnetic field B is applied normally to the plane of the QD with a superconducting solenoid.

Narrow-band FIR-radiation emitted from GaAs/$Al_xGa_{1-x}As$ 2DEG Hall bars[30] or from n-type InSb crystal is used to illuminate the QD sample. In one optical scheme (Fig. 9.6(a)), a 10 μm-wide 2DEG Hall bar ($\mu = 20\,m^2/Vs$) emitter fabricated on a GaAs/$Al_xGa_{1-x}As$ heterostructure crystal is placed within the mixing chamber of the $^3He/^4He$ dilution at a distance of 27 mm from the QD sample. When a current I_{emit} is passed through the emitter, CR transition of non-equilibrium electrons yields narrow-band cyclotron radiation ($\omega_c = eB/m^*$ with $m^* = 0.069\,m_0$, m_0 is the free electron mass) and the radiation is guided to the QD sample through the 25 mm-long pure silicon rod. In the other optical scheme (Fig. 9.6(b)), a GaAs/$Al_xGa_{1-x}As$ 2DEG Hall bar emitter or a n-type InSb ($m^* = 0.014\,m_0$) emitter[34,35] is tuned by a separate superconducting solenoid placed at a distance of about 80 cm from the QD sample. The CR radiation is guided through a metal light pipe of 3 mm Ø bore into the mixing chamber by using a vacuum shielded silicon window and led to the QD sample. The radiation wavelength is independently tunable over a range of $0.1 < \lambda < 2$ mm by scanning the magnetic field, B_{emit}, for the emitter. The spectral bandwidth of the emitted radiation will be specified later. When an emitter is placed in the same magnet (Fig. 9.6(a)), the relative position the emitter to the magnet center is adjusted so as to yield the maximum resonance to the QD.

In either optical scheme, an unwanted radiation in the near- and mid-infrared regions are eliminated by filtering through pure silicon crystals and black polyethylene films. Continuous illumination of the QD for 24 h did not cause appreciable change in the 2DEG density, indicating that the filtered radiation is free from the band gap excitation

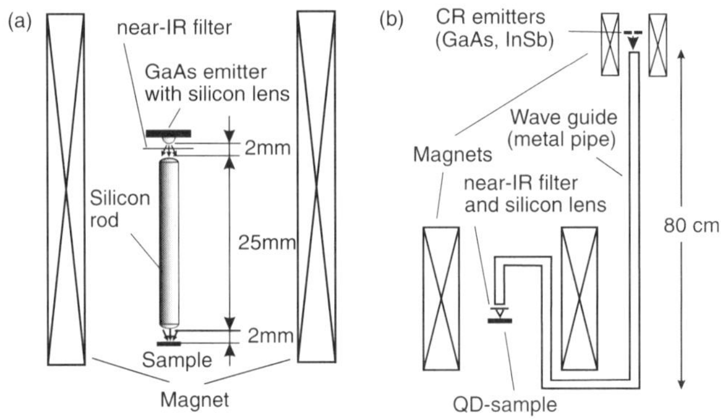

Fig. 9.6. Two types of experimental set-ups: (a) A QD sample is placed in a mixing chamber of a ^{3}He–^{4}He dilution refrigerator together with a GaAs/Al$_x$Ga$_{1-x}$As CR emitter. Both the sample and the emitter are situated in a magnetic field of a superconducting solenoid. (b) A GaAs/Al$_x$Ga$_{1-x}$As CR emitter and a n-InSb CR emitter are installed in a superconducting solenoid ($B_{\rm em}$) that is different from the one (B) for the QD sample.

as well as from the effect of persistent photo-current. Single FIR photons are detected in a B range of 3.4–4.2 T, which corresponds to the wavelength range of $\lambda = 220$–$170\,\mu$m.

The emitted radiation power is roughly proportional to the square of $I_{\rm emit}$. The radiation flux density incident on the area around the QD is approximately as low as 10^{-16}W/mm^2 for a band width of $170 \pm 10\,\mu$m at $I_{\rm emit} = 3\,\mu$A (in the scheme of Fig. 9.6(a)), corresponding to about 10^3 incident photons per second on the effective antenna area of the QD (0.1 mm diameter).

4.4. Ground State

Without FIR illumination, the conductance through the QD exhibits nearly equidistant sharp resonance peaks in the sweep of V_g as shown in the inset of Fig. 9.7(a). Figure 9.7(a) displays the peak positions of the conductance resonance as a function of the magnetic field B. The magnetic field of $B = 4.1$ T corresponds to $\nu = 2$ in this QD, which is smaller than $B = 5.6$ T of $\nu = 2$ in the bulk 2DEG. We find that each trace is featured by tooth-like fine step structures. Two different series of steps, I and II, are distinguished in the close-up representation in Fig. 9.7(b): In series I, steps are relatively large with a nearly unchanged step height, but the step becomes less frequent as B increases (Fig. 9.7(a)). Steps in series II occur at a nearly constant frequency, $\Delta B^{\rm II} = 0.020$ T, while the step height is smaller in a lower B range but gets larger with increasing B. Both types of steps are the shift towards the positive gate voltage when B increases.

Figure 9.8(a) shows step intervals, $\Delta B^{\rm I}$ and $\Delta B^{\rm II}$, against B for both series. The step of series I vanishes at $B \approx 4.1$ T, while series II maintains its interval, $\Delta B^{\rm II} \approx 0.020$ T. The steps of series I and II arise from the tunneling events of LL2$\uparrow$ $\rightarrow$ LL1$\downarrow$ and LL1$\downarrow$ $\rightarrow$ LL1$\uparrow$, respectively. Series I vanishes at $B \approx 4.1$ T because the inner core (LL2$\uparrow$) vanishes ($N_3 = 1 \rightarrow N_3 = 0$) at $B \approx 4.1$ T.

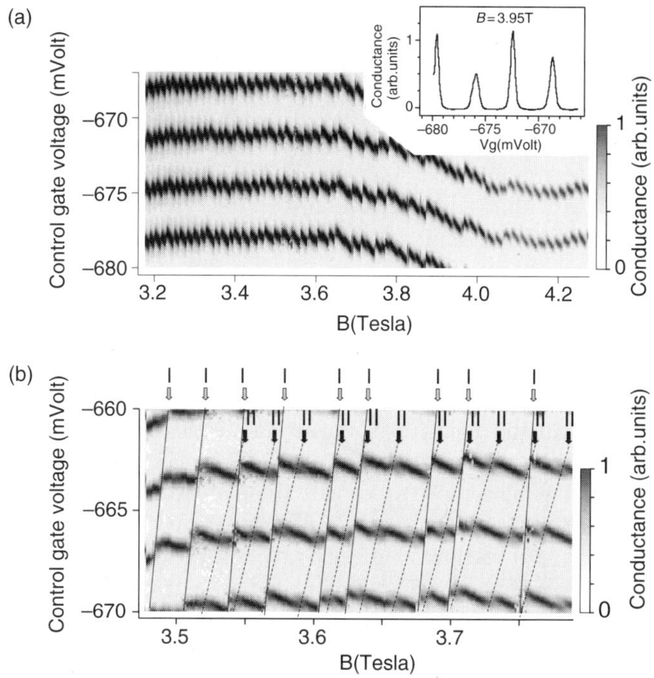

Fig. 9.7. (a) Conductance peak positions as a function of magnetic field. The inset shows an example of conductance oscillation at $B = 3.95\,\mathrm{T}$. (b) Expanded representation of the conductance peak traces.

One can derive N_i in the respective levels as a function of B by counting respective steps. Noting that N_3 decreases by one at each step of series I until it vanishes when $B = 4.1\,\mathrm{T}$ is reached, we derive N_3 versus B as plotted by open circles in Fig. 9.8(b). Equidistant steps of series II indicate that N_1 increases linearly with B (solid triangles in Fig. 9.8(b)). The slope, $N_1 = \pi(D_1/2)^2 B/\Phi_0$ with the flux quantum $\Phi_0 = h/e$, implies that N_1 increases by one each time when Φ_0 is added to the outer metallic ring [33] with the diameter of $D_1 = 0.52\,\mu\mathrm{m}$. We can trace $N_2(B)$ by counting steps of both series and find that it is kept nearly unchanged up to $B = 3.6\,\mathrm{T}$, above which it starts decreasing (black squares in Fig. 9.8(b)).

The basic implication of the data shown in Fig. 9.8(b) is that the electronic structure of a QD in high magnetic fields is indeed split into well-defined levels of LL1$\uparrow$, LL1$\downarrow$, and LL2$\uparrow$. Important for the single-photon detection is that N_3 is small ($N_3 = 33$) already at $B = 3.2\,\mathrm{T}$, which suggests a small size of the inner core and a resulting long lifetime of CR-excited electron–hole pairs in a B-range of $3.2 < B < 4.1\,\mathrm{T}$.

4.5. Photon Detection

Single FIR photons are detected in a B-range of $3.4 \sim 4.2\,\mathrm{T}$. Figure 9.9(a–c) display a conductance peak in one of the samples at $B = 3.67\,\mathrm{T}$ under different illumination intensities ($\lambda \approx 190\,\mu\mathrm{m}$). Here we use the optical scheme shown in Fig. 9.6(a)

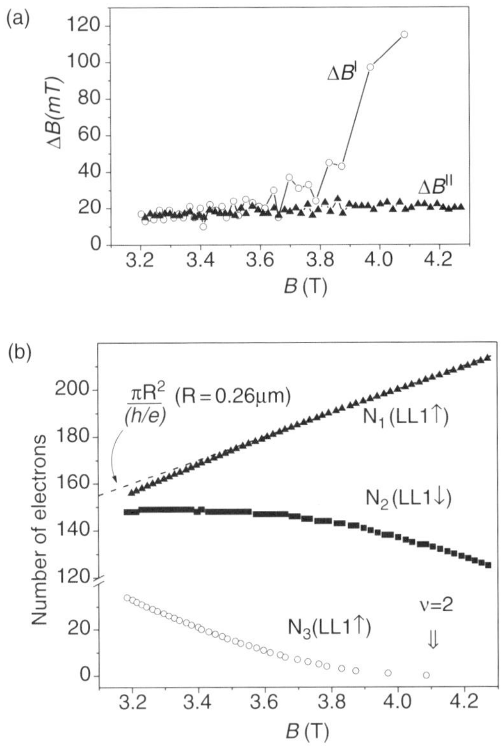

Fig. 9.8. (a) Intervals between successive steps of type I and II, ΔB^{I} and ΔB^{II}, against B. (b) The number of electrons on different Landau levels against B.

with emitters within the mixing chamber. The time constant of measurements is 1 ms and the gate voltage is scanned over 3 min for each curve. When extremely weak FIR illumination is turned on ($I_{\mathrm{em}} = 2\,\mu\mathrm{A}$), the conductance at resonance under the dark (Fig. 9.9(a)) is occasionally switched off, while several spikes of finite conductance show up in a range of V_{g} shifted towards the more negative voltages (Fig. 9.9(b)). When the illumination is intensified ($I_{\mathrm{em}} = 3.5\,\mu\mathrm{A}$), the original conductance resonance, marked by **G** remains as a series of conductance spikes (Fig. 9.9(c)), whilst conductance in the shifted V_{g}-range grows to form a denser array of spikes defining a distinct envelope of another resonance peak centered at the shifted position by $\Delta V_{\mathrm{g}} = -0.55\,\mathrm{mV}$ as marked by **E** (excited state). In addition, less frequent conductance spikes are recognized in a V_{g}-range shifted by $\Delta V_{\mathrm{g}} = -1.1\,\mathrm{mV}$ as marked by **D** (doubly excited state).

The shift, $\Delta V_{\mathrm{g}} = -0.55\,\mathrm{mV}$, is equal in amplitude but opposite in sign to the step of series I (LL2↑ → LL1↓) in Fig. 9.7(b). This indicates that the new conductance resonance **E** arises from the transition LL1↓ → LL2↑ of one electron as discussed in section 4.2. Furthermore, conductance resonance **D** in Fig. 9.9(c) is interpreted as that of the "doubly excited state", in which two electron–hole pairs are excited in the QD.

When V_{g} is fixed at the center of **E**, the conductance is normally off: The conductance is switched on each time when a single FIR-photon is absorbed in the

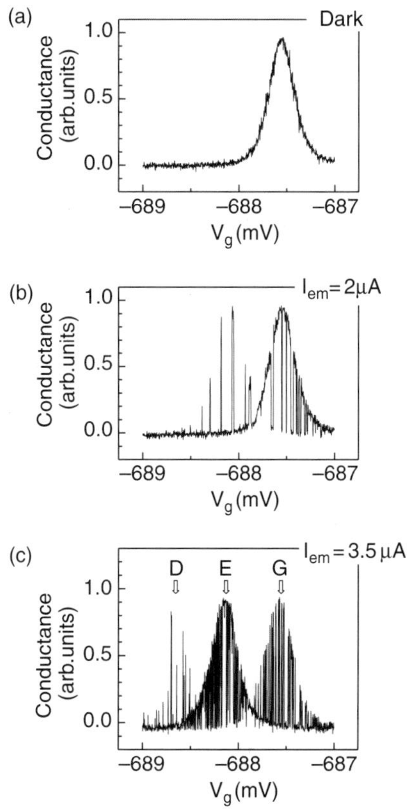

Fig. 9.9. Effects of FIR illumination on a conductance peak at $B = 3.67\,\mathrm{T}$ and $T = 70\,\mathrm{mK}$. (a) Without FIR illumination. (b) Under extremely weak FIR illumination ($I_{em} = 2\,\mu\mathrm{A}$). (c) Under less weak FIR illumination ($I_{em} = 3.5\,\mu\mathrm{A}$).

QD (**G** $\rightarrow$ **E**) as shown by the real time trace in Fig. 9.10(a), where the FIR radiation ($I_{em} = 2\,\mu\mathrm{A}$) is applied in a square wave form of 0.8 Hz. The conductance is reset (**E** $\rightarrow$ **G**) when the excited electron–hole pair vanishes via recombination. Figure 9.10(b) shows that the switching rate increases as the FIR intensity increases ($I_{em} = 3\,\mu\mathrm{A}$). Figure 9.10(c) shows that at still high levels of excitation ($I_{em} = 4\,\mu\mathrm{A}$) the integrated time spent in the excited state levels off, although the switching rate continues to increase. This is because the transition to the "doubly excited state" begins to play a role: If a second electron–hole pair is excited at a time when the previously generated electron–hole pair is still alive in the QD (**E** $\rightarrow$ **D**), it switches off the conductance.

Figure 9.10(d) displays a conductance time trace obtained when V_g is set to the **G** position with $I_{em} = 3\,\mu\mathrm{A}$, where the conductance is normally "on": Excitation event (**G** $\rightarrow$ **E**) switches off the conductance and the recombination (**E** $\rightarrow$ **G**) resets or switches on the conductance. Here, neither transition **E** $\rightarrow$ **D** nor **D** $\rightarrow$ **E** are visible because both of the states (**E** and **D**) correspond to the conductance-off state. Figure 9.10(e) shows the data with V_g fixed at position **D** with $I_{em} = 3\,\mu\mathrm{A}$, where the

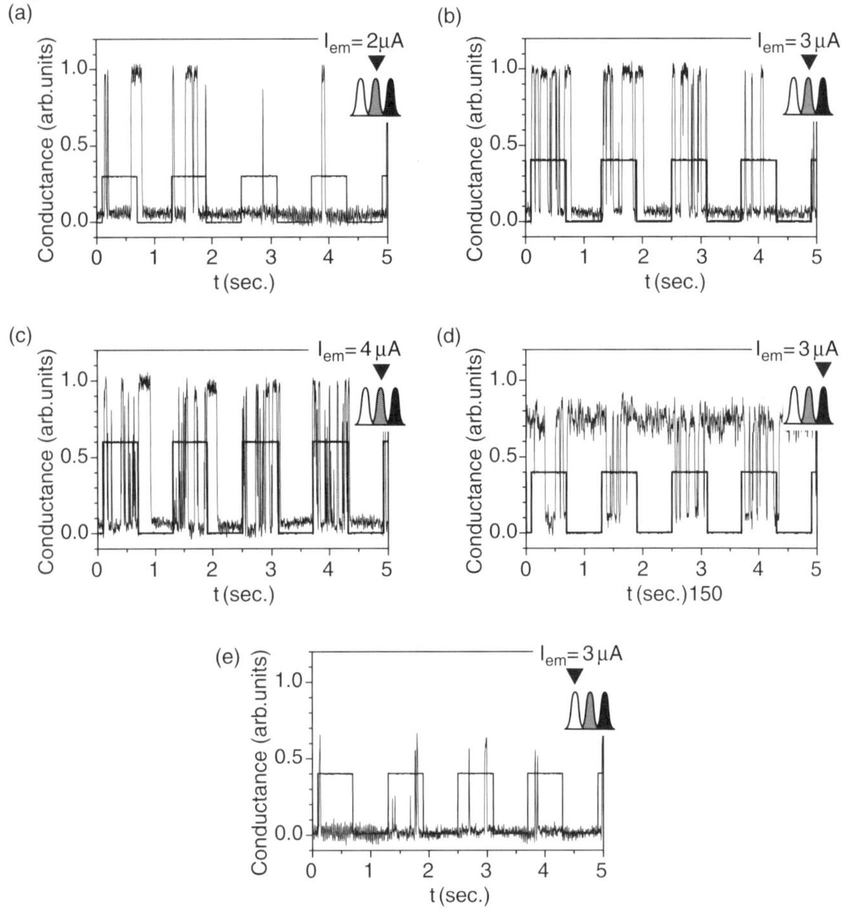

Fig. 9.10. Time traces of the SET conductance under illumination, which is applied in square waves at $f \approx 0.8\,\text{Hz}$. (a) $I_{\text{em}} = 2\,\mu\text{A}$. (b) $I_{\text{em}} = 3\,\mu\text{A}$. (c) $I_{\text{em}} = 4\,\mu\text{A}$. (d) $I_{\text{em}} = 3\,\mu\text{A}$. (e) $I_{\text{em}} = 3\,\mu\text{A}$. Schematically shown peaks with marker in upper right corners of each plot represent position of the trace measurements: V_g is fixed at position **E** in Fig. 9.9(c) for (a)–(c), at position **G** for (d), and at position **D** for (e).

events of switching-on and -off correspond, respectively, to the transitions **E** $\rightarrow$ **D** and **D** $\rightarrow$ **E**. We find that the lifetime of the doubly excited state (**D**) is much shorter than that of the singly excited state (**E**). Still higher multiply excited states (like the triply excited state) are not observable therefore.

Characteristic features are similar to the above for every conductance peak occurring in the B-range of $3.4 \sim 4.15\,\text{T}$, except that the recombination lifetime strongly depends on B as discussed in section 4.7.

Figure 9.11 displays a series of conductance peaks at $B = 3.6\,\text{T}$ for several different levels of FIR illumination (I_{em}). Note that in the measurements presented on Figs 9.11 and 9.12 we use optical scheme of Fig. 9.6(b) with remote emitter, therefore, the

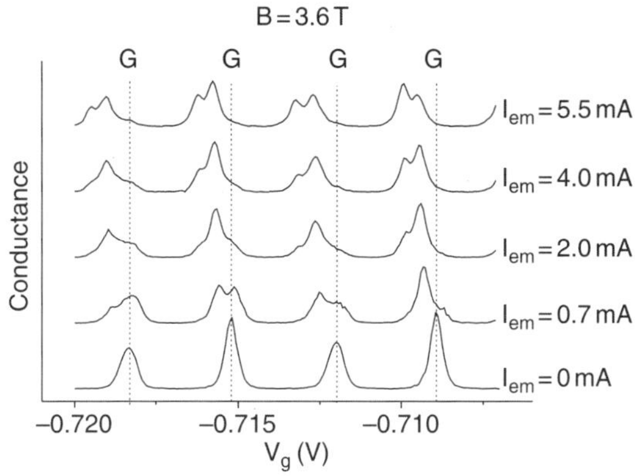

Fig. 9.11. Conductance oscillations taken at $B = 3.60\,\mathrm{T}$ with a long time constant (0.3 s), under illumination up to a high intensity level. The maximal lifetime of the excited state is about 0.1 s. With increasing illumination intensity, ground-state conductance peaks (**G**) decrease their amplitude while a new peak corresponding to the singly excited state (**E**) first grows at a shifted V_g-position. At still higher illumination intensities, yet another peak (**D**) appears at a position shifted twice. Dashed lines mark the positions of **G**.

emitter current I_{em} is much larger than in previous experiments. Individual conductance switches are not seen in the conductance dependencies of Fig. 9.11 because the instrumental time constant is set to be 300 msec, which is much longer than the recombination lifetime of excited states. As the FIR is intensified, each ground-state conductance peak (**G**) reduces its amplitude while two excited-state conductance peaks (**E** and **D**) develop. The integrated line shape of conductance resonance can be well reproduced by a superposition of the separated conductance peaks of **G**, **E**, and **D** (displaced to one another by about 0.55 mV). This is exemplified in Fig. 9.12(a), where the experimental values are indicated by black dots. The fitting procedure at different levels of excitation intensity confirms that the V_g-positions of **E** and **D** are kept unchanged but the relative weight (or the amplitude) of the resonance changes with the excitation intensity. The amplitudes of the respective peaks correspond to the integrated time, which the QD spends in the respective states, or the switching rate multiplied by the recombination lifetime. Figure 9.12(b) displays the amplitudes of the respective peaks against I_{em} (the FIR intensity $\propto I_{em}^2$). With increasing the excitation intensity, the amplitude of peak **E** first develops but levels off at a relatively strong excitation level. Peak **D** becomes appreciable at a higher excitation level but continues to grow in the higher excitation level where peak **E** is already saturated. Ground-state peak **G** diminishes, but the decrease is compensated by the growth of peaks **E** and **D** so that the sum of the peak intensities, $\mathbf{G} + \mathbf{E} + \mathbf{D}$, is kept nearly unchanged.

Figure 9.13 compares the V_g-positions of the conductance peaks under illumination (white circles, $I_{em} = 3.5\,\mu\mathrm{A}$) with those in the ground state **G** (black circles). Here, the instrumental time constant is, again, longer than the recombination lifetime of the excited states. The relevant excited-state is primarily **E**, but several data points also recognized to correspond to **D** ($\Delta V_g \approx -1.1\,\mathrm{mV}$).

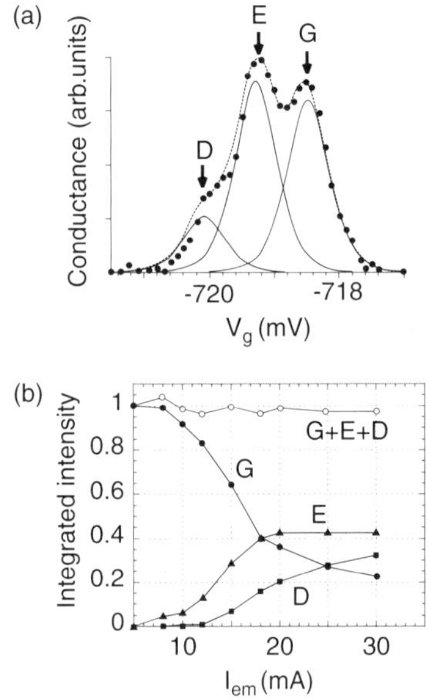

Fig. 9.12. (a) A time-averaged conductance profile under a moderate illumination intensity, taken from Fig. 9.11. The experimental values (solid dots) are reproduced by the dashed line drawn according to the sum of three equally spaced peaks, **G**, **E**, and **D** (solid lines). (b) Variation of the amplitudes of the peaks, **G**, **E**, and **D**, against illumination intensity ($\propto I^2_{\rm emit}$).

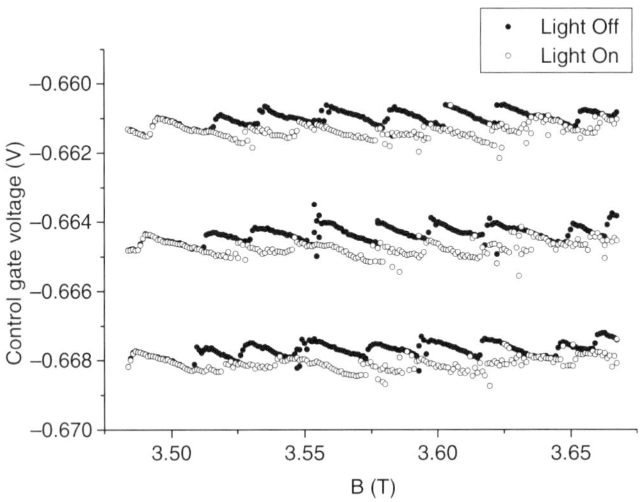

Fig. 9.13. Time-averaged traces of conductance peaks as a function of magnetic field without illumination (solid cycles) and under illumination (open cycles). The illumination intensity is chosen so that peak **E** dominates. Under illumination, peak positions are shifted down by one step of the ground state peak traces.

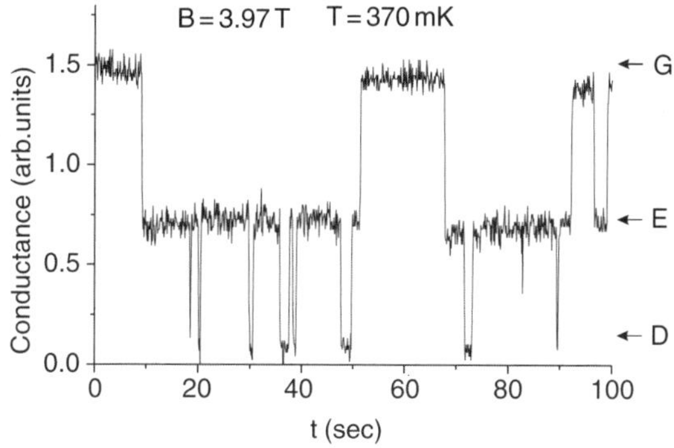

Fig. 9.14. Time trace of conductance at $T = 370\,\text{mK}$ ($B = 3.97\,\text{T}$), where V_g is fixed at the ground-state peak position (**G**). Owing to the thermal broadening of the conductance peak, not only the switches of **G** $\leftrightarrow$ **E**, but also those of **E** $\leftrightarrow$ **D** are visible.

As the temperature is elevated, conductance resonance lines get broader and the recombination lifetime of excited states is reduced. The conductance switches due to CR-absorption are visible up to $T = 400\,\text{mK}$, at which the recombination lifetime is roughly by two order of magnitude smaller than that at $T = 70\,\text{mK}$. Figure 9.14 displays a telegraph-like signal obtained by fixing V_g at a position of **G** at $T = 370\,\text{mK}$ and $B = 3.97\,\text{T}$. (This magnetic field yields the maximum recombination lifetime as will be shown in section 4.7.) The singly and the doubly excited states, **E** and **D**, are now distinguished by different conductance values because of the thermal broadening of conductance resonance.

4.6. Spectral Band of Detection

Spectral measurements are carried out by scanning the wavelength of radiation from a CR emitter in a remote magnet (here, the optical scheme of Fig. 9.6(b) is used) [36]. The radiation spectra of the GaAs/$Al_xGa_{1-x}As$ Hall bar emitter, studied by calibrated Hall-bar detector [30], is well fitted to a Lorentzian curve with a full width at the half-maximum (FWHM) of about $\Delta\omega_{\text{emit}} \approx 4.2\,\text{cm}^{-1}$ at $\omega_{\text{emit}} \approx 55\,\text{cm}^{-1}$.

Figure 9.15 displays how the rate of telegraph-like conductance switching changes as a function of the frequency of incident radiation, ω_{emit} or B_{emit} at $T = 70\,\text{mK}$ and $B = 3.61\,\text{T}$. In the measurements, the radiation power is kept nearly unchanged at such a low level that the average interval between successive events of photon absorption is much longer than the recombination lifetime, τ^*, so that the dead time effect of detection is avoided. The obtained photoresponse is well described by the Lorentzian curve, $1/\{[2(\omega_{\text{emit}} - \omega_R)/\Delta\omega]^2 + 1\}$ with $\omega_R = 53.1\,\text{cm}^{-1}$ and $\Delta\omega = 6.6\,\text{cm}^{-1}$ (FWHM). The absence of background signal suggests that the conductance switching studied is free from dark switching. Noting the spectral width of incident radiation, $\Delta\omega_{\text{emit}} \approx 4.2\,\text{cm}^{-1}$, the true width of excitation spectrum is suggested to be less than $\Delta\omega \approx 2.4\,\text{cm}^{-1}$, much narrower than those found in the averaged transmission spectra of QD arrays. The spectrum is substantially independent of the gate bias condition within a range where the number of electrons in the QD is 280–310.

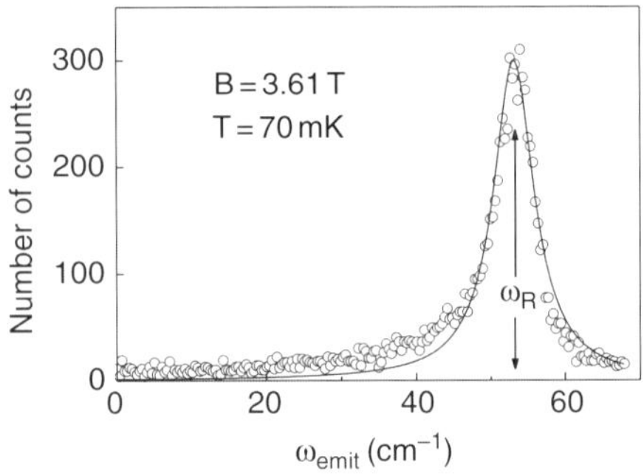

Fig. 9.15. An example of excitation spectrum: The count of the photon-absorption-induced conductance switching, as studied over ten seconds at each point of ω_{emit} (B_{emit}), is shown against ω_{emit} (B_{emit}), which is increased by a step of $\Delta\omega_{\text{emit}} \approx 0.33\,\text{cm}^{-1}$. The data points are well fitted to the Lorentzian curve of the resonance frequency $\omega_{\text{R}} = 53.1\,\text{cm}^{-1}$ and the line width $\Delta\omega = 6.6\,\text{cm}^{-1}$ (FWHM), shown by the solid line.

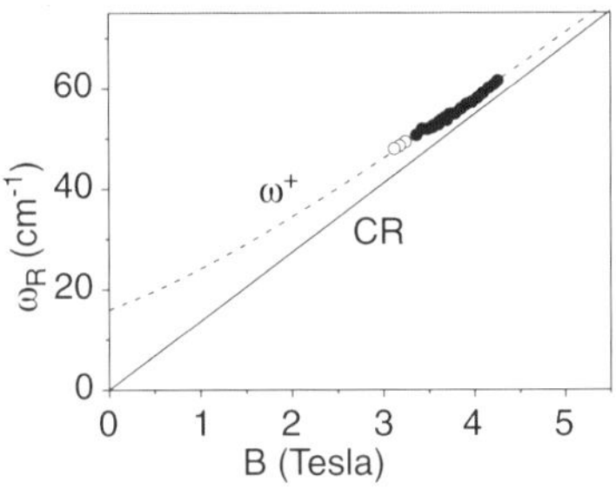

Fig. 9.16. Resonance frequency, ω_{R}, as a function of the magnetic field B applied to the QD. The data points indicated by solid circles are taken via the photon counting method while those represented by open circles via a standard modulation technique. The dashed line shows the dispersion of ω^+ given by Eq. (12) with $\omega_0 = 16\,\text{cm}^{-1}$. The solid line shows the pure CR.

The resonance frequency, ω_{R}, increases with increasing the magnetic field, B, for the QD as shown by the black dots in Fig. 9.16, where additional data obtained through a standard modulation methods are shown by open cycles. (Photon-counting method is not applicable in the range of lower magnetic fields since τ^* becomes shorter than the present instrumental time constant of about 1 msec, as will be seen in section 4.7.)

Figure 9.16 shows that the values of ω_{CR} are slightly higher than those of the bare CR frequency, $\omega_{\text{CR}} = eB/m^*$, indicated by the straight line. The data points are well fitted to a dashed curve drawn according to the upper branch of the dispersion relations

$$\omega^{\pm} = \sqrt{(\omega_{\text{CR}}/2)^2 + \omega_0^2} \pm \omega_{\text{CR}}/2 \tag{12}$$

which is expected for magnetoplasma resonance [37,38], where the best fit to the experimental values is obtained by choosing $\omega_0 = 16\,\text{cm}^{-1}$.

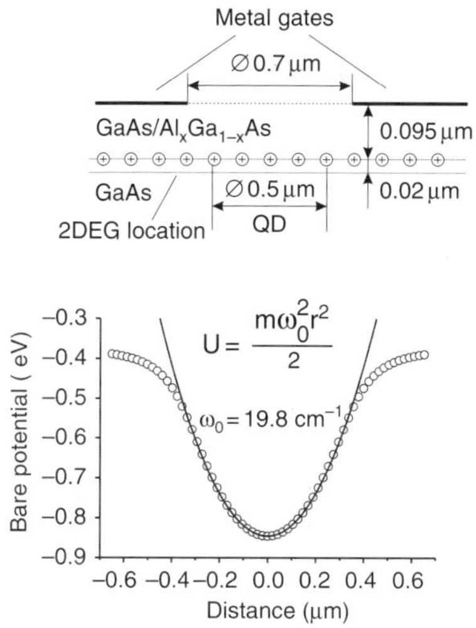

Fig. 9.17. Upper panel: Simplified model of a heterostructure crystal covered by a metal gate with a hole. Lower panel: Open dots indicate calculated values of the bare potential induced by the positive background charge on the plane of the 2DEG. The horizontal scale is common for both panels.

If a hard wall potential is assumed for the confining potential of a QD, a theory [39] predicts $\omega_0 \approx 40\,\mathrm{cm}^{-1}$, more than twice as large as the experimental value. A realistic bare confining potential of the QD considered here can be estimated from a simplified structure depicted in the upper panel of Fig. 9.17. A 2DEG layer of a sheet density n_s is positioned at a depth $h_- = 115$ nm below the crystal surface, and is neutralized by a remote layer of the background positive charge at a depth, $h_+ = 95$ nm. A metal gate with a hole of a 700 nm diameter covers the crystal surface. (The surface charge is ignored for it is compensated by the positive charge induced via band bending of GaAs.) Negatively biasing the gate forms a 2DEG island (QD). The confinement potential formed by the bare positive background charge can be found by taking the prolate spheroidal coordinates ξ, η, φ, which are related to the cylindrical coordinates r, z, φ through $\xi = \sinh u$, $\eta = \cos v$ and $r = R\,(\cosh u)\sin v$, $z = R\,(\sinh u)\cos v$ with parameters u and v being $u > 0, 0 \leq v \leq \pi$. A potential formed by a bounded conducting plane with a circular aperture situated in a uniform electric field E in a half-space $z < 0$ along z-axis (perpendicular to the metal plane) is derived in an analytical form as [40]

$$\varphi = -E\frac{z}{\pi}\left(\arctan\frac{1}{\xi} - \frac{1}{\xi}\right) \quad \text{at } z > 0 \quad \text{and}$$

$$\varphi = -E\frac{z}{\pi}\left(\pi - \arctan\frac{1}{\xi} + \frac{1}{\xi}\right) \quad \text{at } z < 0.$$

Here, the coordinates originate at the aperture center with the z-axis perpendicular to the interface. In our case, the electric field $E = -4\pi n_s e/\varepsilon_{\mathrm{GaAs}}$ is induced by the positive charge sheet below the surface (Fig. 9.17) and an effect of the GaAs–vacuum interface is taken into account by assuming an additional electric field $E' = 4\pi n'_s e/\varepsilon_{\mathrm{GaAs}}$ induced by a mirror charge with a sheet density $n'_s = (\varepsilon - 1)/(\varepsilon + 1)n_s$ at a distance h_+ above the GaAs surface. Open circles in the lower panel of Fig. 9.17 show the potential energy profile for the QD on the plane of 2DEG. In the first order approximation, the potential is parabolic, $U = e\varphi$ as

$$U = \frac{4}{(1+\beta^2)^2}\frac{e^2 n_s}{R\varepsilon_{\mathrm{eff}}}\frac{r^2}{2} + U_0$$

where $\beta = h_-/R$, $\varepsilon_{\mathrm{eff}} = (\varepsilon_{\mathrm{GaAs}} + 1)/2$ with U_0 being a constant not affecting the plasma oscillation frequency. Noting $U = m\omega_p^2 r^2/2$, we derive

$$\omega_p^2 = \frac{4}{(1+\beta^2)^2}\frac{e^2 n_s}{R\varepsilon_{\mathrm{eff}} m^*} \tag{13}$$

Equation (13) yields the characteristic frequency $\omega_p \approx 19.8\,\mathrm{cm}^{-1}$ for the QD. In a pure parabolic potential, generalized Kohn theorem [41,42] predicts that a uniform alternating electric field excites only the center of mass motion at the resonance frequency ω_p.

Thus, $\omega_p \approx 19.8\,\mathrm{cm}^{-1}$ reasonably accounts for the observed resonance frequency $\omega_0 = 16\,\mathrm{cm}^{-1}$. (The discrepancy might arise from a simplified model used to derive Eq. (13).) The confining potential of the QD is suggested to be nearly parabolic. The observed resonance is hence a CR coupled with the Kohn mode plasma oscillation, the resonance frequency of which is not affected by the electron–electron interaction [43].

At this point, a subtle question might arise. That is, a collective center-of-mass motion of all the electrons is excited in our QD, but the resulting signal of conductance switching indicates a single-particle Landau level transition. Quantum mechanically, the state of a collective motion excited via the absorption of one photon may be represented by a superposition of states including different single-particle Landau level excitations: It is impossible to specify which one particular electron is excited in the higher Landau level. However, it is interpreted that, rapidly after the excitation, inelastic scattering processes will drive the QD into a mixed state, in which one particular electron (hole) can be identified in the higher (lower) Landau level. The resonance line width, $\Delta\omega < 2.4\,\mathrm{cm}^{-1}$, implies that the excited collective oscillation has a lifetime of $\tau > 2/\Delta\omega \approx 4.5$ ps, which is suggested to be the relevant inelastic scattering time discussed here. On the final stage, the excited electron (hole) releases its excess energy to the lattice on a time scale of a few nanoseconds. It is only at this final stage that the electrical polarization develops in the QD.

Additional experiments confirm that the lower branch of excitation, ω^- in Eq. (12), does not yield photoresponse. This is because the lower branch excitation does not lead to polarization of a QD.

4.7. Recombination Lifetime

The time duration of switching-off, or the dead time for counting, in the time traces of conductance shown in Fig. 9.10(a) is given by the recombination lifetime, τ, of an

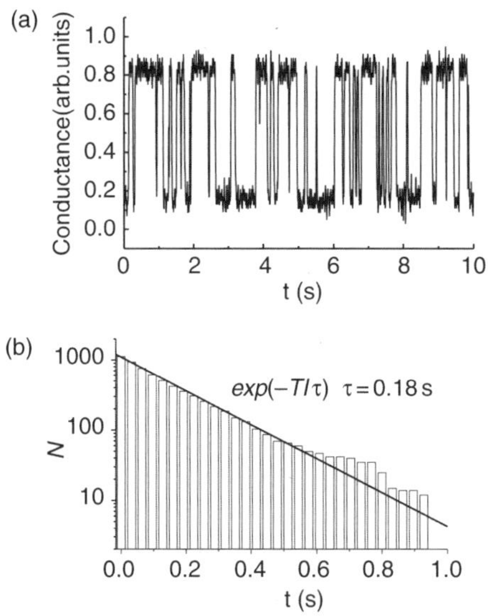

Fig. 9.18. (a) A typical time trace under FIR illumination, taken at $T = 70$ mK and $B = 3.67$ T. (b) Histogram of the recovery time after conductance switching. The number of events with recovery times shorter than time T is represented against T.

excited electron/hole pair within the QD. This lifetime is another important parameter of the photon counter. A histogram of the distribution of the time duration of switching-off is derived from the time trace of Fig. 9.18(a), taken at $T = 70$ mK and $B = 3.67$ T, and is shown in Fig. 9.18(b). The data are well fitted to the exponential decay, $\exp(-t/\tau)$, with a lifetime of $\tau \approx 0.18$ s.

The lifetime, similarly studied in different magnetic fields, is plotted against B by solid cycles in Fig. 9.19(a). The size of the lifetime, τ, spans a wide range of time scale from 1 ms through 1000 s, exhibiting a distinct saw-tooth like behavior. In addition, τ remarkably increases with B to reach about 1000 s at $B \approx 4.0$ T. Above $B \approx 4.0$ T, τ decreases rapidly, down to 1 ms at $B \approx 4.15$ T, at which the inner core (LL2$\uparrow$) is supposed to vanish in the QD (Fig. 9.8(b)). The digits marked on respective saw teeth in Fig. 9.19(a) indicate the number of electrons in LL2$\uparrow$, N_3.

As shown in the inset of Fig. 9.19(a), each saw-tooth in the τ versus B curve correlates with a step of series I ($N_3 \rightarrow N_2$) in the conductance trace (section 4.4): With increasing B, τ rises abruptly by about two orders of magnitude at each transition point ($N_3 \rightarrow N_2$), where one electron tunnels from the inner core (LL2$\uparrow$) to the inner ring (LL1$\downarrow$) in the ground state: τ smoothly decreases with B until the next transition point ($N_3 \rightarrow N_2$) is reached.

As shown in Fig. 9.19(b), τ is reduced by nearly two orders of magnitude as T is elevated from 100 mK up to 300 mK. The T-dependence of the recombination probability is described by the activation-type behavior, $\tau^{-1} \propto \exp(-E^*/kT)$ with $E^* = 100\,\mu$eV, as shown by a dashed line in the inset of Fig. 9.19(b), where τ^{-1} is plotted against T^{-1} for three different magnetic fields.

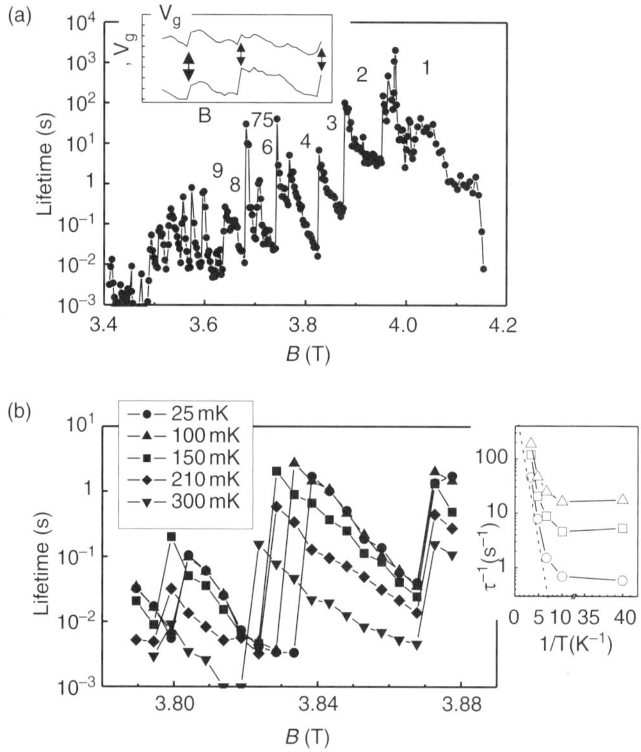

Fig. 9.19. (a) Recombination lifetime τ of an electron-hole pair in the QD as a function of magnetic field B. The inset compares τ with the conductance peak position in the dark. (b) Temperature dependence of τ. The inset shows $\ln \tau^{-1}$ against $1/T$. The dashed strait line is drawn according to $\exp(E^*/kT)$ with $E^* = 100\,\mu\text{eV}$.

As described below, the remarkable length of τ is a consequence of the suppression of scattering due to spin-flip character of the recombination process as well as the large spatial separation between an electron and a hole. The saw-tooth structure of τ, on the other hand, is ascribed to the B-dependence of the internal polarization energy of the QD.

An excess electron and a hole are located in the inner core (LL2$\uparrow$) and the inner ring (LL1$\downarrow$) of the QD (section 4.2). It follows that, at low temperatures, recombination of an electron/hole pair is possible only via a spin-flip process, LL2$\uparrow$ $\rightarrow$ LL1$\downarrow$, as marked by the black dotted arrow α in Fig. 9.20(a). At elevated temperatures, electrons (holes) are thermally excited in LL2$\downarrow$ (LL1$\downarrow$) as $\exp(g\mu_B B/kT)$ where $g\mu_B B$ is the Zeeman energy of electrons in the QD. At such elevated temperatures, spin-conserving processes, LL2$\downarrow$ $\rightarrow$ LL1$\downarrow$ and LL2$\uparrow$ $\rightarrow$ LL1$\uparrow$ marked by β and γ in Fig. 9.20(a), become relevant. Assuming a g-factor to be equal to a bulk value of electrons in GaAs, $g = -0.44$, $g\mu_B B = 100\,\mu\text{eV}$ at $B = 3.8\,\text{T}$. The activation-type behavior of the recombination probability, $\tau^{-1} \propto \exp(-E^*/\text{kT})$ with $E^* = 100\,\mu\text{eV}$ (Fig. 9.19(b)), thus indicates that (i) the recombination process at low temperatures, $T < 100\,\text{mK}$, is indeed limited to the spin-flip process α and (ii) the spin-conserving processes, β and γ,

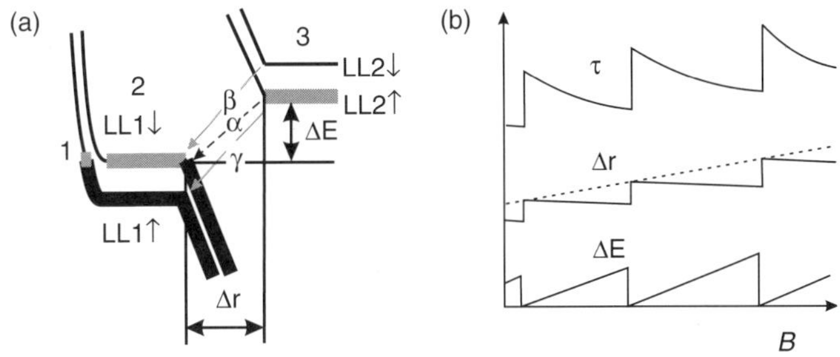

Fig. 9.20. (a) A LL profile with possible recombination processes: (α) LL2↑ → LL1↓, (β) LL2↓ → LL1↑, (γ) LL2↑ → LL1↓. (b) Schematic representation three different physical quantities against magnetic field: ΔE is the internal polarization energy to be released by the recombination process, Δr is the distance between the inner core and the inner ring, and τ is the recombination lifetime.

dominate the recombination at $T > 200\,\text{mK}$ whereas only a few percents of electrons (holes) are thermally excited in the opposite spin states, LL2↓ (LL1↓). The latter is a proof that the transition probability of spin-flip process, α, is far smaller than those of the spin-conserving processes β and γ; viz, the probability is strongly suppressed by the spin-flip character. This is reasonable because the spin-flip process requires the relevance of hyperfine interaction with nuclear spins or spin–orbit interaction along with random potentials.

The recombination process is a higher-order process because two additional constraints are imposed. First, electron–phonon interaction should be involved since an excess energy ΔE stored in the form of internal polarization of the QD has to be released. Secondly, a large angular momentum difference between the initial and the final states has to be overcome, which implies the necessity of random potentials as well as a small overlapping integral of the initial/final-state wave-functions. An approximate form of the recombination probability can be generally written as,

$$\tau^{-1} = A\Delta E^{\text{p}} \exp(-\Delta r^2/4l_{\text{B}}^2) \tag{14}$$

where A is a constant affected by a suppression effect due to spin-flip scattering as well as a strength of random potentials, p (>0) is a power of ΔE, Δr is the distance between the inner core and the inner ring (Fig. 9.20(a)) and l_{B} is the magnetic length. The term ΔE^{p} represents a relevance of phonon scattering [44], and the last term $\exp(-\Delta r^2/4l_{\text{B}}^2)$ is the suppression factor of the overlapping integral [44,45]. In the QD studied, $\Delta r/l_{\text{B}}$ ranges from 3 to 6 yielding a strong suppression, $\exp(-\Delta r^2/4l_{\text{B}}^2) = 10^{-1} \sim 10^{-4}$. The necessity of random potentials is confirmed by noting the wave vector q of a relevant phonon. q is derived from $\Delta E = \hbar c_{\text{s}} q$ with $c_{\text{s}} \approx 5.2 \times 10^3\,\text{m/s}$, the longitudinal sound velocity of GaAs. Since $\Delta E < e^2/C_{23} < 0.8\,\text{meV}$, q is insufficient to satisfy the momentum transfer, $\Delta r/l_{\text{B}}^2$ in the QD with $\Delta r > 50\,\text{nm}$ at $B \geq 3.4\,\text{T}$, M. Stopa (private communication). Thus, the energy is taken by a phonon while the conservation of (angular) momentum is relaxed by random potential scattering.

Equation (14) can be compared with the experimental data if Δr and ΔE are known as a function of B. The dependence of ΔE on B is derived as follows. After an excitation event, most of the absorbed photon energy is rapidly released to the lattice through relaxation (Fig. 9.4), but an energy, ΔE, is stored in the form of internal polarization. Letting $E(N_1, N_2, N_3)$ be the electrostatic energy of the QD with the electron configuration (N_1, N_2, N_3), ΔE is expressed by $\Delta E = E(N_1, N_2 - 1, N_3 + 1) - E(N_1, N_2, N_3)$.

As B increases, ΔE linearly increases. When B reaches a transition point at which the ground state undergoes a transition, $(N_1, N_2, N_3) \rightarrow (N_1, N_2+1, N_3-1)$, the value of ΔE reaches e^2/C_{23} and discontinuously drops to zero, as schematically depicted in the right panel of Fig. 9.20(b) [23]. In the QD studied, $\Delta E \approx \alpha(B - B_0)$ with $\alpha \approx 10$ meV/T, where B_0 is the nearest transition point smaller than B. We note that $\Delta E \gg kT$ except in the close vicinity of transition points, where ΔE should be replaced by kT.

At each transition point, $(N_1, N_2, N_3) \rightarrow (N_1, N_2 + 1, N_3 - 1)$, the inner core shrinks to increase Δr stepwise, as shown in the right panel of Fig. 9.20(b). In between two adjacent transition points, the size of both the inner core and the inner ring is smoothly reduced according to the increasing degeneracy of Landau levels, $l_B^{-2} \sim B$.

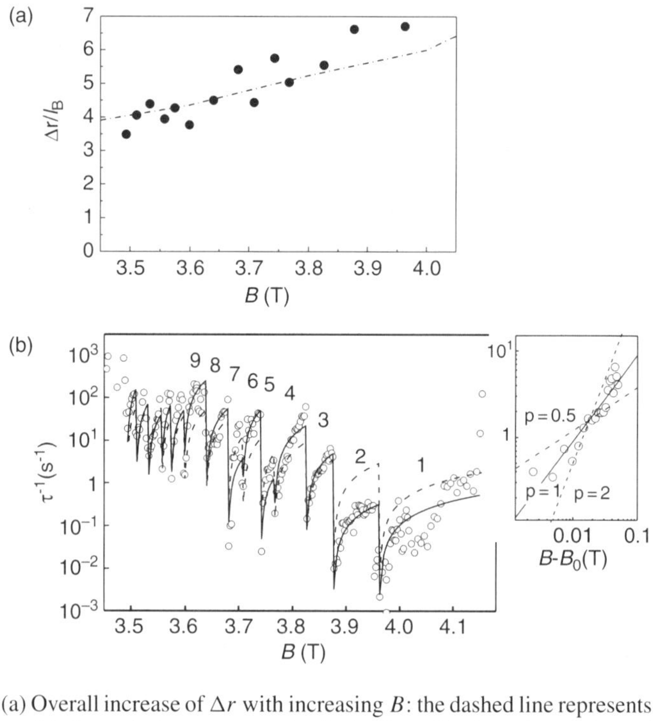

Fig. 9.21. (a) Overall increase of Δr with increasing B: the dashed line represents theoretically derived values, M. Stopa (private communication). The solid dots show values giving the best fit to the experimental data of τ^{-1}. (b) Experimental values of τ^{-1} (open dots) are compared with the values predicted by Eq. (14): The dashed line is obtained with $\Delta r(B)$ shown by the dashed line in Fig. 9.21(a), while the solid line is obtained with $\Delta r(B)$ indicated by the black dots in Fig. (9.21)(a). An inset shows $\ln \tau^{-1}$ versus $\ln(B - B_0)$ for the interval of $N_3 = 3$ ($B_0 = 3.82$ T).

Outer diameter of the inner ring and the inner diameter of the inner ring should both be proportionally to $B^{-1} \sim l_B$, keeping the ratio $\Delta r / l_B$ unchanged in between two adjacent transition points. Overall averaged increase of $\Delta r / l_B$ over a wider B-range are theoretically derived by M. Stopa (private communication), who simulated the QD studied here by considering electrostatic interaction energy without the effect of charge quantization: The smooth dashed line in Fig. 9.21(a) shows the theoretical values of $\Delta r / l_B$. (Since the discreteness of individual electrons is ignored, the increase of $\Delta r / l_B$ is not step-wise.)

As described in the above, the saw-tooth structure of τ arises primarily from the discontinuous drop, $\Delta E = e^2/C_{23} \to 0$, of the internal polarization energy at the transition point followed by a linear increase. The feature is enhanced also by the step-wise increase of Δr at the transition point. The experimental data points of τ (Fig. 9.19(a)) are re-plotted in Fig. 9.21(b). The dashed line in Fig. 9.21(b) indicates theoretical values of Eq. (14), $\tau^{-1} = A \Delta E^{\mathrm{p}} \exp(-\Delta r_0^2/4 l_{\mathrm{B0}}^2)$ with $A = 2 \times 10^4\,\mathrm{s^{-1}\,meV^{-1}}$ and $p = 1$, where values of $\Delta r_0 / l_{\mathrm{B0}}$ are taken from the theoretical curve (Fig. 9.21(b)) at each transition point. The chosen parameter of $p = 1$ is shown to be reasonable in the inset of Fig. 9.21(b), where values of $\ln \tau^{-1}$ are displayed against $\ln(B - B_0)$ for the B-interval of $N_3 = 3$ ($B_0 = 3.82\,\mathrm{T}$). The agreement between the theoretical curve and the experimental data is substantial.

In true QDs, Δr may not smoothly increase with B, but may fluctuate due to random potentials. If $\Delta r / l_B$ is assumed to fluctuate around the theoretical values as shown by the black dots in Fig. 9.21(a), even better fit is obtained as shown by the solid line in Fig. 9.21(b).

5. Detection in the Absence of Magnetic Field

The detector described in section 4 requires the use of strong magnetic fields. In this section, single-photon detection in the absence of magnetic fields is described [27,28].

5.1. *Mechanism of Detection*

The detector is an SET consisting of parallel double QDs as schematically illustrated in Fig. 9.22(a). The first QD (D1) forms an SET, through which tunnel current is transmitted. Adjacent to D1, a second QD (D2) is placed and electrostatically coupled to D1. D2 is coupled to incident SMMW by a planar dipole antenna. As depicted in Fig. 9.22(b), if an electron in D2 gains an excess energy E^* through excitation by SMMW, the excited electron escapes to the electron reservoir through D1 so that the number of electrons, N_2, in D2 decreases by one ($\Delta N_2 = -1$). The electron then rapidly releases its excess energy (via phonon emission or electron–electron interaction) relaxing to the Fermi level, E_{F}. The potential barriers, in turn, prevent the "cold" electron from returning to D2, thereby realizing a long lifetime τ_l of the ionized state of D2. The ionization of D2 ($\Delta N_2 = -1$) decreases the electrochemical potential of D1 by $\Delta \mu_1 \approx e^2 C_{12}/C_1 C_2$ (e is the unit charge) as it directly follows from Eqs (4b) and (5), where $C_{12} \ll C_1$ and $C_{12} \ll C_2$ hold in the experimental condition. This results in the shift of the SET conductance peak by $\Delta \mu_1 / E_{\mathrm{ch}} \approx -C_{12}/C_2 = -(3 \sim 15)\%$ in

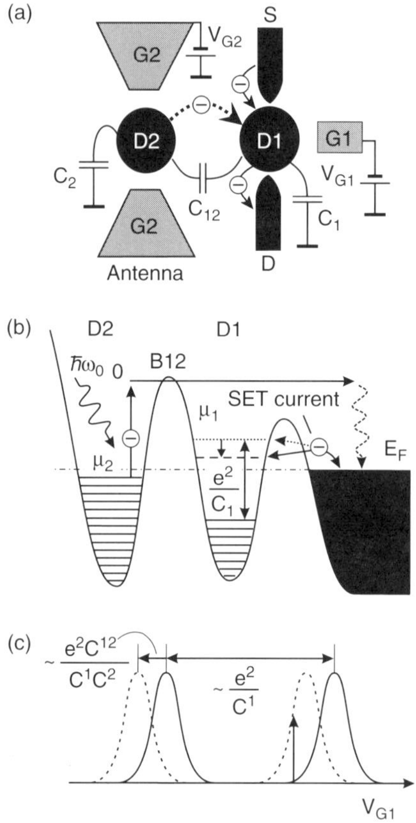

Fig. 9.22. (a) Schematic representation of an SMMW photon detector consisting of parallel double QDs. (b) Ionization of D2 due to SMMW-photon absorption. (c) Conductance peak shift induced by the ionization.

the sweep of V_{G1}, and yields a conductance change as shown in Fig. 9.22(c), where $E_{ch} = e^2/C_1$ is the charging energy of D1.

5.2. Characteristics of Double QD Set

Figure 9.23(a) shows the device fabricated through a standard lithographic technique on a GaAs/Al_xGa_{1-x}As heterostructure crystal with a mobility of 190 m^2/Vs and a sheet carrier density of $n_s = 0.9 \times 10^{15}\,m^{-2}$ at 4.2 K. Areas of light color indicate metal gates deposited on top of the crystal. Negatively biasing the gates depletes the 2DEG below the gates and forms D1, D2, the source (S), the drain (D) and the reservoir (R). The lithographic size of each QD is $0.5 \times 0.5\,\mu m^2$, with about 200 electrons. Gate G12 defines the inter-QD potential barrier, B12. Control gate, G2, controls not only the electrochemical potential of D2, μ_2, but also defines the potential barrier between D2 and the left-hand side reservoir R. Metal leads for G2 and G12 extend over 0.1 mm

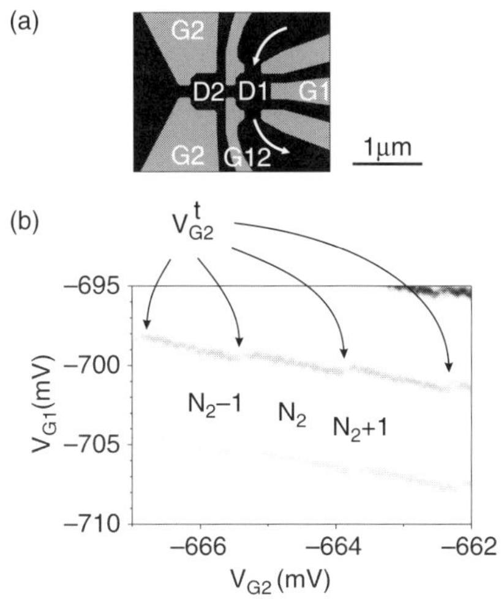

Fig. 9.23. (a) A double-QD photon detector studied. (b) Conductance peak traces on the plane of V_{G1} and V_{G2}.

in length, forming a dipole antenna for D2. The experiments are made at 70 mK by using a ^{3}He–^{4}He dilution refrigerator.

Figure 9.23(b) displays conductance peak traces without SMMW as a function of the bias voltages V_{G1} for G1 and V_{G2} for G2, with a fixed bias voltage for B12, where the dark regions indicate high conductance. As V_{G1} increases at a fixed V_{G2}, the conductance resonance periodically occurs at the positions where $\mu_1 = E_F$: It forms a set of dark traces with a spacing proportional to $E_{ch} = e^2/C_1$. As V_{G2} increases, N_2 increases by one [29] at a transition point, $V_{G2} = V^{t}_{G2}$, where $\mu_2 = E_F$. The change $\Delta N_2 = +1$ causes μ_1 to shift by $+\Delta\mu_1/E_{ch} \approx +C_{12}/C_2 \approx 5\%$. Hence each conductance trace is featured by step-wise small jump in the positive V_{G1}-direction at each transition point V^{t}_{G2} [29].

5.3. *Photon Detection*

The SET sample is illuminated with magnetically tuned SMMWs emitted from a high mobility GaAs/Al$_x$Ga$_{1-x}$As Hall bar described in Fig. 9.6(b). The radiation frequency $f = \omega_c/(2\pi)$ is tuned over the range 0 – 1.5 THz (B_{emit} = 0 T – 4.5 T). Letting P_{em} be the electrical input power fed to the emitter, the incident radiation power with a spectral bandwidth of 5% falling on an antenna area (about 0.2 mm in diameter) is estimated to be on the order of $(10^{-12}\text{–}10^{-14}) \times P_{em}$ for $f = \omega/(2\pi) \approx 500$ GHz ($\lambda \approx 0.6$ mm).

Figure 9.24(a) shows how a typical conductance peak (the solid line on the left hand side) is affected by illumination (on the right hand side). The traces are taken in a sweep of V_{G1} with a fixed value of $V_{G2} = -590.5$ mV. Under illumination, the original conductance peak frequently switches to a shifted new peak indicated by the dashed line drawn for the sake of guiding one's eyes. The switching can be identified to be due to the ionization of D2 ($N_2 \rightarrow N_2 - 1$) through the studies of conductance peak position as shown in Fig. 9.23(b) [27].

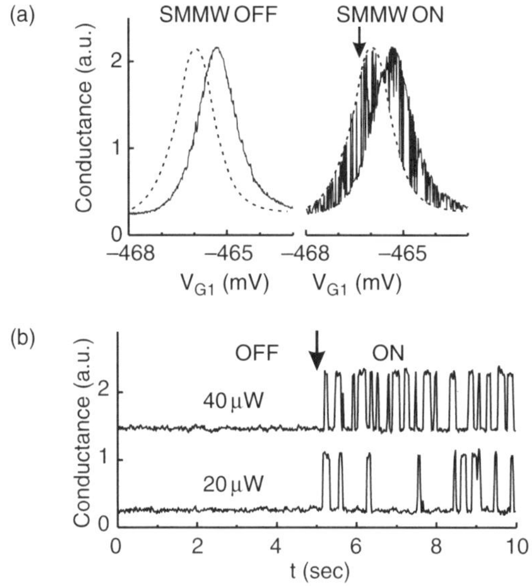

Fig. 9.24. (a) A conductance peak obtained without illumination (left-hand side panel) and under illumination (right-hand side panel) with a radiation frequency of $f \approx 500\,\text{GHz}$ and $P_{\text{em}} = 40\,\mu\text{W}$. Dashed and solid lines represent states wit $N_2 - 1$ and N_2 electrons, respectively, between which the dark switching takes place as $V_{G2} = V_{G2}^{t}$ is approached. (b) Time traces of conductance for $P_{\text{em}} = 20$ and $40\,\mu\text{W}$, both at $V_{G1} = -466.3\,\text{mV}$ and $V_{G2} = -590.5\,\text{mV}$. The arrow marks the instant at which the illumination is turned on.

When V_{G1} is fixed, the photon counting signal manifests itself as telegraph-like conductance switches as displayed in Fig. 9.24(b) for two levels of excitation intensity at $V_{G1} = -466.3\,\text{mV}$ (the position marked by the arrow in Fig. 9.24(a)). No switches are seen without SMMW radiation for five seconds: In this particular condition, the "dark switching" rate is found to be as low as $0.01\,\text{s}^{-1}$, while the recombination lifetime of excited electron–hole pairs is as long as $\tau_L \approx 80\,\text{ms}$.

The photon count is studied as a function of the frequency f of the incident radiation by varying B_{emit} applied to the emitter. The photon count exhibits a distinct resonance structure as shown in Fig. 9.25. The resonance peak frequency, $f_0 \approx 500\,\text{GHz}$ ($\lambda \approx 0.6\,\text{mm}$), is nearly independent of the gate bias condition, and is interpreted as the collective oscillation of electrons in a parabolic confinement potential [26,27], known as the Kohn-mode plasma resonance [38]. The full width at half-maximum (FWHM) of the obtained resonance line is $\Delta f \approx 180\,\text{GHz}$, which implies that the collective plasma oscillation is converted into a single electron excitation after a short lifetime, $\tau = 1/(\pi \Delta f) \approx 1.8\,\text{ps}$.

The data of Figs 9.24 and 9.25 are taken at $V_{G2} = -590.5\,\text{mV}$. The condition is well away from the transition point, $V_{G2}^{t} = -591.05\,\text{mV}$, at which $\mu_2 = E_F$. This is why the dark switching rate is low. As the transition point is approached, the dark switching rate substantially increases. Dependence of detection characteristics on the values of V_{G2} is shown in Fig. 9.26: The rate of dark switching, W_{DS}, is plotted

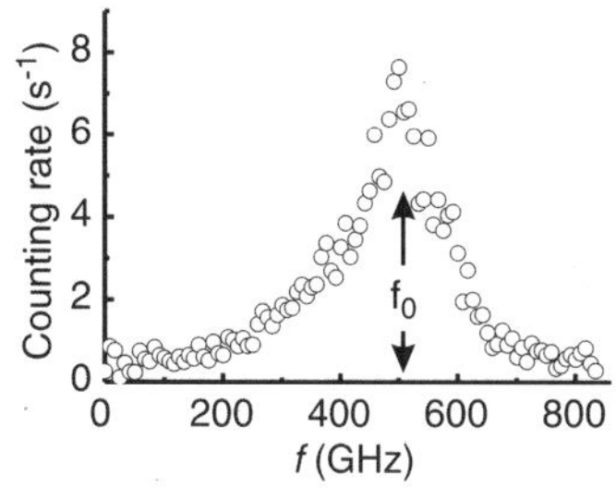

Fig. 9.25. Excitation spectrum in terms of photon count.

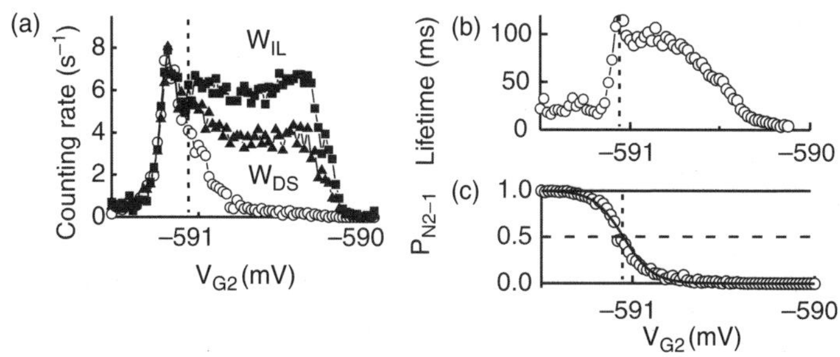

Fig. 9.26. Characteristics of the detector as a function of V_{G2}. The position of $V_{G2}^{t} = -591.05\,\text{mV}$ is indicated by the vertical dashed line in each figure. (a) The rate of conductance switching due to the transition $N_2 \leftrightarrow N_2 - 1$: W_{DS} (open circles), without illumination; W_{IL} (black triangles and black squares), under illumination of $P_{em} = 40$ and $80\,\mu$W. (b) The lifetime of the ionized state, which is the state with N_2 electrons for $V_{G2} < V_{G2}^{t}$ and the state with $N_2 - 1$ electrons for $V_{G2}^{t} < V_{G2}$. (c) The probability, P_{N_2-1}, of finding D2 in the state with $N_2 - 1$ electrons in the dark condition: Experimental data points (white circles) and theoretical values (the solid line: Eq. (15) with $T = 95\,\text{mK}$).

against V_{G2}, and compared with the switching rate, W_{IL}, under illumination at two different levels of radiation intensity in Fig. 9.26(a). The recombination lifetime τ_L is plotted against V_{G2} in Fig. 9.26(b). The photo response, $W_{IL} - W_{DS}$, is visible only on the higher-V_{G2} side to V_{G2}^{t} ($\mu_2 < E_F$), where the response, $W_{IL} - W_{DS}$, is nearly constant up to $V_{G2} \approx -590.4\,\text{mV}$. This characteristics is accounted for by the V_{G2}-dependence of τ_L (Fig. 9.26(b)). The behavior of τ_L against V_{G2} can be explained by considering the energy difference of D2 between the state with $N_2 - 1$ electrons and the one with N_2 electrons, $\Delta E = E_{N_2-1} - E_{N_2} = \alpha e(V_{G2} - V_{G2}^{t})$, where $\alpha = E_{C2}/e\Delta V_{G2} \approx 0.1$ is derived from the experiments [26]. Similarly to the B-dependence of ΔE (Fig. 9.20(b)), ΔE drops to zero at $V_{G2} = V_{G2}^{t}$ and linearly increases with $V_{G2} > V_{G2}^{t}$. Correspondingly, the recombination lifetime, τ_L, increases discontinuously at $V_{G2} = V_{G2}^{t}$ and smoothly decreases with increasing $V_{G2} > V_{G2}^{t}$ until the next transition point is reached.

It is important to clarify physical origin of the dark switching since W_{DS} is a key parameter of photon detectors. In the thermal equilibrium state the probability, P_{N_2-1}, of finding D2 in the state of $N_2 - 1$ electrons should vary with V_{G2} as

$$P_{N_2-1}(V_{G2}) = \left[1 + \exp\{\alpha e(V_{G2} - V_{G2}^{t})/kT\}\right]^{-1} \tag{15}$$

In the experiments, P_{N_2-1} at a given value of V_{G2} can be derived from a conductance time trace taken in the dark condition, exhibiting telegraph-like dark switching. Open circles in Fig. 9.26(c) represent experimentally derived values of $P_{N_2-1}(V_{G2})$, which are in excellent agreement with the theoretical values (the solid line) drawn according to Eq. (15) with $T \approx 95$ mK. The temperature, $T \approx 95$ mK, obtained by the fitting procedure is close to the electron effective temperature, $T \approx 70$ mK, determined by the line shape of Coulomb conductance peak. The dark switching is thus intrinsic and ascribed to the thermal ionization of D2. Hence the best performance of the detector is achieved when the detector is operated at the largest value of V_{G2} that yields negligibly small dark switching rate along with a sufficiently long τ_L.

6. Detector Performance

One of the most important parameters of photon counters is the quantum efficiency, η, which represents the fractional ratio of detected photons to the total incident photons. This quantum efficiency along with the dark switching rate, W_{DS}, determine the effective sensitivity of photon detectors.

In both detector schemes (sections 4 and 5) the incident FIR/SMMW radiation is coupled to small QDs through a planar dipole bow-tie antenna, formed by metal gates. In the described experiments, the antenna is placed on top of GaAs/$Al_xGa_{1-x}As$ heterostructure crystal (dielectric constant: $\varepsilon = 12$) and the radiation is introduced from the front surface of the crystal. When radiation falls from vacuum onto a dielectric medium of the dielectric constant of $\varepsilon \gg 1$, as in the present experiments, most of the radiation power is reflected: The power ratio of the reflected part to the penetrating part is $\varepsilon^{3/2}$, which amounts to $\varepsilon^{3/2} \approx 40$ for $\varepsilon = 12$. Thus, the layout of the present detectors is not optimum and limits the quantum efficiency to be $\eta < 2\%$, though substantial improvement may be straightforward.

Another factor determining the quantum efficiency is the impedance matching of antenna to the QD, $1-[(Z_a - Z_{QD})/(Z_a + Z_{QD})]^2$, where Z_a and Z_{QD} are, respectively, the impedances of an antenna and a QD. Regardless of particular excitation mechanism, the complex impedance of a QD, Z_{QD}, in resonance with incident radiation is purely real value. The impedance is roughly estimated to be $Z_{QD} = 1/(e\mu n_s) \approx 30\,\Omega$ in the experiments described in this chapter, where μ and n_s are, respectively, the mobility and the carrier density of the 2DEG. Since, $Z_a \approx 100\,\Omega$ is expected, the antenna matching is reasonably high, being $1 - [(Z_a - Z_{QD})/(Z_a + Z_{QD})]^2 \approx 0.7$. The total quantum efficiency, η, is estimated to be on the order of $\eta = 1\%$, even in the present non-optimized schemes.

Once η and W_{DS} are known, it is worth while connecting them to an effective sensitivity in terms of the Noise Equivalent Power (NEP). Suppose that a radiation of power P with the photon energy $h\nu$ is incident on a photon detector. A detector catches N photons over time duration T, in addition to the dark switching current-spikes at rate Γ_n. Let us assume that the detector yields short current spikes at times t_k, when positive switches occur. In such a case, the photoresponse current can be simplified to a form

$$i(t) = q_0 \sum_{k=1}^{\infty} \delta(t - t_k) \tag{16}$$

The number of counted photons within time interval T is

$$N = \int_0^T \frac{i(t)}{q_0}\, dt = \int_0^T \sum_{k=1}^{\infty} \delta(t - t_k)\, dt \tag{17}$$

To derive NEP, power detection regime should be considered, where the current is transmitted through a band pass filter with a band width Δf much narrower than the average switching rate $\Gamma = \eta P/h\nu$. $i(t)$ is now averaged over time and yields

$$I(t) = q_0 \Gamma = q_0 \eta P(t)/h\nu \tag{18}$$

This current can be detected if it exceeds the shot noise or the dark switching current, which is written as

$$I_n = q_0 \sqrt{2\Gamma_n \Delta f} \tag{19}$$

Equation (19) is derived by noting that the dark switching current $i_n(t)$ is written in a form similar to Eq. (16) and that the spectral density of the square current deviation $<i_n^2> - <i_n>^2$ is $2q_0^2\Gamma_n$. The minimal detectable power $P_{\min}$ can be found by equating $I_{\min}(t) = I_n(t)$, which gives the final result

$$\mathrm{NEP} = \frac{P_{\min}}{\sqrt{\Delta f}} = \frac{h\nu\sqrt{2\Gamma_n}}{\eta} \tag{20}$$

Equation (20) suggests an NEP on the order of 10^{-21}–10^{-20} W/Hz$^{1/2}$ for the CR detector of FIR (section 4), where $W_{DS} = \Gamma_n = 0.001\,\mathrm{s}^{-1}$ at $B \approx 4.0\,\mathrm{T}$, $\eta \approx 0.01$, and $h\nu \approx 6\,\mathrm{meV}$. Similar sensitivity ($10^{-21}$–$10^{-20}$ W/Hz$^{1/2}$) is suggested also for the double QD detector (section 5), where $\Gamma_n \approx 0.01\,\mathrm{s}^{-1}$ and $\eta \approx 0.01$ with $h\nu \approx 2\,\mathrm{meV}$. These values of NEP are better than the best values of conventional detectors available in the FIR/SMMW region (10^{-18}–10^{-19} W/Hz$^{1/2}$) [46].

We want to stress that η and NEP of the present detectors can be significantly improved by placing a small semiconductor half-spherical lens on top of the crystal surface and passing the radiation through the lens.

7. Conclusion

Two schemes of single-photon detection have been described. In the first-type detector, a semiconductor QD in high magnetic fields is used. At high magnetic fields ($2 < \nu < 3$), different LLs form regions of the inner core (LL2↑), the inner ring (LL1↓) and the outer ring (LL1↑). These regions are equivalent to electrically isolated metallic regions, which form a capacitance network. Absorption of a FIR-photon via a plasma coupled cyclotron resonance in a wavelength range $\lambda = 0.17$–0.20 mm leads to the addition of one extra electron to the inner core (LL2$\uparrow$) and the removal of an electron from the inner ring (LL1$\downarrow$). The FIR-induced electron transfer causes an internal polarization of the QD, which switches on or off the SET conductance through the QD. The lifetime of the internal polarization is long enough for each absorption event to be detected individually in the form of a separated conductance switch, realizing single FIR-photon detection. The lifetime increases with decreasing the size of the core region in the QD, reaching 20 min when just a few electrons remain in the core region.

In this type of detector, the spectral band is tunable by magnetic field, but the tunable range is not very large because the magnetic field should be in a region not far from $\nu = 2$ $(2 < \nu < 3)$. In practice, however, the wavelength range may be expanded by using wafers with different values of sheet electron density, the required magnetic-field region, $2 < \nu < 3$, can be shifted by varying the electron density. Moreover, the electron density can be controlled by an additional gate, so that one device may cover a wider spectral region.

In the second type of single-photon counters, photons at microwave frequencies ($f = 600 \pm 120$ GHz) are detected. The detector uses a single electron transistor formed by two capacitively coupled parallel quantum dots. A point contact separating the double QDs allows prompt escape of an excited electron from one of the QDs to another, but the Coulomb blockade effect prevents the recombination process. The resulting long-lived photo-induced ionization of the QD is detected as a change in the SET current.

The scheme of the second-type photon detection provides us with a wide variety of design capabilities for future developments. For instance, an all-silicon device may be designed. A metal junction SET on top of a semiconductor QD may yield better performance of photon counting. The wavelength range may be largely expanded by utilizing quantum levels in QDs of different sizes.

We have demonstrated that single-photon detectors can be developed in the FIR/SMMW region by using SET. Spectral region of the photon counters is, however, still limited at present. It may not be a fantasy to think of a series of photon detectors which cover the entire region of electromagnetic spectrum spanning the mid-infrared, far-infrared, and SMMWs. Though much work has to be done before seeing them, those photon counters may open up a number of new directions of research.

References

1. I.O. Kulik and R.I. Shekhter, *Zh. Eksp. Teor. Fiz.* **68**, 623 (1975); *Sov. Phys. JETP* **41**, 308 (1975).
2. T.A. Fulton and G.J. Dolan, *Phys. Rev. Lett.* **59**, 109 (1987).
3. M.H. Devoret and H. Grabert, *Single Charge Tunneling* (eds. H. Grabert and M.H. Devoret), Plenum, New York, 1–19 (1992).
4. L.L. Sohn, L.P. Kouwenhoven, and G. Schön, *Mesoscopic Electron Transport* (eds. L.L. Sohn et al.), Kluwer Academic Publishers, Dordrecht, The Netherlands, 105–214 (1997).

5. K.K. Likharev and A.B. Zorin, *J. Low Temp. Phys.* **59**, 347 (1984).
6. D.V. Averin and K.K. Likharev, *J. Low Temp. Phys.* **62**, 345 (1986).
7. P. Lafage, P. Joyez, D. Esteve, C. Urbina, and M.H. Devoret, *Nature* **365**, 422 (1993).
8. Y. Nakamura, C.D. Chen, and J.S. Tsai, *Phys Rev. Lett.* **79**, 2328 (1997).
9. S. Tarucha, D.G. Austing, T. Honda, R.J. van der Hage, and L.P. Kouwenhoven, *Phys. Rev. Lett.* **77**, 3613 (1996).
10. K. Ono, D.G. Austing, Y. Tokura, and S. Tarucha, *Science* **297**, 1313 (2002).
11. M.J. Yoo, T.A. Fulton, H.F. Hess, R.L. Willet, L.N. Dunkleberger, R.J. Chichester, L.N. Pfeiffer, and K.W. West, *Science* **276**, 579 (1997).
12. A. Yacoby, H.F. Hess, T.A. Fulton, L.N. Pfeiffer, and K.W. West, *Solid State Commun.* **111**, 1 (1999).
13. G. Finkelstein, P.I. Glicofridis, R.C. Ashoori, and M. Shayagan, *Science* **289**, 90 (2000).
14. T. Takahashi, A. Fujiwara, K. Yamazaki, H. Namatsu, K. Kurihara, and K. Murase, *Jpn. J. Appl. Phys.* **38**, 2457 (1999).
15. H. Okada and H. Hasegawa, *Physica* B **272**, 123 (1999).
16. J. Motohisa, F. Nakajima, T. Fukui, W.G. van der Wiel, J.M. Elzerman, S. De Franceschi, and L.P. Kouwenhoven, *Appl. Phys. Lett.* **80**, 2797 (2002).
17. Y. Nakamura, Yu.A. Pashkin, and J.S. Tsai, *Nature* **398**, 786 (1999).
18. A.J. Shields, M.P. O'Sullivan, I. Farrer, D.A. Ritchie, M.L. Leadbeater, N.K. Patel, R.A. Hogg, C.E. Norman, N.J. Curson, and M. Pepper, *Jpn. J. Appl. Phys.* Part 1 **40**, 2058 (2001).
19. A. Fujiwara, K. Yamazaki, and Y. Takahashi, *Appl. Phys. Lett.* **80**, 4567 (2002).
20. A.N. Cleland, D. Esteve, C. Urbina, and M.H. Devoret, *Appl. Phys. Lett.* **61**, 2820 (1992).
21. J.M. Hergenrother, M.T. Tuominen, J.G. Lu, D.C. Ralh, and M. Tinkham, *Physica* B **203**, 327 (1994).
22. S. Komiyama, O. Astafiev, V. Antonov, T. Kutsuwa, and H. Hirai, *Nature* **403**, 405 (2000).
23. O. Astafiev, V. Antonov, T. Kutsuwa, and S. Komiyama, *Phys. Rev.* B **62**, 16731 (2000).
24. V. Antonov, O. Astafiev, T. Kutsuwa, H. Hirai, and S. Komiyama, *Physica* E **6**, 367 (2000).
25. S. Komiyama, O. Astafiev, V. Antonov, H. Hirai, and T. Kutsuwa, *Physica* E **7**, 698 (2000).
26. O. Astafiev, S. Komiyama, V. Antonov, and T. Kutsuwa, *Phys. Rev.* B **65**, 085315 (2002).
27. O. Astafiev, S. Komiyama, and T. Kutsuwa, *Appl. Phys. Lett.* **79**, 1199 (2001).
28. O. Astafiev, S. Komiyama, T. Kutsuwa, V. Antonov, Y. Kawaguchi, and K. Hirakawa, *Appl. Phys. Lett.* **80**, 4250 (2002).
29. F. Hofmann, T. Heinzel, D.A. Wharam, J.P. Kotthaus, G. Böhm, W. Klein, G. Tränkle, and G. Weimann, B **51**, 13872 (1995).
30. Y. Kawano, Y. Hisanaga, and S. Komiyama, *Phys. Rev.* B **59**, 12537 (1999).
31. Y. Kawano, S. Komiyama, and K. Hirakawa, *Physica* B **298**, 33 (2001).
32. S. Komiyama and H. Nii, *Physica* B **184**, 7 (1993).
33. C.J.P.M. Harmans and C.T. Foxon, *Phys. Rev.* B **55**, 9746 (1997).
34. K.L.I. Kobayashi, K.F. Komatsubara, and E. Otsuka, *Phys. Rev. Lett.* **30**, 702 (1973).
35. E. Gornik, M. Müller, and F. Gaderer, *Infrared Phys.* **16**, 109 (1976).
36. O. Astafiev, V. Antonov, T. Kutsuwa, and S. Komiyama, *Phys. Rev.* B **65**, 085315 (2002).
37. V. Fock, *Z. Phys.* **47**, 446 (1928).
38. D. Heitmann and J.P. Kotthaus, *Phys. Today* **46**, 56 (1993).
39. A.L. Fetter, *Phys. Rev.* B **33**, 5221 (1986).
40. L.D. Landau and E.M. Lifshits, *Electrodynamics of Continuos Media, 2nd edn* Pergamon Press, Oxford (1963), p. 29.

41. Y.A. Firsov and V.L. Gurevich, *Zh. Esp. Teor. Fiz.* **41**, 512 (1961); *Sov. Phys. JETP* **14**, 367 (1962).
42. L. Brey, N.F. Johnson, and B.I. Halperin, *Phys. Rev.* B **40**, 10647 (1986).
43. R. Krahne, V. Gudmundsson, C. Heyn, and D. Heitmann, *Phys. Rev.* B **63**, 195303 (2001).
44. D.L. Maslov, Y.B. Levinson, and S.M. Badalian, *Phys. Rev.* B **46**, 7002 (1992).
45. S. Komiyama, H. Hirai, M. Ohsawa, Y. Matsuda, S. Sasa, and T. Fuji, *Phys. Rev.* B **45**, 11085 (1992).
46. P.H. Siegel, Terahertz technology, *IEEE Trans. micro. theory tech.* **50**, 910 (2002).

Chapter 10

Quantum-Dot Cellular Automata

C. S. Lent[1], G. L. Snider[1], G. Bernstein[1], W. Porod[1], A. Orlov[1], M. Lieberman[2], T. Fehlner[2], M. Niemier[3], and P. Kogge[3]

[1]*Department of Electrical Engineering, University of Notre Dame, Notre Dame, IN 46556*
[2]*Department of Chemistry and Biochemistry, University of Notre Dame, Notre Dame, IN 46556*
[3]*Department of Computer Science and Engineering, University of Notre Dame, Notre Dame, IN 46556*

1. Introduction

1.1. Needed: A New Device Paradigm for the Nanoscale

The first digital electronic computers were the result of two very good ideas: first, use binary numbers to represent information mathematically; and second, *physically* represent the binary "1" and "0" as the "on" and "off" states of a *current switch.* Konrad Zuse in the 1930s first used electromechanical relays as the current switches, and later changed to vacuum tube triodes. These were eventually replaced by the solid-state version, the semiconductor transistor. Modern CMOS involves a clever use of switches paired so that current flows only when the state of the pair changes. Representing binary information by turning current switch on or off has been one of the most fruitful ideas in the history of technology.

This current-switch paradigm does, however, have serious limitations as device sizes are reduced. As a switch shrinks, it becomes less able to turn the current off and on cleanly. Also, since the current through a single switch is small, it takes longer to charge the interconnect lines between devices. Charge quantization leads to large statistical current fluctuations. Finally, since electrons move from the power supply to ground, considerable energy dissipation occurs.

These limitations have come more fully into view as the shrinking of CMOS technology has continued its remarkable progress. Such fundamental considerations will ultimately limit the device densities attainable with transistors, although the precise trajectory of the microelectronics industry may be determined more by fabrication cost issues. One need not be committed to a particular forecast to see the growing importance of developing alternative approaches that would permit scaling electronics down to the ultimate limits of molecular dimensions.

The quantum-dot cellular automata (QCA) concept [1–15] involves keeping *one* of Zuse's ideas, using a binary representation of information, but replacing the current switch with a cell having a bistable charge configuration. One configuration of charge represents a binary "1," the other a "0," but no current flows into or out of the cell. The field from the charge configuration of one cell alters the charge configuration of the next cell. Remarkably, this basic device–device interaction, coupled with a clocking scheme for modulating the effective barrier between states, is sufficient to support general-purpose computing with very low power dissipation [15]. QCA devices exist and multi-device circuits have been demonstrated at low temperatures [16–34]. Work underway is aimed at developing a room-temperature molecular implementation of the QCA concept [35–41]. The QCA approach, using molecules as structured charge containers, is a more natural match to molecular function than trying to use molecules as current switches.

1.2. The Physical Representation of Information

QCA represents a particular choice of how to represent information physically—use the charge configuration of a cell. It is helpful to compare this choice with alternatives, particularly at the molecular device level. Choices include representing information by:

1. Electronic charge configuration.
2. Nuclear positions.
3. Electronic spin state.
4. Nuclear spin.
5. Collective magnetic moment.
6. Coherent electronic quantum state.
7. Superconducting ground state.

Approach (2) has proven useful for making molecular memory by switching the conformational state of a molecule and sensing the state by resistance changes. This is effective for memory but inherently slow and limited in scaling by the size of the current contacts. Coherent quantum computing (QC) has used a combination of (4) and (6) in the gas phase and (3), (4), and (6) in some solid state proposals. The weakness is the well-known problem of decoherence. Since QC required isolated unitary time-evolution, but also measurement of the output, this problem appears fairly fundamental. Superconductor-based QC devices have combined (6) and (7), but are limited in temperature ($T < T_c$) and size (must have enough carriers to form BCS ground state condensate—not single molecules.). Making necessarily generous assumptions of a long decoherence times, it has been shown that a coherent version of QCA is a computationally complete QCA system [14], but that does not solve the general problems of QC. Direct spin–spin coupling is too weak to support binary (non-QC) versions of QCA based on (3) or (4). Collective magnetic effects, while also limited in scaling, (5), are the basis of conventional memory systems, and a magnetic QCA logic approach is described below. While not small, magnetic systems have the advantage of extremly high coupling energies. For molecular electronics, the combined features of speed, single-molecular size, and robust general-purpose computing argue for using approach (1).

2. The Quantum-Dot Cellular Automata Paradigm

QCA cells. An idealized QCA cell can be viewed as a set of four charge containers, or "dots", positioned at the corners of a square (Fig. 10.1(a)). The cell contains two extra mobile electrons which can quantum-mechanically tunnel between dots but, by design, cannot tunnel between cells. The dots can be realized in a number of ways: electrostatically formed quantum dots in a semiconductor, small metallic islands connected by tunnel junctions, or redox centers in a molecule. The barrier between dots should be high enough so that charge can move only by tunneling and is therefore localized in the dots and not in the connectors. The configuration of charge within the cell is quantified by the cell polarization, which can vary between $P = -1$, representing a binary "0," and $P = +1$, representing a binary "1," as illustrated in the figure.

QCA circuits. Though the potential of the QCA concept extends beyond Boolean circuits, it is important that the approach be rich enough to support any Boolean function. Circuits can be created by putting QCA cells in proximity to each other. A QCA *binary wire* [2] is formed simply by creating a linear array of cells as shown in Fig. 10.1(b). The Coulomb interaction makes nearby cells align in the same state. The corner interaction

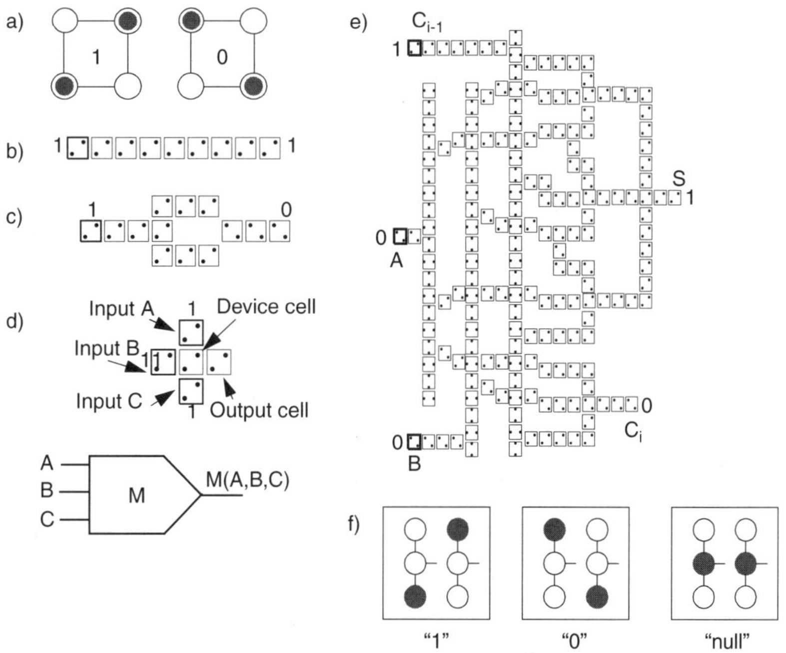

Fig. 10.1. QCA devices formed from cells. (a) Schematic of idealized QCA cells. (b) QCA inverter. (c) QCA majority gate. (d) QCA full adder. (f) Schematic of clocked QCA.

is anti-voting so it can be used to make an inverter (Fig. 10.1(c)). The natural logic *gate* is the three-input majority gate [6] shown in Fig. 10.1(d). We have shown through extensive simulations that larger circuits can be designed using hierarchical design rules. A full *adder* (Fig. 10.1(e)) has been simulated [10] using the full self-consistent Schrödinger equation, verifying that the adder works for all input possibilities.

Clocked QCA. QCA has an architectural future beyond single devices or small-scale circuits. For complex circuits it is useful to be able to clock the cells. Clocking [8,10–12] consists of controlling the activity of the cell by effectively raising and lowering the interdot barriers through a central dot. If barriers are low, the charge resides in the center (see Fig. 10.1(f)) and we say the cell is in a "null" state and holds no information. As the effective interdot barriers are raised, the cell charges localize in either the "1" or "0" state, performing the logical function determined by their layout. High barriers lock the state of each cell; locked cells can be used as input for the next stage of the computation. A large circuit can be broken up into smaller sub-arrays or clocking zones. Since each clocking zone is composed of several cells, it is not necessary to have individual control leads to each cell. Clocking signals for molecular arrays can be distributed by metal lines underlying the array, with a feature size much larger than the cells. Clocked QCA cells exhibit signal power gain, essential to restoring logic levels in real circuits [15,34].

2.1. Key QCA Features

Logical completeness. The three-input majority gate can be considered a programmable AND/OR gate. When combined with the NOT operation, these form a logically complete group. Both a majority gate and an inverter have been fabricated and tested with metal-dot cells.

Restoring logic. The nonlinear response of one cell to the polarization of its neighbors plays the role of voltage gain in conventional devices. This helps restore signal levels down the wire and gives a level of immunity to variations in cell position, orientation, and perfection. This nonlinearity has been confirmed experimentally [28].

Hierarchical design. The electric quadrupole–quadrupole interaction between two cells drops rapidly as the distance between the cells is increased. Near-neighbor and next-near-neighbor interactions strongly dominate, so that hierarchical circuit design is possible. In most instances, one can analyze and design circuit behavior by considering local device groups independently.

Clocking. Clocking groups of cells enables complete control of the direction of information flow. QCA shift registers become the paradigm for moving information from one place to another. Clocking schemes for molecular QCA have also been proposed which use applied fields to move charge to "null" positions. Clocked cells have been experimentally demonstrated [22,32,34].

Memory. The binary information stored in a QCA cell can be held by the clock signal and copied to neighboring cells. Addressable memory cell arrays have been designed and simulated. A single-bit QCA memory has been demonstrated experimentally [32].

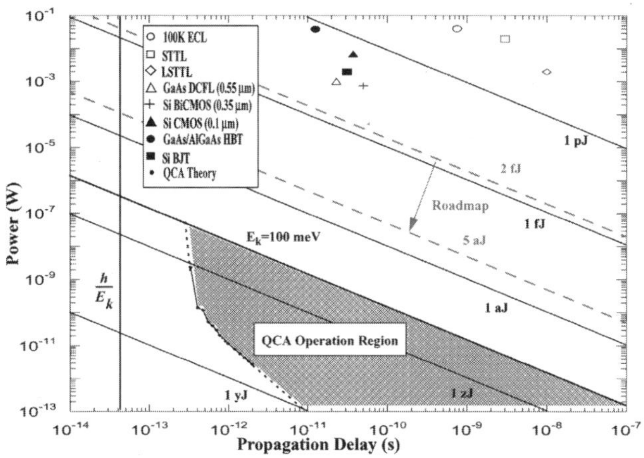

Fig. 10.2. Power-delay product for QCA.

Low power dissipation. By using ramped clock signals, QCA cells can be switched quasi-adiabatically. This means that cells are kept near their instantaneous ground states during the switching process. This idea, developed by Landauer and Keyes 30 years ago [42], enables computation at the theoretical limits of low power dissipation (Fig. 10.2). This is crucial because, at molecular densities, the problem of removing heat dissipated by the devices limits both achievable densities and speed. Fig. 10.2 shows the results of a calculation of QCA power dissipation in molecular circuits. At densities of $10^{13}/\text{cm}^2$ and switching speeds of 10 GHz power dissipation should be less than 10 W/cm^2. It will not melt.

Power gain. When one device causes another to switch, there is inevitable loss of energy to the environment through irreversible processes (phonons, plasmons, molecular vibrations,...). For signals to continue to propagate, this lost signal energy must be restored. In conventional transistors, this energy comes from the power supply. In clocked QCA circuits, this energy comes from the clock itself. QCA circuits can therefore exhibit true signal power gain. This has been shown theoretically [15] and has recently been observed experimentally [34].

3. Experimental Demonstrations of QCA: Metal-Dot Systems

Though the original proposal of QCA [1] considered the cells composed of generic quantum dots populated with two electrons, Lent and Tougaw later suggested [5] the use of metal dots separated by aluminum oxide tunnel junctions (TJ) for QCA fabrication. The Fermi sea of electrons in the metal dots is electrically neutralized by the ions of the crystalline lattice. Hence, for the QCA cell to operate, it must be charged with two excess, noncompensated electrons, and the neutral background may be disregarded.

There are several important advantages of the metal TJ QCA system compared with the more-common semiconductor quantum dot technology. First, many uniform tunneling barriers that form the dots can be produced using metal TJ technology. The fabrication process is relatively simple, with only three major processing steps: electron beam lithography (EBL), development, and metal deposition through a suspended mask [43] with in-situ oxidation. Finally, high yield of the fabricated devices allows the realization of rather complex circuits. In comparison, the fabrication of semiconductor dots requires much more complex processing (typically more than 10 fabrication steps), each tunneling barrier usually requires extra gates to operate, and it is difficult to produce a large number of the quantum dots with similar parameters due to unavoidable influences of impurities and defects.

One disadvantage of the devices produced by the metal TJ process is the smaller charging energy, E_C, of the Al "dots" ($\sim$1 meV) compared to the semiconductor quantum dots ($E_C \sim 10$ meV [44]). This results from the larger barrier (TJ) capacitance, which in turn is defined by the resolution of EBL. To prevent temperature smearing of charge quantization, the condition $E_C \gg k_B T$ (where k_B is Boltzmann's constant and T the absolute temperature) must be satisfied, limiting the operational temperature to below 4 K. (Recently, single electron devices fabricated using metal TJ techniques operating at room temperature were reported [45]. However, these devices are single metal dots, weakly coupled to external leads, which prevents effective coupling between the dots necessary for QCA cells).

It is important to point out that the major advantage of the QCA paradigm compared with conventional FET-based logic is the improvement in performance as the device sizes shrink. As appropriate technologies become available, it should be possible to make complex QCA arrays operating at higher temperatures (>300 K) with much better performance in speed (picosecond switching times) and reliability than existing metal TJ prototypes.

3.1. Fabrication and Electrical Measurement

Fabrication of Al/AlO_x/Al tunnel junctions is accomplished using standard electron-beam lithography procedures with double layer resist (PMMA/MMA) and double angle shadow evaporation [43] of Al on Si or Si/SiO_2 substrates. The bottom electrode metal is oxidized *in situ* by introducing oxygen into a deposition chamber, followed by a second deposition of Al to form the top electrode. The resistance of junctions with sizes from $30 \times 30\,\text{nm}^2$ to $60 \times 60\,\text{nm}^2$ varies between 30 kΩ and 2 MΩ at room temperature, depending on the overlap area and oxidation conditions.

Junction resistances should be large enough to fulfill the condition $R_J > R_Q$, where $R_Q = h/e^2$ (h is Plank's constant, and e is the electron charge) in order to prevent quantum charge fluctuations [46]. Junction capacitance, C_J, is in the range of 100 to 200 aF. The "dots" of the QCA cells are formed by Al islands with a length of 1–3 μm and width of 50–100 nm. To provide a means for electrical measurements, Ti/Au wires and bonding pads are fabricated on the substrate by optical lithography. The samples are then glued into chip carriers and gold wires are bonded to connect the chip carrier to the bonding pads. Measurements are performed in the dilution refrigerator with a base temperature of about 15 mK. To suppress the superconductivity of Al, a magnetic field of 1T is applied to the sample.

Critical to the implementation of QCA is a means of detecting the positions of individual electrons in the output cells. The detector must, therefore, be capable of measuring changes of charge corresponding to the removal or addition of a single electron to the nearby dot. The most natural detection scheme for this application is a single-electron tunneling transistor (SET) [46] situated in close proximity to the dot experiencing the charge change, and coupled to the dot by a non-leaky capacitor. The detailed operation of the SET-electrometer was first presented by Lafarge et al. [47], and experiments detailing the use of electrometers with QCA cells are presented elsewhere [19–21].

3.2. QCA Majority Logic Gate

Binary logic operations in QCA cells are implemented using geometrical arrangements of cells. The fundamental QCA logic device is a three-input majority logic gate (Fig. 10.3), consisting of an arrangement of five standard cells—a central logic cell, three inputs labeled A, B, C, and an output cell. The polarization states of inputs A, B, and C determine that of the logic cell, which is defined by the majority of the three input cells. The output cell polarization follows that of the logic cell. QCA logic gates can be cascaded, so that in a more complex QCA circuit the three inputs would be driven by the outputs of previous gates. Similarly, the output of the majority gate can be connected to drive a subsequent stage of logic gates.

A majority gate can be programmed to act as either an OR gate or an AND gate by fixing one of the three inputs as a program line. If the programming input is a 0 (1), the AND (OR) operation is performed on the remaining two inputs. In our experiments, the cell performing the majority gate function consists of four Al dots, D_1–D_4, connected in a ring by tunnel junctions, as shown in Fig. 10.3(b). In initial biasing of the cell, two excess electrons enter the cell through tunnel junctions, which for simplicity are

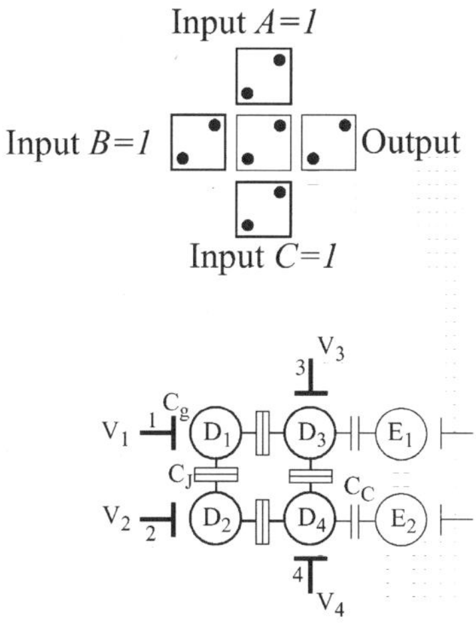

Fig. 10.3. QCA Majority logic gate. (a) The cell in the center is the majority gate, while the surrounding cells are the inputs. (b) Schematic of the QCA majority gate experiment. The cell is defined by dots D_1 – D_4 connected in a ring by tunnel junctions. E_1 and E_2 are the electrometers. External leads and tunnel junctions for the cell and the electrometers are not shown.

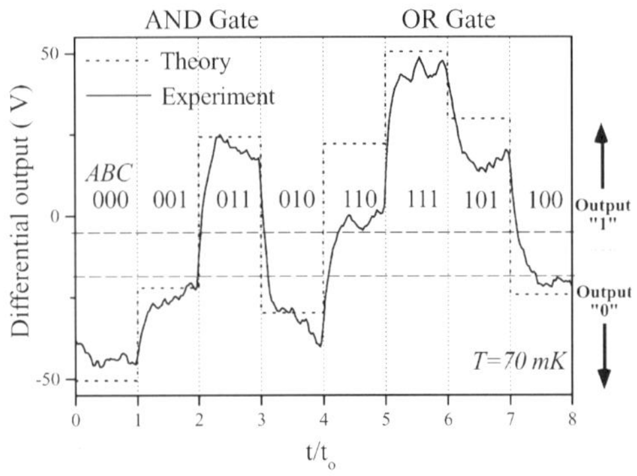

Fig. 10.4. Demonstration of majority gate operation. (a)–(c) Inputs in Gray code. The first and last four inputs illustrate AND and OR operations, respectively. (d) Output characteristic of majority gate where $t_0 = 2_0$ s is the input switching period. The dashed line shows the theory for 70 mK, solid line represents the measured data.

not shown. Each dot is also coupled to a gate, via capacitance C_g, that influences the charge state of its respective dot.

To determine the cell polarization, we measure electrostatic potentials on D_3 and D_4 using electrometers E_1 and E_2. First, the logic cell is biased using gates 1–4 to the "unpolarized" state where logic 1 and 0 are equally probable, and the electrometer outputs are set to 0 V for this condition. This procedure also cancels the effect of the substrate background charge. Differential signals A (between gates 1 and 3), B (between gates 1 and 2), and C (between gates 2 and 4) constitute the inputs to the central cell. The applied bias on a gate mimics the presence of electrons in the input cell, and the amplitudes of are chosen to mimic the potentials due to the polarization of an input cell.

To perform the majority gate experiment the inputs A, B, and C are changed as a function of time to step through all the possible input combinations, and the differential potential between dots D_4 and D_3, $\Phi_{D4} - \Phi_{D3}$, is measured using the electrometers E_2 and E_1. The measured output characteristic, along with theory, is shown in Fig. 10.4. (The transient characteristics are determined by the low cutoff frequency of the data acquisition system, not by the switching time of the cell.) The theoretical results are calculated for the electron temperature in the experiment (70 mK). Though no adjustable parameters are used in the theory, the agreement between the experimental and theoretical results is good. The output high (V_{OH}) and output low (V_{OL}) show a clear separation as required for digital logic. The first and last four input steps are grouped separately, with A as the programming input, to illustrate AND and OR operations. The AND operation is carried out for $A = 0$, for which we see that the output is high only when the remaining two inputs are also high. The OR operation is performed when $A = 1$, for which the output is high when either of the other two inputs is high. This data confirms majority gate operation in QCA, and thus demonstrates a logic gate that requires only two electrons to function.

To achieve better performance for majority gate operation one would need to improve the separation between V_{OH} and V_{OL}. Thermal smearing of the charge states of the dots results in a less than complete polarization of the cell, leading to only small binary level separation. Therefore, the performance of the device could be improved by reducing the total dot capacitance, which will raise the energy of the excited states. For example, when all capacitances are reduced by a factor of 5, the calculated output characteristic shows an increased separation between V_{OH} and V_{OL}. Therefore, future molecular QCA logic gates, for which energy separation between the ground and excited states is many orders of magnitude greater compared to metal TJ prototypes, are expected to yield greater separation in the logic levels at much higher operating temperatures.

3.3. Clocking in QCA

Experiments with QCA cells and majority logic gates [19–31] have demonstrated the basic premise of the QCA paradigm, i.e. coding of information in the position of single electrons. To address issues of metastable states and signal power gain, Lent et al. [8,10,11] suggested a modified version of QCA where the energy is supplied to each cell by means of clock lines, thus achieving major improvements in performance.

Clocking using cyclical manipulation of quantum wells to perform binary operations was first discussed in detail by Keyes and Landauer [42], who credited the original idea to von Neumann from the 1950s. The idea is to use systems that can be transformed continuously and cyclically from a monostable state into a bistable state and back to monostability (MBM clocking scheme). A particle (e.g. an electron) in a time-varying potential in this scheme represents a binary "0" or "1" by the position of the particle (Fig. 10.5). The clocking sequence is RESTORE–SWITCH–HOLD–RESTORE. At the initial phase, RESTORE, the system is in a monostable ground state. During the SWITCH phase, first the input potential is applied, biasing the cell, and then the system is converted from a monostable to a bistable state by altering the potential profile of the well. This work is done by means of an external energy source—the clock. The particular binary state (corresponding to a particle in the left or right well in Fig. 10.5) is selected by the external input, which creates an asymmetry in the emerging double-well and thus forces the particle to choose either the left or right side. The magnitude of the input signal is small, so that the particle cannot be moved by the input signal alone. By the end of the SWITCH phase, binary information becomes stored in the system. During the HOLD stage the information is preserved, and the particle in the well acts as an input for the subsequent stage (e.g. by means of Coulomb interactions).

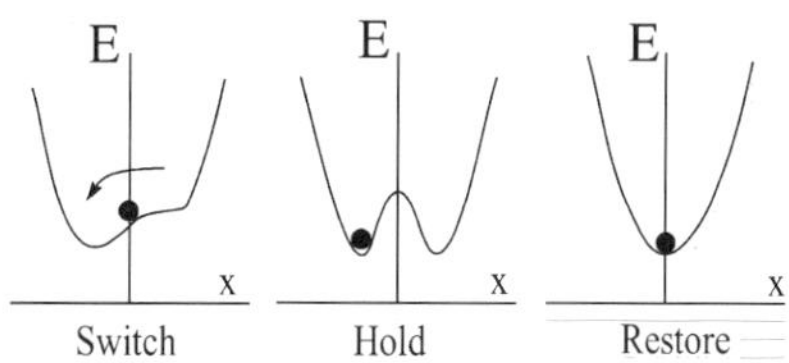

Fig. 10.5. Keyes-Landauer clocking scheme (E-energy, X-coordinate). Black circle represents particle in the potential profile. Binary 1(0) is defined when particle is confined in the left (right) well.

Finally, the system is returned to its initial, monostable state, RESTORE, and is ready to receive a new input.

Using MBM-wells charged with electrons as elementary building blocks, and Coulomb interaction to couple wells, a QCA type architecture could be realized. To implement clocking in QCA, Lent et al. [10] suggested to electrostatically modulate the barriers between quantum dots within the QCA cell. As mentioned previously, the use of single electrons in QCA reduces the power dissipation compared with conventional FETs by many orders of magnitude. Moreover, if the potential profile in Fig. 10.5 changes slowly enough so that the particle remains all the time close to its ground state (quasi-adiabatically), the energy *dissipated* in the process can be lowered to *below kT* ln2 per binary operation (compared to $10^6\,kT$ in contemporary FET-based logic) [15].

Another important advantage of clocked QCA architecture is its inherent ability to realize digital latches. This stems from the fact that the information represented by the position of single electrons within a cell remains stored for as long as the clock signal keeps the barriers high (Hold phase in Fig. 10.5). As a result, the locked electron provides a firm input signal for the next cell. Latches could be used to break a large QCA array into sub-arrays with each sub-array working on different parts of a computational problem (pipelining). Pipelining in QCA resolves the limitation on the size of non-clocked, edge-driven QCA architecture imposed by thermodynamical considerations. By contrast to the edge-driven QCA architecture where the source of energy driving the whole array is the signal inputs, the primary source of energy in clocked QCA is the clock, which acts as a power supply in conventional electronic circuits. In this manner, clocked QCA exhibits power gain needed for logic level restoration [15,34].

For QCA arrays implemented in metal islands coupled by capacitors and oxide tunnel junctions, control over barrier heights is much different from that in semiconductor QCAs because oxide barriers cannot be varied by gate potentials. An implementation of the MBM scheme for the classical Coulomb blockade system with tunnel barriers ("single-electron parametron") was suggested by Likharev et al. in [48] and later refined in [49]. A similar design was used by Toth et al. [11] for clocked control of the switching in metallic QCA cells, which we now discuss.

In this design (Fig. 10.6), a modulated barrier is replaced with an extra dot, D_2, situated between the two end dots, D_1 and D_3. The system, which we call a QCA

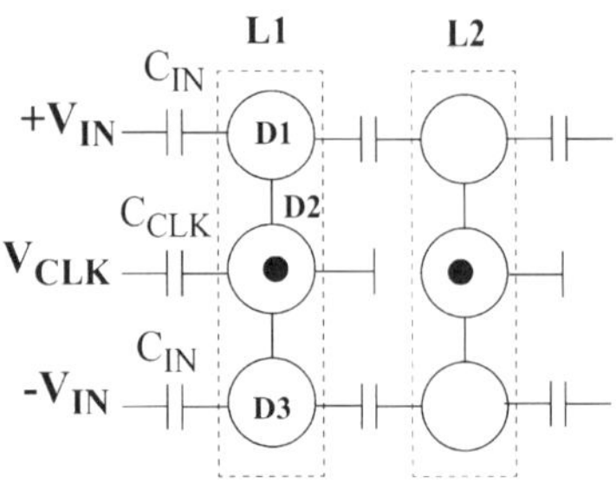

Fig. 10.6. Schematic diagram of a clocked QCA with fixed barriers [21]. The triple dot in the dashed-line box forms a QCA latch. Each latch is pre-charged with one extra electron through the grounded tunnel junction. Two latches form a QCA cell. The null mode is shown.

half-cell, or QCA latch, is precharged with one electron through a TJ connected to ground. Each dot of the device is capacitively coupled to a corresponding gate, and the differential input signal (which comes from either external leads, V_{IN} in Fig. 10.6, or an adjacent latch, e.g. L1 driving L2 in Fig. 10.6) is applied to the top and bottom dots. The clocking signal on the middle gate controls the potential on D_2, varying the effective Coulomb barrier between D_1 and D_3. The three operational modes of such a device, null, active, and locked, are defined by the combination of input and clock biases.

In the null mode, the input signal is zero and clock bias is such that an electron is localized on D_2 (clock LOW). For this combination of gate voltages, the charge state (0,1,0) has the minimal energy configuration. (The numbers in the parentheses are the numbers of excess electrons on D_1, D_2 and D_3, respectively.) In the active mode, a small input signal is first applied to the top and bottom gates. The magnitude of the input signal, ΔV_{IN}, is small, so it cannot lift the Coulomb blockade for tunneling from D_2 to the end dots. The subsequent variation of the clock signal to HIGH forces an excess electron to switch to D_1 or D_3, with the direction of the switching defined by the input signal. As a result of switching, the charge distribution between the dots reaches its minimal energy configuration.

In the locked mode, were an electron to be localized in D_2, the energy of the configuration would be much higher than for the other two configurations in which it is localized on the end dots. As a result, the electron remains trapped on one of the end dots, regardless of the input signal. The excess electron in the locked mode is in the metastable state when the reverse input bias is applied, or in the ground state when input bias stays the same as in the active mode. In either case, the Coulomb barrier created by the clock HIGH prevents an excess electron from switching. Finally, as the clock signal is set back to LOW, the Coulomb barrier is removed, the electron switches back to D_2, and latch returns to the null state.

3.4. *Two Stage QCA Shift Register*

The operation of a basic element in the clocked QCA architecture, a latch, has been successfully demonstrated [32]. Next, the operation of a coupled system where one latch drives the other by means of single electron switching, and powered by the clock, a shift register (SR), is considered. The simplified circuit diagram of the device is shown in Fig. 10.7, and consists of two QCA latches (delineated by dashed lines in Fig. 10.7) and two readout electrometers E_1 and E_2. The two latches are capacitively coupled to each other using lateral capacitors C_C. Each QCA latch consists of three dots D_1–D_3, and D_4–D_6, separated by three tunnel junctions. The multiple tunnel junctions suppress cotunneling [50], increasing the latch hold time. SET electrometers E_1 and E_2 are used to measure the state of each latch.

Operation of the QCA shift register is performed using a two-phase clock (CLK1 and CLK2). The details of the operation of the two-stage SR in the time domain are shown in Fig. 10.8. First, the differential signal V_{IN} corresponding to logical "0" (logical "1") is applied to the inputs V_{IN}^{+}, and V_{IN}^{-} at t_1 (t_7) (Fig. 10.8(a)). As described above, L1 remains in the monostable null state until CLK1 is set HIGH at t_2 (t_8) (Fig. 10.8(b)). When CLK1 is activated, it causes a transition of an electron in L1 (Fig. 10.8(c)). After that the signal input is removed at t_3 (t_9) and the state of L1 no longer depends on the

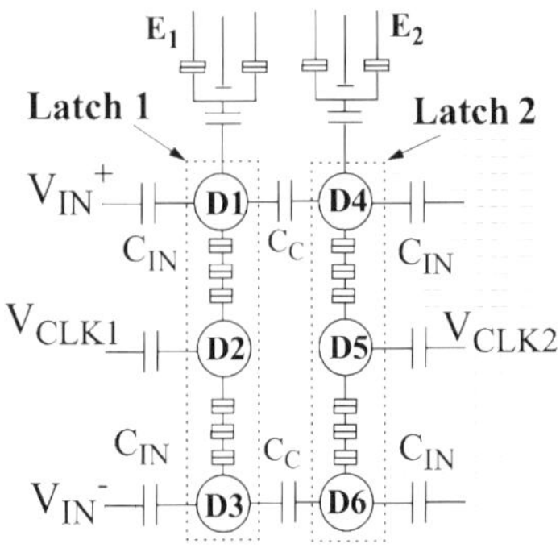

Fig. 10.7. Circuit diagram of a QCA two-stage shift register.

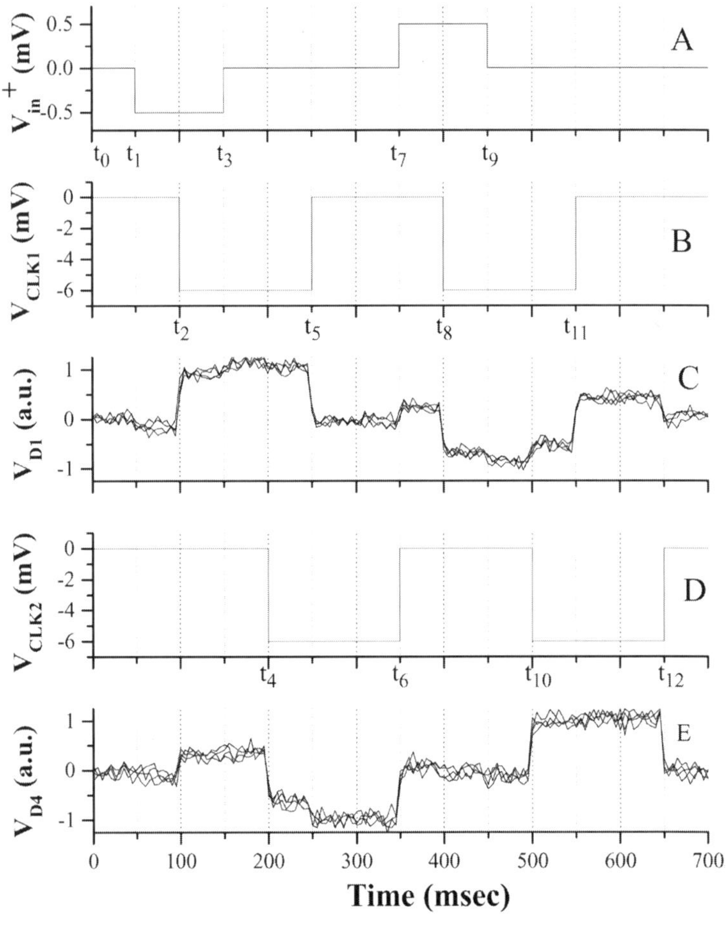

Fig. 10.8. Operation of a QCA shift register. Two phase-shifted clock signals are applied to two capacitively coupled latches to shift binary information from one latch to the next in a sequential manner controlled by the clock. Five successive traces are shown.

input signal. Then CLK2 is applied to L2 at t_4 (t_{10}) (Fig. 10.8(d)), and an electron in L2 switches in the direction determined by the state of L1 (Fig. 10.8(e)). L2 holds the bit after CLK1 is removed at t_5 (t_{11}) for as long as CLK2 is high (until t_6 (t_{12})). The cycle describing the operation of a QCA SR is as follows: monostable → input applied → CLK1 applied and L1 is active → input removed → CLK2 applied and L2 is active → CLK1 is removed. At this time, L1 becomes inactive and is ready to receive new information. The information encoded in the position of a single electron is shifted to L2 and stored there. The QCA SR operates as predicted in [11].

3.5. Experimental Simulation of a Multi-Stage Shift Register

The ultimate utility of a SR would be in large-scale QCA circuits to control the flow of binary information through the circuit. Hence, the performance of a multi-stage SR, especially with regard to preservation of logic levels and vulnerability to digital errors, is of considerable interest. The current device can be used to replicate the propagation of a single bit through such an SR. In a multi-stage SR, a bit is first written into the circuit by the input and then moved along the circuit using each latch as an input to the next (Fig. 10.9(a)). The same situation can be simulated using the two-stage SR by moving the bit back and forth from one latch to the other (Fig. 10.9(b)). Initially, a bit is written into the first latch by the input. Then, using L1 as input, the bit is copied into L2 after which L1 is turned off. Then using L2 as input, the bit is copied back into L1, and L2 is turned off. This process can be repeated a number of times to achieve the same effect as transferring a bit through a long line of latches.

Figure 10.10 shows the timing diagram of the experiment performed for five cycles. Initially, all the signals are zero and the two latches are in the neutral state. Once the input (binary "1") and clock signals are applied to L1, it switches. The input is then removed and the bit is stored in L1. The clock signal is then applied to L2, and it switches using L1 as its input. L1 is then switched off, and the bit is now stored in L2. Instead of applying the clock signal to a third latch in the line, it is applied to L1 which sees L2 as an input and switches accordingly. Then L2 is switched off and the bit is stored once again in L1. This cycle is repeated five times to simulate an SR made of 11 latches. In the second half of the experiment, this scheme is repeated with an input of

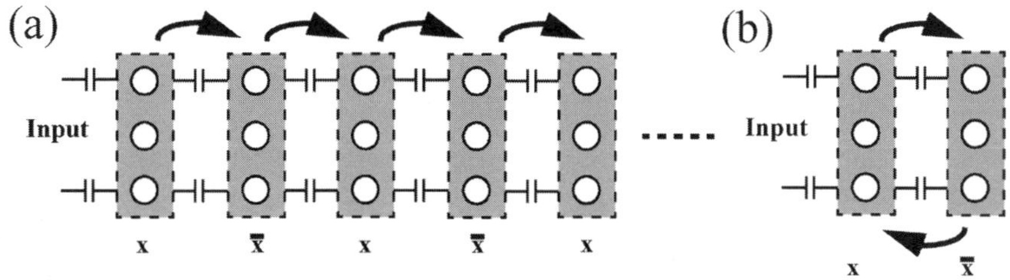

Fig. 10.9. Using a two-stage SR to simulate a multi-stage SR. (a) In a multi-stage SR, a bit is moved sequentially in a single direction from one latch to the next. The bit is inverted at each step. (b) The two-stage SR can be used to simulate a longer SR by moving the bit back and forth from one latch to the other instead of moving it in a single direction.

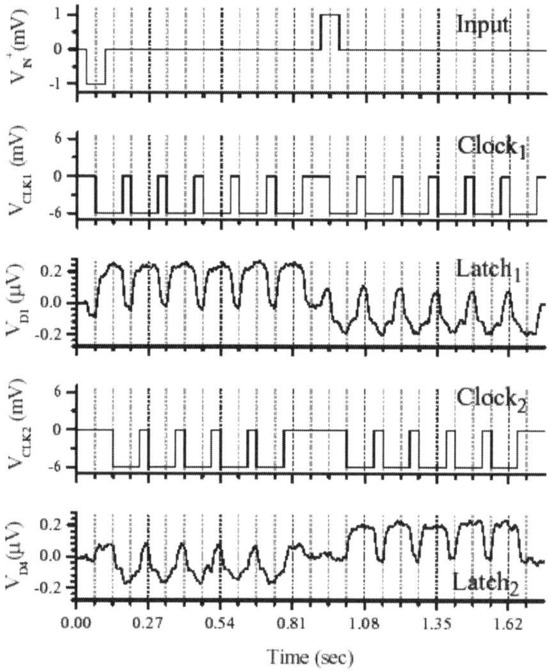

Fig. 10.10. Experiment to simulate a multiple stage SR. The input is applied only at the beginning, to write a bit into the SR. Then the bit is transferred from one latch to the other for five cycles. In the second half of the experiment, the sequence of events is repeated with an input of the reversed polarity.

the opposite sign (binary "0"). This is necessary to verify that bit inversion at the input indeed leads to the bit inversion at the output. The above experiment demonstrates that the direction of the information flow in the circuit is controlled by the sequence of clock signals applied to latches. Another observation that can be made from Fig. 10.10 is that, although the input is applied only once at the beginning of the cycle, we do not see any degradation in the voltage levels as the bit is moved back and forth between the latches, indicating that signal levels would be preserved in multi-stage QCA SRs. The major reason for this stems from the ability of clocked QCA to exhibit power gain [15,34], the clock signal in QCA provides the energy needed for digital level restoration just as a conventional power supply does it for FET circuits.

3.6. Observations

The experiments so far show the feasibility of the QCA architecture, demonstrating functional cells, logic gates, latches and SRs. Future research includes the search for possible QCA cell candidates using metal nanoclusters and molecular complexes, interfacing and assembling of QCA devices. Because the shadow evaporation technique produces capacitances that are still relatively large, these experiments were performed at 70 mK. Molecular devices, by contrast, will operate at room temperature.

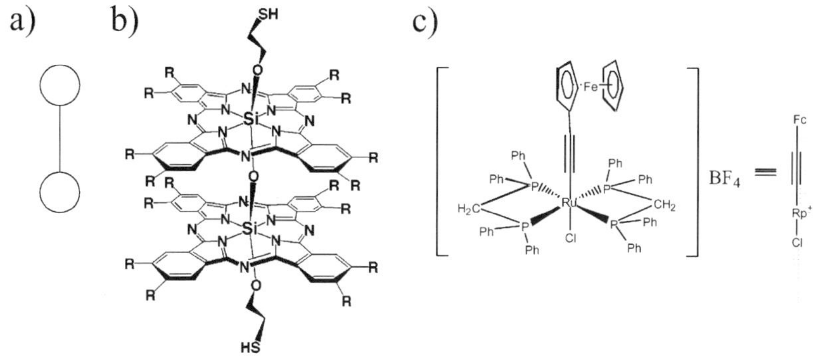

Fig. 10.11. Molecular double-dots. (a) Schematic. (b) Symmetric Silicon phthalocyanine (SiPc) dimer. (c) Dinuclear complex with unsymmetric Fe and Ru centers.

4. Molecular QCA

Creating QCA cells composed of single molecules holds enormous promise for realizing the ultimate limits of electronic device miniaturization and integration. In contrast to present metal-dot cells, the small size of molecules means that Coulomb energies are much larger, so room temperature operation is possible. In contrast to lithographic device fabrication techniques, which always introduce variations in device characteristics, each molecular cell can be made exactly identical using chemical synthesis. Combining molecular QCA with nanoelectronics and molecular/biomolecular sensors opens new possibilities for an integrated molecular technology.

The QCA paradigm is particularly well-matched to the challenge of building single-molecule electronics [35,41]. Encoding binary information in the charge configuration of a molecule is a more natural match to molecular function than pressing molecules into service as current switches. Furthermore the most fundamental obstacle to realizing the full potential of molecular electronics is the need to dramatically lower the power dissipation per device. If we take the ultimate target of molecular electronics to be the development of single-molecule devices, then device densities could approach $10^{14}/\text{cm}^2$ (for 1 nm^2 devices). Any current-switched device at reasonable speeds would melt the chip at those densities. The QCA approach has been shown to be able to reduce power dissipation by many orders of magnitude [15].

QCA molecules. The role of dots in molecular QCA is played by redox centers within the molecule. A redox center can add an electron (be reduced) or loose an electron (be oxidized) without breaking chemical bonds. The redox center is characterized by low-lying nonbonding orbitals. We are particularly interested in two classes of redox centers, those whose nonbonding orbitals are comprised principally of s and p states, and those comprised principally of d states from transition metals. An example of the former is Si-pthalocynanine [51]; an example of the later is the Ru-based Creutz–Taube ion [52]. Molecules with at least two redox centers, between which an electron can move are called mixed-valence compounds, about which a considerable literature exists. It is from the class of mixed-valence compounds that we can seek QCA molecules.

Synthetic approaches based both on transition metals and on π-bonded systems are currently under investigation.

In a pioneering contribution to molecular electronics, Aviram [53] examined a model mixed-valence compound. He considered a particularly simple model system composed of two allyl groups connected by an alkyl bridge. The allyl group is a three-carbon complex which has a low-lying non-bonding π-orbital and acts, in our terms, as a dot. This double-dot system was considered theoretically by Aviram and later by Hush et al. [54]. Aviram proposed a model system in which such a complex could act as the gate of a conducting molecular channel, resulting in a molecular field-effect transistor. Hush and coworkers used the two-allyl molecule to examine the interplay between field effects and the nuclear relaxation that accompanies the charge transfer.

We employ a version of the Aviram double-dot system, shown in Fig. 10.12(a) to illustrate the basic features of molecular QCA. In the singly-charged anion, an electron is localized on one end or the other as shown in the figure, depending on the driver field. The distance between the center charges here is 6.6 Å and switching between the two states corresponds to a change in the molecular dipole moment of 32 Debye. A single molecule can thereby represent a bit of information in a form which can influence its neighbors.

Can one such molecule provide a local driver field strong enough to switch the state of another molecule? To address that question we consider the situation depicted in Fig. 10.12(b). Two Aviram molecules are juxtaposed, parallel to each other at a distance of 7 Å. A driver charge a distance 7 Å away from the first molecule is simply an electron which moves along the axis a distance of 7 Å. This "electron on a rod" mimics the effect of yet another molecule.

The results [55] are shown in Fig. 10.12. The response of the two-molecule cell can be quantified by plotting the induced quadrupole moment as a function of the dipole moment of the single-electron driver [56]. The abrupt switch in the sign of the quadrupole moment corresponds to electrons in each molecule switching positions—the driver switches one molecule which in turns switches the second molecule. The nonlinearity is precisely what is needed for QCA action. Fig. 10.12(c) shows the orbital and potential isosurfaces for the two polarization states. Our calculations show the energy associated with the relaxation of the atomic positions due to switching is much smaller than the switching energy. Figure 10.12(e) shows the energy splitting between the ground-state and the first excited state. The large splitting (about 0.5 eV) suggests that thermal effects would not limit the functioning of even large molecular arrays.

Can QCA molecules be clocked? One strategy for clocking molecular QCA cells is to construct three dots in a "V" shape as shown in Fig. 10.13. We construct a three-dot molecule using the same allyl-based dots as above, again with alkyl bridges between them. An electric field applied perpendicular to the "V" can then raise or lower the potential on the central dot relative to the two top dots. If the mobile charge (here a hole) is in the central position the cell is said to be in the "null" state; occupancy of the two top dots encodes the "1" and "0" state. Figure 10.13(a–c) show these three states by displaying the isopotential surface for each, the orbitals of the two unpaired electrons on the neutral sites, and a schematic representation. Figures 10.13(d) illustrates how the clocking field can either push the hole up into the active state, where it assumes a "1" or "0" position dependent on the state of the neighbors, or down into the "null" state.

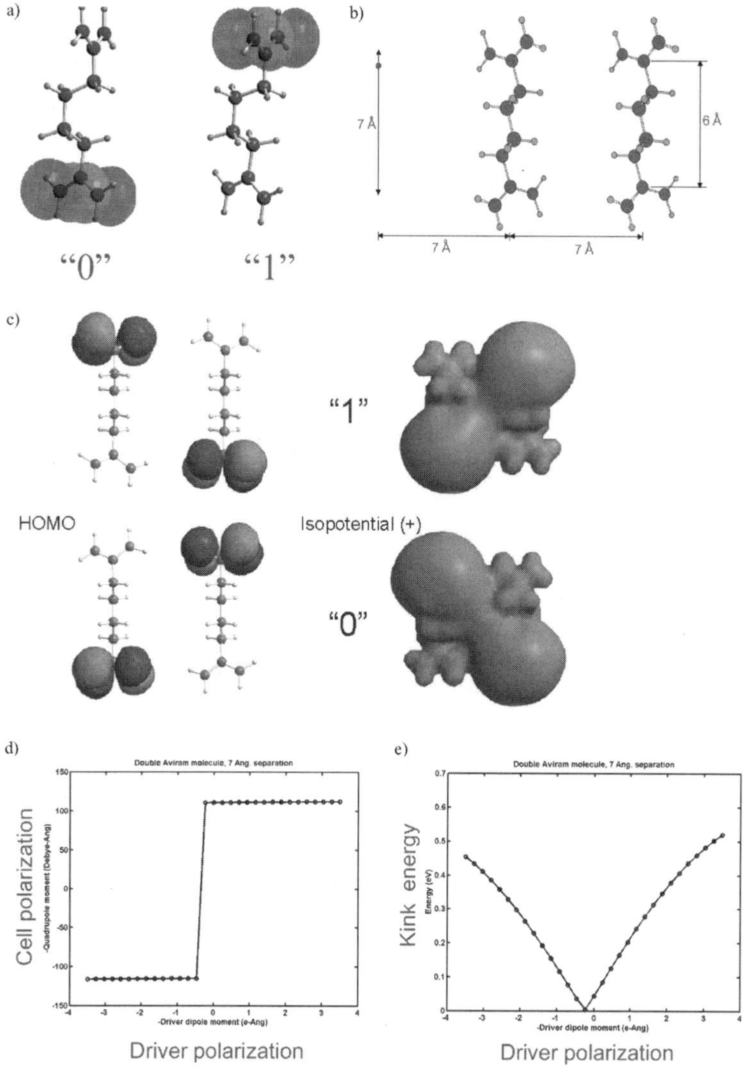

Fig. 10.12. Molecular QCA cells. (a) A single molecule which acts as a molecular double-dot cell. The allyl end-groups act as dots. The extra electron can be in either group. (b) A pair of molecules acts as a full QCA cell. The single-charge driver mimics the effect of a near-by molecule. (c) Orbital and potential isosurfaces for the case of the molecular cell in the "1" and "0" configuration. (d) Cell polarization, as measured by the quadrupole moment of the double molecule. (e) The energy splitting between the ground-state and the first excited states.

The clocking field switches the molecular cell from an active state, in which it's sensitive to the state of its neighbors, to the "null" state, in which it's not. Figure 10.14 shows the calculated response to a driver dipole. The local electric field can switch the cell from a bistable state to a monostable state. Figure 10.15 shows the response of the

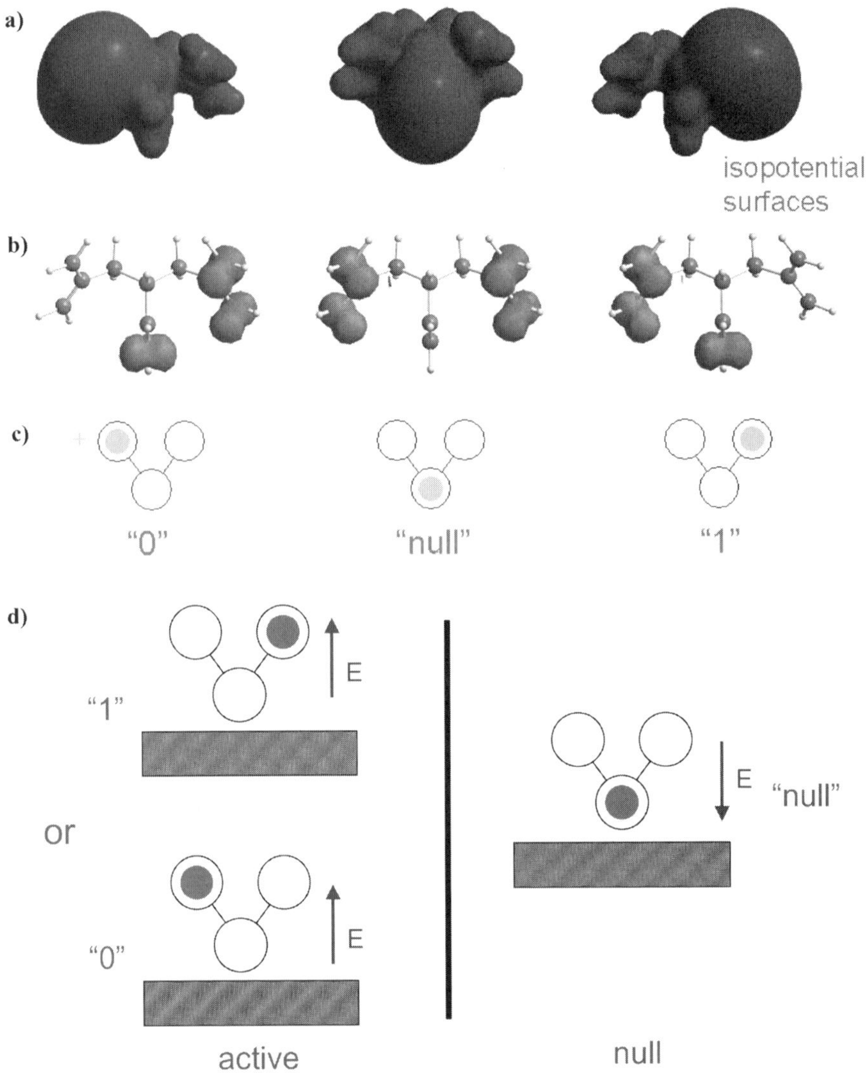

Fig. 10.13. Three-state molecule acts as a clocked QCA cell. (a) The isopotential surfaces of the molecule in each state. (b) Orbitals for the occupied states, which are neutral. (c) Schematic of corresponding hole occupation. (d) Schematic of effect of local clocking field.

cell as a function of both the driver and the clocking field. In this case the zero-field state is the active, bistable state. By altering the geometry or composition of the central redox dot, molecules can be engineered to be "normally-active" or "normally-null." Figure 10.15(b) compares the calculated phase diagram of the three-dot molecule with the measured phase diagram of a thee-dot metal cell. That they are remarkably similar illustrates the transferability of QCA ideas across different implementations.

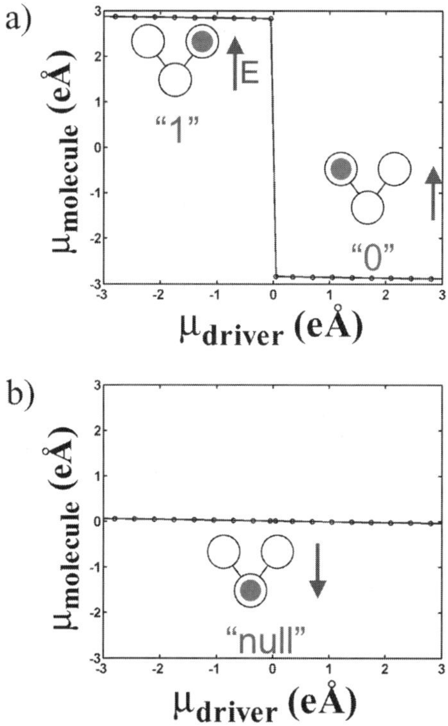

Fig. 10.14. Three-state molecule acts as a clocked QCA cell. (a) When the clocking field is positive the mobile hole is pushed up into one of the active sites. The cell is then sensitive to the state of its neighbors. (b) When the clocking field is negative the mobile hole is pushed down into the center dot and the cell becomes insensitive to the neighbors' configuration.

The clocking field could be varied across the array to make packets of information move across the array surface, being processed as they move [12]. Figure 10.16 illustrates an approach to producing the required clocking field. Buried conducting wires produce the perpendicular clocking field. Four phases of a clocking signal are present as a voltage on each wire; every fourth wire has the same voltage applied to it. This causes the molecules in the molecular layer to be alternately pushed into the active state, or pulled into the null state. Packets of information are formed as a group of cells is switched into it's active state. The area of activity moves across the molecular layer. Computing occurs at the leading edge of this "computational wave" as cells switch into their active state and assume values dependent on the state of the already active cells behind them, which form the input to the local computation. The nature of the local computation is determined by the layout of the cells. Importantly, the conducting clocking wires can be made at a much larger length scale, presumably limited by lithography, than the scale of the molecular circuits.

Figure 10.17 illustrates forms the computational wave could take. The figure shows the calculated electric clocking field intensity produced by particular arrangements of clocking wires (shown in white) below the molecular plane. Light areas of the

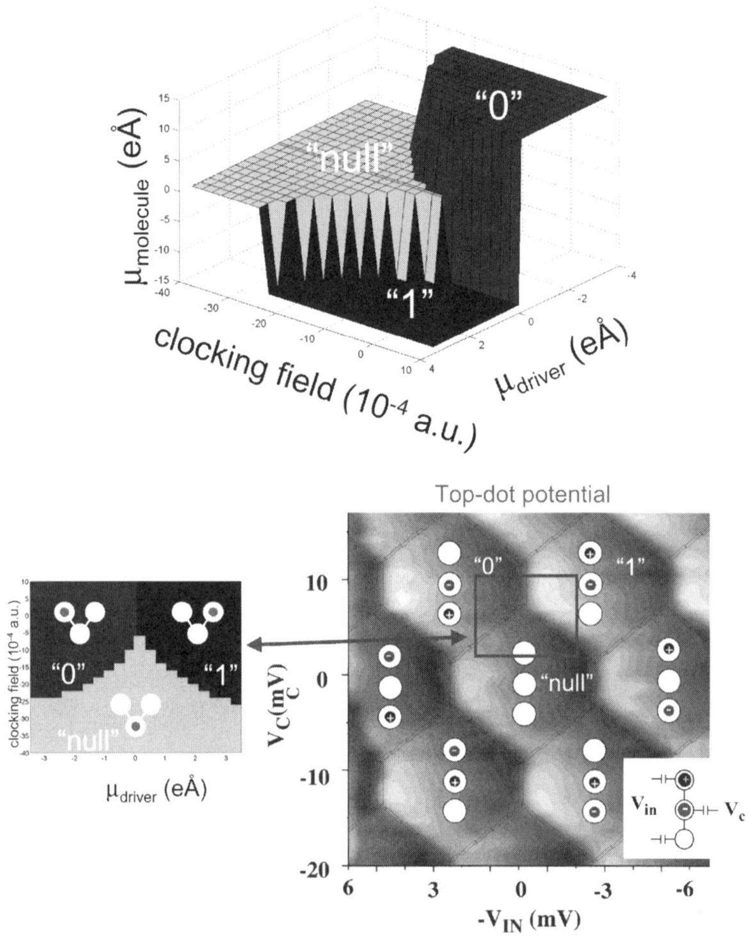

Fig. 10.15. Response of clocked molecule to driver and clocking field. The top figure shows the calculated state, characterized by the induced dipole moment, for the molecule. The lower figure shows the same result as a phase diagram and compares this calculated molecular result with the measured phase diagram of the three-dot metal QCA cell (inset).

gray-scale show higher field intensities, where the cells are pushed into their active states. Figure 10.17(a) shows both a linear path and a connected circulating ring. The linear area allows information to be processed as it moves in one direction. The circulating ring permits both local memory and feedback of signals. Figure 10.17(b) shows the field pattern in a more complex clocked structure. The white arrows show the direction of information motion: up, down, left, and right, in a linear grid, as well as circulating in loops at the intersections of the checkerboard grid. Such a data flow, when combined with the bit-level computations that can be occurring anywhere in the pattern, makes universal general-purpose computation feasible in massively parallel arrays.

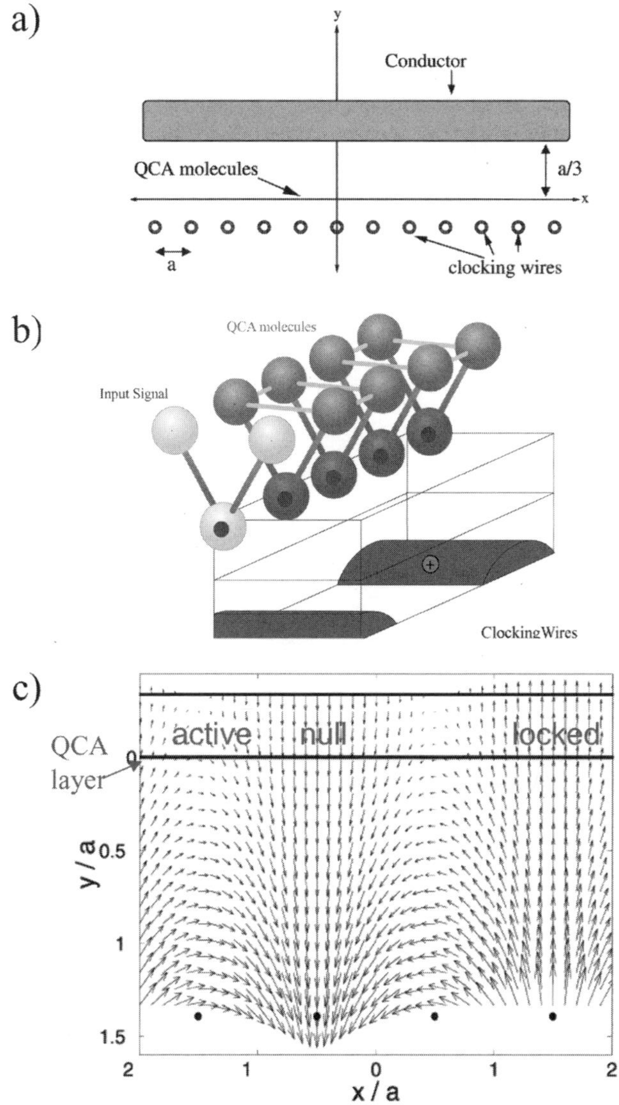

Fig. 10.16. Clocking of QCA molecules. (a) Schematic of buried conductors which act as clocking wires to produce a modulated perpendicular electric field at the surface where the QCA molecules are deposited. (b) The field pushes the molecules between the active and null states. (c) Calculated clocking field from buried wires.

5. Architecture for QCA

All present computational architectures are constructed around the transistor as the fundamental device. Reinventing digital electronics on the basis of a new fundamental element—the QCA cell—requires rethinking the architecture question from the ground up.

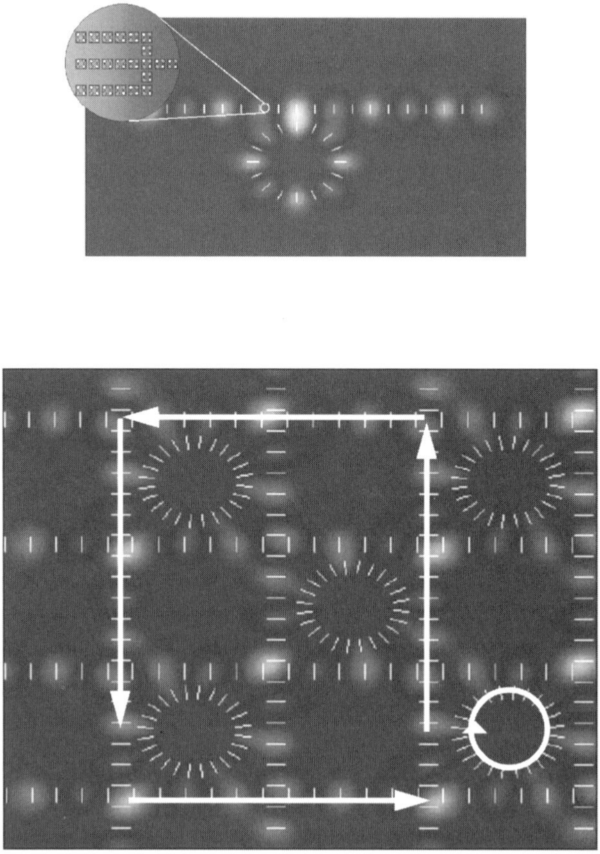

Fig. 10.17. Clocking field patterns produced by buried metal clocking wires. Brighter areas are where the clocking field produces active cells. The pattern determines the flow of information around the circuit. The specific layout of cells (inset) determines the computation performed on the data.

QCA circuits, though transistor-less, are still digital circuits, so it makes sense that the first steps in architectural development have been to engineer functional digital block elements. These include basic gate structures and circuits as complex as adders [10]. These circuits are not critically dependent on a specific clocking scheme; generally a left-to-right clocking is assumed.

An important development has been the design of a simple microprocessor using QCA [57].

The Simple-12 processor was implemented with a clocked QCA design (see Figure 10.18(a)). This involved using physics-level simulators (AQUINAS) for small circuits, and switch-level simulators (Q-BERT) for large circuits. This design, or rather series of improved designs, can be viewed as an exercise in learning how to use QCA effectively. The perspective of computer science and engineering on the problem has been very revealing.

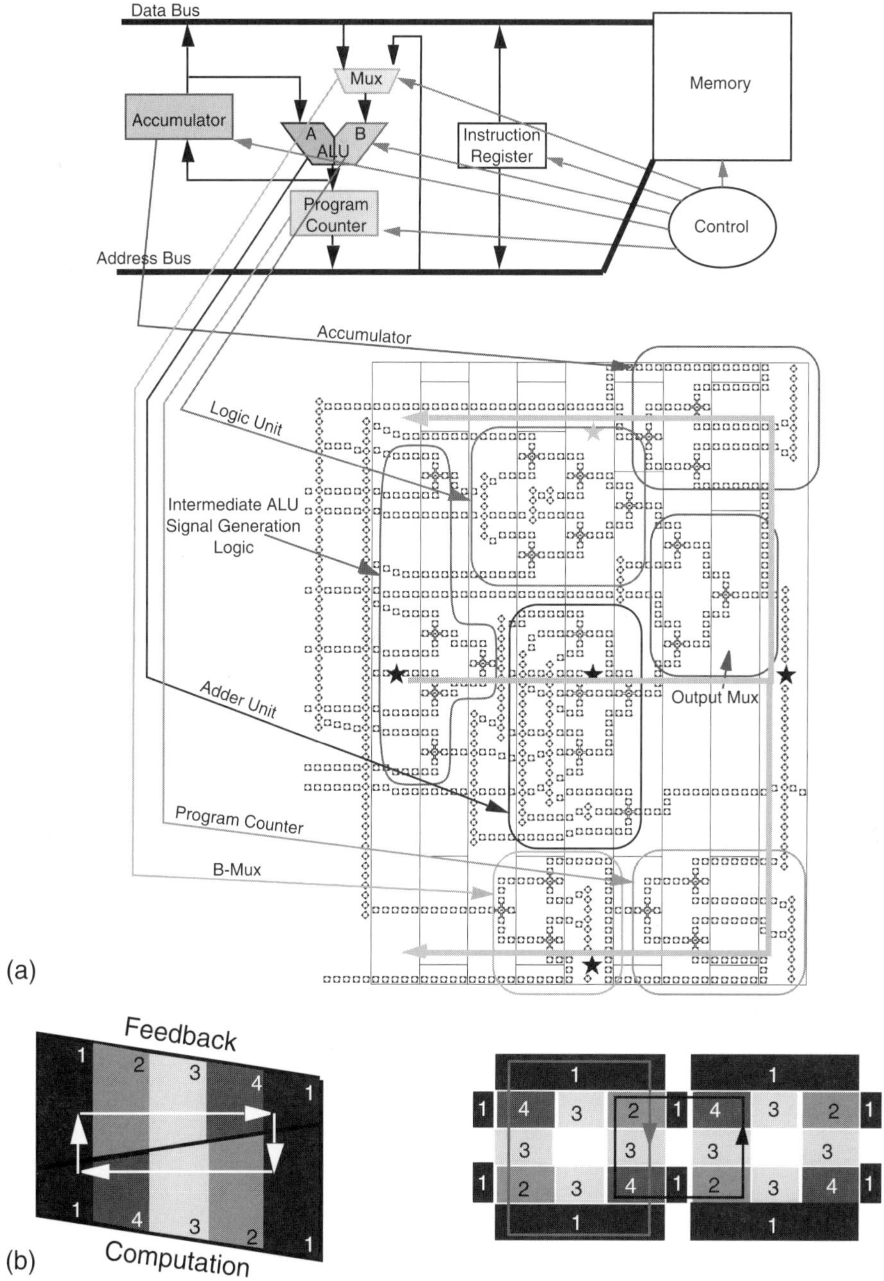

Fig. 10.18. (a) QCA design of Simple-12 microprocessor. (b) Examples of clocking floorplans.

Several themes have emerged:

Processing-in-wire. For clocked QCA, the shift register is the key idea in moving information from place to place. A bit cannot be moved instantly long distances across the circuit, it must be clocked across a QCA wire which acts as a shift register. This disadvantage is balanced by the advantage that a QCA wire can hold many bits of information at once. Once a wire is "filled," a bit is delivered at the output every clock cycle. Furthermore, information can be processed as it is moved. Conventional approaches separate memory from processing. Recent innovations merge the two into "processing-in-memory" (PIM). QCA can be viewed as representing the limit of this sort of integration—"processing-in-wire." Information movement, storage, and processing are now co-mingled [58–60].

Floorplanning first. One of the lessons learned has been that clocking is so important and so linked to the layout that it is helpful to design by starting with a picture of the dataflow. As we have seen in the discussion of molecular clocking fields, the path traversed by the area of the active clocking region can be engineered in many ways. To have feedback, there must be loops in the clocking floorplan. Figure 10.18(b) shows the initial development of the floorplanning notions that, in fact, led to the layout clocking patterns of Figure 10.17. Ongoing work is aimed at developing design rules for clocking floorplans based on the physical constraints of implementing clocks with buried wires.

Layout = timing. The VLSI revolution of Mead and Conway was produced by connecting circuit design with physical layout [61]. Designing with QCA means making the further connection between timing and layout. Clocks act on *local areas* of the circuit and the motion of the active clocking region (the computational wave) around the circuit determines the timing just as the layout of cells determines the function.

Wire crossing plays an important role in any complex design. There are four ways in which QCA wire crossings can occur. (1) Just as in conventional designs, the active area can be physically raised up over another active area in the form of bridges and vias. (2) Wire crossing in the plane can be accomplished by using rotated QCA cells, with symmetry such that no interference occurs between crossed signals (detailed in [10]). (3) Clocking can be extended to eight phases allowing east-west signals to move across a junction at one time and north–south signals to move across at a different time. (4) Logic at the crossing can be added to pass north–south traveling or east–west traveling information in a selectable way.

QCA architecture may provide an implementation for other array-based architectural ideas at a high level. One approach is to use many QCA cells to produce functional super-cells with particular functionality [62–64]. The super-cells can then be laid out in a regular array pattern to form, for example, a cellular neural network (CNN) [65] or conventional cellular automata (CA) architecture. These approaches have been employed for image processing functions [66–69].

Another promising approach is to implement field-programmable gate arrays using QCA cells to form highly functional array elements [70]. Programming areas to keep the clock always low (inactive cells) essentially removes particular areas from the circuit. Information can then be routed differently and particular functions chosen, all by simply changing the clocking signals on specific areas. When combined with new memory-storage paradigms [71], these may yield particularly powerful ultra-dense FPGAs.

An important architectural consideration for any approach to molecular electronics or nanoelectronics is defect tolerance. The nonlinearity inherent in the cell–cell response function (e.g. Fig. 10.12(d)) provides a significant level of tolerance against positional and rotational disorder which are the most likely sort of fabrication defect. Power gain provides further protection. Recent work [72] has shown that a remarkable degree of defect tolerance can be obtained by simply widening the QCA wires to three or more cells. A three-cell wide wire provides essentially triple modular redundancy at each point in the wire. This strategy has been extended to majority gates as well. For a very small cost in density, this strategy provides very strong immunity against fabrication errors. Additionally the excitation energy for a thermally generated mistake, becomes much higher. The kink energy (the cost of making an error) for a single width wire comprised of the molecule in Fig. 10.6 is roughly 0.5 eV, corresponding to a thermal error rate of 10^{-9}. For a three-cell wide wire this becomes about 10^{-25}.

6. Magnetic QCA

In a recent paper, room temperature operation of a magnetic QCA wire based on permanent-magnet dots was demonstrated by a Cambridge University group [73]. The basic idea is to represent each cell in the QCA scheme by a single-domain permalloy magnetic dot, and the information is represented by the orientation of the magnetic moment (Fig. 10.19). In a related paper by the same group, a NOT gate and shift register were also demonstrated [74]. These papers establish the proof-of-principle of magnetic QCA implementations.

Deep submicron—size magnetic dots show single-domain (or near single-domain) magnetic behavior, with nearly parallel magnetization inside the dot, and well-defined switching properties. Dipolar coupling between neighboring dots typically is on the order of several (at room temperature), yielding stable and ordered ground-state configurations. Circular nanomagnets show easy-reversal properties, while anisotropic (shape or crystalline anisotropy) magnets show nearly two-state behavior, making themselves ideal candidates for storing and manipulating binary information.

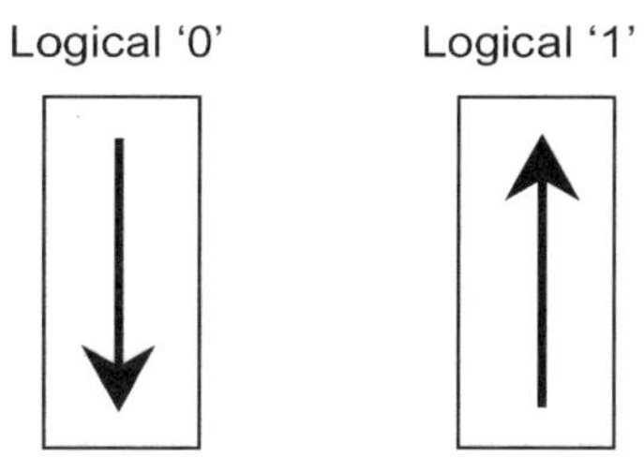

Fig. 10.19. Schematic representation of the magnetization in magnetic pillars.

Nanomagnetic dots exhibit theoretically well-understood dynamics, governed by the Landau–Lifshitz equation [75]. Numerical solutions of these equations, in agreement with measurements, predict nanosecond switching times, low power dissipation, and the possibility of coherent control of magnetization dynamics [76]. Sub-micrometer-diameter electric wires with relatively low currents (few mA) generate sufficiently strong magnetic fields to set the state of nanomagnets (which may be used as an input), and GMR-based sensors are sensitive enough to read out the magnetic state of individual dots. Recent work related to nonvolatile magnetic memories predict feasible integration of magnetic structures and conventional microelectronic devices [77,78].

The above suggests that the basic features needed for an implementation of the general QCA scheme in magnetically-coupled nanostructures appear to be in place. To summarize:

1. Nanometer-sized magnetic dots exhibit single-domain behavior, which may be used to encode (binary as well as analog) information.
2. Magnetic dot-dot coupling in this regime is very strong [79]; interaction energies are on the order of several 100 kT at room temperature.
3. A functional QCA wire based on coupled permalloy dots has been demonstrated to work at room temperature (by the Cambridge group).
4. Input and output to such arrays may be done in ways similar to reading and writing magnetic memory elements (MRAM technology).
5. Logic elements based on such magnetically-coupled nanomagnets would be compatible with the silicon-based microelectronics industry.
6. This proposed investigation into magnetic logic takes advantage of the large experience of the magnetic-recording industry in these material systems and phenomena.

The magnetic coupling [79], i.e. next-neighbor field coupling, of nanomagnets as continuous nonlinear dynamical systems offers a new possibility to build magnetic cellular nonlinear networks. We envision cellular processors composed of nanomagnetic logic coupled to magnetic sensors and memory. These new processors may have important applications where rugged, reliable devices with wide operating temperature range are required.

6.1. Micromagnetic Simulations

In order to investigate the feasibility of these ideas, we have performed simulations of magnetically-coupled magnetic dots [80]. The well-established theory of micromagnetic analysis (structures with size scales between 1–1000 nm) is based on the Landau–Lifshitz equations, which describe the time-evolution of the microscopic magnetization distribution inside and field distribution outside magnetic materials. The theoretical basis is well established, and we summarize here the main features of our simulation approaches:

1. The most accurate (and computationally most expensive) simulations give full three-dimensional distributions of the magnetic fields and the magnetization. We have

performed such simulations in order to study the switching behavior of coupled dots. Most of the micromagnetic simulations were performed with the OOMMF micromagnetic code [81], developed at ITL/NIST (see http://math.nist.gov/oommf/). For other calculations (especially for 3D field calculations) we used the FlexPDE finite element code from PDESolutions Inc: http://www.pdesolutions.com/.

2. In the single-domain regime, one may approximate the magnetization distribution in each dot by its average magnetization; in other words, each dot is described by one magnetization vector, instead of the full distribution. This approximation, known from the Stoner model, significantly reduces the complexity of the simulations. We have employed this approximation, and we have verified its validity against the full three-dimensional theory.
3. Within the single-domain approximation, the partial differential equations (PDE) of the Landau–Lifshitz theory are reduced to ordinary differential equations (ODE). Given this, one can develop circuit models for the resulting ODEs, and one may then employ standard circuit simulation tools (such as SPICE). We already have implemented a SPICE-compatible version of the single-domain-approximation-model equations, which will allow us to co-simulate CMOS devices with magnetically-coupled nanostructures.

We have applied our simulation tools to the study of the switching behavior of coupled magnetic dots, and dots arranged in simple lines and also intersecting lines, which will be the basis for magnetic majority logic gates. This will be discussed in more detail below.

6.2. Switching of Magnetic Pillars

Magnetic pillars with high aspect ratios have been demonstrated in the literature. These pillars have interesting switching properties due to the shape anisotropy of the switching field. The shortest dimension is the "hardest" direction to switch, and the longest is the "softest" direction to switch. One may exploit these properties for selective switching with external fields by appropriate arrangements of pillar geometries and external-field orientations, as will be discussed below.

The ground-state magnetization of such pillars (in the absence of external fields) is always parallel to the easy (soft) axis of the pillar, in our case, perpendicular to the surface. We can assign logic values to these magnetization states, as shown schematically.

6.3. Wire Composed of Magnetic Pillars

The simplest device is the nanomagnet wire, a one-dimensional structure which orders according to the state of the input dot. Figure 10.20 presents plots of micromagnetic simulations [M8], showing how an external pump field is applied at different instances in time to drive the dots from an initial metastable state to the antiferromagnetically-aligned ground state. The input dot is designed to have a larger switching field than the other dots, which assures that it does not switch under the effect of the external pumping field.

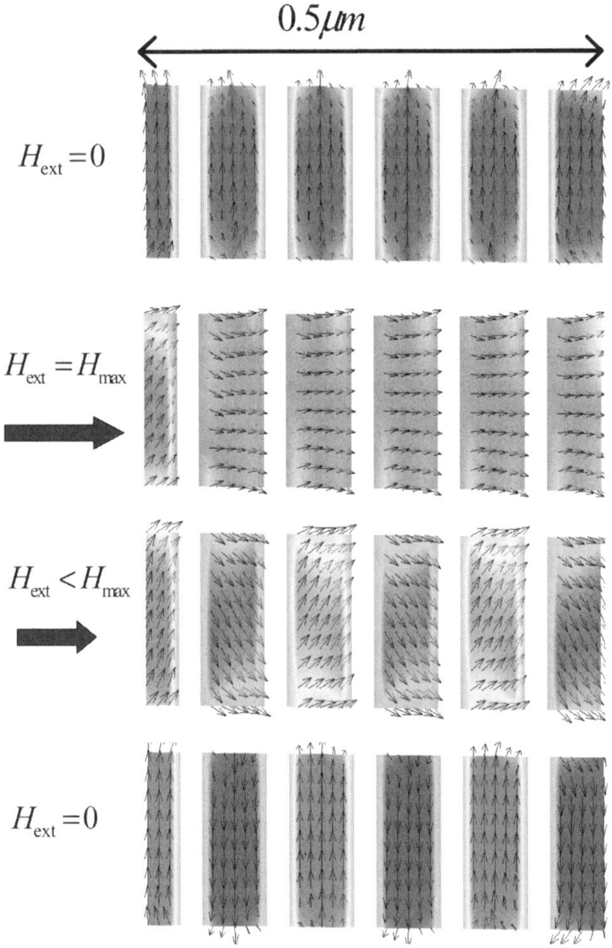

Fig. 10.20. Nanomagnetic simulation of the switching of the nanowire. The four snapshots are different phases of the pumping cycle.

Since metastable states represent deep local minima on the energy landscape of such an array, the external pumping field is always required to drive the system out of a metastable state. The pumping field first erases the "memory" of the structure, as all dots move to their highest energy, but logically neutral state. The external field pumping also ensures the local activity and signal restoration property of the nanoelements.

One can think about the nanowire as an inverter chain: even-dot-length wires simply transmit the state of the input; odd numbers of dots operate as an inverter.

6.4. Majority Gate Composed of Magnetic Pillars

The above scheme can, in principle, be used to realize more complicated structures. In particular, following the QCA work, we have demonstrated the feasibility

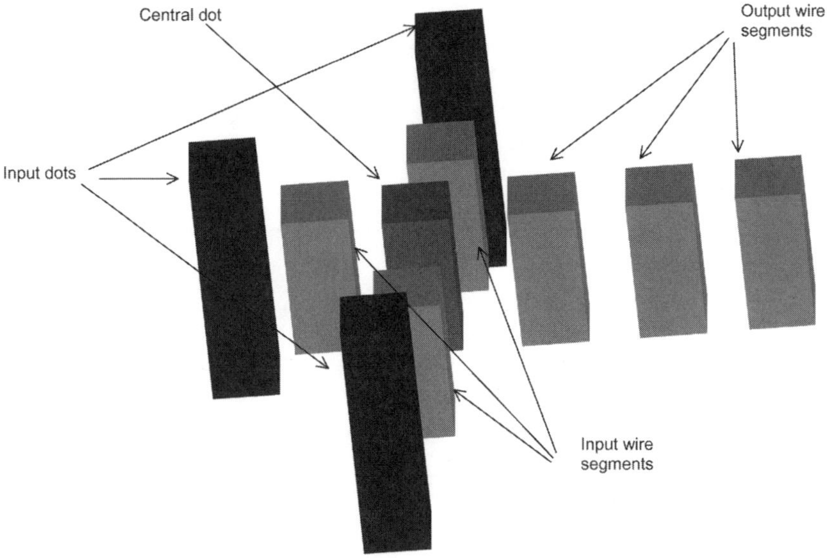

Fig. 10.21. Schematic picture of a majority gate composed of magnetic pillars.

of a majority-logic gate using micromagnetic simulations [81]. Figure 10.21 shows a possible layout of a majority gate implemented with magnetic pillars.

The input (driver) dots (colored blue in the figure) are taller than the others, and thus have larger switching fields which helps preserve their magnetization. In the ground state, the magnetization of the central dot (the majority logic state) should be the opposite of the majority vote of its neighbors. We anticipate that switching of the gate can be accomplished using adiabatic-field pumping, similar to the previously-discussed case of a line. Micromagnetic simulations confirm the operation of the magnetic QCA majority gate [80].

6.5. Experimental Investigation of Nanomagnet Arrays

To validate our models, we have manufactured nanowire-like arrangements of planar permalloy and cobalt dots [82]. A comparison between experimental and simulation results is shown in Fig. 10.22. The line geometry determined by atomic force microscopy (AFM) is shown in part (a), and part (b) shows the micromagnetic simulations using the OOMMF code. Parts (c) and (d) show magnetic force microscopy (MFM) images, both as expected from simulations of the field gradients (c), and the experimental MFM image (d) which clearly exhibits the expected coupling in the wire.

Switching-field variations, which arise from the imperfect shape or rough surface of the dots, will introduce faults in longer lines, and we find a typical correlation length of about 5–10 dots. This result is consistent with the observation of other groups, and suggests that design for fault-tolerance is a crucial issue in nanomagnetic networks. A simple example of fault-tolerant design is non-reciprocal structures, which show

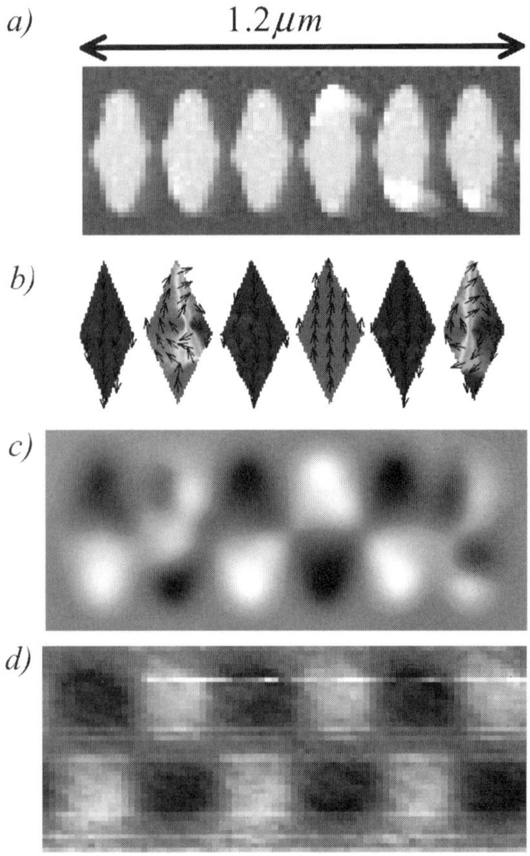

Fig. 10.22. (a) AFM topographic picture of a Cobalt nano-structure, (b) magnetization distribution (c) MFM image as expected from simulations, and (d) recorded MFM image.

much less sensitivity against defects than the ones lacking definite signal propagation direction.

In this section, we demonstrated that it is, in principle, possible to construct a family of logic devices from coupled ferromagnetic dots and pillars, using electric circuits only as the interfaces of the computing array. The magnetic devices are characterized by low power dissipation (a few ten or hundred kT per bit operation) and relatively high speed (on the order of a hundred MHz, as estimated from time-dependent simulations) [82]. These magnetic structures take advantage of the technological base developed by the magnetic data storage industry. The computing architecture naturally incorporates memory elements in the logic circuit, which might be very well suited for certain applications. We are currently working on the experimental realization of the proposed devices. While the realization of complex circuits might pose formidable difficulties, we do not see any fundamental obstacle of building a magnetic QCA-type logic device of high complexity.

Acknowledgments

This work was supported in part by the Office of Naval Research, DARPA, the National Science Foundation, the Semiconductor Research Corporation, NASA, and the W. M. Keck Foundation.

For the theoretical work, Craig Lent acknowledges with gratitude the help of P. Douglas Tougaw, John Timler, and Geza Toth. For the experimental work, Gregory Snider acknowledges A. N. Korotkov (UC Riverside) for suggesting bit-shifting experiments, and K. Yadavalli for technical support. For work on magnetic systems, Wolfgang Porod acknowledges fruitful discussions with Prof. A. Csurgay. The micromagnetic simulations were performed by G. Csaba, and the MFM experiments were done by A. Imre on structures provided by Prof. V. Metlushko (University of Illinois at Chicago).

References

1. C.S. Lent, P.D. Tougaw, W. Porod, and G.H. Bernstein, Quantum cellular automata, *Nanotechnology* **4**(1), 49–57 (1993).
2. C.S. Lent and P.D. Tougaw, Lines of interacting quantum-dot cells: A binary wire, *J. Appl. Phys.* **74**(10), 6227–6233 (1993).
3. P.D. Tougaw, C.S. Lent, and W. Porod, Bistable saturation in coupled quantum-dot cells, *J. Appl. Phys.* **74**(5), 3558–3566 (1993).
4. C.S. Lent, P.D. Tougaw, and W. Porod, Bistable saturation in coupled quantum dots for quantum cellular automata, *Appl. Phys. Lett.* **62**(7), 714–716 (1993).
5. C.S. Lent and P.D. Tougaw, Bistable saturation due to single electron charging in rings of tunnel junctions, *J. Appl. Phys.* **75**(8), 4077–4080 (1994).
6. P.D. Tougaw and C.S. Lent, Logical devices implemented using quantum cellular automata, *J. Appl. Phys.* **75**(3), 1818–1825 (1994).
7. P.D. Tougaw and C.S. Lent, Effect of stray charge on quantum cellular-automata, *Japan J. Appl. Phys.* **34**(8B), 4373–4375 (1995).
8. C.S. Lent, P.D. Tougaw, and W. Porod, Quantum cellular automata: The physics of computing with arrays of quantum dot molecules, in *Proceedings Workshop on Physics and Computation—PhysComp '94*, 5–13 (1994).
9. G. Toth, P.D. Tougaw, and C.S. Lent, Dynamic behavior of quantum cellular automata, *J. Appl. Phys.* **80**(8), 4722–4736 (1996).
10. C.S. Lent and P.D. Tougaw, A device architecture for computing with quantum dots, *Proc. IEEE* **85**(4), 541–557 (1997).
11. G. Toth and C.S. Lent, Quasiadiabatic switching for metal-island quantum-dot cellular automata, *J. Appl. Phys.* **85**(5), 2977–2984 (1999).
12. K. Hennessy and C.S. Lent, Clocking of molecular quantum-dot cellular automata, *J. Vac. Sci. Technol.* B 19(5), 1752–1755 (2001).
13. G. Toth and C.S. Lent, Role of correlation in the operation of quantum-dot cellular automata, *J. Appl. Phys.* **89**(12), 7943–7953 (2001).
14. G. Toth and C.S. Lent, Quantum computing with quantum-dot cellular automata, *Phys. Rev.* A **63**(5), 052315 (2001).
15. J. Timler and C.S. Lent, Power gain and dissipation in quantum-dot cellular automata, *J. Appl. Phys.* **91**(2), 823–831 (2002).

16. G. Bazan, A.O. Orlov, G.L. Snider, and G.H. Bernstein, Charge detector realization for AlGaAs/GaAs quantum-dot cellular automata, *J. Vac. Sci. Technol.* B **14**(6), 4046–4050 (1996).
17. G.H. Bernstein, G. Bazan, M.H. Chen, C.S. Lent, J.L. Merz, A.O. Orlov, W. Porod, G.L. Snider, and P.D. Tougaw, Practical issues in the realization of quantum-dot cellular automata, *Superlatt. Microstruct.* **20**(4), 447–459 (1996).
18. G.H. Bernstein, I. Amlani, A.O. Orlov, and G.L. Snider, Si-compatible quantum-dot cellular automata, in *Silicon Nanoelectronics Workshop* (1997). (Workshop Abstracts, 38–39).
19. A.O. Orlov, I. Amlani, G.H. Bernstein, C.S. Lent, and G.L. Snider, Realization of a functional cell for quantum-dot cellular automata, *Science* **277**(5328), 928–930 (1997).
20. I. Amlani, A.O. Orlov, G.L. Snider, C.S. Lent, and G.H. Bernstein, External charge state detection of a double-dot system, *Appl. Phys. Lett.* **71**(12), 1730–1732 (1997).
21. I. Amlani, A.O. Orlov, G.L. Snider, and G.H. Bernstein, Differential charge detection for quantum-dot cellular automata, *J. Vac. Sci. Technol.* B **15**(6), 2832–2835 (1997).
22. I. Amlani, A.O. Orlov, G.L. Snider, C.S. Lent, and G.H. Bernstein, Demonstration of a six-dot quantum cellular automata system, *Appl. Phys. Lett.* **72**(17), 2179–2181 (1998).
23. I. Amlani, A.O. Orlov, G.L. Snider, C.S. Lent, and G.H. Bernstein, Demonstration of a functional quantum-dot cellular automata cell, *J. Vac. Sci. Technol.* B **16**(6), 3795–3799 (1998).
24. A.O. Orlov, I. Amlani, G. Toth, C.S. Lent, G.H. Bernstein, and G.L. Snider, Correlated electron transport in coupled metal double dots, *Appl. Phys. Lett.* **73**(19), 2787–2789 (1998).
25. G.L. Snider, A.O. Orlov, I. Amlani, G.H. Bernstein, C.S. Lent, J.L. Merz, and W. Porod, Experimental demonstration of quantum-dot cellular automata, *Semicond. Sci. Technol.* **13**(8A), A130–A134, Suppl. S (1998).
26. I. Amlani, A.O. Orlov, G. Toth, C.S. Lent, G.H. Bernstein, and G.L. Snider, Digital logic gate using quantum-dot cellular automata, *Science* **284**(5412), 289–291 (1999).
27. G.H. Bernstein, I. Amlani, A.O. Orlov, C.S. Lent, and G.L. Snider, Observation of switching in a quantum-dot cellular automata cell, *Nanotechnology* **10**(2), 166–173 (1999).
28. A.O. Orlov, I. Amlani, G. Toth, C.S. Lent, G.H. Bernstein, and G.L. Snider, Experimental demonstration of a binary wire for quantum-dot cellular automata, *Appl. Phys. Lett.* **74**(19), 2875–2877 (1999).
29. G.L. Snider, A.O. Orlov, I. Amlani, G.H. Bernstein, C.S. Lent, J.L. Merz, and W. Porod, Quantum-dot cellular automata: Line and majority logic gate, *Japan J. Appl. Phys.* **38**(12B), 7227–7229 (1999).
30. G. Toth, A.O. Orlov, I. Amlani, C.S. Lent, G.H. Bernstein, and G.L. Snider, Conductance suppression due to correlated electron transport in coupled double quantum dots, *Phys. Rev.* B **60**(24), 16906–16912 (1999).
31. I. Amlani, A.O. Orlov, R.K. Kummamuru, G.H. Bernstein, C.S. Lent, and G.L. Snider, Experimental demonstration of a leadless quantum-dot cellular automata cell, *Appl. Phys. Lett.* **77**(5), 738–740 (2000).
32. A.O. Orlov, I. Amlani, R.K. Kummamuru, R. Ramasubramaniam, G. Toth, C.S. Lent, G.H. Bernstein, and G.L. Snider, Experimental demonstration of clocked single-electron switching in quantum-dot cellular automata, *Appl. Phys. Lett.* **77**(2), 295–297 (2000).
33. T. Junno, S.B. Carlsson, H.Q. Xu, L. Samuelson, A.O. Orlov, and G.L. Snider, Single-electron tunneling effects in a metallic double dot device, *Appl. Phys. Lett.* **80**(4), 667–669 (2002).
34. R.K. Kummamuru, J. Timler, G. Toth, C.S. Lent, R. Ramasubramaniam, A.O. Orlov, G.H. Bernstein, and G.L. Snider, Power gain in a quantum-dot cellular automata latch, *Appl. Phys. Lett.* **81**(7), 1332–1334 (2002).

35. C. Lent, Bypassing the transistor paradigm, *Science* **228**(5471), 1597–1598 (2000).
36. O.G. Wiest and S. Braun-Sand, Quantum dot cellular automata: Encoding information in molecules, *Abstr. Pap. Am. Chem. Soc.* **219**, 112-COMP (Part 1) (2000).
37. C.J. Russo and C.S. Lent, Quantum mechanical calculation of QCA molecule properties, in *Device Research Conference* (2001) (Conference Digest Cat. No.01TH8561, 165).
38. M. Lieberman, Z.Y. Li, F. Peiris, and G.S. Snider, Molecular quantum-dot cellular automata: Surface attachment and STM spectroscopy, *Abstr. Pap. Am. Chem. Soc.* **221**, 10-COLL (Part 1) (2001).
39. M. Lieberman, Y.L. Wang, and S. Chellamma, Toward molecular implementations of quantum-dot cellular automata, *Abstr. Pap. Am. Chem. Soc.* **221**, 271-COLL (Part 1) (2001).
40. Q.L. Hang, Y.L. Wang, M. Lieberman, and G.H. Bernstein, Molecular patterning through high-resolution polymethylmethacrylate masks, *Appl. Phys. Lett.* **80**(22), 4220–4222 (2002).
41. M. Lieberman, S. Chellamma, B. Varughese, Y.L. Wang, C. Lent, G.H. Bernstein, G. Snider, and F.C. Peiris, Quantum-dot cellular automata at a molecular scale, *Ann. NY Acad. Sci.* **960**, 225–239 (2002).
42. R.W. Keyes and R. Landauer, Minimal energy dissipation in logic, *IBM J. Res. Dev.* **14**, 152 (1970).
43. T.A. Fulton and G.H. Dolan, Observation of single-electron charging effects in small tunnel junctions, *Phys. Rev. Lett.* **59**, 109 (1987).
44. L.P. Kouwenhoven, C.M. Marcus, P.L. McEuen, S. Tarucha, R.M. Westervelt, and N. Wingreen, Electron transport in quantum dots, in *Proceedings of the NATO Advanced Study Institute on Mesoscopic Electron Transport* (1997) (Kluwer Series E345).
45. Yu, A. Pashkin, Y. Nakamura, and J.S. Tsai, Room-temperature Al single-electron transistor made by electron-beam lithography, *Appl. Phys. Lett.* **76**, 2256 (2000).
46. D.V. Averin and K.K. Likharev, Single electronics, in *Mesoscopic Phenomena in Solids* (ed. B.L. Altshuler, P.A. Lee, and R.A. Webb), Elsevier (1991).
47. P. Lafarge, H. Pothier, E.R. Williams, D. Esteve, C. Urbina, and M.H. Devoret, *Z. Phys.* B **85**, 327 (1991).
48. K.K. Likharev and A.N. Korotkov, Single-electron parametron: Reversible computation in a discrete state system, *Science* **273**, 763 (1996).
49. A. Korotkov and K.K. Likharev, Single-electron-parametron-based logic devices, *J. Appl. Phys.* **84**, 6114 (1998).
50. D.V. Averin and A.A. Odintsov, Macroscopic quantum tunneling of the electric charge in small tunnel junctions, *Phys. Lett.* A **140**, 251 (1989).
51. Z. Li and M. Lieberman, Axial reactivity of soluble silicon (IV) phthalocyanines, *Inorg. Chem.* **40**, 932–939 (2001).
52. K.D. Demadis, C.M. Hartshorn, and T.J. Meyer, The localized-to-delocalized transition in mixed-valence chemistry, *Chem. Rev.* **101**, 2655–2685 (2001).
53. A. Aviram, Molecules for memory, logic, and amplification, *J. Am. Chem. Soc.* **110**(17), 5687–5692 (1988).
54. N.S. Hush, A.T. Wong, G.B. Bacskay, and J.R. Reimers, Electron and energy-transfer through bridged systems.6. Molecular switches—the critical-field in electric-field activated bistable molecules, *J. Am. Chem. Soc.* **112**(11), 4192–4197 (1990).
55. Gaussian 98, Revision A.11, M.J. Frisch, G.W. Trucks, H.B. Schlegel, G.E. Scuseria, M.A. Robb, J.R. Cheeseman, V.G. Zakrzewski, J.A. Montgomery, Jr., R.E. Stratmann, J.C. Burant, S. Dapprich, J.M. Millam, A.D. Daniels, K.N. Kudin, M.C. Strain, O. Farkas, J. Tomasi, V. Barone, M. Cossi, R. Cammi, B. Mennucci, C. Pomelli, C. Adamo, S. Clifford, J. Ochterski, G.A. Petersson, P.Y. Ayala, Q. Cui, K. Morokuma, P. Salvador, J.J. Dannenberg,

D.K. Malick, A.D. Rabuck, K. Raghavachari, J.B. Foresman, J. Cioslowski, J.V. Ortiz, A.G. Baboul, B.B. Stefanov, G. Liu, A. Liashenko, P. Piskorz, I. Komaromi, R. Gomperts, R.L. Martin, D.J. Fox, T. Keith, M.A. Al-Laham, C.Y. Peng, A. Nanayakkara, M. Challacombe, P.M.W. Gill, B. Johnson, W. Chen, M.W. Wong, J.L. Andres, C. Gonzalez, M. Head-Gordon, E.S. Replogle, and J.A. Pople, Gaussian, Inc., Pittsburgh, PA, (2001).

56. For a charged molecule the dipole moment depends on the choice of coordinate origin. We take the origin at the midpoint of the line segment connecting the top allyl dots. We take the direction of the dipole moment to be from negative to positive charge, following A. J. Stone, *The Theory of Intermolecular Forces*. Oxford University Press, Oxford (1996). For the calculation of quadrupole moments below we take the coordinate origin at the center of the square formed by the 4 upper dots and calculate Qxy, the moment in the plane of the dots.
57. M.T. Niemier, M.J. Kontz, and P.M. Kogge, A design of and design tools for a novel quantum dot based microprocessor, Proceedings 2000. Design Automation Conference. IEEE Cat. No. 00CH37106. 2000: 227–32.
58. M.T. Niemier and P.M. Kogge, Logic in wire: Using quantum dots to implement a microprocessor, in *International Conference on Electronics, Circuits, and Systems (ICECS '99)* Cyprus, (1999).
59. M.T. Niemier and P.M. Kogge, Logic in wire: Using quantum dots to implement really dense processing logic, in *Proceedings of the Third Petaflops Workshop* (1999) (Associated with Frontiers of Massively Parallel Processing, Annapolis, MD).
60. M.T. Niemier and P.M. Kogge, Problems in designing with QCAs: layout=timing, *Int. J. Circuit Theory Appl.* **29**(1), 49–62 (2001).
61. C.A. Mead and L. Conway, *An Introduction to VLSI Systems*. Addison Wesley, (1980).
62. A. Gin, S. Williams, H.Y. Meng, and P.D. Tougaw, Hierarchical design of quantum-dot cellular automata devices, *J. Appl. Phys.* **85**(7), 3713–3720 (1999).
63. A. Gin, P.D. Tougaw, and S. Williams, An alternative geometry for quantum-dot cellular automata, *J. Appl. Phys.* **85**(12), 8281–8286 (1999).
64. J.R. Pasky, L. Henry, and P.D. Tougaw, Regular arrays of quantum-dot cellular automata macrocells, *J. Appl. Phys.* **87**(12), 8604–8609 (2000).
65. C.S. Lent, P.D. Tougaw, Y. Brazhnik, W.W. Weng, W. Porod, R.W. Liu, and Y.F. Huang, Quantum cellular neural networks, *Superlatt. Microstruct.* **20**(4), 473–478 (1996).
66. A.I. Csurgay, W. Porod, and C.S. Lent, Signal processing with near-neighbor-coupled time-varying quantum-dot arrays, *IEEE Trans. Circuits Syst. I—Fundam. Theory Appl.* **47**(8), 1212–1223 (2000).
67. A. Csurgay and W. Porod, Equivalent circuit representation of arrays composed of Coulomb-coupled nanoscale devices: Modelling, simulation and realizability, *Int. J. Circuit Theory Appl.* **29**(1), 3–35 (2001).
68. J.L. Cardenas-Barrera, K.N. Platoniotis, and A.N. Venetsanopoulos, QCA implementation of a multichannel filter for image processing, *Math. Probl. Eng.* **8**(1), 87–99 (2002).
69. M. Helsingius, P. Kuosmanen, and J. Astola, Quantum-dot cells and their suitability for nonlinear signal processing, in *Proceedings of the IEEE EURASIP Workshop on Nonlinear Signal and Image Processing NSIP'99*, 659–663 vol. 2 (1999).
70. M.T. Niemier, A.F. Rodrigues, and P.M. Kogge, A potentially implementable FPGA for quantum dot cellular automata, in *1st Workshop on Non-Silicon Computation (NSC-1)* (Held in Conjunction with 8th International Symposium on High Performance Computer Architecture (HPCA-8), Boston, MS) (February 3, 2002).
71. S.E. Frost, A.F. Rodrigues, A.W. Janiszewski, R.T. Rausch, and P.M. Kogge, Memory in motion: A study of storage structures in QCA, in *1st Workshop on Non-Silicon Computation*

(NSC-1) (Held in Conjunction with 8th International Symposium on High Performance Computer Architecture (HPCA-8), Boston, MS) (February 3, 2002).
72. A. Fijany and B.N. Toomarian, New design for quantum dots cellular automata to obtain fault tolerant logic gates, *J. Nanopart. Res.* **3**(1), 27–37 (2001).
73. R.P. Cowburn and M.E. Welland, Room temperature magnetic cellular automata, *Science* **287**, 1466 (2000).
74. D.A. Allwood, Gang Xiong, M.D. Cooke, C.C. Faulkner, D. Atkinson, N. Vernier, and R.P. Cowburn, Sub micrometer ferromagnetic NOT gate and shift register, *Science* **296**, 2003–2006 (2002).
75. A. Aharoni, *Introduction to the Theory of Ferromagnetism*. Oxford University Press, Oxford (2000).
76. T.M. Crawford, P. Kabos, and T.J. Silva, Coherent control of precessional dynamics in thin film permalloy, *Appl. Phys. Lett.* **76**, 15 (2000).
77. G.A. Prinz, Magnetoelectronics, *Science* **282**, 1160 (1998) and references therein.
78. J.-G. Zhu,Y. Zheng, and G. A. Prinz, Ultrahigh density vertical magnetoresistive random access memory, *J. Appl. Phys.* **87**(9) (2000).
79. H. Koo, M. Dryer, V.V. Metlushko, and R.D. Gomez, Systematic investigation of the magnetic interaction of small permalloy particles, *IEEE Trans. Magn.* **37**, 2049 (2001).
80. G. Csaba and W. Porod, Simulation of field-coupled architectures based on magnetic dot arrays, *J. Comput. Electron.* **1**, 87–91 (2002).
81. M.J. Donahue and D.G. Porter, *Interagency Report NISTIR 6376*, OOMMF User's Guide, Version 1.0. http://math.nist.gov/oommf/
82. G. Csaba, A. Imre, G.H. Bernstein, V. Metlushko, and W. Porod, Nanocomputing by field-coupled nanomagnets, *IEEE Trans. Nanotechnol.* **1**, 209–219 (2002).

Chapter 11

Carbon Nanotubes for Nanoscale Spin-Electronics

B. Alphenaar[1], S. Chakraborty[1] and K. Tsukagoshi[2]

[1]*Department of Electrical and Computer Engineering, University of Louisville, Louisville, KY 40292*

[2]*The Institute of Physical and Chemical Research (RIKEN), Hirosawa 2-1, Wako, Saitawa 351-0198, Japan*

1. Introduction

During the last ten years, spin electronics has grown rapidly, with the development of spin electronic magnetic tunnel junctions (MTJs) [1], giant magnetoresistive (GMR) devices, and spin valves [2] for applications as nonvolatile magnetic memory and magnetic read heads. These devices are based on a two-terminal ferromagnetic/non-ferromagnetic layer structure in which the magnetic moments of the ferromagnetic layers rotate independently of each other. Due to spin scattering at the ferromagnetic/non-ferromagnetic interface, the resistance is high when the ferromagnetic moments are anti-parallel, and low when the ferromagnetic moments are parallel, resulting in a bi-state spin resistor. The percent change in resistance is $\Delta R/R_a = (R_a - R_p)/R_a$, where R_a and R_p are the resistances in the anti-parallel and parallel configurations, respectively.

The challenge now facing spin electronics is to find a way to enhance the functionality of the spin beyond spin-scattering in two-terminal resistors [3]. An individual electron spin state could, for example, be thought of as a bit of information, to be transmitted between a spin source and a spin detector. Spin manipulation during transmission would provide the possibility of constructing spin logic or computational devices—including three-terminal spin transistors and quantum computational bits based on spin entanglement. The advantages of spin-based logic devices include a decrease in power consumption, and an increase in switching speed when compared with charge based devices. Such advanced spin applications are currently limited, however, by the difficulty of transmitting spin signals over long enough distances to allow for spin manipulation.

In this chapter we review initial work on one possible solution to this problem—the use of carbon nanotubes to transmit electron spin between ferromagnetic contacts. Nanotubes are easily synthesized with extremely small diameters (1–3 nm) and relatively long lengths (up to 100 microns), allowing them to be integrated within ultra-high density circuits [4]. Considering their small size, they are extremely strong and stable, and in comparison with other molecular scale wires and devices, they are easy to contact to form low resistance interconnects. Carbon nanotubes already form the basis of a number of charge based electronic devices. Metallic nanotubes connected to semiconducting nanotubes form molecular scale rectifiers [5], while gated nanotubes produce field effect transistors [6] and single electron transistors [7]. For spin-electronic applications, the large electron scattering lengths in nanotubes make them candidates for low-loss spin transmission lines. Spin signals can be amplified or suppressed by gating the nanotube to change the electron density. High ballistic current flow also produces anomalously large magnetic moments in carbon nanotube circuits [8]. This is useful for magnetic memory and switching applications to amplify and sense local magnetic fields. Finally, the possibility for studying the interaction of electron spin with Coulomb charging [9] and Luttinger liquid [10] behavior makes nanotubes an important physicist's tool for studying spin transport in low-dimensional systems.

1.1. Overview of Spin Injection and Detection

In a ferromagnetic material (such as Co, Ni, or Fe) a majority of electron states at the Fermi energy point in a preferred spin direction (see Fig. 11.1(a)). Ferromagnets thus provide a source of spin-polarized electron current. In 1971, Tedrow and Meservey did the first work showing that a ferromagnet can be used to inject spin into a non-ferromagnetic material [11]. In these experiments, the conductance of a ferromagnetic/superconducting junction is measured as a function of bias, while a magnetic field applied parallel to the superconductor produces spin splitting in the quasi-particle density of states. Conductance peaks are observed for each spin level, and the difference in the heights of the peaks corresponds to the difference in populations of the spin states within the ferromagnet. The results of these experiments provide a measure of the spin polarization of ferromagnetic materials, and are still widely used.

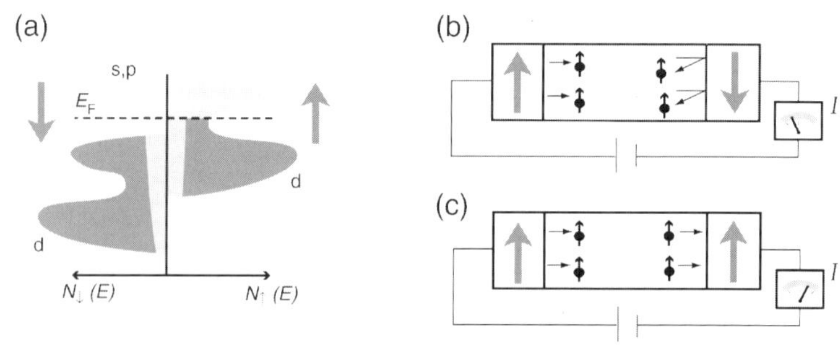

Fig. 11.1. (a) Band structure for a ferromagnetic material showing the imbalance in the density of states at the Fermi energy. (b) Schematic representation of a F/N/F spin-electronic device in the anti-parallel and (c) parallel configurations (after [3]).

Work proceeded toward the development of spin-electronic devices in which spin was both injected and detected by ferromagnetic contacts. The model device structure (shown in Fig. 11.1(b) and (c)) consists of two ferromagnetic contacts separated by a non-ferromagnetic transmission layer. The behavior of this device is often explained using the following intuitive argument [3]. A bias placed across the F/N/F structure injects spin-polarized electrons from the left ferromagnetic contact into the non-ferromagnetic material. The electrons travel through the non-ferromagnetic layer, remaining polarized for some spin scattering length l_s. Electron transmission into the right ferromagnetic contact depends upon the orientation of the incoming electron spin. The transmission rate is higher for spins pointing parallel to the majority spin population than for spins pointing anti-parallel, due to the higher population of majority spin states in the right contact. The total device resistance then depends upon the polarization of the right contact with respect to the left contact, (assuming the thickness of the non-ferromagnetic material is less than the spin scattering length). The resistance is high when the ferromagnetic moments are anti-parallel (Fig. 11.1(b)) and low when the ferromagnetic moments are parallel (Fig. 11.1(c)). The spin orientation follows the magnetization of the ferromagnetic materials and can be rotated by applying an external magnetic field.

Julliere [12] demonstrated the first device of this kind using iron and cobalt ferromagnetic contacts separated by a thin germanium oxide barrier. Since in this case the non-ferromagnetic region is a tunnel barrier, this type of device has become known as a magnetic tunnel junction or MTJ. Transport is presumed to be solely due to tunneling across the barrier so that the conductance is proportional to the product of the density of states in either ferromagnetic contact. The difference between the tunnel resistance in the parallel (R_p) and anti-parallel (R_a) states is then given by:

$$\Delta R/R_a = (R_a - R_p)/R_a = 2P_1P_2/(1 + P_1P_2) \qquad (1)$$

Here, P_1 and P_2 are the percentages of conduction electrons polarized in the majority spin band in the two ferromagnetic contacts ($P_{Co} = 34\%$ and $P_{Fe} = 44\%$) [1]. Julliere observed $\Delta R/R_a = 14\%$ at 4.2 K, which dropped off to zero as the temperature increased to 300 K. This is much smaller than the theoretically predicted value of 27%. However, in 1995, improvements in interface conditions finally allowed much larger room temperature values of $\Delta R/R_a$ to be observed [13], and MTJs now routinely produce resistance changes of 40% or more [14].

Further work showed that spin polarized current can survive over relatively long distances in non-ferromagnetic metals. Johnson and Silsbee [15] performed a series of influential experiments on the injection of spin from a permalloy (Py) contact into aluminum. The voltage difference was measured between a Py detector contact and a non-ferromagnetic contact at a point downstream from the Py injector. These experiments showed that spin polarization could be detected even after the electrons traveled distances of 100 microns or more. Further work by Johnson [16] explored the possibility of a three terminal Py/Au/Py spin injection device. In this case, the non-ferromagnetic Au layer acted as a base for the injection of spin polarized current between the ferromagnetic emitter and collector.

Interest in semiconductor spin-electronic devices was motivated by the theoretical work of Datta and Das [17]. They proposed an electron wave analog of an electro-optic modulator in which spin polarized electrons are injected into a narrow-gap semiconductor by a ferromagnetic contact. While moving through the semiconductor, the electron spin precesses (at a rate that depends on the strength of the spin–orbit coupling in the semiconductor) and is eventually detected by a second ferromagnetic contact. The total detected signal depends upon the final direction of the electron spin with respect to the magnetic moment in the ferromagnetic detector. An external gate voltage can be used to control the amount of spin–orbit coupling, and thus the degree of spin rotation. There have been a number of attempts to produce the spin-electronic device proposed by Datta and Das [18,19]. Typically, permalloy (Py) contacts are made to an InAs channel, and the resistance is measured as a function of magnetic field. $\Delta R/R_a$ of about 1% is observed as the magnetic moments of the contacts switch from the parallel to the anti-parallel state. In addition, Hu et al. [19], observe a change in the sign of the signal as a function of gate voltage, that seems to indicate the rotation of the spin due to a change in the spin–orbit coupling constant.

1.2. Problems with Electrical Spin Detection

Electrical detection of spin transported over large distances (as described above) has provoked a great deal of controversy, and the role of the spin in these experiments is still under debate. In these devices the percentage change $\Delta R/R_a$ between the anti-parallel and the parallel ferromagnetic moment configurations is often disappointingly small, and is typically less than 1% . For such small changes, it is difficult to distinguish between the influence of spin injection and the influence of the stray magnetic field of the ferromagnetic contacts [20]. For example, the resistance of a low-density semiconductor is strongly dependent on applied field, due to the Hall effect and weak localization. At the same time, the ferromagnetic contacts can produce a fairly high field, as much as 1 T near the contact surface. When the alignment of the ferromagnetic contacts flips between the anti-parallel and the parallel states, the total stray field due to the contacts changes, and the channel resistance is modified. (This effect has even been incorporated in a non-volatile memory device design [21].) Spin-injection requires that the contacts be very close to the transport path, and also to each other, enhancing the influence of the stray field. It is then difficult to separate the spurious stray field effect from the spin injection phenomenon.

The stray field effects can be eliminated from the measurement by removing the ferromagnetic detector from the current path. The voltage difference between a non-ferromagnetic contact and a ferromagnetic contact probing the same region is measured, so that the voltage drop in the non-ferromagnetic material has no effect. This strategy was first used by Johnson and Silsbee [15] (as described above) and recently re-employed by Jedema et al. [22] (using an Al non-ferromagnetic region) and Filip et al. [23] (using an InAs non-ferromagnetic region). In this configuration, Jedema et al. measure a resistance change of 0.5 mΩ (about 0.2% change) while Filip et al. measure no noticeable resistance change between the parallel and anti-parallel states. These results suggest that a substantial portion of the resistance change measured in the standard two-terminal arrangements are due to stray field effects rather than spin injection.

If the spin transport lengths in metals and semiconductors are larger than the contact separation, why then are the spin mediated resistance changes much smaller than those observed in MTJs? The problem lies in the non-judicial application of the intuitive model described in Fig. 11.1. The model assumes that the energy distribution of the spin polarized electron population is the same in the non-ferromagnetic layer as it is in the ferromagnetic injector contact, and remains unchanged as the electrons travel between the contacts. This requires that electrons are injected and detected across tunnel barriers, and travel ballistically through the non-ferromagnetic layer, without any momentum scattering events. It also assumes that the resistance is determined solely by tunneling resistances at the interfaces, and neglects the resistance of the contacts or transmission channel.

These conditions are not met if the F/N/F device consists of transmissive ferromagnetic contacts made to a semiconductor or metal non-ferromagnetic region. In this case, electron transport is diffusive (electrons undergo many momentum scattering events between the two ferromagnetic contacts) and the bias drops across the entire sample and not predominately at the contact interfaces. The diffusive picture for a F/N/F device was first described by Valet and Fert [24], and has recently been re-investigated by Schmidt et al. [25] and Filip et al. [23] In the simplest approximation, the resistance of a F/N/F structure is given by sum of the resistances of the individual elements. Each ferromagnetic contact is assumed to contain two conducting channels. One corresponds to the spin up electrons and has a resistance R_{up}, and the other corresponds to the spin down electrons, and has a resistance R_{down}. (We take up to be the majority spin direction, so that $R_{\mathrm{up}} < R_{\mathrm{down}}$.) The resistance of the non-ferromagnetic region is R_{n}, for both spin directions. It is then easy to show that the difference in the resistance ΔR between the parallel and anti-parallel states is given by [24]:

$$\Delta R = \frac{(R_{\mathrm{down}} - R_{\mathrm{up}})^2}{2(R_{\mathrm{up}} + R_{\mathrm{down}} + R_{\mathrm{n}})} \tag{2}$$

The resistance change is inversely proportional to the resistance of the normal region, R_{n}, and will quickly go to zero as the ratio between R_{n} and $(R_{\mathrm{down}} - R_{\mathrm{up}})$ increases. For large changes to occur, R_{n} must be comparable to $(R_{\mathrm{down}} - R_{\mathrm{up}})$. As recent experiments have shown, this is difficult to achieve in standard configurations, and R_{n} is typically 10 to 100 times larger than $(R_{\mathrm{down}} - R_{\mathrm{up}})$. The diffusive picture thus does not predict large spin mediated resistance changes in F/N/F structures (as are observed in MTJs). For this, the electron wavefunctions must remain coherent over the length of the non-ferromagnetic region.

1.3. Optical Spin Injection

For the sake of completeness, we point out that in addition to electrical spin injection, it is also possible to inject and detect spin using optical techniques. In GaAs, (and other optically active semiconductors) circularly polarized light can be used to excite spin polarized carriers. The diffusion of the carriers from the injection point can be detected by their influence on the reflectivity. Kikkawa and Awschalom [26] used this technique to demonstrate that the spin-scattering time in GaAs can be as large as 1 μs, giving a spin scattering length of up to 1 cm. In addition, the recombination of spin polarized carriers results in the emission of circularly polarized light. Alvardo and Renaud [27] used this technique to detect the spin polarization of electrons injected from a ferromagnetic STM tip. Also, Fielderling et al. [28] and Ohno et al. [29] have injected spin from magnetically doped semiconductors into an LED type structure, and the spin polarization was determined from the polarization of the emitted light. Optical spin detection in semiconductors is a very promising technique that has generated considerable interest [30]. Nevertheless, conventional logic and memory devices are electrical rather than optical, so that all-electrical spin-injection devices still appear more useful for device applications than those using optical spin-injection.

2. Spin Transport in Carbon Nanotubes

Work continues at a number of institutions to develop new spin electronic devices with increased spin functionality [3]. A difficult question in light of the discussion above is how to increase the length of the spin transmission channel to allow spin manipulation without decreasing the electrical spin detection signal to imperceptible levels. As shown in Table 11.1, spin transmission lengths in two-terminal spin resistors are limited to only a few nanometers. In MTJs, the tunnel resistance increases exponentially with increasing spin transmission length, while in metal or semiconductor devices, the increasing background resistance of the transmission layer washes out the spin resistance change. To maintain high spin resistance changes over long transmission lengths requires that electrons travel without loss of spin or momentum. In other words, the spin transmission medium must transmit electrons *ballistically*, with a minimal number of electron scattering events.

The carbon nanotube is a spin transmission medium that provides a solution to this problem and that potentially enables spin manipulation during transmission [4]. Because carbon nanotubes are one-dimensional conductors with stable, low-defect crystalline structures, electron scattering mechanisms are relatively weak. Electrons travel long distances through nanotubes without experiencing any kind of scattering (10 microns or more), so that spin mediated resistance changes are easily measurable. Large spin transmission lengths also mean that a third terminal can be attached to the ferromagnetically contacted carbon nanotube to create a nanotube spin transistor.

Below, we discuss the progress so far in the fabrication and measurement of ferromagnetically contacted carbon nanotubes for spin electronic devices. Work in this area began a short time ago, and has yielded promising evidence for spin transport

Table 11.1. Properties of F/N/F device structures

Transmission channel	Devices	Conduction mechanism	Maximum length
Dielectric	Magnetic tunnel, junction	Tunneling	1–2 nm
Bulk metal or semiconductor	Spin valve, GMR device	Diffusive transport	3–5 nm
Carbon nanotube	Spin transistor, nanohelix sensor	Ballistic transport	>0.25 μm

effects. As with normal nanotube electronics, the challenge remains to understand and control the carbon nanotube electrical properties, while also dealing with the substantial complications presented by the nanotube/ferromagnetic interfaces. An improved experimental and theoretical understanding will hopefully allow for the observation of large, clean spin-modulated resistances.

2.1. Fabrication of Ferromagnetic/Nanotube Devices

The basic device structure used for studying spin transport through carbon nanotubes is shown in Fig. 11.2. It consists of a ferromagnetically contacted single wall nanotube (SWNT) or multi-wall nanotube (MWNT) on an insulating substrate. Figure 11.2(a) shows a scanning electron micrograph of a MWNT with cobalt contacts on a SiO_2/Si substrate. The fabrication procedure follows that described for non-ferromagnetically contacted nanotubes [4]. Nanotubes are dispersed by sonication in a solvent such as dicholoethane or methanol. The nanotube/solvent mixture is then dropped onto a SiO_2/Si substrate and allowed to dry. The solvent/nanotube concentration ratio and sonication time is varied to produce well-separated nanotubes, randomly distributed on the sample surface. The dispersed nanotubes are located with respect to pre-defined alignment marks using AFM or SEM (AFM is preferred to avoid electron beam damage to the nanotubes). Contact pads are defined in resist using electron beam lithography, after which a ferromagnetic film is deposited using thermal or electron beam evaporation. (Sputter deposition cannot be used because the high-energy sputtered particles damage the nanotube surface.) Large area leads and bond-pads are similarly defined. Using non-ferromagnetic leads and bond-pads (such as AuTi) to connect the ferromagnetic contacts to external wires is preferred since it reduces the magnetic dependent resistance contribution of the ferromagnetic contacts. The silicon substrate can be doped to act as a field effect gate, or alternatively an in-plane metal gate can be deposited.

To minimize the gap between the contacts (and thus the scattering length needed to observe spin polarized transport) it is easiest to define the two contacts in a single lithography step. This results in both contacts being made of the same ferromagnetic material. One alternative is to use a shadow evaporation technique in which the contact material is deposited at a glancing angle to the sample surface. This shifts the position of the deposition with respect to the resist pattern. If a second contact material is then

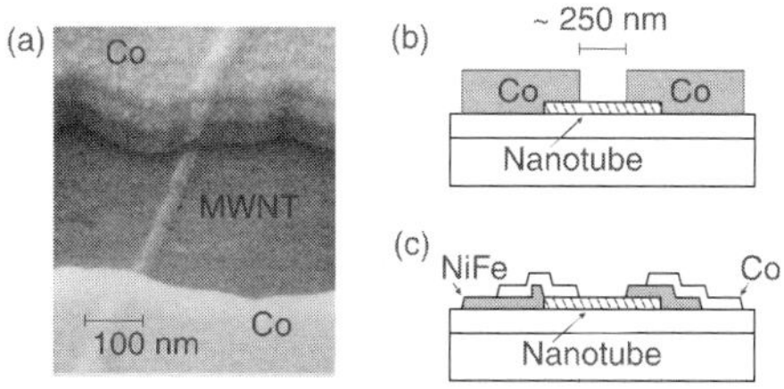

Fig. 11.2. (a) Scanning electron microscope image of a Co contacted multi-wall nanotube. The diameter of the nanotube is 30 nm and the conducting channel length is 250 nm. (b) Schematic diagram of a typical device. The substrate is a semi-insulating silicon wafer covered by an insulating layer of SiO_2. (c) Device structure fabricated using the shadow evaporation technique described in the text.

deposited at the opposite glancing angle, it will be shifted in the opposite direction, forming a stepped contact pattern as shown in Fig. 11.2(c). Here, the left contact material is cobalt and the right contact material is permalloy, and the shortest transport path through the nanotube is from the cobalt to the permalloy. For this to be the dominant transport path requires that the contact resistance be relatively uniform over the contact area, which is, as we describe in section 2.4, not necessarily a valid assumption.

Because of their small size, SWNTs are more difficult to locate and contact than MWNTs. An alternative method is to grow SWNTs on the surface at specific sites using CVD [31]. Nanometer scale catalyst particles are first deposited and patterned using standard lithography. Subsequent CVD growth results in the formation of SWNTs spreading from the catalyst areas. A ferromagnetic layer is then deposited on the catalyst islands, forming contacts to a number of SWNTs in parallel. It is usually assumed in such devices that conduction is dominated by a single nanotube.

Although it is clear from measurements of MTJs that the ferromagnetic/non-ferromagnetic interface can have a strong influence on the spin transport, there has not yet been a systematic study of the influence of interface properties or the contact deposition method on the spin electronic properties of ferromagnetically contacted nanotubes. Surface contamination on the nanotube is likely to be a major source of variations in contact resistance. Organic contamination can be partially removed by O_2 plasma etching prior to contact deposition, taking care not to damage the nanotube with excess plasma exposure. Thermal annealing after contact deposition can also be used to reduce the contact resistance. A 10 min anneal at 400°C in a 5% H_2/95% Ar gas mixture can reduce the contact resistance by two orders of magnitude or more. The mechanism for this reduction is not yet understood. An additional complicating factor is caused by the contact domain structure. Increasing the magnetic film thickness and altering the size and shape of the contact can improve the domain size and thus also enhance the effective spin polarization.

2.2. *Magnetoresistance Switching in Multi-Wall Nanotubes*

Tsukagoshi et al. [32] reported the first electrical measurements of ferromagnetically contacted nanotubes in 1999. Crude MWNTs were used, synthesized from graphite rods by the arc discharge method under a helium atmosphere. This produces nanotubes typically 10–40 nm in diameter and 1–10 microns in length. Cobalt contacts 60–100 nm in thickness were deposited by thermal evaporation and magnetoresistance measurements performed at 4.2 K in a helium bath cryostat with the magnetic field from a superconducting magnet directed in the plane of the substrate, and perpendicular to the current path. The resistance was measured using an ac lock-in technique with an excitation voltage of 100 μV.

Figure 11.3 shows the two-terminal resistance of three different Co-contacted nanotubes as a function of magnetic field. The field is first swept from -100 to 100 mT (solid line) and then back to -100 mT (dashed line). In each trace a resistance peak appears as the magnetic field moves through 0 T. There is a large hysteresis in the peak position between positive and negative sweep directions. The resistance switching appears to be correlated with the switching of the magnetization in the ferromagnetic contacts. At high magnetic field, the contact magnetizations align parallel to the

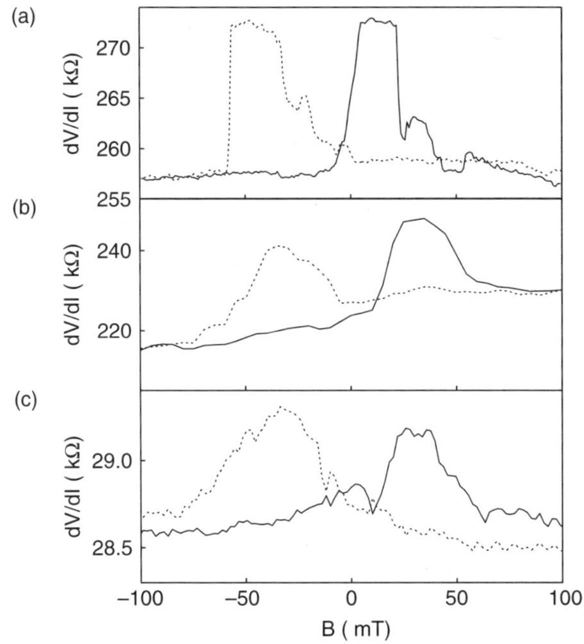

Fig. 11.3. Two-terminal differential resistance as a function of magnetic field for three different cobalt contacted multi-wall nanotube devices. The dashed (solid) line shows the down (up) sweep direction. The magnetic field is directed parallel to the substrate and perpendicular to the current path. The temperature is 4.2 K.

magnetic field direction, and the resistance is low. As the field is swept through 0 T, the magnetization polarities change sign as the applied magnetic field is increased above the coercive field of the magnetic contact. If the two contact magnetizations switch at slightly different magnetic fields they become anti-parallel near $B = 0\,\text{T}$. When this occurs the majority spin states are misaligned, and the device resistance increases above the low resistance parallel magnetization state, producing the observed resistance peaks. The maximum resistance change observed between the parallel and anti-parallel states is $\Delta R/R_a = 9\%$ at 4.2 K. Among the devices measured, however, there are large variations in both the resistance change and the total resistance. Recently, two other groups have reported magnetoresistance switching in ferromagnetically contacted MWNTs. Orgassa et al. [33] observed resistance switching in Co/MWNT/Ni-Fe devices fabricated using a shadow evaporation technique while Zhao et al. [34] reported anomalously high $\Delta R/R_a$ in cobalt contacted MWNTs (for further discussion, see section 2.5).

These results suggest that spin injected from the ferromagnetic contacts travels coherently through the nanotube to modify the nanotube device resistance, in the same way as described for the generic F/N/F device in Fig. 11.1. The F/nanotube/F device fits the requirements of the ideal model description fairly well. The electron scattering length for the MWNT is approximately equal to the contact separation ($\approx$200 nm) allowing for quasi-ballistic electron transport while the ferromagnets most likely form tunneling contacts to the nanotubes, whose resistance will dominate the total device resistance. It is instructive to calculate the lower limit of the nanotube spin scattering length that would be required in order to produce the observed results. We can make a rough estimate for the spin scattering length by using a modified version of Julliere's model for spin injection across a tunnel barrier. We assume that electrons are injected from contact 1 into the nanotube with a polarization P_1 and are detected by contact 2 with spin polarization P_2. As the electrons travel through the nanotube, the spin polarization decays exponentially with a characteristic spin scattering length l_s, before reaching the second contact, a distance l away. The percent resistance change between the parallel and anti-parallel states is then given by:

$$\frac{\Delta R}{R_a} = \frac{2P_1 \exp(-l/l_s) P_2}{1 + P_1 \exp(-l/l_s) P_2} \tag{3}$$

For cobalt, the polarization has been estimated to be 34%, so that the maximum possible resistance change in a cobalt contacted nanotube is 21%. In the nanotube device measurements, $\Delta R/R_a$ reaches 9%, giving a value for the spin scattering length of 262 nm.

Although the spin scattering length calculated here seems reasonable, there is good reason to believe that the actual length is much larger than this. There are few spin scattering centers in carbon nanotubes and carbon is too light an atom for appreciable spin–orbit scattering to occur. Results from ESR measurements [35] suggest that the low temperature spin scattering time is 10^{-8} s in MWNTs which assuming a Fermi velocity of 10^6 m/s gives an electron spin scattering length of approximately 1 cm. There are thus most likely other factors limiting $\Delta R/R_a$ that are more important than spin scattering in the nanotube. First, the percentage of polarized spin injected into the nanotube will depend upon the interface quality as well as spin scattering at the interface and could be much lower than 21%. Second, momentum scattering events will act to re-distribute the spin population and reduce the spin polarization of the conduction electrons near the Fermi energy. Finally, due to the high contact resistance, an electron must on average undergo multiple transits of the nanotube before it is eventually transmitted into the detector contact. The expectation value for the total transport length is $\langle l \rangle = 1 + (2r_1r_2/(1 - r_1r_2))$ where r_1 and r_2 are the reflection coefficients for contacts 1 and 2. For high resistance contacts the reflection coefficient can be as high as 0.95 giving a value of approximately 10 microns for the total transport length.

2.3. *Influence of Stray Magnetic Fields*

Low temperature measurements of non-ferromagnetically contacted MWNTs show that the MWNT resistance itself changes with magnetic field. This background magnetoresistance is thought to be due to a weak-localization effect caused by the influence of the magnetic field on phase coherent pathways in the nanotube [36]. We must be careful then, in our interpretation of the ferromagnetically contacted nanotube measurements to take into account the influence from stray magnetic fields. As the magnetization of the contacts flips between the anti-parallel and the parallel configurations, the stray field from the contacts changes and this will in turn change the measured resistance. To approximate the magnitude of the resistance change, we first calculate the change in

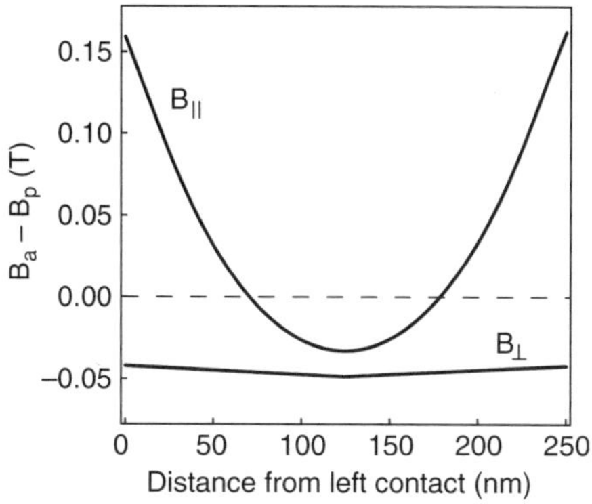

Fig. 11.4. Calculation of the difference in the anti-parallel (B_a) and parallel (B_p) configurations of the stray magnetic field in a cobalt contacted nanotube (see text). The field difference is shown as a function of distance from the left contact, for the magnetic moments aligned ($B_\|$) and transverse ($B_\perp$) to the current path. The contact separation is 250 nm, the contact thickness is 100 nm and the contact width is 2 microns.

the stray field between the parallel and anti-parallel configurations. The stray magnetic field is calculated using:

$$\vec{H}(\vec{r}) = \vec{\nabla} \int d\vec{r}' \frac{\vec{\nabla}\vec{M}(\vec{r}')}{|\vec{r} - \vec{r}'|} \tag{4}$$

where $M(r)$ is the magnetization at a point r in the contact.We model the contacts as two semi-infinite sheets, each of uniform magnetization, and separated by a small gap. The magnetization is taken to be the saturation magnetization of cobalt (1400 kA/m). We calculate the stray field change for two different configurations: magnetic moments transverse to the current path ($B_\perp$)—corresponding to the configuration used in the experiments above—and magnetic moments parallel to the current path ($B_\|$). Figure 11.4 shows the stray field change in the two configurations as a function of the position between the two contacts for a contact separation of 250 nm, a contact width of 2 microns and a contact thickness of 100 nm. The maximum fields are $B_\perp = 0.05$ T (in the center of the sample) and $B_\| = 0.15$ T (near the contacts). From non-ferromagnetic contact measurements we find that the resistance of a MWNT typically decreases by about 1% as the applied field increases from 0 to 0.1 T at 4.2 K. This means that the maximum resistance change due to stray field effects in the transverse configuration is 0.5% , far smaller than the 9% change that we observe. (We note that a direct experimental comparison of the magnitude of the switching in the two configurations would be a useful way to estimate the stray field effects.)

One way to differentiate the background magnetoresistance from the magnetoresistance switching is to compare the temperature dependence of the two effects. Such a comparison is shown in Fig. 11.5 for a cobalt contacted MWNT device. Resistance measurements were made from −1 to +1 T so that the decrease in resistance due to the background weak localization and the magnetoresistance switching around zero magnetic field can both be observed. In Fig. 11.5(a) the magnitude of the zero field switching is shown to be relatively independent of temperature below 20 K. Above

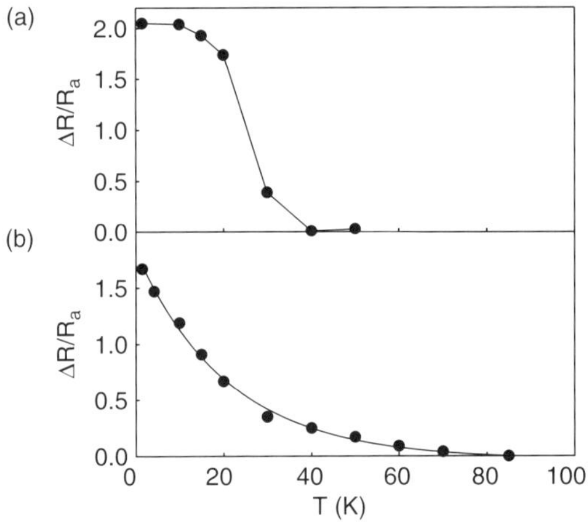

Fig. 11.5. Comparison of the temperature dependence of the (a) magnetoresistance switch and the (b) background magnetoresistance for a cobalt contacted multi-wall nanotube.

20 K, however, the magnitude decreases, and drops sharply to zero with increasing temperature. Orgassa et al. [33] observe a similar temperature dependence in FeNi/MWNT/Co devices, with the magnetoresistance ratio abruptly disappearing above 30 K. In contrast, the background magnetoresistance shown in Fig. 11.5(b), decreases continuously with increasing temperature, with approximately an inverse power law dependence. This agrees with measurements on non-ferromagnetically contacted MWNTs, and with weak localization theory, which predicts a $T^{-1/2}$ dependence for the magnetoconductance [36].

The difference in the temperature dependence of the two types of magnetoresistance provides further evidence that the low-field magnetoresistance switching is not caused by simple stray field effects. However, a detailed explanation for the temperature dependence of the magnetoresistance switching is not yet available. It appears that the sharp decrease in the magnetoresistance at a particular temperature is due to a transition in the contact from a ferromagnetic to a paramagnetic or anti-ferromagnetic state. Although the Curie temperature for cobalt is over 1,000°C, the material at the interface is most likely a mixture of cobalt and cobalt oxide, with a much-reduced ferromagnetic transition temperature. Depending on the quality of the interface, the transition temperature will vary, resulting in a large variation in the temperature dependence among devices.

Improvements in the nanotube interface should make it possible to obtain a better understanding of the spin scattering mechanisms in the carbon nanotube. For a MTJ, the reduction in $\Delta R/R_a$ with temperature is thought to be mainly due to the excitation of magnons in the ferromagnetic material, which effectively reduce the spin polarization [1]. This results in a $T^{3/2}$ dependence. In metals, on the other hand, spin relaxation is thought to occur by the Elliot–Yafet mechanism, in which scattering from a spin conserving potential (due to impurities, defects or phonons) can induce transitions between spin up and down states when a spin–orbit interaction is present [37]. From this, it follows that the spin relaxation rate is proportional to T at high temperatures, a dependency that is observed in monovalent (e.g. Na and K) metals. However, in polyvalent metals (e.g. Al and Mg), spin scattering "hot-spots" exist where the energy separation between neighboring bands is small. The spin-scattering rate is thus enhanced by orders of magnitude. A theory for spin scattering in carbon nanotubes has not yet been developed, but, it is possible that nanotubes also contain such spin hot spots, or some other spin enhancement mechanisms. This implies that a different spin scattering rate should be observable in nanotubes with different chirality.

2.4. *Ferromagnetic/Nanotube Interface Effects*

The temperature dependence shown in Fig. 11.5 suggests that the contact interface plays an extremely important role in determining the magnetoresistance of ferromagnetically contacted carbon nanotubes. The two terminal resistance of a ferromagnetically contacted MWNT is typically 200 kΩ or more at 4.2 K while the MWNT resistance measured with transmissive non-ferromagnetic contacts can be as low as 12 kΩ. The resistance due to spin-polarized transport is thus dominated by transmission across the contact interface, with the nanotube acting as a low-resistance transport channel or resonant cavity for the electron spin [34]. The contact resistance is determined by the energy and momentum overlap of nanotube and contact electron states, which is strongly dependent on the surface conditions. It has also been shown that the spin polarization can change in magnitude or even change direction according to the local surface potential [38]. Since the contact area between the ferromagnet and the nanotube is so small, local fluctuations in interface conditions do not average out, and can have a dramatic influence on individual device behavior.

Local fluctuations are in part caused by variations in the domain configuration at the interface. The cobalt surface domains at zero magnetic field (probed using MFM) are approximately the same width as the diameter of a MWNT (50 nm) [39]. Fluctuations in individual domains can thus completely change the spin polarization of the injected electrons, and it is these fluctuations that most likely cause the misalignment in contact magnetization that results in the magnetoresistance switching that is observed. Measurements of nanotubes contacted by two different ferromagnetic materials show no larger or more reproducible magnetoresistance than nanotubes with two contacts made from the same ferromagnetic material. An example is shown in Fig. 11.6 where the resistance of a NiFe/MWNT/Co device is plotted as a function of magnetic field. The contact resistance is lower than for nanotubes contacted with a single ferromagnetic layer, presumably because the multi-layered contact improves step-coverage. However, the magnetoresistance ratio is also lower, and there is no evidence that the magnetoresistance ratio improves by using materials with different coercivities (see also [33]). Local fluctuations can also be caused by contaminants on the surface. Of particular concern is the presence of cobalt oxide on the nanotube interface that forms during the deposition process. Cobalt oxidizes very easily even at room temperature in air [40], so it is likely that the interface is at least partially oxidized. Cobalt oxide is an antiferromagnet, but mixtures of cobalt oxide and cobalt show ferromagnetic behavior, but with a lowered Curie temperature [41]. As described above, its presence will have a detrimental effect on the magnetoresistance ratio.

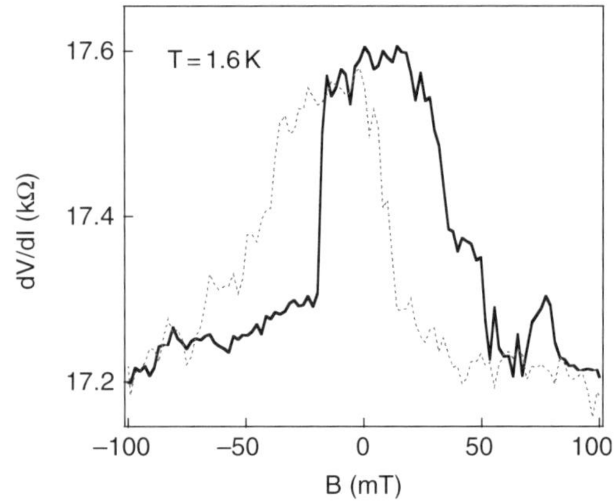

Fig. 11.6. Two-terminal differential resistance as a function of magnetic field for a NiFe/MWNT/Co device measured at 1.6 K. The device was made using a shadow evaporation technique producing a device layout as shown in Fig. 11.2(c).

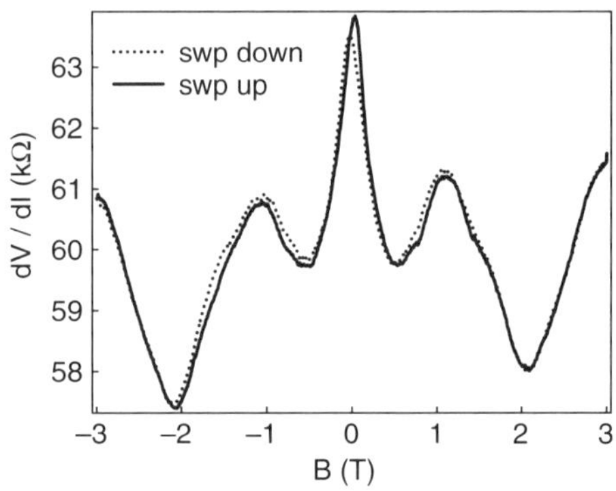

Fig. 11.7. Two-terminal differential resistance as a function of magnetic field for a multi-wall nanotube having one Co contact and one Pt/Au contact. The dashed (solid) line shows the down (up) sweep direction. The magnetic field is directed parallel to the substrate, perpendicular to the current path, and the temperature is 4.2 K.

It is possible that fluctuations in the domain configuration of the ferromagnetic contacts produce a much larger local magnetic field variation than is indicated by the calculation above for the homogeneous contact. It is difficult to estimate the magnitude of this effect; however, we can experimentally test this possibility by performing magnetoresistance measurements on devices having only a single ferromagnetic contact. Resistance changes due to the local stray field should be observed in a MWNT having only a single ferromagnetic contact, while resistance changes due to spin injection require two ferromagnetic contacts (a spin injector and detector). We fabricated five MWNT devices contacted by one ferromagnetic contact (Co), and one non-ferromagnetic contact (Pt/Au) [42]. Figure 11.7 shows results from a typical device. A small hysteresis in the magnetoresistance is observed around $B = 0$, presumably produced by the stray field from the contact. However, no hysteretic switching similar to that in Fig. 11.3 is observed in this, or any of the single ferromagnetic contact devices. Although this evidence is suggestive, there is of course still some possibility that if more single ferromagnetic contact devices were made, one of them would show switching due to changes in the domain configuration. This is because of the inherently large sample-to-sample variations in carbon nanotube device characteristics.

Surprisingly, even though there is no evidence for magnetoresistance switching in single ferromagnetic contacted nanotubes, there is evidence that the domain configuration can have an influence on the resistance. As can be seen in Fig. 11.7, the magnetoresistance of the single ferromagnetic contact device is asymmetric with magnetic field: $R(-B) \neq R(B)$. The asymmetry can survive to fields as high as 6 T, and has been observed in all reported double ferromagnetic contact measurements. Since the nanotube has no known properties that are asymmetric with magnetic field, the asymmetry must be caused by the magnetic contacts. Note though that it is not obvious that the asymmetry is caused by a change in the stray field because the magnetoresistance peaks (whose position depends on the applied field) do not shift with the magnetic field direction. It is likely that the domain configuration is pinned due to the surface potential in a configuration that is independent of magnetic field. If the total magnetization of the contact is finite in the pinned configuration, the energy of the electron states will shift asymmetrically with increasing magnetic field, thus changing the contact resistance. A more detailed analysis of this result is now in progress.

2.5. *Negative Magnetoresistance Switching*

Figure 11.8 shows the magnetoresistance of a cobalt contacted MWNT as reported recently by Zhao et al. [34]. In this case the resistance is observed to decrease near zero field, rather than increase, in conflict with our expectations based on the spin injection model described above. Inverse magnetoresistance switching in MWNTs has also been reported by Orgassa et al. [33], and has occasionally been observed by the authors as well. As described by Zhao et al., the inverse switching will occur if the electron spin populations near the interface of the two contacts are polarized in the opposite sense when the magnetizations are parallel and in the same sense when the magnetizations are anti-parallel. Although initially counterintuitive, recent theory suggests that this can in fact be the case [43]. At the Fermi energy, cobalt contains negatively spin-polarized d-electrons and positively spin-polarized s-electrons. The spin polarization injected into the non-ferromagnetic region is determined by the relative coupling strength of the two electron populations across the ferromagnetic/non-ferromagnetic interface. Since the coupling strength is strongly dependent on the interface conditions, it is possible for the spin polarization to be either positive or negative. Experiments on MTJs show that the injected spin polarization is in fact dependent on the material at the interface. A thin Ru interface layer has been observed to change the sign of the spin polarization injected from a cobalt contact leading to an inverse magnetoresistance change [44]. MTJs made with $SrTiO_3$ or Ta_2O_5 tunnel barriers also show an inverse magnetoresistance change at low bias [45,46]. For the ferromagnetically contacted nanotube then, it is easy to imagine a situation where the interface conditions at the two ferromagnetic contacts differ, to make the sign of the spin polarization opposite

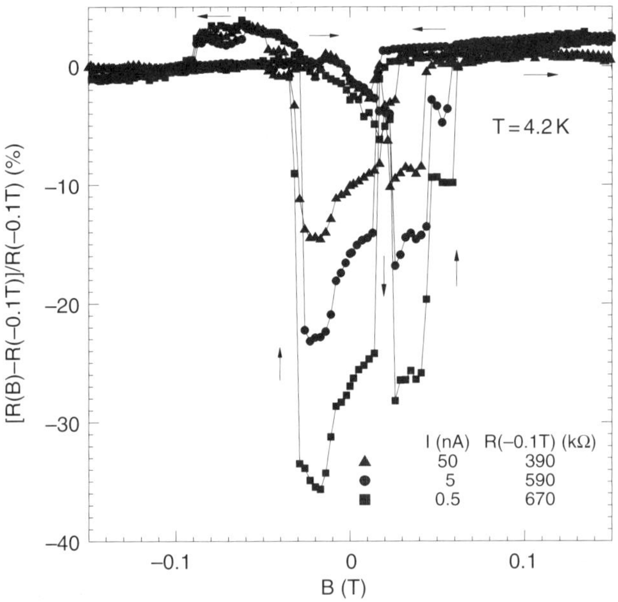

Fig. 11.8. Magnetoresistance ratio of a cobalt contacted multi-wall nanotube as a function of magnetic field for a series of different biases as reported by Zhao et al. (see [34]).

when the magnetization of the contacts are parallel, and equal when the magnetization of the contacts are anti-parallel, thus leading to a negative magnetoresistance switch.

A clue to the type of interface condition necessary to produce a change from normal to inverse magnetoresistance switching is provided by measurements of two of our nanotube devices in which an unusual temperature dependence of the magnetoresistance is observed. Figure 11.9(a) shows the resistance of a single Co contacted MWNT device at three different temperatures. At low temperature, the magnetoresistance switch is normal, but as the temperature increases, the magnitude of the magnetoresistance switch first decreases, and then changes to the inverse direction. Figure 11.9(b) plots $\Delta R/R_a$ for this device over a wide range of temperatures. The change from normal to inverse magnetoresistance occurs at about 20 K in this device, while in the second anomalous device that we found the changeover occurs at around 10 K.

It appears that the properties of the ferromagnetic interface layer are also dependent on temperature. We hypothesize (as described in section 2.4) that the interface consists of a Co and Co–O mixture, which undergoes a ferromagnetic–paramagnetic transition at relatively low temperatures. The region closest to the nanotube contains the highest percentage of Co–O, and undergoes the transition at correspondingly lower temperatures. The result is that the thickness of the interfacial paramagnetic layer increases with increasing temperature. Theoretically, it has been shown that the polarization of spin injected from cobalt through a tunnel barrier changes sign as the thickness of the tunnel barrier increases [38], while in MTJs it has been observed that the direction of the magnetoresistance switch changes from normal to inverse by increasing the thickness of a interfacial paramagnetic layer [47]. In our case, the polarization change occurs at only

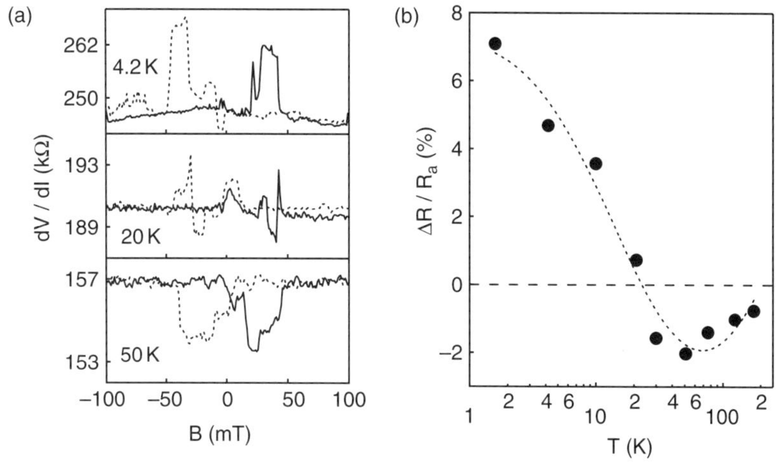

Fig. 11.9. (a) Two-terminal differential resistance as a function of magnetic field for a cobalt contacted multi-wall nanotube device at three different temperatures. (b) Magnetoresistance ratio as a function of temperature for the same device. The dashed line is a guide to the eye.

one of the ferromagnetic interfaces, resulting in the observed changeover from normal to inverse magnetoresistance switching. Missing from this discussion is an analysis of the influence of the carbon nanotube electron state distribution on the sign of the injected spin. A detailed model of the type described in [38] needs to be developed for ferromagnetic/nanotube junctions to determine if the sign of the polarization depends upon the nanotube chirality.

2.6. Ferromagnetically Contacted Single Wall Nanotubes

Single wall nanotubes (SWNTs) have a number of advantages over multi-wall nanotubes for spin transport measurements. The SWNT is a simpler and cleaner system than the MWNT. Inter-layer scattering does not exist, and the transport is truly one-dimensional. In contrast to the MWNT, the SWNT resistance has little or no magnetic field dependence. This makes it possible to ignore the influence of stray fields from the contacts in interpreting measurements of ferromagnetically contacted SWNTs. SWNTs also differ from MWNTs in that SWNTs can be influenced by Coulomb charging at relatively high temperatures and behave as quasi-one-dimensional quantum dots [9]. The conductance of a SWNT shows sharp peaks as a function of gate voltage that correspond to the single electron charging levels of the nanotube. The SWNT is thus a promising system for exploring the influence of Coulomb charging on spin-injection. The possibility of relatively high temperature operation (Coulomb charging effects survive in nanotubes up to 100 K) makes the system more appealing for device applications.

The disadvantage of SWNTs is that their small size makes it difficult to make reliable low-resistance contacts, particularly when using ferromagnetic materials. SWNT devices tend to be extremely sensitive to thermal cycling and stray charge and have short lifetimes. Because of this, there is only a small amount of data available for

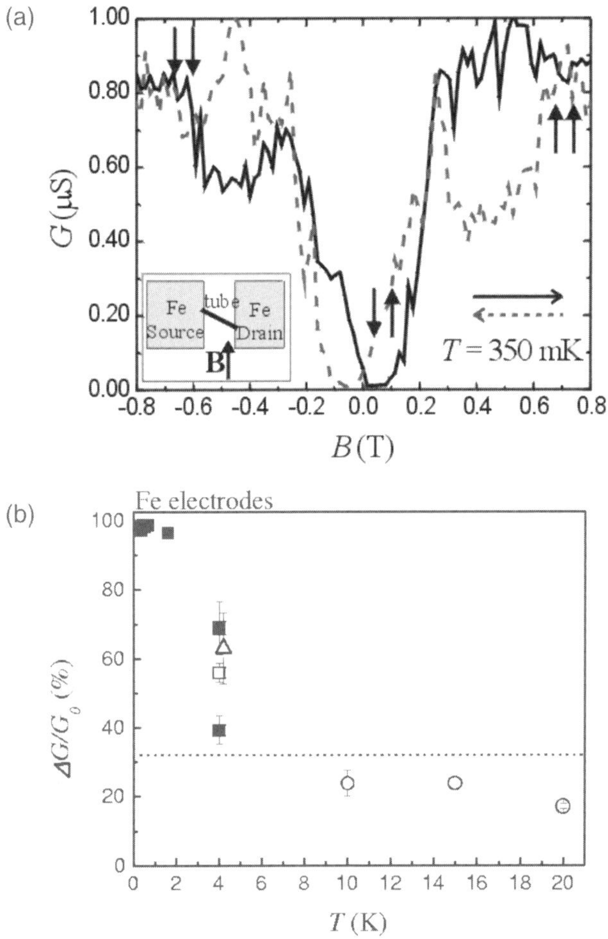

Fig. 11.10. (a) Two-terminal differential conductance as a function of temperature for an iron contacted single wall nanotube device measured at 350 mK. (b) Magnetoconductance ratio as a function of temperature composed from data taken at three different cool-downs. The dotted line indicates the highest possible value according to Julliere's model [12]. From Jensen et al. [48].

ferromagnetically contacted SWNTs. Recently, Jensen et al. [48] successfully fabricated ferromagnetically contacted SWNTs using a CVD process to grow nanotubes from a ferromagnetic precursor. The CVD technique makes the SWNTs relatively easy to locate and to contact. Low temperature magnetoconductance measurements of an iron contacted SWNT are shown in Fig. 11.10(a). A clear conductance switch is observed with a magnetoconductance ratio $\Delta G/G_0$ equal to the surprisingly large value of 100% . As shown in Fig. 11.10(b), $\Delta G/G_0$ is strongly dependent on temperature, and increases sharply below temperatures of 4.2 K.

It is not completely understood why the magnetoconductance ratio is so large in this device. Taking a spin polarization for Fe of 44%, Julliere's model predicts a maximum conductance change of $\Delta G/G_0 = 43\%$. Recent measurements of cobalt contacted MWNTs have also observed a magnetoresistance ratio larger than predicted by Julliere's model (see Fig. 11.8). This was attributed to the influence of Coulomb blockade on spin transport through the nanotube [34]. It has been demonstrated both theoretically [49] and experimentally [50] that Coulomb blockade leads to an enhancement in the magnetoresistance for F/F/F junctions, in which a small ferromagnetic island is connected via tunnel barriers to two ferromagnetic contacts. This occurs when co-tunneling becomes the dominate transport mechanism at low temperatures, and enhances the difference between spin-up and spin-down transmission probabilities. However, it is not clear if Coulomb blockade has the same effect for F/N/F junctions, in which the central region is a non-ferromagnetic island (as is the case for the F/nanotube/F devices). Immamura et al. [51] propose that a different charging energy exists for the two spin directions in the F/N/F junction leading to an enhancement in the magnetoresistance ratio. But Coulomb charging also clearly increases the dwell time of the electron on the island, and thus the amount of time that an electron spin must remain polarized for spin transport to occur. Brataas and Wang [52] have used a linear response model to show that at sufficiently low temperatures the dwell time on the small non-ferromagnetic island is larger than the spin-flip relaxation time, resulting in a decrease in the magnetoresistance of a Coulomb blockaded F/N/F junction.

From our discussion in section 2.4, it is clear that the population of spin polarized electrons injected from the ferromagnetic contact strongly depend upon the conditions of the interface to the non-ferromagnetic material. Due to improvements in interface conditions, MTJs can now routinely be made that show magnetoresistance ratios higher than those predicted by Julliere's model [1,53]. Nevertheless, the huge enhancement in the magnetoresistance ratio observed in SWNTs suggests that the polarization can be especially high for spin injection into nanotubes. It is hoped that further experiment and theory will provide clues for understanding this result.

2.7. Theoretical Models

For the development of a spin-electronic carbon nanotube device, it is useful to determine the maximum possible spin-mediated resistance change, and the conditions under which the maximum can be observed. In a carbon nanotube, the elastic and phase scattering lengths are extremely long, on the order of 10 μm or more at 4.2 K. This means that the carbon nanotube is a ballistic conductor, and electron wave functions should be considered as coherent over the length of the nanotube. Mehrez et al. [54] have reported calculations of the resistance of a ferromagnetically contacted SWNT as a function of the angle between the moments in the two contacts. They apply a non-equilibrium Green's function technique, using tight bonding theory to model the electron potential. They predict that the $\Delta R/R_a$ can be as high as 20%, but that also $\Delta R/R_a$ depends strongly on the nanotube length and chirality.

In a recent publication [42], we reported preliminary calculations performed by M. Wagner on the transmission through a F/SWNT/F structure, taking into account the coherence of the electron wavefunction within the nanotube. A two-band model is used for the conduction electrons in the ferromagnets to describe the d and s–p bands that are partially filled in Fe, Co, or Ni. The d band is characterized by a nonparabolic

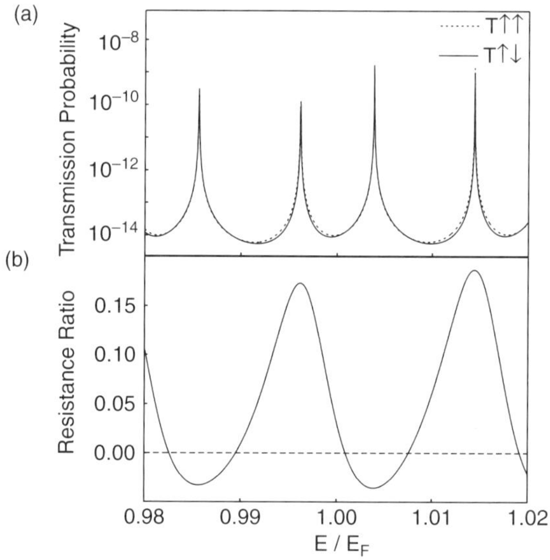

Fig. 11.11. (a) Transmission probability for the parallel ($T_{\uparrow\uparrow}$) and anti-parallel ($T_{\downarrow\uparrow}$) configurations calculated for a ferromagnet/SWNT/ferromagnet device using the ballistic model described in the text. (b) The resistance ratio $\partial R = (T_{\uparrow\downarrow} - T_{\uparrow\uparrow})/(T_{\uparrow\downarrow} + T_{\uparrow\uparrow})$ calculated using the results in (a) (from [42]).

dispersion relation and is assumed to be completely spin polarized, while the s–p band is free-electron like, with randomly oriented spin. A (5,5) armchair SWNT is modeled, in which the dispersion relation consists of a pair of bands that meet degenerately at the Fermi energy, and allow for metallic conduction. The contacts between the ferromagnet and the nanotube are modeled by delta function potential tunnel barriers. Figure 11.11(a) shows the one-dimensional transmission coefficients for the parallel ($T_{\uparrow\uparrow}$) and the anti-parallel ($T_{\downarrow\uparrow}$) configurations as a function of electron energy. Sharp peaks are observed, indicating resonances due to partial backscattering at the carbon nanotube/ferromagnet interface.

If we assume that the current is proportional to the transmission probability (i.e. the zero temperature linear response result), the resistance ratio can be defined as $\partial R = (T_{\uparrow\downarrow} - T_{\uparrow\uparrow})/(T_{\uparrow\downarrow} + T_{\uparrow\uparrow})$. (More realistically, we should integrate over the lateral density of states perpendicular to the tunneling direction to determine the current.) In Fig. 11.11(b), the resistance ratio δR is plotted as a function of energy using the results from Fig. 11.11(a). The exact value for δR is dependent on a number of parameters, not all of which are known experimentally. These include the precise form of the ferromagnet dispersion relations, the spin polarization and distribution of spin among the two bands, and the nanotube chirality. Nevertheless, using reasonable parameters our model shows that it is possible to obtain a reasonably high value for δR of 20%. In addition, large oscillations in δR are predicted as a function of energy. These might be observable by measuring the nanotube magnetoresistance as a function of bias on a capacitively coupled gate, thus forming a new kind of nanotube spin transistor.

2.8. Luttinger Liquid Behavior

Theoretical models for spin transport up to now, all assume that a Fermi liquid model can be applied. Recent experimental results [55] suggest however that due to exchange interactions in one-dimension, the Fermi liquid picture no longer holds for carbon nanotubes, and they instead can be described as Lüttinger liquids, with dramatically different conduction properties. (Although the results for SWNTs have been more widely publicized, transport in MWNTs also appears to be influenced by Lüttinger liquid behavior [56].) One of the most startling predictions of Lüttinger liquid theory is that the charge and the spin can be treated as if they are carried by separate particles. Recent theory suggests that because of this, the F/nanotube/F device will not only show a large change in resistance as a function of spin polarization, but that the current will oscillate as a function of applied bias [10]. Besides the obvious physical implications this observation would have in confirming the Lüttinger liquid model, it would also make for an extremely interesting three-terminal spin-electronic device. It should also be possible to test the existence of spin-charge separation in the nanotube by comparing the temperature dependence of the transport scattering length (determined from the resistivity) with the temperature dependence of the spin scattering length (determined from the spin dependent resistance change) [57].

3. Conclusions

Carbon nanotubes are an extremely efficient charge and spin transmission medium due to the absence of electron scattering mechanisms. This allows for manipulation of intrinsic electron spin transmission rates, and for the possibility of developed three-terminal spin electronic devices. The work described here, while preliminary, provides a basis for further exploration of spin electronics using carbon nanotubes. This will hopefully lead to a more thorough understanding of spin transport mechanisms in carbon nanotubes and spin injection across nanotube/ferromagnetic interfaces.

Acknowledgments

The authors would like to acknowledge M. Wagner for his help with the theoretical analysis and modeling. We also thank D. Lane for her editorial assistance in the preparation of this manuscript.

References

1. J.S. Moodera, J. Nassar, and G. Mathon, Spin tunneling in ferromagnetic junctions, *Annu. Rev. Mater. Sci.* **29**, 381 (1998).
2. S.S.P. Parkin, Giant magnetoresistance in magnetic nanostructures, *Annu. Rev. Mater. Sci.* **25,** 357 (1995).
3. G.A. Prinz, Magnetoelectronics, *Science* **282**, 1660 (1998); J. Gregg, W. Allen, N. Viart, R. Kirschman, C. Sirisathitkul, J.P. Schille, M. Gester, S. Thompson, P. Sparks, V. DaCosta,

K. Ounadjela, and M. Skvarla, The art of spin electronics, *J. Mag. Mag. Mat.* **175**, 1 (1997); J. de Boeck and G. Borghs, Magnetoelectronics, *Phys. World* **12**, 27 (1999).

4. R. Saito, G. Dresselhaus, and M.S. Dresselhaus, *Physical Properties of Carbon Nanotubes* (Imperial College Press, Singapore, 1998); C. Dekker, Carbon nanotubes as molecular quantum wires, *Phys. Today* **52**, 22 (1999).
5. Z. Yao, H.W.C. Postma, L. Balents, and C. Dekker, Crossed nanotube intramolecular junctions, *Nature* **402**, 273 (1999); M.S. Fuhrer, J. Nygard, L. Shih, M. Forero, Y.G. Yoon, M.S.C. Mazzoni, H.J. Choi, J. Ihm, S.G. Louie, A. Zettl, and P.L. McEuen, Crossed nanotube junctions, *Science* **288**, 494 (2000).
6. S.J. Tans, A.R.M. Verschueren, and C. Dekker, Room temperature transistor based on single walled carbon nanotube, *Nature* **393**, 49 (1998); R. Martel, T. Schmidt, H.R. Shea, T. Hertel, and P. Avouris, Single- and multi-wall carbon nanotube field-effect transistors, *Appl. Phys. Lett.* **73**, 2447 (1998).
7. S.J. Tans, M.H. Devoret, H.J. Dai, A. Thess, R.E. Smalley, L.J. Geerligs, and C. Dekker, Individual single-wall carbon nanotubes as quantum wires, *Nature* **386**, 474 (1997); M. Bockrath, D.H. Cobden, P.L. McEuen, N.G. Chopra, A. Zettl, A. Thess, and R.E. Smalley, Single-electron transport in ropes of carbon nanotubes, *Science* **275**, 1922 (1997).
8. L. Liu, G.Y. Guo, C.S. Jayanthi, and S.Y. Wu, Colossal paramagnetic moments in metallic carbon nanotori, *Phys. Rev. Lett.* **88**, 217206 (2002).
9. D.H. Cobden, M. Bockrath, P.L. McEuen, A.G. Rinzler, and R.E. Smalley, Spin splitting and even–odd effects in carbon nanotubes, *Phys. Rev. Lett.* **81**, 681 (1998); S.J. Tans, M.H. Devoret, R.J.A. Groeneveld, and C. Dekker, Electron–electron correlations in carbon nanotubes, *Nature* **394**, 761 (1998).
10. L. Balents and R. Egger, Spin transport of interacting quantum wires and carbon nanotubes, *Phys. Rev. Lett.* **85**, 3464 (2000).
11. P.M. Tedrow and R. Meservey, Direct observation of spin-state mixing in superconductors, *Phys. Rev. Lett.* **27**, 919 (1971).
12. M. Julliere, Tunneling between ferromagnetic films, *Phys. Lett.* **54A**, 225 (1975).
13. J.S. Moodera, L.R. Kinder, T.M. Wong, and R. Meservey, Large magnetoresistance at room temperature in ferromagnetic thin film tunnel junctions, *Phys. Rev. Lett.* **74**, 3273 (1995).
14. T. Miyazaki, TMR for next generation hard disk and memory, *Nikkei Electron.* **771**, 165 (2000).
15. M. Johnson and R.H. Silsbee, Spin-injection experiment, *Phys. Rev.* B **37**, 5326 (1988).
16. M. Johnson, Spin accumulation in gold films, *Phys. Rev. Lett.* **70**, 2142 (1993).
17. S. Datta and B. Das, Electronic analog of the electro-optic modulator, *Appl. Phys. Lett.* **56**, 665 (1990).
18. A. Cabbibo, J.R. Childress, S.J. Pearton, F. Ren, and J.M. Kuo, Fabrication of spin-current field-effect transistor structures, *J. Vac. Sci. Tech.* A **15**, 1215 (1997); S. Gardelis, C.G. Smith, C.H.W. Barnes, E.H. Linfield, and D.A. Ritchie, Spin-valve effects in a semiconductor field effect transistor: a spintronic device, *Phys. Rev.* B **60**, 7764 (1999).
19. C.M. Hu, J. Nitta, A. Jensen, J.B. Hansen, and H. Takayanagi, Spin polarized transport in a two dimensional electron gas with inter-digital-ferromagnetic ohmic contacts (preprint).
20. F.G. Monzon, H.X. Tang and M.L. Roukes, Magnetoelectronic phenomena at a ferromagnet-semiconductor interface, *Phys. Rev. Lett.* **84**, 5022 (2000); B.J. van Wees, Comment on observation of spin injection at a ferromagnet-semiconductor interface, *Phys. Rev. Lett.* **84**, 5023 (2000).
21. M. Johnson, B.R. Bennett, M.J. Yang, M.M. Miller, and B.V. Shanabrook, Hybrid hall effect device, *Appl. Phys. Lett.* **71**, 974 (1997).
22. F.J. Jedema, A.T. Filip, and B.J. van Wees, Observation of spin injection and accumulation at room temperature in an all-metal mesoscopic spin valve, *Nature* **410**, 345 (2001).

23. A.T. Filip, B.H. Hoving, F.J. Jedema, and B.J. van Wees, Experimental search for the electrical spin injection in a semiconductor, *Phys. Rev.* B **62**, 9996 (2000).
24. T. Valet and A. Fert, Theory of the perpendicular magnetoresistance in magnetic multilayers, *Phys. Rev.* B **48**, 7099 (1993).
25. G. Schmidt, A basic obstacle for electrical spin injection from a ferromagnetic metal into a diffusive semiconductor, *Phys. Rev.* B **62**, R4790 (2000).
26. J.M. Kikkawa and D.D. Awschalom, Lateral drag of spin coherence in gallium arsenide, *Nature* **397**, 139 (1999).
27. S. Alvardo and M. Renaud, Observation of spin polarized tunneling from a ferromagnet into GaAs, *Phys. Rev. Lett.* **68**, 1387 (1991).
28. R. Fiederling, M. Keim, G. Reuscher, W. Ossau, G. Schmidt, A. Waag, and L.W. Molenkamp, Injection and detection of a spin-polarized current in a light-emitting diode, *Nature* **402**, 787 (1999).
29. Y. Ohno, D.K. Young, B. Beschoten, F. Matsukura, H. Ohno, and D.D. Awschalom, Electrical spin injection in a ferromagnetic semiconductor heterostructure, *Nature* **402**, 790 (1999).
30. R. Fitzgerald, Magnetic semiconductors enable efficient electrical spin injection, *Phys. Today* **53**, 21 (2000).
31. For a review see H. Dai, Carbon nanotubes: Opportunities and challenges, *Surf. Sci.* **500**, 218 (2002).
32. K. Tsukagoshi, B.W. Alphenaar, and H. Ago, Coherent transport of electron spin in a ferromagnetically contacted carbon nanotube, *Nature* **401**, 572 (1999).
33. D. Orgassa, G.J. Mankey, and H. Fujiwara, Spin injection into carbon nanotubes and a possible application in spin-resolved scanning tunneling microscopy, *Nanotechnology*, **12**, 281 (2001).
34. B. Zhao, I. Monch, T. Muhl, H. Vinzelberg, and C.M. Schneider, Spin-dependent transport in multiwalled carbon nanotubes, *J. Appl. Phys.* **91**, 7026 (2002); B. Zhao, I. Monch, H. Vinzelberg, and C.M. Schneider, Spin-coherent transport in ferromagnetically contacted carbon nanotubes, *Appl. Phys. Lett.* **80**, 3144 (2002).
35. O. Chauvet, G. Baumgartner, M. Carrard, W. Basca, D. Ugarte, W.A. de Heer, and L. Forro, ESR study of potasium-doped aligned carbon nanotubes, *Phys. Rev.* B **53**, 13996 (1996); A.S. Kotosonov and D.V. Shilo, Electron spin resonance study of carbon nanotubes, *Carbon* **36**, 1649 (1998).
36. C. Schonenberger, A. Bachtold, C. Strunk, J.P. Salvetat, and L. Forro, Interference and interaction in multi-wall carbon nanotubes, *Appl. Phys.* A **69**, 283 (1999).
37. J. Fabian and S.D. Sarma, Spin relaxation of conduction electrons, *J. Vac. Sci. Technol.* B **17**, 1708 (1999).
38. I.I. Oleinik, E.Y. Tsymbal, and D.G. Pettifor, Structural and electronic properties of $Co/Al_2O_3/Co$ magnetic tunnel junction from first principles, *Phys. Rev.* B **62**, 3952 (2000).
39. U. Rüdiger, J. Yu, L. Thomas, S.S.P. Parkin, and A.D.M. Kent, Magnetoresistance, micromagnetism, and domain-wall scattering in epitaxial hcp Co films, *Phys. Rev.* B **59**, 11914 (1999).
40. J.H. Thomas, Electrical properties of thin cobalt oxide films on cobalt, *Thin Solid Films* **44**, 155 (1977).
41. D.V. Dimitrov, G.C. Hadjipanayis, V. Papaefthymiou, and A. Simopoulos, Magnetic properties and microstructure of Fe-O and Co-O thin films, *IEEE Trans. Mag.* **33,** 4363 (1997).
42. B.W. Alphenaar, K. Tsukagoshi, and M. Wagner, Spin transport in nanotubes, *J. Appl. Phys.* **89**, 6863 (2001).

43. E.Y. Tsymbal and D.G. Pettifor, Modeling of spin polarized electron tunneling from 3d ferromagnets, *J. Phys.* C **9**, L411 (1997).
44. P. LeClair, B. Hoex, H. Wieldraaijer, J.T. Kohlhepp, H.J.M. Swagten, and W.J.M. de Jonge, Sign reversal of spin polarization in Co/Ru/Al_2O_3/Co magnetic tunnel junctions, *Phys. Rev.* B **64**, 100406 (2001).
45. J.M. De Teresa, A. Barthelemy, A. Fert, J.P. Contour, R. Lyonnet, F. Montaigne, P. Seneor, and A. Vaures, *Phys. Rev. Lett.* **82**, 4288 (1999).
46. M. Sharma, S.X. Wang, and J.H. Nickel, *Phys. Rev. Lett.* **82**, 616 (1999).
47. J.S. Moodera, J. Nowak, L.R. Kinder, P.M. Tedrow, R.J.M. van de Veerdonk, B.A. Smits, M. van Kampen, H.J.M. Swagten, and W.J.M. de Jonge, Quantum well states in spin-dependent tunnel structures, *Phys. Rev. Lett.* **83**, 3029 (1999).
48. A. Jensen, J. Nygaard, J. Borggreen, and B.W. Alphenaar, Spin transport in single wall carbon nanotubes with ferromagnetic contacts, unpublished.
49. S. Takahashi and S. Maekawa, Effect of coulomb blockade on magnetoresistance in ferromagnetic tunnel junctions, *Phys. Rev. Lett.* **80**, 1758 (1998).
50. K. Ono, H. Shimada, and Y. Ootuka, Enhanced magnetic valve effect and magneto-coulomb oscillations in ferromagnetic single electron transistor, *J. Phys. Soc. Jap.* **66**, 1261 (1997); L.F. Schep, A. Fert, F. Fettar, P. Holody, S.F. Lee, J.L. Maurice, F. Petroff, and A. Vaures, Spin dependent tunneling with coulomb blockade, *Phys. Rev.* B **56**, R5747 (1997).
51. H. Imamura, S. Takahashi, and S. Maekawa, Spin dependent coulomb blockade in ferromagnet/normal-metal/ferromagnet double tunnel junctions, *Phys. Rev.* B **59**, 6017 (1999).
52. A. Brataas and X.H. Wang, Linear-response conductance and magnetoresistance of ferromagnetic single-electron transistors, *Phys. Rev.* B **64**, 104434 (2001).
53. S.S.P. Parkin, *Cambridge University Condensed Matter Seminar* (January 14, 2000).
54. H. Mehrez, J. Taylor, H. Guo, J. Wang, and C. Roland, Carbon nanotube based magnetic tunnel junctions, *Phys. Rev. Lett.* **84**, 2682 (2000).
55. C. Kane, L. Balents, and M.P.A. Fischer, Coulomb interactions and mesoscopic effects in carbon nanotubes, *Phys. Rev. Lett.* **79**, 5086 (1997); R. Egger and A.O. Gogolin, Effective low-energy theory for correlated carbon nanotubes, *Phys. Rev. Lett.* **79**, 5082 (1997).
56. A. Bachtold, M. de Jonge, K. Grove-Rasmussen, P.L. McEuen, M. Buitelaar, and C. Schonenberger, Supression of tunneling into multiwall carbon nanotubes, *Phys. Rev. Lett.* **87**, 166801 (2001).
57. Q.M. Si, Probing spin-charge transport using spin transport, *Physica* C **341**,1519 (2000).

Index